Student Solutions Manual for

The Allyn J. Washington Series in

Basic Technical Mathematics

Fifth Edition

Basic Technical Mathematics

Basic Technical Mathematics with Calculus

Basic Technical Mathematics with Calculus, Metric Version

Frances Bazen Wilbanks

Southern Bell Telephone Company

Anne Zeigler

Florence Darlington Technical College

In Consultation with Allyn J. Washington

Addison-Wesley Publishing Company
Reading, Massachusetts • Menlo Park, California • New York
Don Mills, Ontario • Wokingham, England • Amsterdam • Bonn
Sydney • Singapore • Tokyo • Madrid • San Juan • Milan • Paris

Sponsoring Editor: Lisa J. Moller
Assistant Editor: Mary Ann Telatnik
Production Editor: Sharon Montooth
Cover Artist: Victoria Ann Philp

ISBN 0-8053-8891-5

    6 7 8 9 10 -AL- 95 94 93 92

Preface

We are pleased to bring you this Student Solutions Manual as a supplement to the fourth edition of the Basic Technical Mathematics series by Allyn J. Washington. We have written and revised this manual in close consultation with Professor Washington in order to produce a supplement fully compatible with his texts. Every other odd problem in Basic Technical Mathematics, fifth edition; Basic Technical Mathematics with Calculus, fifth edition; and Basic Technical Mathematics with Calculus, Metric Version, fifth edition has been solved.

Our goal is to instill a more thorough understanding of the topics in the Technical Mathematics series, and thereby help you gain confidence in your mathematical skills. Therefore, this manual provides detailed, step-by-step solutions including many supplementary graphs not shown in the answer sections of the main texts.

Answers in the solutions manual may, in some cases, vary slightly from the answers in the main text as these were worked by a computer that carried nine significant digits. The problems in this solutions manual were worked with a hand-held calculator and were rounded when proceeding from one step to another.

Frances Bazen Wilbanks
Anne Zeigler

Contents

CHAPTER 1

Fundamental Concepts and Operations

1. 3 is an integer (whole number); 3 is rational (may be written as a ratio of integers, 3/1); 3 is real (not the square root of a negative number). $-\pi$ is irrational; it is not the ratio of integers; $-\pi$ is real.

5. $|3| = 3$ (3 is 3 units from the origin); $\left|\frac{7}{2}\right| = \frac{7}{2}$

9. $6 < 8$; six is less than eight (the arrow opens toward the larger number).

13. $-4 < -3$; -4 is to the left of -3 on the number line.

17. The reciprocal of $3 = \frac{3}{1}$ is $\frac{1}{3}$. The reciprocal of $-2 = -\frac{2}{1}$ is $-\frac{1}{2}$.

21.

25.

-18	$-\lvert -3 \rvert$	-1	$\sqrt{5}$	π	$\lvert -8 \rvert$	9
-18	-3	-1	2.24	3.14	8	9

29. (a) $x > 0$; to the right of the origin

    (b) $x < -4$; to the left of -4

33. One kilobyte = 1000 bytes; each byte contains n bits
    $N = (a \text{ bits/byte})(1000 \text{ bytes/kilobyte})(n \text{ kilobytes}) = 1000an$

1. $6 + 5 = (+6) + (+5)$
   $= +(6 + 5) = +11$
   $= 11$

5. $16 - 7 = (+16) - (+7)$
    $= (+16) + (-7) = +(16 - 7)$
    $= +9 = 9$

9. $(8)(-3) = (+8)(-3)$
    $= -(8 \times 3) = -24$

13. $\frac{-9}{+3} = -(\frac{9}{3}) = -3$

17. $(-2)(+4)(-5) = -(2 \times 4)(-5)$
    $= (-8)(-5) = +(8 \times 5)$
    $= +40 = 40$

21. $9 - 0 = 9;\ a \pm 0 = a$

25. $8 - 3(-4) = (+8) - [-(3 \times 4)]$
    $= (+8) - (-12) = (+8) + (+12)$
    $= +20 = 20$

29. $\frac{(+3)(-6)(-2)}{0 - 4} = \frac{-(3 \times 6)(-2)}{0 - (+4)}$

    $= \frac{(-18)(-2)}{0 + (-4)} = \frac{+(18 \times 2)}{-4}$

    $= \frac{+36}{-4} = -\frac{36}{4}$

    $= -9$

33. $(-7) - \frac{-14}{2} - 3(2) = (-7) - (-7) - (+6)$
    $= (-7) + (+7) + (-6)$
    $= 0 + (-6)$
    $= -6$

37. $(6)(7) = (7)(6)$; commutative law of multiplication.

41. $3 + (5 + 9) = (3 + 5) + 9$; associative law of addition.

45. Positive. You can always pair up numbers, and the product of each pair is always positive.

49. $100\ m + 200\ m = 200\ m + 100\ m$; commutative law of addition

Exercises 1-4, page 13

1. If your calculator uses algebraic logic, the numbers are entered as shown:
    3.16, ⊞ 53.9, ⊡ 117.8, = 3.6175552. Rounded off to three significant digits, the result ≈ 3.62.

If your calculator does not use algebraic logic, it may be necessary to perform the division first.   53.9, ⊡ 117.8, ⊟ , ⊞ 3.16 ⊟ 3.6175552 ≈ 3.62.

5. If your calculator uses algebraic logic, the numbers are entered as shown:
23.962, ⊠ , 0.01537, ⊡ , 〔 10.965, ⊟ 8.249, 〕 , ⊟ 0.1356023 ≈ 0.1356.
If your calculator does not use algebraic logic, it is necessary to evaluate the numerator and denominator first.
23.962, ⊠ , 0.01537, ⊟ 0.3682959
10.965, ⊟ , 8.249, ⊟ 2.716
0.3682959, ⊡ , 2.716, ⊟ 0.1356023 ≈ 0.1356

9. 37.962, ⊞ 5.049, ⊟ 43.011
5.049, ⊞ 37.962, ⊟ 43.011   Commutative law of addition

13. 1.46, ⊞/⊟ , ⊠ , 5.62, ⊟ −8.2052.
If your calculator does not have ⊞/⊟ :
1.46, ⊠ , 5.62, ⊟ −8.2052
Result is −8.21.   (Product of two real numbers of unlike sign is the negative of their product.)

17. 37.2, ⊞ , 2.49, ⊟ 14.939759.   Rounded to three significant digits, the quotient is 14.9.

21. The reciprocal of 2.8 is (a) 1, ⊡ , 2.8, ⊟ 0.3571429 = 0.36;
(b) 2.8, $\boxed{\frac{1}{x}}$ = 0.3571429 = 0.36

25. (a) $\frac{8}{33}$ = 8, ⊡ , 33, ⊟ 0.24$\overline{24}$
(b) π ≈ 3.1415927

29. 460, ⊞ , 180, ⊟ 2.6

Exercises 1-5, page 19

1. $x^3 x^4 = x^{3+4} = x^7$ Eq. (1-1)

5. $\frac{m^5}{m^3} = m^{5-3} = m^2$ Eq. (1-2)

9. $(a^2)^4 = a^{2\times4} = a^8$ Eq. (1-3)

13. $(2n)^3 = 2^3 n^3 = 8n^3$ Eq. (1-4)

17. $(\frac{2}{b})^3 = \frac{2^3}{b^3} = \frac{8}{b^3}$ Eq. (1-4)

21. $7^0 = 1$ Eq. (1-5)

25. $6^{-1} = \frac{1}{6^1} = \frac{1}{6}$ Eq. (1-6)

29. $(-t^2)^7 = (-1t^2)^7$
$= (-1)^7(t^2)^7 = -1t^{14} = -t^{14}$
Eq. (1-4), Eq. (1-3)

33. $(4xa^{-2})^0 = 1$ Eq. (1-4)

37. $\dfrac{2a^4}{(2a)^4} = \dfrac{2a^4}{(2a)(2a)(2a)(2a)} = \dfrac{2a^4}{16a^4} = \dfrac{1}{8}$

41. $(5^0 x^2 a^{-1})^{-1} = (5^0)^{-1}(x^2)^{-1}(a^{-1})^{-1}$

$= 5^{0(-1)} x^{2(-1)} a^{-1(-1)} = 5^0 x^{-2} a^1$

$= (1)\left(\dfrac{1}{x^2}\right)(a) = \dfrac{a}{x^2}$ Eq. (1-4), Eq. (1-3)

45. $(-8gs^3)^2 = (-8)^2 (g)^2 (s^3)^2$

$= 64g^2 s^6$ Eq. (1-4), Eq. (1-3)

49. $7(-4) - (-5)^2 = -28 - 25$
$= -53$

53. 2.38, ⊠ , 60.7, $\boxed{x^2}$, ⊟ , 2540, ⊟ 6229.09 or
60.7, ⊠ , 60.7, ⊠ , 2.38, ⊟ 8769.0862, ⊟ , 2540, ⊟ 6229.09 = 6230

57. $\pi\left(\dfrac{r}{2}\right)^3\left(\dfrac{4}{3\pi r^2}\right) = \left(\dfrac{\pi}{1}\right)^3\left(\dfrac{r}{2^3}\right)\left(\dfrac{4}{3\pi r^2}\right) = \dfrac{4\pi r^3}{24\pi r^2} = \dfrac{r}{6}$

Exercises 1-6, page 23

1. $4.5 \times 10^4 = 45,000$; Move decimal point 4 places to right by adding 3
zeros.

5. $3.23 \times 10^0 = 3.23$; Move the decimal point zero places.

9. $40\ 000 = 4 \times 10^4$ 13. 6; The number is between 1 and 10.
     ↑   ↑ 4 places   It may be written 6×10^0

17. $(28\ 000)(2\ 000\ 000\ 000) = 2.8 \times 10^4 \times 2 \times 10^9 = 5.6 \times 10^{13}$

21. (29M.) $(1\ 280)(865\ 000)(43.8) = 1.28 \times 10^3 \times 8.65 \times 10^5 \times 4.38 \times 10 =$
              1.28, EE , 3, ⊠ , 8.65, EE , 5, ⊠ , 4.38, EE , 1, ⊟ 4.8485 10,
              which means 4.85×10^{10}, rounded off.

21M. 3200 g $= 3.2 \times 10^3$ g $= 3.2$ kg

25. 3.642, EE , 8, $\boxed{+/-}$, ⊠ , 2.736, EE , 5, ⊟ 9.965×10^{-3}, rounded off.

25M. 5.3 Mm $= 5.3 \times 10^6$ m

29. (33M.)   6 500 000 kW $= 6.5 \times 10^6$ kW
              ↑   ↑
              6 places

33. (37M.) $2 \times 10^5 = 2\ 00\ 000$
                           ↑   ↑
                           5 places

37. (41M.) $2.4 \times 10^{-43} = 0.000\ 000\ 000\ 000\ 000\ 000\ 000\ 000\ 000\ 000\ 000\ 000\ 24$

41. $1 \text{ mi}^2 = (161\ 000 \text{ cm})^2 = (1.61 \times 10^5)^2 = 1.61^2 \times 10^{10} \text{ cm}^2$
$= 2.59 \times 10^{10} \text{ cm}^2$

45M. $25\ 700 \text{ km}^2 = 25\ 700\,(10^3 \text{ m})^2 = 25\ 700\,(10^5 \text{ cm})^2$
$= 2.57 \times 10^4 (10^5 \text{ cm})^2 = 2.57 \times 10^{14} \text{ cm}^2$

Exercises 1-7, page 26

1. $\sqrt{25} = \sqrt{5^2} = 5$ 5. $-\sqrt{49} = -7$ 9. $\sqrt[3]{125} = 5$

13. $(\sqrt{5})^2 = \sqrt{5}\sqrt{5} = \sqrt{25} = 5$ 17. $\sqrt{18} = \sqrt{9 \times 2} = 3\sqrt{2}$

21. $\sqrt{44} = \sqrt{4 \times 11} = 2\sqrt{11}$ 25. $2\sqrt{84} = 2\sqrt{4 \times 21} = 2(2)\sqrt{21} = 4\sqrt{21}$

29. $\sqrt{36 + 64} = \sqrt{100} = 10$

33. $\sqrt{85.4}$; 85.4, $\boxed{\text{INV}}$, $\boxed{x^2}$, $\boxed{=}$ 9.24$\overline{12}$ = 9.24 or 85.4, $\boxed{\sqrt{}}$, = 9.24$\overline{12}$ = 9.24

37. $t = 1.11\sqrt{L} = 1.11\sqrt{1.75} = 1.11(1.323) = 1.468 = 1.47$ s

37M. $2.01\sqrt{L} = 2.01\sqrt{1.75}$

1.75, $\boxed{\sqrt{}}$, $\boxed{\times}$, 2.01 $\boxed{=}$ 2.66

41. $C^2 = A^2 + B^2$; A = 15.2 in, B = 11.4 in
$C^2 = (15.2 \text{ in})^2 + (11.4 \text{ in})^2 = 231.04 \text{ in}^2 + 129.96 \text{ in}^2 = 361 \text{ in}^2$
$C = \sqrt{361} = 19.0$ in

41M. $C^2 = A^2 + B^2$; A = 38.6 cm, B = 29.0 cm
$C^2 = (38.6 \text{ cm})^2 + (29.0 \text{ cm}^2) = 1489.96 \text{ cm}^2 + 841.00 \text{ cm}^2 = 2331 \text{ cm}^2$
$C = \sqrt{2331} = 48.3$ cm

Exercises 1-8, page 31

1. $5x + 7x - 4x = x(5 + 7 - 4)$
$= x(8) = 8x$; distributive law

5. $2a - 2c - 2 + 3c - a = 2a - a + 3c - 2c - 2 = a(2 - 1) + c(3 - 2) - 2$
$= a(1) + c(1) - 2 = a + c - 2$

9. $s + (4 + 3s) = s + 4 + 3s$
$= 4s + 4$

13. $2 - 3 - (4 - 5a) = 2 - 3 + (-4) + (+5a)$
$= 2 - 3 - 4 + 5a = 5a - 5$

17. $-(t - 2u) + (3u - t) = -t + 2u + 3u - t$
$= -2t + 5u$

21. $-7(6 - 3c) - 2(c + 4) = -42 + 21c - 2c - 8$
$$= 19c - 50$$

25. $2[4 - (t^2 - 5)] = 2[4 - t^2 + 5]$
$$= 2[9 - t^2] = 18 - 2t^2$$
$$= -2t^2 + 18$$

29. $a\sqrt{xy} - [3 - (a\sqrt{xy} + 4)] = a\sqrt{xy} - [3 - a\sqrt{xy} - 4]$
$$= a\sqrt{xy} - [-1 - a\sqrt{xy}] = a\sqrt{xy} + 1 + a\sqrt{xy}$$
$$= 2a\sqrt{xy} + 1$$

33. $5p - (q - 2p) - [3q - (p - q)] = 5p - q + 2p - [3q - p + q]$
$$= 7p - q - 3q + p - q$$
$$= 8p - 5q$$

37. $3a - [6 - (a + 3)] = 3a - [6 - a - 3]$
$$= 3a - 6 + a + 3$$
$$= 4a - 3$$

41. $3D - (D - d) = 3D - D + d = D(3 - 1) + d = D(2) + d = 2D + d$

<u>Exercises 1-9</u>, page 33

1. $(a^2)(ax) = (a^2)(a^1)(x)$
$$= a^3 x$$

5. $(2ax^2)^2(-2ax) = (4a^2x^4)(-2ax)$
$$= -8a^3x^5$$

9. $a^2(x + y) = (a^2)(x) + (a^2)(y)$
$$= a^2x + a^2y$$

13. $5m(m^2n + 3mn) = (5m)(m^2n) + (5m)(3mn)$
$$= 5m^3n + 15m^2n$$

17. $ab^2c^4(ac - bc - ab) = (ab^2c^4)(ac) - (ab^2c^4)(bc) - (ab^2c^4)(ab)$
$$= a^2b^2c^5 - ab^3c^5 - a^2b^3c^4$$

21. $(x - 3)(x + 5) = x(x) + x(5) + (-3)(x) + (-3)(5)$
$$= x^2 + 5x - 3x - 15$$
$$= x^2 + 2x - 15$$

25. $(2a - b)(3a - 2b) = 2a(3a) + 2a(-2b) + (-b)(3a) + (-b)(-2b)$
$$= 6a^2 - 4ab - 3ab + 2b^2$$
$$= 6a^2 - 7ab + 2b^2$$

29. $(x^2 - 1)(2x + 5) = x^2(2x) + x^2(5) + (-1)(2x) + (-1)(5)$
$$= 2x^3 + 5x^2 + (-2x) + (-5) = 2x^3 + 5x^2 - 2x - 5$$

33. $(x + 1)(x^2 - 3x + 2) = x(x^2) + x(-3x) + x(2) + 1(x^2) + 1(-3x) + 1(2)$
$$= x^3 - 2x^2 - x + 2$$

37. $2(a + 1)(a - 9) = 2(a^2 - 9a + a - 9)$
$$= 2(a^2 - 8a - 9)$$
$$= 2a^2 - 16a - 18$$

41. $(2x - 5)^2 = (2x - 5)(2x - 5)$
$$= 4x^2 - 10x - 10x + 25$$
$$= 4x^2 - 20x + 25$$

45. $(xyz - 2)^2 = (xyz - 2)(xyz - 2)$
$$= x^2y^2z^2 - 2xyz - 2xyz + 4$$
$$= x^2y^2z^2 - 4xyz + 4$$

49. $(2 + x)(3 - x)(x - 1) = (2 + x)[3(x) + 3(-1) + (-x)(x) + (-x)(-1)]$
$$= (2 + x)[3x - 3 - x^2 + x] = (2 + x)(-x^2 + 4x - 3)$$
$$= 2(-x^2) + 2(4x) + 2(-3) + x(-x^2) + x(4x) + x(-3)$$
$$= -x^3 + 2x^2 + 5x - 6$$

53. $(x + 2)(x - 1) = x^2 - x + 2x - 2 = x^2 + x - 2$

Exercises 1-10, page 37

1. $\dfrac{8x^3y^2}{-2xy} = -4x^{3-1}y^{2-1}$
$$= -4x^2y$$

5. $\dfrac{(15x^2)(4bx)(2y)}{30bxy} = \dfrac{(15)(4)(2)bx^3y}{30bxy}$
$$= 4b^{1-1}x^{3-1}y^{1-1}$$
$$= 4x^2$$

9. $\dfrac{a^2x + 4xy}{x} = \dfrac{a^2x}{x} + \dfrac{4xy}{x}$
$$= a^2 + 4y$$

13. $\dfrac{4pq^3 + 8p^2q^2 - 16pq^5}{4pq^2} = \dfrac{4pq^3}{4pq^2} + \dfrac{8p^2q^2}{4pq^2} - \dfrac{16pq^5}{4pq^2}$
$$= q + 2p - 4q^3$$

17. $\dfrac{3ab^2 - 6ab^3 + 9a^3b}{9a^2b^2} = \dfrac{3ab^2}{9a^2b^2} - \dfrac{6ab^3}{9a^2b^2} + \dfrac{9a^3b}{9a^2b^2}$
$$= \dfrac{1}{3a} - \dfrac{2b}{3a} + \dfrac{a}{b}$$

21.
$$\begin{array}{r} 2x + 1 \\ x + 3 \overline{)\,2x^2 + 7x + 3} \\ \underline{2x^2 + 6x} \\ x + 3 \\ \underline{x + 3} \end{array}$$

25.
$$\begin{array}{r} 4x^2 - x - 1 \\ 2x - 3 \overline{)\,8x^3 - 14x^2 + x + 0} \\ \underline{8x^3 - 12x^2} \\ -2x^2 + x \\ \underline{-2x^2 + 3x} \\ -2x + 0 \\ \underline{-2x + 3} \\ -3 \text{ Rem.} \end{array}$$

29.
$$\begin{array}{r} x^2 + x - 6 \\ x + 2 \overline{)\,x^3 + 3x^2 - 4x - 12} \\ \underline{x^3 + 2x^2} \\ x^2 - 4x \\ \underline{x^2 + 2x} \\ -6x - 12 \\ \underline{-6x - 12} \end{array}$$

33.
$$\begin{array}{r} x^2 - 2x + 4 \\ x + 2 \overline{)\,x^3 + 0x^2 + 0x + 8} \\ \underline{x^3 + 2x^2} \\ -2x^2 + 0x \\ \underline{-2x^2 - 4x} \\ 4x + 8 \\ \underline{4x + 8} \end{array}$$

37. $\dfrac{8A^5 + 4^3\mu^2E^2 - A\mu^4E^4}{8A^4} = \dfrac{8A^5}{8A^4} + \dfrac{4A^3\mu^2E^2}{8A^4} - \dfrac{A\mu^4E^4}{8A^4} = A + \dfrac{\mu^2E^2}{2A} - \dfrac{\mu^4E^4}{8A^3}$

Exercises 1-11, page 40

1. $x - 2 = 7$
    $x = 7 + 2$
    $x = 9$

5. $\dfrac{t}{2} = 5$
    $t = 10$

9. $3t + 5 = -4$
    $3t = -9$
    $t = -3$

13. $3x + 7 = x$
     $2x + 7 = 0$
     $2x = -7$
     $x = -\dfrac{7}{2}$

17. $6 - (r - 4) = 2r$
     $6 - r + 4 = 2r$
     $6 + 4 = 2r + r$
     $10 = 3r$
     $r = \dfrac{10}{3}$

21. $x - 5(x - 2) = 2$
     $x - 5x + 10 = 2$
     $-4x = -8$
     $x = 2$

25. $5.8 - 0.3(x - 6.0) = 0.5x$
$5.8 - 0.3x + 1.8 = 0.5x$
$7.6 - 0.3x = 0.5x$
$-0.3x = 0.5x - 7.6$
$-0.8x = -7.6$
$x = 9.5$

33. $15(5.5 + v) = 24(5.5 - v)$
$82.5 + 15v = 132 - 24v$
$15v + 24v = 132 - 82.5$
$39v = 49.5$
$v = 1.3$

29. $2(x - 3) + 1 = 2x - 5$
$2x - 6 + 1 = 2x - 5$
$2x - 5 = 2x - 5$ which is true for all x

Exercises 1-12, page 43

1. $ax = b;$ $x = \dfrac{b}{a}$

5. $ax + 6 = 2ax - c$
$-ax + 6 = -c$
$-ax = -c - 6$
$ax = c + 6$
$x = \dfrac{c + 6}{a}$

9. $\theta = kA + \lambda$
$\theta - kA = \lambda$

13. $P = 2\pi Tf$
$\dfrac{P}{2\pi f} = \dfrac{2\pi Tf}{2\pi f}$
$T = \dfrac{P}{2\pi f}$

17. $p = \dfrac{\pi^2 EI}{L^2}$
$pL^2 = \pi^2 EI$
$\dfrac{pL^2}{\pi^2 I} = E$

21. $C_0^2 = C_1^2\,(1 + 2v)$
$\dfrac{C_0^2}{C_1^2} = 1 + 2v$
$\dfrac{C_0^2}{C_1^2} - 1 = 2v$
$\dfrac{C_0^2}{2C_1^2} - \dfrac{1}{2} = \dfrac{C_0^2 - C_1^2}{2C_1^2} = v$

25. $Q_1 = P(Q_2 - Q_1)$
$\dfrac{Q_1}{P} = Q_2 - Q_1$
$Q_2 = \dfrac{Q_1}{P} + Q_1$

29. $L = \pi(r_1 + r_2) + 2x_1 + 2_2$
$L - 2x_1 - x_2 = \pi(r_1 + r_2)$
$\dfrac{L - 2x_1 - x_2}{\pi} = r_1 + r_2$
$r_1 = \dfrac{L - 2x_1 - x_2}{\pi} - r_2$

33. $p = \pi r + 2r + 2h$
$p - \pi r - 2r = 2h$
$h = \dfrac{p - \pi r - 2r}{2}$
$= \dfrac{172 - \pi(22.5) - 2(22.5)}{2}$
$= \dfrac{172 - 70.7 - 45.0}{2}$
$= \dfrac{56.3}{2} = 28.2$

33M. $p = \pi r + 2r + 2h$
       $p - \pi r - 2r = 2h$

   $h = \dfrac{p - \pi r - 2r}{2}$

   $= \dfrac{344 - \pi(45.0) - 2(45.0)}{2}$

   $= \dfrac{344 - 141.4 - 90.0}{2} = 56.3$

Exercises 1-13, page 48

1. Let x = cost of one program, $x + 114$ = cost of other program
    $x + x + 114 = 390$; $2x + 114 = 390$; $2x = 390 - 114 = 276$; $x = \$138$
                                                          $x + 114 = \$252$

5. Let x = number of acres of land purchased at \$20,000 per acre
    $70 - x$ = number of acres of land purchased at \$10,000 per acre

    $x(20,000) + (70 - x)(10,000) = 900,000$
    $20,000x + 700,000 - 10,000x = 900,000$
    $10,000x + 700,000 = 900,000$
    $10,000x = 200,000$
    $x = 20$ acres
    $70 - x = 50$ acres

5M. $x = 20$ ha, $70 - x = 50$ ha

9. Let x = first current, $2x$ = second current, and $x + 9.2$ = third current

    $x + 2x + x + 9.2 = 0$
    $4x + 9.2 = 0$
    $4x = 0 - 9.2 = -9.2$
    $x = -2.3\ \mu A$, $2x = -4.6\mu A$, $x + 9.2 = 6.9\ \mu A$

13. Let x = length of main pipeline
    $x + 2.6$ = length of smaller pipeline
    $x + 3(x + 2.6) = 35.4$
    $x + 3x + 7.8 = 35.4$
    $4x + 7.8 = 35.4$
    $4x = 27.6$
    $x = 6.9$ km
    $x + 2.6 = 9.5$ km

17. Let x = speed of slower jet; $x + 400$ = speed of faster jet
    $1.5x$ = distance traveled by slower jet
    $1.5(x + 400)$ = distance traveled by faster jet

    $1.5x + 1.5(x + 400) = 5400$
    $1.5x + 1.5x + 600 = 5400$
    $3.0x + 600 = 5400$
    $3.0x = 5400 - 600 = 4800$
    $x = 1600$; $x + 400 = 2000$

21. To make 8ℓ of solution that is 15/16 gasoline, $x\ell$ of gasoline must be added to the existing amount of $(8 - x)\,\ell$ which is 3/4 gasoline. The quantity of gasoline total is described by combining the existing with that which is added to produce the 15/16 of 8ℓ.

$$x + \frac{3}{4}(8 - x) = \frac{15}{16}(8); \quad x + 6 - \frac{3}{4}x = 7\frac{1}{2}; \quad \frac{1}{4}x = 1\frac{1}{2}; \quad x = 6\,\ell$$

Review Exercises for Chapter 1, page 50

1. $(-2) + (-5) - (+3) = (-2) + (-5) + (-3) = -10$

5. $-5 - 2(-6) + \dfrac{-15}{+3} = -5 + 12 - 5$
    $$= -10 + 12 = 2$$

9. $\sqrt{16} - \sqrt{64} = 4 - 8 = -4$ 13. $(-2rt^2)^2 = (-2)^2 r^2 (t^2)^2$
    $$= 4r^2 t^4$$

17. $(x^0 y^{-1} z^3)^2 = (x^0)^2 (y^{-1})^2 (z^3)^2$
     $$= 1 y^{-2} z^6 = \frac{1}{y^2} z^6$$
     $$= \frac{z^6}{y^2}$$

21. $\sqrt{45} = \sqrt{(9)(5)} = \sqrt{9}\sqrt{5} = 3\sqrt{5}$

25. 37.38, $\boxed{-}$, 16.92, $\boxed{\times}$, 1.067, $\boxed{x^2}$, $\boxed{=}$ 18.12; or
     16.92, $\boxed{\times}$, 1.067, $\boxed{\times}$, 1.067, $\boxed{=}$ 19.263234;
     37.38, $\boxed{-}$, 19.263234, $\boxed{=}$ 18.12

29. $a - 3ab - 2a + ab = (a - 2a) + (ab - 3ab) = -a + (-2ab) = -a - 2ab$

33. $(2x - 1)(x + 5) = 2x^2 + 10x - x - 5 = 2x^2 + 9x - 5$

37. $\dfrac{2h^3 k^2 - 6h^4 k^5}{2h^2 k} = \dfrac{2h^3 k^2}{2h^2 k} - \dfrac{6h^4 k^5}{2h^2 k} = hk - 3h^2 k^4$

41. $2xy - \{3z - [5xy - (7z - 6xy)]\}$
     $= 2xy - \{3z - [5xy - 7z + 6xy]\}$
     $= 2xy - \{3z - [11xy - 7z]\}$
     $= 2xy - \{3z - 11xy + 7z\}$
     $= 2xy - \{10z - 11xy\}$
     $= 2xy - 10z + 11xy$
     $= 13xy - 10z$

45. $-3y(x - 4y)^2$
     $= -3y(x - 4y)(x - 4y)$
     $= -3y(x^2 - 4xy - 4xy + 16y^2)$
     $= -3y(x^2 - 8xy + 16y^2)$
     $= -3x^2 y + 24xy^2 - 48y^3$

49. $\dfrac{12p^3q^2 - 4p^4q + 6pq^5}{2p^4q}$

$= \dfrac{12p^3q^2}{2p^4q} - \dfrac{4p^4q}{2p^4q} + \dfrac{6pq^5}{2p^4q}$

$= \dfrac{6q}{p} - 2 + \dfrac{3q^4}{p^3}$

53.

$$\begin{array}{r} x^2 - 2x + 3 \\ 3x - 1{\overline{)3x^3 - 7x^2 + 11x - 3}} \\ \underline{3x^3 - x^2} \\ -6x^2 + 11x \\ \underline{-6x^2 + 2x} \\ 9x - 3 \\ \underline{9x - 3} \end{array}$$

57. $-3\{(r + s - t) - 2[(3r - 2s) - (t - 2s)]\}$
    $= -3\{r + s - t - 2[3r - 2s - t + 2s]\}$
    $= -3\{r + s - t - 2[3r - t]\} = -3\{r + s - t - 6r + 2t\}$
    $= -3\{-5r + s + t\} = 15r - 3s - 3t$

61. $3s + 8 = 5s$
     $-2s + 8 = 0$
      $-2s = -8$
        $s = 4$

65. $6x - 5 = 3(x - 4)$
      $6x - 5 = 3x - 12$
      $3x - 5 = -12$
       $3x = -7$
        $x = -\dfrac{7}{3}$

69. $3t - 2(7 - t) = 5(2t + 1)$
     $3t - 14 + 2t = 10t + 5$
       $5t - 14 = 10t + 5$
           $5t = 10t + 19$
          $-5t = 19$
            $t = -\dfrac{19}{5}$

73. $25,000 = 2.5(10,000) = 2.5 \times 10^4$ mi/h

73M. $40\ 000 = 4(10\ 000) = 4 \times 10^4$ km/h

77. $0.000\ 001\ 2$ cm^2 = 1.2×10^6 cm^2
      ↑   ↑
      6 spaces

81. $3s + 2 = 5a$
      $3s = 5a - 2$
       $s = \dfrac{5a - 2}{3}$

85. $R = n^2z$

      $\dfrac{R}{n^2} = z$

89. $m = dV (1 - e)$

      $\dfrac{m}{dV} = 1 - e$

      $\dfrac{m}{dV} - 1 = -e;\ e = 1 - \dfrac{m}{dV}$

93. $E = \dfrac{J - K}{1 - S}$

      $E (1 - S) = J - K$
      $E (1 - S) - J = -K$
      $K = J - E (1 - S)$

97. $aL(T_2 - T_1);\ a = 0.670 \times 10^{-5}$°F; $L = 75.0$ ft; $T_2 = 75.6$°F;
      $T_1 = 39.8$°F
      $0.670 \times 10^{-5} \times 75.0(75.6 - 39.8) = 50.25 \times 10^{-5}(35.8) = 1798.95 \times 10^{-5}$
      $= 1.80 \times 10^{-2}$ ft

97M. $aL(T_2 - T_1)$; $a = 1.21 \times 10^{-5} °C$; $L = 75.0$ m; $T_2 = 34.2°C$;
$T_1 = 14.3°C$
$1.21 \times 10^{-5} \times 75.0(34.2 - 14.3) = 90.75 \times 10^{-5}(19.9) = 1805.925 \times 10^{-5}$
$$= 1.81 \times 10^{-2} \text{ m}$$

101. $1 - r^2 - 4r(1 - r) = 1 - r^2 - 4r + 4r^2 = 3r^2 - 4r + 1$

105. x = capacity of 1st computer
$4x$ = capacity of 2nd computer

$x + 4x = 81\ 920$
$5x = 81\ 920$
$x = 16\ 384$

109. x = vertical strands, 135 cm long
$x + 30$ = horizontal strands, each 95 cm long
$135(x) + 95(x + 30) = 14\ 810$
$135x + 95x + 2850 = 14810$
$230x = 14810 - 2850 = 11960$
$x = 52$
$x + 30 = 82$

113. Let x = liters of 0.50% mixture
$1000 - x$ = liters of 0.75% mixture

$0.50\%(x) + 0.75\%(1000 - x) = 0.65\%(1000)$
$0.0050(x) + 0.0075(1000 - x) = 0.0065(1000)$
$0.0050x + 7.5 - 0.0075x = 6.5$
$- 0.0025x = -1.5$
$x = 600;\ 1000 - x = 400$

CHAPTER 2

Functions and Graphs

1. From geometry, $A = \pi r^2$ 5. From geometry, $A = \ell w$; $A = \ell(5) = 5\ell$

9. $f(1) = 2(1) + 1 = 2 + 1 = 3$; $f(-1) = 2(-1) + 1 = -2 + 1 = -1$

13. $f(3) = 3^2 - 9(3) = 9 - 27 = -18$;
    $f(-5) = (-5)^2 - 9(-5) = 25 + 45 = 70$

17. $g\left(-\dfrac{1}{2}\right) = a\left(-\dfrac{1}{2}\right)^2 - a^2\left(-\dfrac{1}{2}\right) = a\left(\dfrac{1}{4}\right) + a^2\left(\dfrac{1}{2}\right) = \dfrac{1}{4}\,a + \dfrac{1}{2}\,a^2$

    $g(a) = a(a^2) - a^2(a) = a^3 - a^3 = 0$

21. $f(x) = 5x^2 - 3x$

    $f(3.86) = 5, \boxed{\text{x}}\,,\ 3.86, \boxed{x^2}\,, \boxed{-}\,,\ 3, \boxed{\text{x}}\,,\ 3.86,\ \boxed{=}\quad 62.918 \approx 62.9$

    $f(-6.92) = 5, \boxed{\text{x}}\,,\ 6.92, \boxed{x^2}\,, \boxed{-}\,,\ 3, \boxed{\text{x}}\,,\ 6.92, \boxed{+/-}\,, \boxed{=}\ 260.192 \approx 260$

25. $f(x) = x^2 + 2$; square the value of the independent variable and add 2

29. From "y is equal to the square root of x" we know that $y = \sqrt{x}$.
    Since "y is a function of x" implies $y = f(x)$, we may show this
    function as $f(x) =$

33. $s = 1.75 - 9.8t^2$; $f(t) = 1.75 - 9.8t^2$
    $$f(1.3) = 1.75 - 9.8(1.3)^2 = 1.75 - 9.8(1.69)$$
    $$= 1.75 - 16.56 = -14.8$$

Exercises 2-2, page 62

1. $f(x) = x + 5$;   Since $x + 5$ is defined for all real numbers, the domain is the set of real numbers. The range is also the set of real numbers since there is a real value for x such that $x + 5$ is equal to any real number.

5. $f(s) = \dfrac{2}{s^2}$; $\dfrac{2}{s^2}$ is not defined for any x such that $s^2 = 0$, so $s = 0$ cannot belong to the domain. Therefore the domain is the set of real numbers except 0. Since s^2 will be positive for all values in the domain, the fraction will also be positive. The range of $f(s)$ is the set of positive real numbers.

9. $Y(y) = \dfrac{y + 1}{\sqrt{y - 2}}$;   Since division by 0 is not defined, the domain must be restricted to exclude any value for which $\sqrt{y - 2}$ is equal to 0. This excludes $y = 2$ from the domain. Also, since square roots are not defined for negative values, y must be restricted to those values of y greater than 2, and the domain becomes the set of real numbers greater than 2.

13. $F(t) = 3t - t^2$
$$F(-2) = 3(-2) - (-2)^2$$
$$= -6 - (+4)$$
$$= -6 - 4$$
$$= -10$$

$$F\left(\tfrac{1}{3}\right) = 3\left(\tfrac{1}{3}\right) - \left(\tfrac{1}{3}\right)^2 = 1 - \tfrac{1}{9} = \tfrac{8}{9}$$

17. $F(x) = \begin{cases} x + 1, & x < 1 \\ \sqrt{x + 3}, & x \geq 1 \end{cases}$ for $x = -2, f(x) = x + 1 = -2 + 1 = -1$
for $x = 2, f(x) = \sqrt{x + 3} = \sqrt{2 + 3} = \sqrt{5}$

21. $d = f(t) = 40(2) + 55(t)$

25. $w = f(h) = 110 + 0.5(h - 1000)$
$$= 0.5h - 390$$

29. $0.10(x) + 0.40(y) = 1200$
$0.40y = 1200 - 0.10x$

$$y = \frac{1200 - 0.10x}{0.40} = 3000 - 0.25x$$

$$y = f(x) = 3000 - 0.25x$$

33. $f = \dfrac{1}{2\pi\sqrt{C}}$;   since the function is undefined for any value of C that makes $2\pi\sqrt{C} = 0$, the domain must exclude C = 0. Therefore, the domain is the set of all real numbers greater than 0 since C cannot be a negative value.

Exercises 2-3, page 66

1. A = (2,1) B = (-1,2) C = (-2,-3)

5.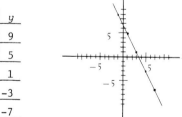

9.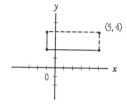

13. Abscissas are x-coordinates; all points $(1,y)$ where y is any real number, are points on a vertical line formed by varying the value of y.

17. When the abscissa equals the ordinate, then $x = y$. These pairs lie on a line formed of points made by varying x. These points form a line bisecting quadrants I and III.

21. All points in Quadrant I and IV have $x > 0$.

25. All points that have $y < 0$ lie below the x-axis. Of these, those points that have $x = 0$ lie on the y-axis.

Exercises 2-4, page 71

1. $y = 3x$

x	y
-2	-6
-1	-3
0	0
1	3
2	6

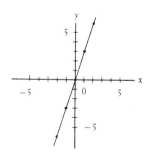

5. $y = 7 - 2x$

x	y
-1	9
1	5
3	1
5	-3
7	-7

9. $y = x^2$

x	y
-4	16
-3	9
-2	4
0	0
2	4
3	9
4	16

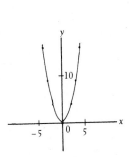

13. $y = \frac{1}{2}x^2 + 2$

x	y
-4	10
-2	4
0	2
2	4
4	10

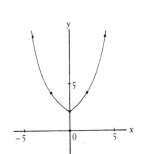

17. $y = x^2 - 3x + 1$

x	y
-3.0	19.0
-2.0	11.0
0.0	1.0
1.5	-1.3
2.0	-1.0
4.0	5.0
5.0	11.0
6.0	19.0

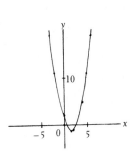

21. $y = x^3 - x^2$

x	y
-2.0	-12.0
-1.0	-2.0
0.0	0.0
0.5	-0.1
1.0	0.0
2.0	4.0
3.0	18.0

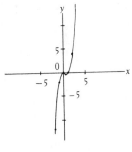

25. $y = \frac{1}{x}$

x	y
-4.0	-0.3
-2.0	-0.5
-1.0	-1.0
-0.5	-2.0
-0.3	-3.3
0.3	3.3
0.5	2.0
1.0	1.0
2.0	0.5
4.0	0.3

29. $y = \sqrt{x}$

x	y
0	0
1	1
4	2
9	3

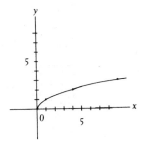

33. $V = 50\ 000 - 0.2m$ for $m \leq 100\ 000$

m (thousands)	V(thousands)
100	30
75	35
50	40
25	45
0	50

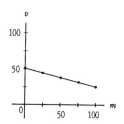

37. $r = 0.42v^2$

v (mi/h)	r(ft)
10	42
20	168
30	378
40	672
50	1050

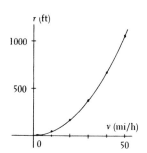

41. $2\ell + 2w = 60$
     $2\ell = 60 - 2w$
     $\ell = 30 - w$
     (ℓ and w must be
     greater than 0
     and less than 30)

$A = \ell \times w$
$\quad = (30 - w)(w)$
$\quad = 30w - w^2$

w	A
0	0
5	125
10	200
15	225
20	200
25	125
30	0

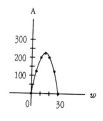

45. $y = x$ $y = |x|$

x	y
-10	-10
-5	-5
0	0
5	5
10	10

x	y
-10	10
-5	5
0	0
5	5
10	10

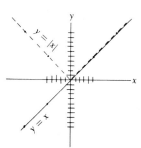

49. This is the graph of a function. Any vertical line will intercept
     the graph in only one point.

Exercises 2-5, page 75

1.

Week	1	2	3	4	5	6	7	8	9	10
Prod.	765	780	840	850	880	840	845	820	760	810

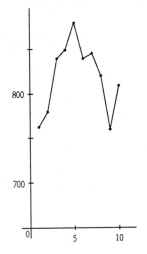

5.

(cm)	0.0	2.0	4.0	6.0	8.0	10.0	12.0
(H)	0.77	0.75	0.61	0.49	0.38	0.25	0.17

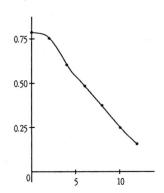

9. (a) Estimate 0.3 of the interval between 4 and 5 on the t-axis and mark
    it. Draw a line vertically from this point to the graph. From the point
    where it intersects the graph, draw a horizontal line to the T-axis,
    which it crosses at 132. Therefore, T = 132.1°C.
    (b) Draw a line horizontally from T = 145.0°C to the point where it
    intersects the graph. Draw a vertical line from this point to the
    t-axis, which it crosses between 0 and 1. Estimate t at 0.7 min.

13.

$$2.0 \begin{bmatrix} 1.2 \begin{bmatrix} 8.0 & 0.38 \\ 9.2 & ? \\ 10.0 & 0.25 \end{bmatrix} x \end{bmatrix} 0.13$$

$$\frac{1.2}{2.0} = \frac{x}{0.13} \qquad 0.38 - 0.08 = 0.30$$

$$2.0x = 0.156$$
$$x = 0.078 \ or \ 0.08$$

17.

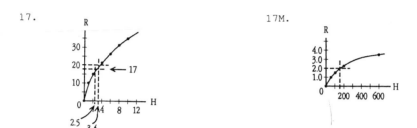

17M.

21. $A = 36 \ m^2;$ A is between 30 and 40. Using linear interpolation,

$$0.07 \begin{bmatrix} x \begin{bmatrix} 0.30 & 30 \\ & 36 \end{bmatrix} 6 \\ 0.37 & 40 \end{bmatrix} 10$$

$$\frac{x}{0.07} = \frac{6}{10}$$
$$10x = 0.42$$
$$x = 0.042$$

Therefore, $f = 0.30 + 0.042 = 0.342 \approx 0.34$

25.

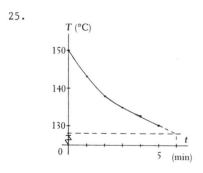

As shown, the graph is extended using a straight line segment. The estimated value of the temperature is 130.3°C.

Exercises 2-6, page 79

1. $5x - 10$; $f(x) = 5x - 10$; let $f(x) = 0$; then $5x - 10 = 0$; $5x = 10$; $x = 2$

5. Let $y = 2x - 7$

x	y
2	-3
4	1
6	5

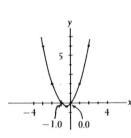

$y = 0$ when $x = 3.5$

Check: Let $f(x) = 0$
$2x - 7 = 0$
$2(3.5) - 7 = 0$

9. Let $y = x^2 + x$

x	y
-3.0	6.0
-2.0	2.0
-1.0	0.0
-0.5	-0.3
0.0	0.0
1.0	2.0
2.0	6.0

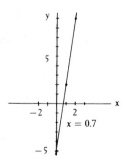

$y = 0$ when $x = 0.0$ and -1.0

Check: $x^2 + x = 0$
$(-1)^2 + (-1) = 1 + (-1) = 0$
$(0)^2 + 0 = 0 + 0 = 0$

13. Let $y = 7x - 5$

x	y
0	-5
1	2
2	9

$y = 0$ when x is approximately 0.7

17. Let $y = x^2 + x - 5$

x	y
-3	1
-2	-3
-1	-5
0	-5
1	-3
2	1

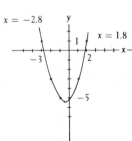

$y = 0$ when x is approximately -2.8 and when x is approximately 1.8

21. Let $y = x^3 - 4x$

x	y
-3	-15
-1	3
0	0
1	-3
3	15

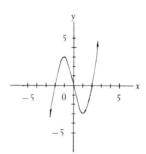

$y = 0$ when $x = -2.0, 0.0, 2.0$

25. $\sqrt{2x + 2} = 3$; $(\sqrt{2x + 2})^2 = 3^2$; $2x + 2 = 9$; $2x - 7 = 0$;   let $y = 2x - 7$

x	y
-2	-11
0	-7
2	-3
4	1

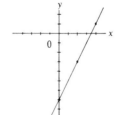

$y = 0$ when x is
approximately 3.5

29. $i = 0.01v - 0.06$

v	i
0	-0.06
5	-0.01
10	0.04
15	0.09
20	0.14

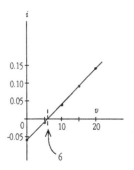

33. $1 = w + 12$; $A = l \times w = (w + 12)(w) = w^2 + 12w = 520$

$$w \approx 18, \quad l \approx 30$$

w	A
0	0
5	85
10	220
15	405
20	640
25	925

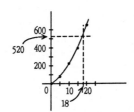

Review, Exercises for Chapter 2, page 80

1. $V = \pi r^2 h$; $V = \pi r^2(8)$; $V = 8\pi r^2$

5. $f(3) = 7(3) - 5 = 21 - 5 = 16$
    $f(-6) = 7(-6) - 5 = -42 - 5 = -47$

9. $H(-4) = \sqrt{1 - 2(-4)} = \sqrt{1 + 8} = \sqrt{9} = 3$
    $H(2h) = \sqrt{1 - 2(2h)} = \sqrt{1 - 4h}$

13. $f(5.87) = 8.07$, ⊟ , 2, ⊠ , 5.87, ⊟ −3.67
     $f(-4.29) = 8.07$, ⊟ , 2, ⊠ , 4.29, +/− , ⊟ 16.7

17. x^4 is a valid quantity for all real values of x, so the domain is
     the set of real numbers. Since x^4 is never negative, the least
     value of x^4 is 0, and for $x^4 + 1$ is 1. The range of $x^4 + 1$ is all
     real numbers equal to or greater than 1.

21. $y = 4x + 2$

x	y
-2	-6
0	2
2	10

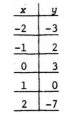

25. $y = 3 - x - 2x^2$

x	y
-2	-3
-1	2
0	3
1	0
2	-7

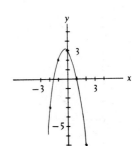

29. $y = 2 - x^4$

x	y
-1.5	-3.0
-1.0	1.0
0.0	2.0
1.0	1.0
1.5	-3.0

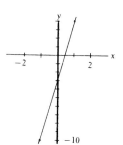

33. From exercise 21, $y = 0$ when x is approximately -0.5

37. From exercise 25, $y = 0$ when x is approximately -1.5 and 1.0

41. From exercise 29, $y = 0$ when x is approximately -1.2 and 1.2.

45. Let $y = 7x - 3$

x	y
-1	-10
0	-3
1	4

$y = 0$ when x is approximately 0.4

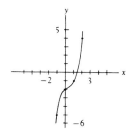

49. Let $y = x^3 - x^2 + x - 2$

x	y
-1	-5
0	-2
1	-1
2	4

$y = 0$ when x is approximately 1.4

53. If a and b have opposite signs, $A(a,b)$ will lie in quadrant diagonally opposite from $B(b,a)$.

57. $F_2 = 0.75F_1 + 45$

F_1 (N)	F_2 (N)
0	45.0
10	52.5
20	60.0

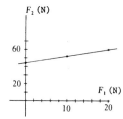

61. $P = 1.5 \times 10^{-6} i^3 - 0.77$

i	P
80	-0.002
90	0.32
100	0.73
110	1.23
120	1.82
130	2.53
140	3.35

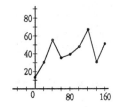

65.

D	d
0	15
20	32
40	56
60	33
80	29
100	47
120	68
140	31
160	52

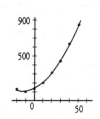

69. $S = 135 + 4.9T + 0.19T^2$

T	S
-20	113
-10	105
0	135
10	203
20	309
30	453
40	635
50	855

CHAPTER 3

The Trigonometric Functions

1. (a) 60°
    (b) 120°
    (c) −90°

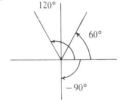

5. 45° + 360° = 405°
    45° − 360° = −315°

9. 70°30' + 360° = 430°30'
    70°30' − 360° = −289°30'

13. To change 0.265 rad to degrees, multiply by 57.30, which gives 15.2. 0.265 rad = 15.2°.

17. $15°12' = 15 \frac{12}{60}^{\circ}$

     $= 15 \frac{1}{5}^{\circ} = 15.2°$

21. $301°16' = 301 \frac{16}{60}^{\circ}$

     $= 301.27°$

25. $0.5 = \frac{5}{10} = \frac{30}{60}$, so 47.5° = 47°30'

29. $5.62° = 5°37'$ since $0.62 = \frac{31}{50}$ and $\frac{31}{50} = \frac{37}{60}$, or 0.62(60') = 37'

33. (4,2)

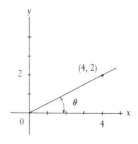

37. (−7,5)

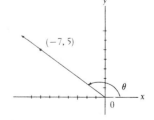

41. $21°42'36" = 21°42 \dfrac{36'}{60}$

$\qquad\qquad = 21°42.6' = 21 \dfrac{42.6°}{60}$

$\qquad\qquad = 21.710°$

Exercises 3-2, page 91

1. $r = \sqrt{6^2 + 8^2} = \sqrt{100} = 10$ (Pythagorean Theorem)

$\sin \theta = \dfrac{y}{r} = \dfrac{8}{10} = \dfrac{4}{5}$ $\qquad$ $\csc \theta = \dfrac{r}{y} = \dfrac{10}{8} = \dfrac{5}{4}$

$\cos \theta = \dfrac{x}{r} = \dfrac{6}{10} = \dfrac{3}{5}$ $\qquad$ $\sec \theta = \dfrac{r}{x} = \dfrac{10}{6} = \dfrac{5}{3}$

$\tan \theta = \dfrac{y}{x} = \dfrac{8}{6} = \dfrac{4}{3}$ $\qquad$ $\cot \theta = \dfrac{x}{y} = \dfrac{6}{8} = \dfrac{3}{4}$

5. $r = \sqrt{9^2 + 40^2} = \sqrt{1681} = 41$

$\sin \theta = \dfrac{y}{r} = \dfrac{40}{41}$ $\qquad$ $\csc \theta = \dfrac{r}{y} = \dfrac{41}{40}$

$\cos \theta = \dfrac{x}{r} = \dfrac{9}{41}$ $\qquad$ $\sec \theta = \dfrac{r}{x} = \dfrac{41}{9}$

$\tan \theta = \dfrac{y}{x} = \dfrac{40}{9}$ $\qquad$ $\cot \theta = \dfrac{x}{y} = \dfrac{9}{40}$

9. $r = \sqrt{1^2 + 1^2} = \sqrt{2}$

$\sin \theta = \dfrac{y}{r} = \dfrac{1}{\sqrt{2}}$ $\qquad$ $\csc \theta = \dfrac{r}{y} = \dfrac{\sqrt{2}}{1} = \sqrt{2}$

$\cos \theta = \dfrac{x}{r} = \dfrac{1}{\sqrt{2}}$ $\qquad$ $\sec \theta = \dfrac{r}{x} = \dfrac{\sqrt{2}}{1} = \sqrt{2}$

$\tan \theta = \dfrac{y}{x} = \dfrac{1}{1} = 1$ $\qquad$ $\cot \theta = \dfrac{x}{y} = \dfrac{1}{1} = 1$

13. $r = \sqrt{3.25^2 + 5.15^2} = \sqrt{37.085} = 6.090$

$\sin \theta = \dfrac{y}{r} = \dfrac{5.15}{6.090} = 0.846$

$\cos \theta = \dfrac{x}{r} = \dfrac{3.25}{6.090} = 0.534$

$\tan \theta = \dfrac{y}{x} = \dfrac{5.15}{3.25} = 1.58$

$\csc \theta = \dfrac{r}{y} = \dfrac{6.090}{5.15} = 1.18$

$\sec \theta = \dfrac{r}{x} = \dfrac{6.090}{3.25} = 1.87$

$\cot \theta = \dfrac{x}{y} = \dfrac{3.25}{5.15} = 0.631$

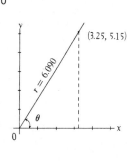

17. $\cos \theta = \dfrac{12}{13} = \dfrac{x}{r}$; $r^2 = x^2 + y^2$;

$13^2 = 12^2 + y^2$; $y^2 = 25$; $y = \pm 5$

$\sin \theta = \dfrac{y}{r} = \dfrac{5}{13}$ for acute θ

$\cot \theta = \dfrac{x}{y} = \dfrac{12}{5}$ for acute θ

21. $\sin \theta = 0.750 = \dfrac{0.750}{1} = \dfrac{y}{r}$; $r^2 = x^2 + y^2$; $1^2 = x^2 + 0.750^2$;

$x = \sqrt{0.4375} = \pm 0.6614$

$\cot \theta = \dfrac{x}{y} = \dfrac{0.6614}{0.750} = 0.882$ for acute θ

$\csc \theta = \dfrac{r}{y} = \dfrac{1}{0.750} = 1.33$ for θ in Quadrant I or Quadrant II

25. $(3,4)$; $x = 3$, $y = 4$, $r = \sqrt{9 + 16} = 5$; $\sin \theta = \dfrac{4}{5}$, $\tan \theta = \dfrac{4}{3}$

$(6,8)$; $x = 6$, $y = 8$, $r = \sqrt{36 + 64} = 10$; $\sin \theta = \dfrac{8}{10} = \dfrac{4}{5}$,

$\tan \theta = \dfrac{8}{6} = \dfrac{4}{3}$

$(4.5,6)$; $x = 4.5$, $y = 6$, $r = \sqrt{20.25 + 36} = 7.5$; $\sin \theta = \dfrac{6}{7.5} = 0.8 = \dfrac{4}{5}$

$\tan \theta = \dfrac{6}{4.5} = \dfrac{4}{3}$

29. $\cos \theta = \dfrac{x}{r}$; $\dfrac{r}{x} = \sec \theta$

Exercises 3-3, page 95

1.

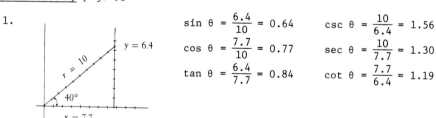

$\sin \theta = \dfrac{6.4}{10} = 0.64$ $\csc \theta = \dfrac{10}{6.4} = 1.56$

$\cos \theta = \dfrac{7.7}{10} = 0.77$ $\sec \theta = \dfrac{10}{7.7} = 1.30$

$\tan \theta = \dfrac{6.4}{7.7} = 0.84$ $\cot \theta = \dfrac{7.7}{6.4} = 1.19$

5. (Set calculator for degrees.)
22.4, [SIN]. The display shows 0.38107038.
sin 22.4° = 0.381

9. 15.71, COS . The display shows 0.9626445.
    cos 15.71° = 0.9626

13. 67.78, TAN , 1/x . The display shows 0.4084997.
     cot 67.78° = 0.4085

17. 49.3, sin $\frac{1}{x}$. The display shows 1.3190274. Round to 1.32.

21. 0.3261, INV , COS . The display shows 70.967768.
     If cos θ = 0.3261, then θ = 70.97°.
     (Calculator may use ARCCOS or COS⁻¹ instead of INV , COS .)

25. 0.207 INV , TAN . The display shows 11.7695053.
     If tan θ = 0.207, then θ = 11.70.
     (Calculator may use ARCTAN or TAN⁻¹ instead of INV , TAN .)

29. 1.245, 1/x , INV , SIN . The display shows 53.438012.
     If csc θ = 1.245, then θ = 53.44°.
     (Calculator may use ARCSIN or SIN⁻¹ instead of INV , SIN .)

41. Scan down "Degrees" column to 19°0', then read across to column labeled
     "sin θ" at top of column and read "0.3256."

45. Read down column labeled "tan θ" at top of column to find "0.8441";
     read "40°10' in left column labeled "degrees."

49. $1.0° \begin{bmatrix} 0.8° \begin{bmatrix} \tan 28.0° = 0.532 \\ \tan 28.8° = \quad ? \\ \tan 29.0° = 0.554 \end{bmatrix} x \end{bmatrix} 0.022$ $\frac{0.8}{1.0} = \frac{x}{0.022}$

     $x = 0.8(0.022) = 0.0176$

     tan 28.8° = 0.532 + 0.0176 = 0.5496 or 0.550.

53. $-1.0° \begin{bmatrix} x' \begin{bmatrix} \cos 73.0° = 0.292 \\ \cos θ \quad = 0.296 \\ \cos 72.0° = 0.309 \end{bmatrix} 0.004 \end{bmatrix} 0.017$ $\frac{x}{-1.0} = \frac{0.004}{0.017}$

     cos θ = 73.0° - 0.2° = 72.8°. $x = -0.23$

57. d = 70 + 30 cos θ = 70 + 30 cos 54.5°.
     54.5, COS X , 30, = , + , 70, = , 87dB

Exercises 3-4, page 100

1.

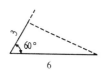

Once the given parts are in place only one side remains to include. Only one length and one direction of line will close up the space and form a triangle.

5.

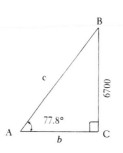

since $\sin A = \dfrac{a}{c}$

$$c = \dfrac{a}{\sin A}$$

$$= \dfrac{6700}{\sin 77.8°}$$

$$= 6850$$

$B = 90.0° - 77.8°$

$= 12.2°$

$\tan B = \dfrac{b}{a}$

$b = a \tan B$

$= 6700 \tan 12.2°$

$= 1450$

9.

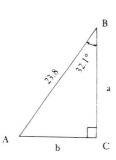

$\sin B = \dfrac{b}{c}$

$b = c \sin B$

$= 23.8 \sin 32.1°$

$= 12.6$

$A = 90.0° - 32.1°$

$= 57.9°$

$\sin A = \dfrac{a}{c}$

$a = c \sin A$

$= 23.8 \sin 57.9°$

$= 20.2$

13.

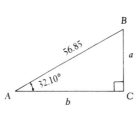

$\sin A = \dfrac{a}{c}$

$a = c \sin A$

$= 56.85 \sin 32.10°$

$= 30.21$

$B = 90.00° - 32.10°$

$= 57.90°$

$\cos A = \dfrac{b}{c}$

$b = c \cos A$

$= 56.85 \cos 32.10°$

$= 48.16$

17.

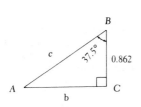

$\tan B = \dfrac{b}{a}$

$\quad b = a \tan B$

$\quad\quad = 0.862 \tan 37.5°$

$\quad\quad = 0.661$

$\cos B = \dfrac{a}{c}$

$\quad c = \dfrac{a}{\cos B}$

$\quad\quad = \dfrac{0.862}{\cos 37.5°}$

$\quad\quad = 1.09$

$A = 90.0° - 37.5°$

$\quad = 52.5°$

21.

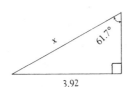

$\tan A = \dfrac{a}{b}$

$\quad A = \arctan \dfrac{a}{b}$

$\quad\quad = \arctan \dfrac{591.87}{264.93}$

$\quad\quad = \arctan 2.23406$

$\quad\quad = 65.89°$

$\tan B = \dfrac{b}{a}$

$\quad B = \arctan \dfrac{b}{a}$

$\quad\quad = \arctan \dfrac{264.93}{591.87}$

$\quad\quad = \arctan 0.447615$

$\quad\quad = 24.11°$

$c = \sqrt{(264.93)^2 + (591.87)^2} = 648.46$

25.

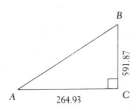

$\sin 61.7° = \dfrac{3.92}{x}$

$\quad\quad\quad x = \dfrac{3.92}{\sin 61.7°}$

$\quad\quad\quad\quad = 4.45$

29.

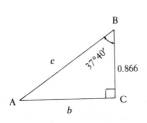

$\sin A = \dfrac{a}{c}$

$\quad c = \dfrac{a}{\sin A}$

$\quad\quad = \dfrac{0.886}{\sin 52°20'}$

$\quad\quad = \dfrac{0.886}{0.7916}$

$\quad\quad = 1.12$

$\tan B = \dfrac{b}{a}$

$\quad b = a \tan B$

$\quad\quad = 0.886(\tan 37°40')$

$\quad\quad = 0.886(0.7720)$

$\quad\quad = 0.684$

$A = 90°0' - 37°40'$

$\quad = 52°20'$

33. $C = 90°,\quad B = 90° - A,\quad a = c \sin A$ since $\sin A = \dfrac{a}{c}$, and $b = c \cos A$

Exercises 3-5, page 103

1.

$$\sin 54.0° = \frac{h}{120}$$

$$h = 120 \sin 54.0° = 120(0.809) = 97$$

5.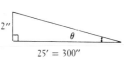

$$\tan \theta = \frac{2}{300} = 0.00667$$

$$\theta = \arctan 0.00667 = 0.38°$$

5M.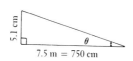

$$\tan \theta = \frac{5.1}{750} = 0.0068$$

$$\theta = \arctan 0.0068 = 0.39°$$

9.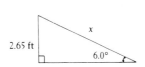

$$\sin 6.0° = \frac{2.65}{x}$$

$$x = \frac{2.65}{\sin 6.0°} = 25.4 \text{ ft}$$

9M.

$$\sin 6.0° = \frac{80.0}{x}$$

$$x = \frac{80.0}{\sin 6.0°} = 765.3 \text{ cm}$$

13.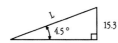

$$\sin 4.5° = \frac{15.3}{L}$$

$$L = \frac{15.3}{\sin 4.5°} = 195 \text{ ft}$$

13M.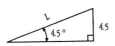

$$\sin 4.5° = \frac{4.5}{L}$$

$$L = \frac{4.5}{\sin 4.5°} = 57.4 \text{ m}$$

17.

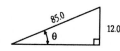

$$\sin \theta = \frac{12.0}{85.0} = 0.1412$$

$$\theta = 8.1°$$

21.

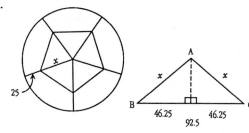

$$\angle A = 360° + 5 = 72°$$

$$\angle B = \angle C = \frac{180° - 72°}{2} = 54°$$

$$\cos C = \frac{46.25}{x}$$

$$r = x + 25$$

$$x = \frac{46.25}{\cos C} = \frac{46.25}{0.588} = 78.65$$

$$C = 2\pi(x + 25) = 78.65, \boxed{+}, 25, \boxed{=}, \boxed{\times}, 2, \boxed{\times}, \boxed{\pi}, \boxed{=} 651$$

$$\cos 31.8° = \frac{12.8\pi}{l}$$

$$l = \frac{12.8\pi}{\cos 31.8°}$$

$$\frac{1}{2} C = \frac{1}{2} \left[2\pi (11.8 + 1) \right]$$

$$= 12.8\pi$$

$$l = 12.8, \boxed{\times}, \boxed{\pi}, \boxed{=}, \boxed{+}, 31.8, \boxed{\cos}, \boxed{=} \quad 47.3$$

Review Exercises for Chapter 3, page 105

1. Smallest positive angle coterminal with 17.0° is 360° + 17.0° = 377.0°.
    Smallest numerical negative angle coterminal with 17.0° is −360° + 17.0°
    = −343.0°.

5. $31°54' = 31\dfrac{54°}{60} = 31.9°$

9. $17.5° = 17\dfrac{5}{10}°$

    $= 17\dfrac{30°}{60} = 17°30'$

13. $x = 24$, $y = 7$

$r = \sqrt{24^2 + 7^2} = \sqrt{576 + 49}$

$\quad = \sqrt{625} = 25$

$\sin \theta = \dfrac{y}{r} = \dfrac{7}{25}$ $\qquad$ $\csc \theta = \dfrac{r}{y} = \dfrac{25}{7}$

$\cos \theta = \dfrac{x}{r} = \dfrac{24}{25}$ $\qquad$ $\sec \theta = \dfrac{r}{x} = \dfrac{25}{24}$

$\tan \theta = \dfrac{y}{x} = \dfrac{7}{24}$ $\qquad$ $\cot \theta = \dfrac{x}{y} = \dfrac{24}{7}$

17. $\sin \theta = \dfrac{y}{r} = \dfrac{5}{13}$; $y = 5$; $r = 13$;

$r = \sqrt{x^2 + 5^2} = \sqrt{x^2 + 25} \; = 13$

$\qquad\qquad\qquad x^2 + 25 = 13^2 = 169$

$\qquad\qquad\qquad\qquad x^2 = 169 - 25 = 144$

$\qquad\qquad\qquad\qquad\; x = 12$

$\cos \theta = \dfrac{x}{r} = \dfrac{12}{13} = 0.923$

$\cot \theta = \dfrac{x}{y} = \dfrac{12}{5} = 2.40$

21. 72.1°, [SIN] gives display of 0.95159441
     so sin 72.1° = 0.952

25. 18.4°, [cos], [1/x], = 1.05

29. 0.950, [inv], [cos], = 18.20

33. 4.713, [1/x], [sin], = 12.25°

37.

$\tan 17.0° = \dfrac{a}{6.00}$ $\qquad\qquad$ $\cos 17.0° = \dfrac{6.00}{c}$

$\qquad a = 6.00 \tan 17.0°$ $\qquad\qquad$ $c = \dfrac{6.00}{\cos 17.0°}$

$\qquad a = 6.00(0.3057)$ $\qquad\qquad\;\; c = \dfrac{6.00}{0.9563}$

$\qquad a = 1.83$ $\qquad\qquad\qquad\qquad c = 6.27$

$B = 90.0° - 17.0°$

$\quad = 73.0°$

41.

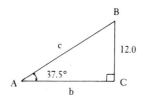

$\tan 37.5° = \dfrac{12.0}{b}$ $\sin 37.5° = \dfrac{12.0}{c}$

$b = \dfrac{12.0}{\tan 37.5°}$ $c = \dfrac{12.0}{\sin 37.5°}$

$b = \dfrac{12.0}{0.7673}$ $c = \dfrac{12.0}{0.6088}$

$b = 15.6$ $c = 19.7$

$B = 90.0° - 37.5°$
$= 52.5°$

45.

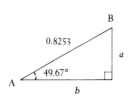

$\sin 49.67° = \dfrac{a}{0.8253}$

$a = 0.8253 \sin 49.67°$

$a = 0.8253(0.7623)$

$a = 0.6292$

$\cos 49.67° = \dfrac{b}{0.8253}$

$b = 0.8253 \cos 49.67°$

$b = 0.8253(0.6472)$

$b = 0.5341$

$B = 90.00° - 49.67°$
$= 40.33°$

49. $F_y = F \cos \theta$
$F_y = 56.0 \cos 37.5°$
$F_y = 56.0(0.7934)$
$F_y = 44.4$ N

53. $\tan \theta = \dfrac{v^2}{gr} = \dfrac{(80.7)^2}{(32.2)(950)}$ = 80.7, $\boxed{x^2}$, $\boxed{+}$, 32.2, $\boxed{+}$, 950, $\boxed{=}$, $\boxed{\text{INV}}$, $\boxed{\tan}$ = 12°

53M. $\tan \theta = \dfrac{(24.2)^2}{(9.80)(285)}$ = 24.2, $\boxed{x^2}$, $\boxed{+}$, 9.8, $\boxed{+}$, 285, $\boxed{=}$, $\boxed{\text{INV}}$, $\boxed{\tan}$ = 12°

57. $90° - 65° = 25°$ $\tan 25° = \dfrac{2.0}{2.5 + x}$

$x + 2.5 = \dfrac{2.0}{\tan 25°}$

$x = \dfrac{2.0}{\tan 25°} - 2.5 =$ 2 $\boxed{+}$ 25 $\boxed{\tan}$ $\boxed{=}$ $\boxed{-}$ 2.5
$= 1.8$, or 56°

57M. $\tan 25° = \dfrac{0.60}{0.75 + x}$

$x + 0.75 = \dfrac{0.60}{\tan 25°}$; $x = 6.0$, $\boxed{+}$, 25, $\boxed{\tan}$, $\boxed{=}$, $\boxed{-}$, 0.75, $\boxed{=}$, 0.54 or 56°

73.

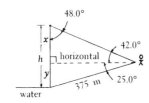

$$\sin 25.0° = \frac{x}{375}; \quad x = 375 \sin 25° = 158.48$$

$$\cos 25.0° = \frac{\text{horizontal}}{375}; \quad \text{horizontal} = 339.87$$

$$\tan 42.0° = \frac{y}{339.87}; \quad y = 339.87 \tan 42.0° = 306.02$$

$$h = x + y = 158.48 + 306.02 \approx 464 \text{ m}$$

CHAPTER 4

Systems of Linear Equations; Determinants

1. The coordinates of the point $(3,1)$ satisfy the equation since $2(3) + 3(1)$ $= 6 + 3 = 9$. The coordinates of the second point do not satisfy the equation, since $2(5) + 3(\frac{1}{3}) = 10 + 1 = 11 \neq 9$.

5. (a) $5(1) - y = 6$; $5 - y = 6$; $-y = 1$; $y = -1$
   (b) $5(-2) - y = 6$; $-10 - y = 6$; $-y = 16$; $y = -16$

9. If the values $x = 4$, $y = -1$ satisfy both equations, they are a solution. $4 - (-1) = 4 + 1 = 5$; $2(4) + (-1) = 8 - 1 = 7$. Therefore the given values are a solution.

13. $2(\frac{1}{2}) - 5(-\frac{1}{5}) = 1 + 1 = 2 \neq 0$; $4(\frac{1}{2}) + 10(-\frac{1}{5}) = 2 - 2 = 0 \neq 4$. The given values are not a solution.

17. If $p = 260$ and $w = 40$, then $260 + 40 = 300$ and $260 - 40 = 220$, and this is a solution.

1. By taking $(3,8)$ as (x_2,y_2) and $(1,0)$ as (x_1,y_1), $m = \dfrac{8 - 0}{3 - 1} = \dfrac{8}{2} = 4$

5. By taking $(-2,-5)$ as (x_2,y_2) and $(5,-3)$ as (x_1,y_1), $m = \dfrac{-5 - (-3)}{-2 - 5}$
   $= \dfrac{-5 + 3}{-7} = \dfrac{2}{7}$

9. $m = 2$, $b = -1$
   Plot the y-intercept point $(0,-1)$. Since the slope is $2/1$, from this point, go up 2 units and over 1 unit to the right, and plot a second point. Sketch the line between the two points.

$(0, -1)$

13. $m = \frac{1}{2}$, $b = 0$
     Plot the y-intercept, point (0,0).
     Since the slope is $\frac{1}{2}$, from this point
     go up 1 unit, and over 2 units to the
     right, and plot a second point.
     Sketch the line between the two
     points.

17. $y = -2x + 1$
     Since the equation is solved for y,
     the coefficient of x is the slope.
     Therefore, the slope of the line is
     -2/1. Since the b-term is the y-
     intercept, the y-intercept of this
     line is (0,1). Plot the y-intercept.
     From this point go down 2 units, and
     over 1 unit to the right. Draw the
     line between these points.

21. $5x - 2y = 4$; $-2y = -5x + 4$; $y =$
     $\frac{5}{2}x - 2$
     The coefficient of x, $\frac{5}{2}$, is the slope.
     The y-intercept is (0,-2). Plot the
     y-intercept, (0,-2). From this point,
     go up 5 units, and over 2 units to the
     right. Draw the line between the
     points.

25. $x + 2y = 4$; let $x = 0$; $0 + 2y = 4$;
     $y = 2$
     Therefore, the y-intercept is (0,2).
     Let $y = 0$; $x + 2(0) = 4$; $x = 4$.
     Therefore, the x-intercept is (4,0).
     Plot the points (0,2) and (4,0). A
     third point is found as a check.
     Let $x = 6$; $6 + 2y = 4$; $2y = -2$;
     $y = -1$. Therefore, the point (6,-1)
     should lie on the line.

29. $y = 3x + 6$; let $x = 0$; $y = 3(0) + 6$
     $= 6$.
     Therefore, the y-intercept is (0,6).
     Let $y = 0$; $0 = 3x + 6$; $3x = -6$;
     $x = -2$; therefore, the x-intercept is
     (-2,0). Plot these points. As a
     check, let $x = 1$; $y = 3(1) + 6 = 9$.
     The point (1,9) should be on the line.

33. $d = 0.2\ell + 1.2$; y-intercept is
     $(0,1.2)$; $m = 0.2 = 1/5$
     Plot $(0,1.2)$. From this point,
     go up 1 unit, and over 5 units
     to the right. Draw the line
     between the points.

Exercises 4-3, page 119

1. $y = -x + 4$; $y = x - 2$
    The slope of the first line is -1, and the
    y-intercept is 4. The slope of the second
    line is 1, and the y-intercept is -2. From
    the graph, the point of intersection is
    $(3.0,1.0)$. Thus the solution of the system
    of equations is $x = 3.0$, $y = 1.0$.

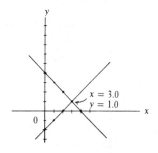

5. $3x + 2y = 6$; $x - 3y = 3$
    The intercepts of the first line are $(0,3)$
    and $(2,0)$. A third point is $(-2,6)$. The
    intercepts of the second line are $(0,-1)$
    and $(3,0)$. A third point is $(6,1)$. From
    the graph, the point of intersection is
    $(2.2,-0.3)$. Thus the solution to the
    system of equations is $x = 2.2$ and
    $y = -0.3$.

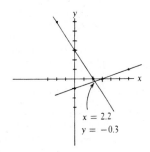

9. $x - 4y = 8$; $2x + 5y = 10$
    The intercepts of the first line are $(0,-2)$
    and $(8,0)$. Another point is $(10,0.5)$.
    The intercepts of the second line are
    $(0,2)$ and $(5,0)$. Another point is
    $(10,-2)$. From the graph, the point of
    intersection is $(6.2,-0.5)$, and the
    solution to the system of equations is
    $x = 6.2$ and $y = -0.5$.

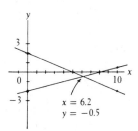

13. $x - 4y = 6$; $2y = x + 4$
The intercepts of the first line
are $(0,-1.5)$ and $(6,0)$. The intercepts of the second line are $(0,2)$
and $(-4,0)$. From the graph, the
point of intersection is $(-14.0,-5.0)$,
and the solution to the system of
equations is $x = -14.0$ and $y = -5.0$.

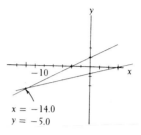

$x = -14.0$
$y = -5.0$

17. $x = 4y + 2$; $3y = 2x + 3$; $y = \frac{2}{3}x + 1$
The intercepts of the first line
are $(0,-\frac{1}{2})$ and $(2,0)$. The slope
of the second line is $\frac{2}{3}$, and the
y-intercept is 1.

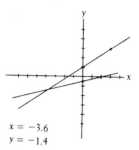

$x = -3.6$
$y = -1.4$

21. $x - 5y = 10$; $2x - 10y = 20$
The intercepts for the first line are
$(0,-2)$ and $(10,0)$. The intercepts for
the second line are $(0,-2)$ and $(10,0)$.
The lines are the same, and have all
points in common. The system is
dependent.

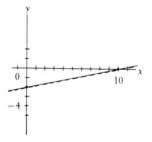

25. $5x = y + 3$; $4x = 2y - 3$
The intercepts for the first line are
$(0.0,-3.0)$ and $(0.6,0.0)$. The intercepts for the second line are $(0.0,1.5)$
and $(-0.75,0.0)$. From the graph, the
point of intersection is $(1.5,4.5)$, and
the solution to the system of equations
is $x = 1.5$ and $y = 4.5$.

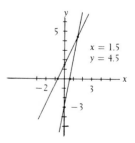

$x = 1.5$
$y = 4.5$

29. $0.8T_1 - 0.6T_2 = 12$; $0.6T_1 + 0.8T_2 = 68$

The intercepts for the first line are
$(0,15)$ and $(-20,0)$. A third point is
$(40,45)$.   The intercepts for the second
line are $(1,113.3)$ and $(85,0)$. From the
graph the point of intersection is $(47,50)$.

Exercises 4-4, page 125

1. (1) $x = y + 3$
    (2) $x - 2y = 5$
         $y + 3 - 2y = 5$ the value of x from (1) substituted for x in (2)
         $-y = 2; \quad y = -2$
         $x = -2 + 3 = +1$ the value of y substituted for y in (1)
    The solution is $x = 1, y = -2$.

5. (1) $x + y = -5; \quad x = -5 - y$
    (2) $2x - y = 2$
         $2(-5 - y) - y = 2$ the value of x from (1) substituted for x in (2)
         $-10 - 2y - y = 2; \quad -10 - 3y = 2; \quad -3y = 12; \quad y = -4$
         $x + (-4) = -5; \quad x = -1$ the value of y substituted in (1)
    The solution is $x = -1, y = -4$.

9. (1) $3x + 2y = 7; \quad 3x = 7 - 2y; \quad x = \frac{7}{3} - \frac{2}{3}y$
    (2) $2y = 9x + 11; \quad 2y = 9(\frac{7}{3} - \frac{2}{3}y) + 11; \quad 2y = 32 - 6y$
         $8y = 32; \quad y = 4; \quad 3x + 2(4) = 7; \quad 3x + 8 = 7; \quad 3x = -1; \quad x = -\frac{1}{3}$
    The solution is $x = -\frac{1}{3}, y = 4$.

13. (1) $x + 2y = 5$
    (2) $\underline{x - 2y = 1}$
         $2x \quad = 6$ add (1) and (2)
         $x = 3$
         $3 + 2y = 5$ substitute for x in (1)
         $2y = 2$
         $y = 1$
    The solution is $x = 3, y = 1$.

17. (1) $2x + 3y = 8$
    (2) $x = 2y - 3$
    (3) $x - 2y = -3$ (2) put in standard form
    (4) $-2x + 4y = 6$ multiply (3) by -2
    (5) $\underline{2x + 3y = 8}$ recopy (1)
         $7y = 14$ add (4) and (5)
         $y = 2$
         $x = 2(2) - 3 = 1$ substitute for y in (2)
    The solution is $x = 1, y = 2$.

21. (1) $2x - 3y - 4 = 0$
     (2) $3x + 2 = 2y$
     (3) $2x - 3y = 4$ (1) put in standard form
     (4) $3x - 2y = -2$ (2) put in standard form
     (5) $4x - 6y = 8$ (3) multiplied by 2
     (6) $\underline{-9x + 6y = 6}$ (4) multiplied by -3
            $-5x\qquad = 14$ (5) and (6) added

            $x = -\dfrac{14}{5}$

            $2(-\dfrac{14}{5}) - 3y = 4$ substitute for x in (1)

            $-\dfrac{28}{5} - 3y = \dfrac{20}{5};\quad -3y = \dfrac{48}{5};\quad y = -\dfrac{48}{15} = -\dfrac{16}{5}$

     The solution is $x = -\dfrac{14}{5},\ y = -\dfrac{16}{5}$.

25. (1) $2x - y = 5$
     (2) $6x + 2y = -5$
     (3) $\underline{4x - 2y = 10}$ multiply (1) by 2
            $10x\qquad = 5$ add (2) and (3)

            $x = \dfrac{1}{2}$

            $2(\dfrac{1}{2}) - y = 5$ substitute for x in (1)

            $1 - y = 5;\quad -y = 4;\quad y = -4$

     The solution is $x = \dfrac{1}{2},\ y = -4$.

29. (1) $3x - 6y = 15$
     (2) $\underline{4x - 8y = 20}$
     (3) $12x - 24y = 60$ multiply (1) by 4
     (4) $\underline{-12x + 24y = -60}$ multiply (2) by -3
            $0 = 0$ add (3) and (4)

     The solution is dependent; i.e., the two equations represent the same line.

33. (1) $v_1 + v_2 = 15$

     (2) $v_1 - v_2 = 3$
     (3) $2v_1\qquad = 18$ Add (1) and (2)
            $v_1 = 9$
            $9 + v_2 = 15$ Substitute for v_1 in (1)
            $v_2 = 6$
     The solution is $v_1 = 9,\ v_2 = 6$.

37. $t_2 = t_1 - 12$ 37M. $t_2 = t_1 - 12$

          $2000t_1 = 3200t_2$ $600t_1 = 960t_2$

          $2000t_1 = 3200(t_1 - 12)$ $600t_1 = 960(t_1 - 12)$

          $2000t_1 = 3200t_1 - 38400$ $600t_1 = 960t_1 - 11520$

          $-1200t_1 = -38400$ $-360t_1 = -11520$

          $t_1 = 32$ $t_1 = 32$

          $t_2 = 32 - 12 = 20$ $t_2 = 32 - 12 = 20$

41. Let x = number of mL of 5% solution;
     y = number of mL of 25% solution

   (1) $x + y = 1000$
   (2) $0.05x + 0.25y = 0.10(1000);$ $0.05x + 0.25y = 100$
   (3) $y = 1000 - x$
   (4) $0.05x + 0.25(1000 - x) = 100$
   (5) $0.05x + 250 - 0.25x = 100$

   $-0.20x = 100 - 250 = -150;$ $x = \dfrac{-150}{-0.20} = 750$ mL 5% solution, 250 mL 25%

Exercises 4-5, page 131

1. $\begin{vmatrix} 2 & 4 \\ 3 & 1 \end{vmatrix} = 2(1) - 3(4) = 2 - 12 = -10$

5. $\begin{vmatrix} 8 & -10 \\ 0 & 4 \end{vmatrix} = 8(4) - 0(-10) = 32 - 0 = 32$

9. $\begin{vmatrix} 0.7 & -1.3 \\ 0.1 & 1.1 \end{vmatrix} = 0.7(1.1) - 0.1(-1.3) = 0.77 + 0.13 = 0.9$

13. $x + 2y = 5$
     $x - 2y = 1$

$x = \dfrac{\begin{vmatrix} 5 & 2 \\ 1 & -2 \end{vmatrix}}{\begin{vmatrix} 1 & 2 \\ 1 & -2 \end{vmatrix}} = \dfrac{5(-2) - 1(2)}{1(-2) - 1(2)} = \dfrac{-10 - 2}{-2 - 2} = \dfrac{-12}{-4} = 3$

$y = \dfrac{\begin{vmatrix} 1 & 5 \\ 1 & 1 \end{vmatrix}}{\begin{vmatrix} 1 & 2 \\ 1 & -2 \end{vmatrix}} = \dfrac{1(1) - 1(5)}{1(-2) - 1(2)} = \dfrac{1 - 5}{-2 - 2} = \dfrac{-4}{-4} = 1$

17. $2x + 3y = 8$
     $x = 2y - 3;$ $x - 2y = -3$

$x = \dfrac{\begin{vmatrix} 8 & 3 \\ -3 & -2 \end{vmatrix}}{\begin{vmatrix} 2 & 3 \\ 1 & -2 \end{vmatrix}} = \dfrac{8(-2) - (-3)(3)}{2(-2) - 1(3)} = \dfrac{-16 + 9}{-4 - 3} = \dfrac{-7}{-7} = 1$

$y = \dfrac{\begin{vmatrix} 2 & 8 \\ 1 & -3 \end{vmatrix}}{\begin{vmatrix} 2 & 3 \\ 1 & -2 \end{vmatrix}} = \dfrac{2(-3) - 1(8)}{-7} = \dfrac{-6 - 8}{-7} = \dfrac{-14}{-7} = 2$

21. $2x - 3y - 4 = 0;$ $2x - 3y = 4$
     $3x + 2 = 2y;$ $3x - 2y = -2$

$$x = \frac{\begin{vmatrix} 4 & -3 \\ -2 & -2 \end{vmatrix}}{\begin{vmatrix} 2 & -3 \\ 3 & -2 \end{vmatrix}} = \frac{4(-2) - (-2)(-3)}{2(-2) - 3(-3)} = \frac{-8 - 6}{-4 + 9} = \frac{-14}{5} = -\frac{14}{5}$$

$$y = \frac{\begin{vmatrix} 2 & 4 \\ 3 & -2 \end{vmatrix}}{\begin{vmatrix} 2 & -3 \\ 3 & -2 \end{vmatrix}} = \frac{2(-2) - 3(4)}{5} = \frac{-4 - 12}{5} = \frac{-16}{5} = -\frac{16}{5}$$

25. $2x - y = 5$
     $6x + 2y = -5$

$$x = \frac{\begin{vmatrix} 5 & -1 \\ -5 & 2 \end{vmatrix}}{\begin{vmatrix} 2 & -1 \\ 6 & 2 \end{vmatrix}} = \frac{5(2) - (-5)(-1)}{2(2) - 6(-1)} = \frac{10 - 5}{4 + 6} = \frac{5}{10} = \frac{1}{2}$$

$$y = \frac{\begin{vmatrix} 2 & 5 \\ 6 & -5 \end{vmatrix}}{\begin{vmatrix} 2 & -1 \\ 6 & 2 \end{vmatrix}} = \frac{2(-5) - 6(5)}{10} = \frac{-10 - 30}{10} = -4$$

29. $3x - 6y = 15$
     $4x - 8y = 20$

$$x = \frac{\begin{vmatrix} 15 & -6 \\ 20 & -8 \end{vmatrix}}{\begin{vmatrix} 3 & -6 \\ 4 & -8 \end{vmatrix}} = \begin{matrix} 15(-8) - 20(-6) = 0 \\ 3(-8) - 4(-6) = 0 \end{matrix}$$ The equations are dependent.

33. $F_1 + F_2 = 21$
     $2F_1 = 5F_2$ or $2F_1 - 5F_2 = 0$

$$F_1 = \frac{\begin{vmatrix} 21 & 1 \\ 0 & -5 \end{vmatrix}}{\begin{vmatrix} 1 & 1 \\ 2 & -5 \end{vmatrix}} = \frac{21(-5) - 0(1)}{1(-5) - 2(1)}$$ $$F_2 = \frac{\begin{vmatrix} 1 & 21 \\ 2 & 0 \end{vmatrix}}{\begin{vmatrix} 1 & 1 \\ 2 & -5 \end{vmatrix}} = \frac{1(0) - (2)(21)}{-7} = \frac{-42}{-7} = 6$$

$$= \frac{-105}{-7} = 15$$

37.   24 min = 0.4 hr
         $t_2 = t_1 - 0.4$ or $t_1 - t_2 = 0.4$

      $42t_1 = 50t_2$ or $42t_1 - 50t_2 = 0$

$$t_1 = \frac{\begin{vmatrix} 0.4 & -1 \\ 0 & -50 \end{vmatrix}}{\begin{vmatrix} 1 & -1 \\ 42 & -50 \end{vmatrix}} = \frac{0.4(-50) - (-1)(0)}{1(-50) - (42)(-1)}$$

$$= \frac{-20}{-8} = 2.5 \text{ hr}$$

$$t_2 = \frac{\begin{vmatrix} 1 & 0.4 \\ 42 & 0 \end{vmatrix}}{\begin{vmatrix} 1 & -1 \\ 42 & -50 \end{vmatrix}} = \frac{1(0) - (42)(0.4)}{-8} = \frac{-16.8}{-8} = 2.1 \text{ hr}$$

41. $0.03x + 0.08y = 0.06(2) = 0.12$
      $x + y = 2$

$$x = \frac{\begin{vmatrix} 0.12 & 0.08 \\ 2 & 1 \end{vmatrix}}{\begin{vmatrix} 0.03 & 0.08 \\ 1 & 1 \end{vmatrix}} = \frac{0.12(1) - (0.08)(2)}{0.03(1) - (0.08)(1)} = \frac{-0.04}{-0.05} = 0.8 \text{ L}$$

$$y = \frac{\begin{vmatrix} 0.03 & 0.12 \\ 1 & 2 \end{vmatrix}}{\begin{vmatrix} 0.03 & 0.08 \\ 1 & 1 \end{vmatrix}} = \frac{0.03(2) - (1)(0.12)}{-0.05} = \frac{-0.06}{-0.05} = 1.2 \text{ L}$$

Exercises 4-6, page 137

1. (1) $x + y + z = 2$
    (2) $x \quad - z = 1$
    (3) $x + y \quad = 1$
    (4) $\quad -z = 1 - x; \quad z = x - 1$ equation (2) solved for z
    (5) $x + y + (x - 1) = 2; \quad 2x + y = 3$ (4) substituted in (1)
    (6) $\underline{\quad\quad\quad\quad -x - y = -1}$ (3) multiplied by −1
    (7) $\quad\quad\quad\quad x \quad = 2$ (5) and (6) added
    (8) $2 - z = 1; \quad -z = -1; \quad z = 1$ (7) substituted in (2)
    (9) $2 + y = 1; \quad y = -1$ (7) substituted in (3)

    The solution is $x = 2$, $y = -1$, $z = 1$.

5.   (1) $5x + 6y - 3z = 6$
      (2) $4x - 7y - 2z = -3$
      (3) $\underline{3x + y - 7z = 1}$
      (4) $10x + 12y - 6z = 12$ (1) multiplied by 2
      (5) $\underline{-12x + 21y + 6z = 9}$ (2) multiplied by -3
      (6) $-2x + 33y = 21$ (4) and (5) added

      (7) $35x + 42y - 21z = 42$ (1) multiplied by 7
      (8) $\underline{-9x - 3y + 21z = -3}$ (3) multiplied by -3
      (9) $26x + 39y = 39$ (7) and (8) added
      (10) $\underline{-26x + 429y = 273}$ (6) multiplied by 13
      (11) $468y = 312$ (9) and (10) added

      (12) $y = \dfrac{2}{3}$

      (13) $-2x + 33\left(\dfrac{2}{3}\right) = 21$
               $-2x = -1; \quad x = \dfrac{1}{2}$ (12) substituted in (6)
      (14) $5\left(\dfrac{1}{2}\right) + 6\left(\dfrac{2}{3}\right) - 3z = 6$ (12) and (13) substituted in (1)
               $z = \dfrac{1}{6}$

      The solution is $x = \dfrac{1}{2}$, $y = \dfrac{2}{3}$, $z = \dfrac{1}{6}$.

9.   (1) $3x - 7y + 3z = 6$
      (2) $3x + 3y + 6z = 1$
      (3) $5x - 5y + 2z = 5$
      (4) $-6x + 14y - 6z = -12$ (1) multiplied by -2
               $\underline{3x + 3y + 6z = 1}$ (2)
      (5) $-3x + 17y = -11$ (4) and (2) added
      (6) $-15x + 15y - 6z = -15$ (3) multiplied by -3
               $\underline{3x + 3y + 6z = 1}$ (2)
      (7) $-12x + 18y = -14$ (6) and (2) added
      (8) $\underline{12x - 68y = 44}$ (5) multiplied by -4
      (9) $-50y = 30$ (7) and (8) added
      (10) $y = -\dfrac{3}{5}$

      (11) $-3x + 17\left(-\dfrac{3}{5}\right) = -11$ (10) substituted in (5)

      (12) $-3x - \dfrac{51}{5} = -11$

      (13) $-3x = -\dfrac{4}{5}$

      (14) $x = \dfrac{4}{15}$

      (15) $3\left(\dfrac{4}{15}\right) - 7\left(-\dfrac{3}{5}\right) + 3z = 6$ (10) and (15) substituted in (1)

      (16) $\dfrac{4}{5} + \dfrac{21}{5} + 3z = 6; \quad z = \dfrac{1}{3}$

      The solution is $x = \dfrac{4}{15}$, $y = -\dfrac{3}{5}$, $z = \dfrac{1}{3}$.

13. (1) $2x + 3y - 5z = 7$
    (2) $4x - 3y - 2z = 1$
    (3) $8x - y + 4z = 3$
    (4) $\underline{8x - 6y - 4z = 2}$ (2) multiplied by 2
    (5) $16x - 7y \quad = 5$ (3) and (4) added
    (6) $4x + 6y - 10z = 14$ (1) multiplied by 2
    (7) $\underline{-20x + 15y + 10z = -5}$ (2) multiplied by -5
    (8) $-16x + 21y \quad = 9$ (6) and (7) added
    (9) $\underline{48x - 21y \quad = 15}$ (5) multiplied by 3
    (10) $32x \quad = 24$ (8) and (9) added

$$x = \frac{3}{4}$$

    (11) $16\left(\frac{3}{4}\right) - 7y = 5; \quad y = 1$ (10) substituted in (5)

    (12) $4\left(\frac{3}{4}\right) - 3(1) - 2z = 1; \quad z = -\frac{1}{2}$ (10) and (11) substituted in (2)

The solution is $x = \frac{3}{4}$, $y = 1$, $z = -\frac{1}{2}$.

17. (1) $P + M + I = 1150$
    (2) $P = 4I - 100$
    (3) $P = 6M + 50$
    (4) $I = \frac{1}{4}P + 25$ (2) solved for I

    (5) $M = \frac{1}{6}P - \frac{50}{6}$ (3) solved for M

    (6) $P + (\frac{1}{6}P - \frac{50}{6}) + (\frac{1}{4}P + 25) = 1150$

            (4) & (5) substituted in (1)
    (7) $P = 800$ (6) solved for P
    (8) $I = 225$ (7) substituted in (4) and solved
    (9) $M = 125$ (7) substituted in (5) and solved

21. (1) $A + B + C = 180$
    (2) $A + B = 90 - A$ or $2A + B = 90$
    (3) $A + 2B = 180 - A - B$ or $2A + 3B = 180$
    (4) $2B = 90$ (2) subtracted from (3)
    (5) $B = 45.0$
    (6) $A = 22.5$ (5) substituted in (2)
    (7) $C = 112.5$ (5) and (6) substituted in (1)

25. (1) $x - 2y - 3z = 2$
    (2) $x - 4y - 13z = 14$
    (3) $-3x + 5y + 4z = 0$

    (4) $-x + 4y + 13z = -14$ (2) multiplied by -1
         $\underline{x - 2y - 3z = 2}$ (1)
    (5) $2y + 10z = -12$ (4) and (1) added
    (6) $y + 5z = -6$ (5) divided by 2

    (7) $3x - 6y - 9z = 6$ (1) multiplied by 3
         $\underline{-3x + 5y + 4z = 0}$ (3)
    (8) $-y - 5z = 6$ (7) and (3) added
    (9) $y + 5z = -6$ (8) multiplied by -1

(6) and (9) are dependent. There are an unlimited number of solutions.
One solution is found by choosing $z = 0$, then $y = -6$, and, from (1),
$x - 2(-6) - 3(0) = 2$ and $x = 2 - 12 + 0 = -10$. Another solution is
if $z = 1$, $y = -11$, $x = -17$.

<u>Exercises 4-7,</u> page 143

1. $\begin{vmatrix} 5 & 4 & -1 \\ -2 & -6 & 8 \\ 7 & 1 & 1 \end{vmatrix}\begin{matrix} 5 & 4 \\ -2 & -6 \\ 7 & 1 \end{matrix}$ = $(-30) + (+224) + (+2) - (+42) - (+40) - (-8) = 122$

5. $\begin{vmatrix} -3 & -4 & -8 \\ 5 & -1 & 0 \\ 2 & 10 & -1 \end{vmatrix}\begin{matrix} -3 & -4 \\ 5 & -1 \\ 2 & 10 \end{matrix}$ = $(-3) + 0 + (-400) - (+16) - (0) - (+20) = -439$

9. $\begin{vmatrix} 5 & 4 & -5 \\ -3 & 2 & -1 \\ 7 & 1 & 3 \end{vmatrix}\begin{matrix} 5 & 4 \\ -3 & 2 \\ 7 & 1 \end{matrix}$ = $(+30) + (-28) + (+15) - (-70) - (-5) - (-36) = 128$

13. $x = \dfrac{\begin{vmatrix} 4 & 3 & 1 \\ -3 & 0 & -1 \\ -5 & -2 & 2 \end{vmatrix}\begin{matrix} 4 & 3 \\ -3 & 0 \\ -5 & -2 \end{matrix}}{\begin{vmatrix} 2 & 3 & 1 \\ 3 & 0 & -1 \\ 1 & -2 & 2 \end{vmatrix}\begin{matrix} 2 & 3 \\ 3 & 0 \\ 1 & -2 \end{matrix}} = \dfrac{0 + 15 + 6 - 0 - 8 + 18}{0 - 3 - 6 - 0 - 4 - 18} = \dfrac{31}{-31} = -1$

$y = \dfrac{\begin{vmatrix} 2 & 4 & 1 \\ 3 & -3 & -1 \\ 1 & -5 & 2 \end{vmatrix}\begin{matrix} 2 & 4 \\ 3 & -3 \\ 1 & -5 \end{matrix}}{-31} = \dfrac{-12 - 4 - 15 + 3 - 10 - 24}{-31} = \dfrac{-62}{-31} = 2$

$z = \dfrac{\begin{vmatrix} 2 & 3 & 4 \\ 3 & 0 & -3 \\ 1 & -2 & -5 \end{vmatrix}\begin{matrix} 2 & 3 \\ 3 & 0 \\ 1 & -2 \end{matrix}}{-31} = \dfrac{0 - 9 - 24 - 0 - 12 + 45}{-31} = \dfrac{0}{-31} = 0$

17. $x = \dfrac{\begin{vmatrix} 2 & 3 & 1 \\ -1 & 2 & 3 \\ 0 & -3 & 1 \end{vmatrix}\begin{matrix} 2 & 3 \\ -1 & 2 \\ 0 & -3 \end{matrix}}{\begin{vmatrix} 2 & 3 & 1 \\ -1 & 2 & 3 \\ -3 & -3 & 1 \end{vmatrix}\begin{matrix} 2 & 3 \\ -1 & 2 \\ -3 & -3 \end{matrix}} = \dfrac{4 + 0 + 3 - 0 + 18 + 3}{4 - 27 + 3 + 6 + 18 + 3} = \dfrac{28}{7} = 4$

$y = \dfrac{\begin{vmatrix} 2 & 2 & 1 \\ -1 & -1 & 3 \\ -3 & 0 & 1 \end{vmatrix}\begin{matrix} 2 & 2 \\ -1 & -1 \\ -3 & 0 \end{matrix}}{7} = \dfrac{-2 - 18 + 0 - 3 - 0 + 2}{7} = \dfrac{-21}{7} = -3$

$z = \dfrac{\begin{vmatrix} 2 & 3 & 2 \\ -1 & 2 & -1 \\ -3 & -3 & 0 \end{vmatrix}\begin{matrix} 2 & 3 \\ -1 & 2 \\ -3 & -3 \end{matrix}}{7} = \dfrac{0 + 9 + 6 + 12 - 6 - 0}{7} = \dfrac{21}{7} = 3$

21.

$$x = \dfrac{\begin{vmatrix} 5 & -2 & 3 \\ -1 & 1 & -2 \\ 0 & -1 & -3 \\ 2 & -2 & 3 \\ 2 & 1 & -2 \\ 4 & -1 & -3 \end{vmatrix}\begin{matrix} 5 & -2 \\ -1 & 1 \\ 0 & -1 \\ 2 & -2 \\ 2 & 1 \\ 4 & -1 \end{matrix}}{} = \dfrac{-15 + 0 + 3 - 0 - 10 + 6}{-6 + 16 - 6 - 12 - 4 - 12} = \dfrac{-16}{-24} = \dfrac{2}{3}$$

$$y = \dfrac{\begin{vmatrix} 2 & 5 & 3 \\ 2 & -1 & -2 \\ 4 & 0 & -3 \end{vmatrix}\begin{matrix} 2 & 5 \\ 2 & -1 \\ 4 & 0 \end{matrix}}{-24} = \dfrac{6 - 40 + 0 + 12 - 0 + 30}{-24} = \dfrac{8}{-24} = -\dfrac{1}{3}$$

$$z = \dfrac{\begin{vmatrix} 2 & -2 & 5 \\ 2 & 1 & -1 \\ 4 & -1 & 0 \end{vmatrix}\begin{matrix} 2 & -2 \\ 2 & 1 \\ 4 & -1 \end{matrix}}{-24} = \dfrac{0 + 8 - 10 - 20 - 2 - 0}{-24} = \dfrac{-24}{-24} = 1$$

25.

$$x = \dfrac{\begin{vmatrix} 0 & 2 & 2 \\ -1 & 6 & -3 \\ -8 & -3 & 6 \\ 1 & 2 & 2 \\ 2 & 6 & -3 \\ 4 & -3 & 6 \end{vmatrix}\begin{matrix} 0 & 2 \\ -1 & 6 \\ -8 & -3 \\ 1 & 2 \\ 2 & 6 \\ 4 & -3 \end{matrix}}{} = \dfrac{0 + 48 + 6 + 96 - 0 + 12}{36 - 24 - 12 - 48 - 9 - 24} = \dfrac{162}{-81} = -2$$

$$y = \dfrac{\begin{vmatrix} 1 & 0 & 2 \\ 2 & -1 & -3 \\ 4 & -8 & 6 \end{vmatrix}\begin{matrix} 1 & 0 \\ 2 & -1 \\ 4 & -8 \end{matrix}}{-81} = \dfrac{-6 + 0 - 32 + 8 - 24 - 0}{-81} = \dfrac{-54}{-81} = \dfrac{2}{3}$$

$$z = \dfrac{\begin{vmatrix} 1 & 2 & 0 \\ 2 & 6 & -1 \\ 4 & -3 & -8 \end{vmatrix}\begin{matrix} 1 & 2 \\ 2 & 6 \\ 4 & -3 \end{matrix}}{-81} = \dfrac{-48 - 8 + 0 - 0 - 3 + 32}{-81} = \dfrac{-27}{-81} = \dfrac{1}{3}$$

29.

$$A = \dfrac{\begin{vmatrix} 80 & 0 & -0.6 \\ 0 & 1 & -0.8 \\ 0 & 0 & -10 \\ 1 & 0 & -0.6 \\ 0 & 1 & -0.8 \\ 6 & 0 & -10 \end{vmatrix}\begin{matrix} 80 & 0 \\ 0 & 1 \\ 0 & 0 \\ 1 & 0 \\ 0 & 1 \\ 6 & 0 \end{matrix}}{} = \dfrac{-800 + 0 + 0 - 0 - 0 - 0}{-10 + 0 + 0 + 3.6 - 0 - 0} = \dfrac{-800}{-6.4} = 125 \text{ N}$$

$$B = \dfrac{\begin{vmatrix} 1 & 80 & -0.6 \\ 0 & 0 & -0.8 \\ 6 & 0 & -10 \end{vmatrix}\begin{matrix} 1 & 80 \\ 0 & 0 \\ 6 & 0 \end{matrix}}{-6.4} = \dfrac{0 - 384 + 0 - 0 - 0 - 0}{-6.4} = 60 \text{ N}$$

$$F = \dfrac{\begin{vmatrix} 1 & 0 & 80 \\ 0 & 1 & 0 \\ 6 & 0 & 0 \end{vmatrix}\begin{matrix} 1 & 0 \\ 0 & 1 \\ 6 & 0 \end{matrix}}{-6.4} = \dfrac{0 + 0 + 0 - 480 - 0 - 0}{-6.4} = 75 \text{ N}$$

<u>Review</u>   Exercises for Chapter 4, page 145

1. $\begin{vmatrix} -2 & 5 \\ 3 & 1 \end{vmatrix} = -2 - 15 = -17$

5. By taking $(4,-8)$ as (x_2, y_2) and $(2,0)$ as (x_1, y_1), $m = \dfrac{-8 - 0}{4 - 2} = \dfrac{-8}{2} = -4$

9. $y = -2x + 4$; the slope is -2, and the y-intercept is $(0,4)$. Plot the y-intercept. Since the slope is $-\dfrac{2}{1}$, from this point, go down 2 units and over 1 unit to the right, and plot a second point. Sketch the line between the points.

13. $y = 2x - 4$

    $y = -\dfrac{3}{2}x + 3$

    The slope for the first line is 2, and the y-intercept is -4. The slope for the second line is $-\dfrac{3}{2}$ and the y-intercept is 3. From the graph, the point of intersection is $(2,0)$, and the solution is $x = 2.0$ and $y = 0.0$.

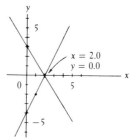

17. The intercepts and a third point for the first line are $(0.0,-7.0)$, $(2.0,0.0)$, and $(3.0,3.5)$. The intercepts and a third point for the second line are $(0.0,4.0)$, $(1.0,0.0)$, and $(2.0,-4.0)$. From the graph the point of intersection is $(1.5,-1.9)$, and the solution is $x = 1.5$, $y = -1.9$.

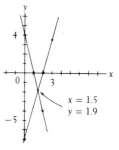

21. (1) $x + 2y = 5$
    (2) $x + 3y = 7$
    (3) $-y = -2$ (2) subtracted from (1)
    (4) $y = 2$
    (5) $x + 2(2) = 5$ (4) substituted in (1)
    (6) $x = 1$
    The solution is $x = 1$, $y = 2$.

25. (1) $3x + 4y = 6$
     (2) $9x + 8y = 11$
     (3) $\underline{-6x - 8y = -12}$ (1) multiplied by -2
              $3x\qquad = -1$ (2) and (3) added

                  $x = -\dfrac{1}{3}$

     (5) $3(-\tfrac{1}{3}) + 4y = 6$ (4) substituted in (1)

     (6) $y = \dfrac{7}{4}$

     The solution is $x = -\dfrac{1}{3}$, $y = \dfrac{7}{4}$.

29. (1) $7x = 2y - 6$
     (2) $7y = 12 - 4x$
     (3) $7x - 2y = -6$ (1) put in standard form
     (4) $4x + 7y = 12$ (2) put in standard form
     (5) $49x - 14y = -42$ (3) multiplied by 7
     (6) $\underline{8x + 14y = 24}$ (4) multiplied by 2
     (7) $57x\qquad = -18$ (5) and (6) added

                  $x = -\dfrac{6}{19}$

     (8) $4(-\tfrac{6}{19}) + 7y = 12$ (7) substituted in (4)

     (9) $y = \dfrac{36}{19}$

     The solution is $x = -\dfrac{6}{19}$, $y = \dfrac{36}{19}$.

33. $x + 2y = 5;\quad x + 3y = 7$

$$x = \dfrac{\begin{vmatrix} 5 & 2 \\ 7 & 3 \end{vmatrix}}{\begin{vmatrix} 1 & 2 \\ 1 & 3 \end{vmatrix}} = \dfrac{5(3) - 7(2)}{1(3) - 1(2)} = \dfrac{1}{1} = 1 \qquad y = \dfrac{\begin{vmatrix} 1 & 5 \\ 1 & 7 \end{vmatrix}}{\begin{vmatrix} 1 & 2 \\ 1 & 3 \end{vmatrix}} = \dfrac{1(7) - 1(5)}{1(3) - 1(2)} = \dfrac{2}{1} = 2$$

37. $3x + 4y = 6$
     $9x + 8y = 11$

$$x = \dfrac{\begin{vmatrix} 6 & 4 \\ 11 & 8 \end{vmatrix}}{\begin{vmatrix} 3 & 4 \\ 9 & 8 \end{vmatrix}} = \dfrac{6(8) - 11(4)}{3(8) - 9(4)} = \dfrac{4}{-12} = -\dfrac{1}{3}$$

$$y = \dfrac{\begin{vmatrix} 3 & 6 \\ 9 & 11 \end{vmatrix}}{\begin{vmatrix} 3 & 4 \\ 9 & 8 \end{vmatrix}} = \dfrac{3(11) - 9(6)}{-12} = \dfrac{-21}{-12} = \dfrac{7}{4}$$

41. $7x = 2y - 6$; $7x - 2y = -6$
    $7y = 12 - 4x$; $4x + 7y = 12$

$$x = \frac{\begin{vmatrix} -6 & -2 \\ 12 & 7 \end{vmatrix}}{\begin{vmatrix} 7 & -2 \\ 4 & 7 \end{vmatrix}} = \frac{-6(7) - 12(-2)}{7(7) - 4(-2)} = \frac{-18}{57} = -\frac{6}{19}$$

$$y = \frac{\begin{vmatrix} 7 & -6 \\ 4 & 12 \end{vmatrix}}{\begin{vmatrix} 7 & -2 \\ 4 & 7 \end{vmatrix}} = \frac{84 - 4(-6)}{57} = \frac{84 + 24}{57} = \frac{108}{57} = \frac{36}{19}$$

45. $\begin{vmatrix} 4 & -1 & 8 \\ -1 & 6 & -2 \\ 2 & 1 & -1 \end{vmatrix} \begin{matrix} 4 & -1 \\ -1 & 6 \\ 2 & 1 \end{matrix} = (-24) + (+4) + (-8) - (+96) - (-8) - (-1) = -115$

49. (1) $2x + y + z = 4$
     (2) $x - 2y - z = 3$
     (3) $3x + 3y - 2z = 1$
     (4) $3x - y = 7$ (1) and (2) added

     (5) $-2x + 4y + 2z = -6$ (2) multiplied by -2
           $\underline{3x + 3y - 2z = 1}$ (3)
     (6) $x + 7y = -5$ (5) and (3) added
     (7) $\underline{21x - 7y = 49}$ (4) multiplied by 7
     (8) $22x = 44$ (6) and (7) added
     (9) $x = 2$
    (10) $3(2) - y = 7$; $y = -1$ (9) substituted in (4)
    (11) $2 - 2(-1) - z = 3$ (10) and (9) substituted in (2)
    (12) $z = 1$
    The solution is $x = 2$, $y = -1$, and $z = 1$.

53. (1) $2r + s + 2t = 8$
     (2) $3r - 2s - 4t = 5$
     (3) $\underline{-2r + 3s + 4t = -3}$
     (4) $r + s = 2$ (2) and (3) added
     (5) $4r + 2s + 4t = 16$ (1) multiplied by 2
           $\underline{3r - 2s - 4t = 5}$ (2)
     (6) $7r = 21$; $r = 3$ (5) and (2) added
     (7) $3 + s = 2$; $s = -1$ (6) substituted in (4)
     (8) $2(3) + (-1) + 2t = 8$; $t = \frac{3}{2}$ (6) and (7) substituted in (1)

     The solution is $r = 3$, $s = -1$, and $t = \frac{3}{2}$.

57. $2x + y + z = 4;$ $x - 2y - z = 3;$ $3x + 3y - 2z = 1$

$$x = \frac{\begin{vmatrix} 4 & 1 & 1 \\ 3 & -2 & -1 \\ 1 & 3 & -2 \\ 2 & 1 & 1 \\ 1 & -2 & -1 \\ 3 & 3 & -2 \end{vmatrix}\begin{matrix} 4 & 1 \\ 3 & -2 \\ 1 & 3 \\ 2 & 1 \\ 1 & -2 \\ 3 & 3 \end{matrix}}{} = \frac{(+16) + (-1) + (+9) - (-2) - (-12) - (-6)}{(+8) + (-3) + (+3) - (-6) - (-6) - (-2)} = \frac{44}{22} = 2$$

$$y = \frac{\begin{vmatrix} 2 & 4 & 1 \\ 1 & 3 & -1 \\ 3 & 1 & -2 \end{vmatrix}\begin{matrix} 2 & 4 \\ 1 & 3 \\ 3 & 1 \end{matrix}}{22} = \frac{(-12) + (-12) + (+1) - (+9) - (-2) - (-8)}{22} = \frac{-22}{22} = -1$$

$$z = \frac{\begin{vmatrix} 2 & 1 & 4 \\ 1 & -2 & 3 \\ 3 & 3 & 1 \end{vmatrix}\begin{matrix} 2 & 1 \\ 1 & -2 \\ 3 & 3 \end{matrix}}{22} = \frac{(-4) + (+9) + (+12) - (-24) - (+18) - (+1)}{22} = \frac{22}{22} = 1$$

61. $2r + s + 2t = 8;$ $3r - 2s - 4t = 5;$ $-2r + 3s + 4t = -3$

$$r = \frac{\begin{vmatrix} 8 & 1 & 2 \\ 5 & -2 & -4 \\ -3 & 3 & 4 \\ 2 & 1 & 2 \\ 3 & -2 & -4 \\ -2 & 3 & 4 \end{vmatrix}\begin{matrix} 8 & 1 \\ 5 & -2 \\ -3 & 3 \\ 2 & 1 \\ 3 & -2 \\ -2 & 3 \end{matrix}}{} = \frac{(-64) + (+12) + (+30) - (+12) - (-96) - (+20)}{(-16) + (+8) + (+18) - (+8) - (-24) - (+12)} = \frac{42}{14}$$
$$= 3$$

$$s = \frac{\begin{vmatrix} 2 & 8 & 2 \\ 3 & 5 & -4 \\ -2 & -3 & 4 \end{vmatrix}\begin{matrix} 2 & 8 \\ 3 & 5 \\ -2 & -3 \end{matrix}}{14} = \frac{(+40) + (+64) + (-18) - (-20) - (+24) - (+96)}{14} = \frac{-14}{14}$$
$$= -1$$

$$t = \frac{\begin{vmatrix} 2 & 1 & 8 \\ 3 & -2 & 5 \\ -2 & 3 & -3 \end{vmatrix}\begin{matrix} 2 & 1 \\ 3 & -2 \\ -2 & 3 \end{matrix}}{14} = \frac{(+12) + (-10) + (+72) - (+32) - (+30) - (-9)}{14} = \frac{21}{14}$$
$$= \frac{3}{2}$$

65. Let $u = \frac{1}{x}$ and $v = \frac{1}{y}$

(1) $u - v = \frac{1}{2}$

(2) $u + v = \frac{1}{4}$

(3) $2u = \frac{3}{4};$ $u = \frac{3}{8}$ (1) and (2) added

(4) $\frac{3}{8} - v = \frac{1}{2};$ $v = -\frac{1}{8}$ (3) substituted in (1)

The solution is $u = \frac{1}{x} = \frac{3}{8},$ $x = \frac{8}{3};$ and $v = \frac{1}{y} = -\frac{1}{8},$ $y = -\frac{8}{1} = -8.$

69. In order to be dependent, the equations must be multiples of each other.
     We multiply the second equation by 3, the coefficient of x in the first
     equation, and obtain $3x + 6y = 6$. The coefficient of y is 6. Therefore,
     $-a = 6$ or $a = -6$.

73.

$$F_1 = \frac{\begin{vmatrix} 280 & 2 & 0 & 280 & 2 \\ 0 & 0 & -1 & 0 & 0 \\ 600 & -4 & 0 & 600 & -4 \\ 1 & 2 & 0 & 1 & 2 \\ 0.866 & 0 & -1 & 0.866 & 0 \\ 3 & -4 & 0 & 3 & -4 \end{vmatrix}} = \frac{0 + (-1200) + 0 - (0 - 1120 + 0)}{0 + (-6) + 0 - (0 + 4 + 0)}$$

$$= \frac{-2320}{-6 - 4} = \frac{-2320}{-10} = 232$$

$$F_2 = \frac{\begin{vmatrix} 1 & 280 & 0 & 1 & 280 \\ 0.866 & 0 & -1 & 0.866 & 0 \\ 3 & 600 & 0 & 3 & 600 \end{vmatrix}}{-10} = \frac{0 - 840 + 0 - (0 - 600 + 0)}{-10} = 24$$

$$F_3 = \frac{\begin{vmatrix} 1 & 2 & 280 & 1 & 2 \\ 0.866 & 0 & 0 & 0.866 & 0 \\ 3 & -4 & 600 & 3 & -4 \end{vmatrix}}{-10} = \frac{0 + 0 - 969.92 - (0 + 0 - 1039.2)}{-10}$$

$$= 201$$

77. Let x = speed of space shuttle and y = launch speed of satellite

     (1) $x + y = 24\ 200$

     (2) $x - y = 21\ 400$

     (3) $2x = 45\ 600$

              $x = 22\ 600$

     $22\ 800 + y = 24\ 600$

              $y = 1400$

81.

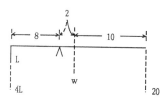

$8(L) = 2(w)$

$8(4L) = 2(w) + 12(20)$

$32L = 2w + 240$

$24L = 240$

$L = 10$

$8(10) = 2w;\ w = 40$

81M.

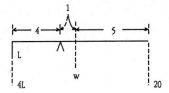

$$4(L) = 1(w)$$
$$4(4L) = 1(w) + 6(20)$$
$$16L = w + 120$$
$$16L = 4L + 120; \quad L = 10$$
$$4(10) = w = 40$$

CHAPTER 5

Factoring and Fractions

1. $40(x - y) = 40x - 40y$ special product (5-1)

5. $(y + 6)(y - 6) = y^2 - 36$ special product (5-2)

9. $(4x - 5y)(4x + 5y) = 16x^2 + 20xy - 20xy - 25y^2$
$$= 16x^2 - 25y^2$$

13. $(5f + 4)^2 = (5f)^2 + 2(5f)(4) + 4^2$
$$= 25f^2 + 40f + 16$$ special product (5-3)

17. $(x - 1)^2 = x^2 + 2(-x) + 1$
$$= x^2 - 2x + 1$$ special product (5-3)

21. $(4x - 2y)^2 = 16x^2 - 16xy + 4y^2$ special product (5-4)

25. $(x + 1)(x + 5) = x^2 + (x + 5x) + 5$
$$= x^2 + 6x + 5$$ special product (5-5)

29. $(3x - 1)(2x + 5) = 6x^2 + (15x - 2x) - 5$
$$= 6x^2 + 13x - 5$$ special product (5-6)

33. $(5v - 3)(4v + 5) = 20v^2 + (25v - 12v) - 15$
$$= 20v^2 + 13v - 15$$ special product (5-6)

37. $2(x - 2)(x + 2) = 2(x^2 - 4)$
$$= 2x^2 - 8$$

41. $6a(x + 2b)^2 = 6a(x^2 + 4bx + 4b^2)$
$$= 6ax^2 + 24abx + 24ab^2$$

45. $4a(2a - 3)^2 = 4a(4a^2 - 12a + 9) = 16a^3 - 48a^2 + 36a$

49. $(3 - x - y)^2 = [(3-x) - y]^2 = (3 - x)^2 - 2(y)(3 - x) + y^2 =$
$9 - 6x + x^2 - 6y + 2xy + y^2 = x^2 + y^2 + 2xy - 6x - 6y + 9$

53. $(2x + 5t)^3 = 8x^3 + 12x^2(5t) + 75t^2(2x) + 125t^3$
$= 8x^3 + 60x^2t + 150xt^2 + 125t^3$

57. $(x + 2)(x^2 - 2x + 4) = x^3 + 2^3$
$= x^3 + 8$ special product (5-9)

61. $P_1(P_0c + G) = P_1P_0c + P_1G$

65. $\frac{1}{2}\pi(R + r)(R - r) = \frac{1}{2}\pi(R^2 - r^2) = \frac{1}{2}\pi R^2 - \frac{1}{2}\pi r^2$

Exercises 5-2, page 156

1. $6x + 6y = 6(x + y)$ (factor out a common factor of 6)

5. $3x^2 - 9x = 3x(x - 3)$ (factor out the common factor of $3x$)

9. $12n^2 + 6n = 6n(2n + 1)$ (factor out the common factor of $6n$)

11. $2x + 4y - 8z = 2(x + 2y - 4z)$ (factor out the common factor of 2)

13. $3ab^2 - 6ab + 12ab^3 = 3ab(b - 2 + 4b^2)$ (factor out a common factor of 3)

15. $12pq^2 - 8pq - 28pq^3 = 4pq(3q - 2 - 7q^2)$ (factor out the common factor of $4pq$)

17. $2a^2 - 2b^2 + 4c^2 - 6d^2 = 2(a^2 - b^2 + 2c^2 - 3d^2)$ (factor out a common factor of 2)

19. $x^2 - 4 = (x + 2)(x - 2)$ (because $-2x + 2x = 0x = 0$)

21. $100 - y^2 = 10^2 - y^2 = (10 + y)(10 - y)$ (because $x^2 - y^2 = (x + y)(x - y)$)

25. $81s^2 - 25t^2 = (9s + 5t)(9s - 5t)$ (because $-45st + 45st = 0st = 0$)

29. $(x + y)^2 - 9 = [(x + y) + 3][(x + y) - 3] = (x + y + 3)(x + y - 3)$

33. $3x^2 - 27z^2 = 3(x^2 - 9z^2) = 3(x + 3z)(x - 3z)$

37. $x^4 - 16 = (x^2 + 4)(x^2 - 4)$
$= (x^2 + 4)(x + 2)(x - 2)$

41. $2a - b = ab + 3$
$2a - ab = 3 + b$
$a(2 - b) = 3 + b$

$a = \frac{3 + b}{2 - b}$

45. $3x - 3y + bx - by = (3x - 3y) + (bx - by)$
$= 3(x - y) + b(x - y) = (x - y)(3 + b)$
$= (3 + b)(x - y)$

49. $x^3 + 3x^2 - 4x - 12 = (x^3 + 3x^2) - (4x + 12)$
$= x^2(x + 3) - 4(x + 3)$
$= (x + 3)(x^2 - 4)$
$= (x + 3)(x + 2)(x - 2)$

53. $Rv + Rv^2 + Rv^3 = Rv(1 + v + v^2)$

57. $PbL^2 - Pb^3 = Pb(L^2 - b^2) = Pb(L + b)(L - b)$

Exercises 5-3, page 162

1. $x^2 + 5x + 4 = (x + 1)(x + 4)$ (because $+5x = 4x + 1x$)

5. $t^2 + 5t - 24 = (t + 8)(t - 3)$ (because $+8t + (-3t) = 5t$)

7. $x^2 + 2x + 1 = (x + 1)(x + 1)$
$= (x + 1)^2$ (because $2x = 1x + 1x$)

9. $x^2 - 4xy + 4y^2 = (x - 2y)(x - 2y) = (x - 2y)^2$

11. $3x^2 - 5x - 2 = (3x + 1)(x - 2)$ (because $-5x = -6x + x$)

13. $3y^2 - 8y - 3 = (3y + 1)(y - 3)$ (because $-9y + y = -8y$)

17. $3f^2 - 16f + 5 = (3f - 1)(f - 5)$ (because $-15f - f = -16f$)

21. $3t^2 - 7tu + 4u^2 = (3t - 4u)(t - u)$ (because $-7tu = -3tu - 4tu$)

25. $9x^2 + 7xy - 2y^2 = (x + y)(9x - 2y)$

29. $4x^2 - 12x + 9 = (2x - 3)(2x - 3)$
$= (2x - 3)^2$ see (5-4)

31. $9t^2 - 15t + 4 = (3t - 4)(3t - 1)$ (because $-15t = -12t - 3t$)

33. $8b^2 + 31b - 4 = (8b - 1)(b + 4)$ (because $32b - b = 31b$)

35. $4p^2 - 25pq + 6q^2 = (4p - q)(p - 6q)$ (because $-25pq = -24pq - pq$)

37. $12x^2 + 47xy - 4y^2 = (12x - y)(x + 4y)$ (because $48xy - xy = 47xy$)

39. $2x^2 - 14x + 12 = 2(x^2 - 7x + 6)$
$= 2(x - 6)(x - 1) = 2(x - 1)(x - 6)$

41. $4x^2 + 14x - 8 = 2(2x^2 + 7x - 4)$
$= 2(2x - 1)(x + 4)$

45. $a^2 + 2ab + b^2 - 4 = (a^2 + 2ab + b^2) - 4 = (a + b)^2 - 4$
$= [(a + b) + 2][(a + b) - 2]$

49. $x^3 + 3x^2 + 3x + 1 = (x + 1)^3$ special product (5-7)

53. $4s^2 + 16s + 12 = 4(s^2 + 4s + 3) = 4(s + 3)(s + 1)$

57. $wx^4 - 5wLx^3 + 6wL^2x^2 = wx^2(x^2 - 5Lx + 6L^2) = wx^2(x - 2L)(x - 3L)$

Exercises 5-4, page 166

1. $\dfrac{2}{3} = \dfrac{2(7)}{3(7)} = \dfrac{14}{21}$

5. $\dfrac{2}{x + 3} = \dfrac{2(x - 2)}{(x + 3)(x - 2)} = \dfrac{2x - 4}{x^2 + x - 6}$

9. $\dfrac{28}{44} = \dfrac{28 \div 4}{44 \div 4} = \dfrac{7}{11}$

13. $\dfrac{2(x - 1)}{(x - 1)(x + 1)} = \dfrac{2(x - 1) \div (x - 1)}{(x - 1)(x + 1) \div (x - 1)} = \dfrac{2}{x + 1}$

17. $\dfrac{2a}{8a} = \dfrac{1(2a) \div 2a}{4(2a) \div 2a} = \dfrac{1}{4}$

21. $\dfrac{a + b}{5a^2 + 5ab} = \dfrac{1(a + b) \div (a + b)}{5a(a + b) \div (a + b)} = \dfrac{1}{5a}$

25. $\dfrac{4x^2 + 1}{(2x + 1)(2x - 1)}$; $4x^2$'s and 1's do <u>not</u> cancel and the fraction cannot be reduced.

29. $\dfrac{2y + 3}{4y^3 + 6y^2} = \dfrac{2y + 3}{2y^2(2y + 3)}$

$= \dfrac{(2y + 3)(1)}{(2y + 3)(2y^2)} = \dfrac{1}{2y^2}$

33. $\dfrac{2x^2 + 5x - 3}{x^2 + 11x + 24} = \dfrac{(2x - 1)(x + 3) \div (x + 3)}{(x + 3)(x + 8) \div (x + 3)}$

$= \dfrac{2x - 1}{x + 8}$

37. $\dfrac{x^4 - 16}{x + 2} = \dfrac{(x^2 - 4)(x^2 + 4)}{x + 2}$

$= \dfrac{(x + 2)(x - 2)(x^2 + 4)}{x + 2} = (x^2 + 4)(x - 2)$

39. $\dfrac{x^2y^4 - x^4y^2}{y^2 - 2xy + x^2} = \dfrac{x^2y^2(y^2 - x^2)}{(y - x)(y - x)}$

$= \dfrac{x^2y^2(y + x)(y - x) \div (y - x)}{(y - x)(y - x) \div (y - x)} = \dfrac{x^2y^2(y + x)}{(y - x)}$

41. $\dfrac{(x-1)(3+x)}{(3-x)(1-x)} = \dfrac{(x-1)(3+x)}{(3-x)(-1)(x-1)}$

$= \dfrac{(3+x)}{-(3-x)} = \dfrac{x+3}{x-3}$

43. $\dfrac{y-x}{2x-2y} = \dfrac{-1(-y+x)}{2(x-y)} = \dfrac{-1(x-y) \div (x-y)}{2(x-y) \div (x-y)} = -\dfrac{1}{2}$

45. $\dfrac{2x^2 - 9x + 4}{4x - x^2} = \dfrac{(2x-1)(x-4)}{x(4-x)}$

$= \dfrac{(2x-1)(-1)(4-x) \div (4-x)}{x(4-x) \div (4-x)} = \dfrac{-(2x-1)}{x}$

49. $\dfrac{x^3 + y^3}{2x + 2y} = \dfrac{(x+y)(x^2 - xy + y^2) \div (x+y)}{2(x+y) \div (x+y)} = \dfrac{x^2 - xy + y^2}{2}$

53. (a) $\dfrac{x^2(x+2)}{x^2+4}$ does not reduce since $x^2 + 4$ does not factor.

(b) $\dfrac{x^4 + 4x^2}{x^4 - 16} = \dfrac{x^2(x^2+4) \div (x^2+4)}{(x^2+4)(x^2-4) \div (x^2+4)}$

$= \dfrac{x^2}{x^2 - 4}$

57. $\dfrac{8\pi r^3}{6\pi^2 r^2} = \dfrac{4(2\pi r^2)(r) + (2\pi^2 r)}{3(2\pi r^2)(\pi) + (2\pi r^2)} = \dfrac{4r}{3\pi}$

Exercises 5-5, page 170

1. $\dfrac{3}{8} \times \dfrac{2}{7} = \dfrac{(3)(2)}{(8)(7)} = \dfrac{3(1)}{4(7)}$

$= \dfrac{3}{28}$ (divide out common factor of 2)

5. $\dfrac{2}{9} \div \dfrac{4}{7} = \dfrac{2}{9} \times \dfrac{7}{4} = \dfrac{(2)(7)}{(9)(4)}$

$= \dfrac{1(7)}{9(2)} = \dfrac{7}{18}$ (divide out common factor of 2)

9. $\dfrac{4x+12}{5} \times \dfrac{15t}{3x+9} = \dfrac{4(x+3)(5)(3)(t)}{5(3)(x+3)}$

$= \dfrac{4(t)}{1} = 4t$ divide out common factors of 5, 3, and $(x+3)$

13. $\dfrac{2a + 8}{15} \div \dfrac{a^2 + 8a + 16}{25} = \dfrac{2(a + 4)}{3(5)} \times \dfrac{5(5)}{(a + 4)(a + 4)} = \dfrac{2(a + 4)(5)(5)}{3(5)(a + 4)(a + 4)}$

$$= \dfrac{2(5)}{3(a + 4)} = \dfrac{10}{3(a + 4)}$$

Common factors of 5 and $(a + 4)$ were divided out.

17. $\dfrac{3ax^2 - 9ax}{10x^2 + 5x} \times \dfrac{2x^2 + x}{a^2x - 3a^2} = \dfrac{3ax(x - 3)(x)(2x + 1)}{5x(2x + 1)(a^2)(x - 3)} = \dfrac{3(x)}{5(a)} = \dfrac{3x}{5a}$

Common factors are a, x, $(x - 3)$, and $(2x + 1)$.

21. $\dfrac{ax + x^2}{2b - cx} \div \dfrac{a^2 + 2ax + x^2}{2bx - cx^2} = \dfrac{x(a + x)}{2b - cx} \times \dfrac{x(2b - cx)}{(a + x)(a + x)}$

$$= \dfrac{x(a + x)(x)(2b - cx)}{(2b - cx)(a + x)(a + x)} = \dfrac{x^2}{a + x}$$

Common factors of $(a + x)$ and $(2b - cx)$ were divided out.

25. $\dfrac{x^2 - 6x + 5}{4x^2 - 17x - 15} \times \dfrac{6x + 21}{2x^2 + 5x - 7} = \dfrac{(x - 5)(x - 1)}{(4x + 3)(x - 5)} \times \dfrac{3(2x + 7)}{(2x + 7)(x - 1)}$

$$= \dfrac{3}{4x + 3}$$

Common factors of $(x - 5)$, $(x - 1)$, and $(2x + 7)$ were divided out.

29. $\dfrac{7x^2}{3a} \div (\dfrac{a}{x} \times \dfrac{a^2x}{x^2}) = \dfrac{7x^2}{3a} \div \dfrac{a^3x}{x^3}$

$$= \dfrac{7x^2}{3a} \times \dfrac{x^3}{a^3x} = \dfrac{7x^5}{3a^4x}$$

$$= \dfrac{7x^4}{3a^4} \quad \text{common factor of } x$$

33. $\dfrac{x^3 - y^3}{2x^2 - 2y^2} \times \dfrac{x^2 + 2xy + y^2}{x^2 + xy + y^2} = \dfrac{(x - y)(x^2 + xy + y^2)(x + y)(x + y)}{2(x + y)(x - y)(x^2 + xy + y^2)}$

$$= \dfrac{x + y}{2} = \dfrac{1}{2}(x + y)$$

Common factors of $(x - y)$, $(x + y)$, and $(x^2 + xy + y^2)$ were divided out.

37. $\dfrac{n^2a^2}{v^2} \div \dfrac{n - an}{v} = \dfrac{n^2a^2}{v^2} \times = \dfrac{v}{n - an} = \dfrac{n^2a^2v}{v^2(n)(1 - a)} = \dfrac{na^2}{v(1 - a)}$

Exercises 5-6, page 176

1. $\dfrac{3}{5} + \dfrac{6}{5} = \dfrac{3 + 6}{5} = \dfrac{9}{5}$ 5. $\dfrac{1}{2} + \dfrac{3}{4} = \dfrac{1(2)}{2(2)} + \dfrac{3}{4} = \dfrac{2 + 3}{4} = \dfrac{5}{4}$

9. $\dfrac{a}{x} - \dfrac{b}{x^2} = \dfrac{a(x)}{x(x)} - \dfrac{b}{x^2} = \dfrac{ax - b}{x^2}$

13. L.C.D. $= 2(5)(a) = 10a$

$$\frac{2}{5a} + \frac{1}{a} - \frac{a}{10} = \frac{2}{5a} + \frac{1}{a} - \frac{a}{2(5)}$$

$$= \frac{2(2)}{5a(2)} + \frac{1(2)(5)}{a(2)(5)} - \frac{a(a)}{10(a)}$$

$$= \frac{4 + 10 - a^2}{10a} = \frac{14 - a^2}{10a}$$

17. L.C.D. $= 2(2x - 1)$

$$\frac{3}{2x - 1} + \frac{1}{4x - 2} = \frac{3}{2x - 1} + \frac{1}{2(2x - 1)}$$

$$= \frac{3(2)}{(2x - 1)(2)} + \frac{1}{2(2x - 1)}$$

$$= \frac{6 + 1}{2(2x - 1)} = \frac{7}{2(2x - 1)}$$

21. L.C.D. $= 2(s - 3)(2) = 4(s - 3)$

$$\frac{s}{2s - 6} + \frac{1}{4} - \frac{3s}{4s - 12} = \frac{s}{2(s - 3)} + \frac{1}{2(2)} - \frac{3s}{2(2)(s - 3)}$$

$$= \frac{s(2)}{2(s - 3)(2)} + \frac{1(s - 3)}{2(2)(s - 3)} - \frac{3s}{2(2)(s - 3)}$$

$$= \frac{2s + s - 3 - 3s}{4(s - 3)} = \frac{-3}{4(s - 3)}$$

25. L.C.D. $= (x - 4)(x - 4) = (x - 4)^2$

$$\frac{3}{x^2 - 8x + 16} - \frac{2}{4 - x} = \frac{3}{(x - 4)(x - 4)} - \frac{2}{(-1)(x - 4)}$$

$$= \frac{3}{(x - 4)(x - 4)} + \frac{2(x - 4)}{(x - 4)(x - 4)}$$

$$= \frac{3 + 2(x - 4)}{(x - 4)(x - 4)} = \frac{2x - 5}{(x - 4)^2}$$

29. L.C.D. $= (3x - 1)(x - 4)$

$$\frac{x - 1}{3x^2 - 13x + 4} - \frac{3x + 1}{4 - x} = \frac{(x - 1)}{(3x - 1)(x - 4)} - \frac{(3x + 1)(3x - 1)}{(-1)(x - 4)(3x - 1)}$$

$$= \frac{(x - 1)}{(3x - 1)(x - 4)} + \frac{(3x + 1)(3x - 1)}{(x - 4)(3x - 1)}$$

$$= \frac{(x - 1) + (9x^2 - 1)}{(3x - 1)(x - 4)} = \frac{9x^2 + x - 2}{(3x - 1)(x - 4)}$$

33. $\dfrac{1 + \dfrac{1}{x}}{1 - \dfrac{1}{x}} = \dfrac{\dfrac{x + 1}{x}}{\dfrac{x - 1}{x}} = \dfrac{x + 1}{x} \times \dfrac{x}{x - 1} = \dfrac{x + 1}{x - 1}$

Common factor of x was divided out.

37. $\dfrac{\dfrac{3}{x}+\dfrac{1}{x^2+x}}{\dfrac{1}{x+1}-\dfrac{1}{x-1}}$ Before performing the division, perform the addition in the numerator and denominator.

$$\frac{3}{x}+\frac{1}{x(x+1)}=\frac{3(x+1)+1}{x(x+1)}$$

$$=\frac{3x+4}{x(x+1)};\quad \frac{(x-1)-(x+1)}{(x+1)(x-1)}=\frac{x-1-x-1}{(x+1)(x-1)}$$

$$=\frac{-2}{(x+1)(x-1)}$$

Replace the numerator and denominator and invert the divisor:

$$\frac{3x+4}{x(x+1)}\cdot\frac{(x+1)(x-1)}{-2}=\frac{(3x+4)(x-1)}{-2x}$$

$$=\frac{3x^2+x-4}{-2x}=\frac{4-x-3x^2}{2x}$$

An alternate method of solution is to find the L.C.D. of the fractions in the numerator and denominator; i.e., $x(x+1)(x-1)$. Multiply the numerator and denominator by the L.C.D.

$$\frac{3(x+1)(x-1)+1(x-1)}{x(x-1)-x(x+1)}=\frac{3x^2-3+x-1}{x^2-x-x^2-x}$$

$$=\frac{3x^2+x-4}{-2x}=\frac{4-x-3x^2}{2x}$$

$$=-\frac{(3x+4)(x-1)}{2x}$$

41. $f(x)=\dfrac{x}{x+1}$, $f(x+h)=\dfrac{x+h}{x+h+1}$

$$f(x+h)-f(x)=\frac{x+h}{x+h+1}-\frac{x}{x+1}$$

$$=\frac{(x+h)(x+1)}{(x+h+1)(x+1)}-\frac{(x)(x+h+1)}{(x+1)(x+h+1)}$$

$$=\frac{x^2+x+hx+h-(x^2+xh+x)}{(x+1)(x+h+1)}$$

$$=\frac{h}{(x+1)(x+h+1)}$$

45. $(\tan\theta)(\cot\theta)+(\sin\theta)^2-\cos\theta$

$$\left(\frac{y}{x}\right)\left(\frac{x}{y}\right)+\left(\frac{y}{r}\right)^2-\frac{x}{r}=\frac{xy}{xy}+\frac{y^2}{r^2}-\frac{x}{r}$$

$$=1+\frac{y^2}{r^2}-\frac{x}{r}=\frac{r^2}{r^2}+\frac{y^2}{r^2}-\frac{x(r)}{r(r)}$$

$$=\frac{y^2-rx+r^2}{r^2}$$

49. $\dfrac{3}{4\pi} - \dfrac{3H_0}{4\pi H} = \dfrac{3H - 3H_0}{4\pi H} = \dfrac{3(H - H_0)}{4\pi H}$

Exercises 5-7, page 180

Each problem should be checked by substituting the solution into the equation to check that each side simplifies to the same value.

1. $\dfrac{x}{2} + 6 = 2x$ L.C.D. = 2; multiply each term by the L.C.D.

$$\dfrac{2(x)}{2} + 2(6) = 2(2x)$$
$$x + 12 = 4x$$
$$12 = 3x$$
$$x = 4$$

5. $\dfrac{1}{2} - \dfrac{t - 5}{6} = \dfrac{3}{4}$ L.C.D. = 12; multiply each term by the L.C.D.

$$\dfrac{12(1)}{2} - \dfrac{12(t - 5)}{6} = \dfrac{12(3)}{4}$$
$$6 - 2(t - 5) = 9$$
$$-2t + 16 = 9$$
$$-2t = -7$$
$$t = \dfrac{7}{2}$$

9. $\dfrac{3}{x} + 2 = \dfrac{5}{3}$ L.C.D. = 3x; multiply each term by the L.C.D.

$$\dfrac{3x(3)}{x} + 3x(2) = \dfrac{3x(5)}{3}$$
$$9 + 6x = 5x$$
$$x = -9$$

13. $\dfrac{2y}{y - 1} = 5$ L.C.D. = (y - 1); multiply each term by the L.C.D.

$$\dfrac{(y - 1)(2y)}{y - 1} = (y - 1)(5)$$
$$2y = 5y - 5$$
$$-3y = -5$$
$$y = \dfrac{5}{3}$$

17. $\dfrac{5}{2x + 4} + \dfrac{3}{x + 2} = 2$ L.C.D. is 2(x + 2) = 2x + 4; multiply each term by the L.C.D.

$$\dfrac{(2x + 4)(5)}{2x + 4} + \dfrac{(2x + 4)(3)}{x + 2} = (2x + 4)(2)$$
$$5 + 6 = 4x + 8$$
$$3 = 4x$$
$$x = \dfrac{3}{4}$$

21. $\dfrac{1}{x} + \dfrac{3}{2x} = \dfrac{2}{x+1}$ L.C.D. = $(2x)(x+1)$; multiply each term by the L.C.D.

$$\dfrac{(2x)(x+1)(1)}{x} + \dfrac{(2x)(x+1)(3)}{2x} = \dfrac{(2x)(x+1)(2)}{x+1}$$

$$2x + 2 + 3x + 3 = 4x$$
$$5x + 5 = 4x$$
$$x = -5$$

25. $\dfrac{1}{x^2 - x} - \dfrac{1}{x} = \dfrac{1}{x-1}$ L.C.D. is $(x)(x-1)$; multiply each term by the L.C.D.

$$\dfrac{(x)(x-1)(1)}{(x)(x-1)} - \dfrac{(x)(x-1)(1)}{x} = \dfrac{(x)(x-1)(1)}{x-1}$$

$$1 - (x-1) = x$$
$$2 = 2x$$
$$x = 1$$

Checking: Substituting 1 into the first and third terms yields $\dfrac{1}{1-1}$ or $\dfrac{1}{0}$, which is undefined. Therefore, $x = 1$ is not a solution. This equation does not have a solution.

29. $2 - \dfrac{1}{b} + \dfrac{3}{c} = 0$ L.C.D. is $(b)(c)$; multiply each term by the L.C.D.

$$(b)(c)(2) - \dfrac{(b)(c)(1)}{b} + \dfrac{(b)(c)(3)}{c} = (b)(c)(0)$$

$$2bc - c + 3b = 0$$
$$2bc - c = -3b$$
$$c(2b - 1) = -3b$$
$$c = \dfrac{-3b}{2b-1} = \dfrac{3b}{1-2b}$$

33. $n = n_1 - \dfrac{n_1 v}{V}$

$nV = n_1 V - n_1 v$ multiply each term by V

$n_1 v = n_1 V - nV$ add $n_1 v$ subtract nV

$v = \dfrac{n_1 V - nV}{n_1}$ divide by n_1

37. $z = \dfrac{1}{g_m} - \dfrac{jx}{g_m R}$

$$zg_m R = R - jx$$
$$zg_m R - R = -jx$$
$$R(zg_m - 1) = -jx$$
$$R = \dfrac{-jx}{zg_m - 1} = \dfrac{jx}{1 - zg_m}$$

41. $R = \dfrac{L_1}{kA_1} + \dfrac{L_2}{kA_2}$

   $kA_1A_2R = A_2L_1 + A_1L_2$ multiply each term by kA_1A_2

   $A_2L_1 = kA_1A_2R - A_1L_2$ subtract A_1L_2

   $L_1 = \dfrac{kA_1A_2R - A_1L_2}{A_2}$ divide by A_2

45. Since work = rate(time), rate = $\dfrac{\text{work}}{\text{time}}$, where work is one job.

   Let t = time required. For first pipe, $r = \dfrac{1}{4}$; second pipe is $r = \dfrac{1}{6}$.

   $\dfrac{1}{4}t + \dfrac{1}{6}t = 1$; $\dfrac{5}{12}t = 1$; $t = \dfrac{12}{5} = 2.4h$

49. Rate with wind = $450 + w$; $t = \dfrac{s}{r} = \dfrac{2580}{450 + w}$

   Rate against wind = $450 - w$; $t = \dfrac{1800}{450 - w}$

   $\dfrac{2580}{450 + w} = \dfrac{1800}{450 - w}$

   $\dfrac{2580(450 - w)}{(450 + w)(450 - w)} = \dfrac{1800(450 + w)}{(450 + w)(450 - w)}$

   $2580(450 - w) = 1800(450 + w)$
   $2580(450) - 2580w = 1800(450) + 1800w$
   $2580(450) - 1800(450) = 1800w + 2580w$
   $780(450) = 4380w$
   $351\ 000 = 4380w$
   $80 = w$

Review, Exercises for Chapter 5, page 181

1. $3a(4x + 5a) = 12ax + 15a^2$ special product (5-1)

5. $(2a + 1)^2 = (2a)^2 + 2(2a)(1) + (1)^2$
        $= 4a^2 + 4a + 1$ special product (5-3)

9. $(2x + 5)(x - 9) = 2x^2 + (-18x + 5x) - 45$
             $= 2x^2 - 13x - 45$ special product (5-6)

13. $3s + 9t = 3(s + 3t)$

17. $x^2 - 144 = (x + 12)(x - 12)$

21. $9t^2 - 6t + 1 = (3t - 1)^2$

25. $x^2 + x - 56 = (x + 8)(x - 7)$ (because $8x - 7x = +x$)

29. $2x^2 - x - 36 = (2x - 9)(x + 4)$ (because $+8x - 9x = -x$)

33. $10b^2 + 23b - 5 = (5b - 1)(2b + 5)$ (because $-2b + 25b = +23b$)

37. $x^3 + 9x^2 + 27x + 27 = (x + 3)^3$ special product (5-7)

41. $ab^2 - 3b^2 + a - 3 = b^2(a - 3) + (a - 3)$
$$= (a - 3)(b^2 + 1)$$

45. $\dfrac{48ax^3y^6}{9a^3xy^6} = \dfrac{16x^2}{3a^2}$ Common factors of 3, a, x, and y^6 were divided out.

49. $\dfrac{4x + 4y}{35x^2} \times \dfrac{28x}{x^2 - y^2} = \dfrac{4(x + y)(28x)}{35x^2(x + y)(x - y)}$

$$= \dfrac{16}{5x(x - y)}$$ Common factors of $(x + y)$, 7, and x were divided out.

53. $\dfrac{\dfrac{3x}{7x^2 + 13x - 2}}{\dfrac{6x^2}{x^2 + 4x + 4}} = \dfrac{\dfrac{3x}{(7x - 1)(x + 2)}}{\dfrac{6x^2}{(x + 2)^2}}$

$$= \dfrac{3x}{(7x - 1)(x + 2)} \times \dfrac{(x + 2)^2}{6x^2} = \dfrac{x + 2}{2x(7x - 1)}$$

Common factors of 3, x, and $(x + 2)$ were divided out.

57. $\dfrac{4}{9x} - \dfrac{5}{12x^2} = \dfrac{4(4x)}{9x(4x)} - \dfrac{5(3)}{12x^2(3)}$

$$= \dfrac{16x - 15}{36x^2}$$

61. $\dfrac{a + 1}{a + 2} - \dfrac{a + 3}{a} = \dfrac{(a + 1)(a)}{(a + 2)(a)} - \dfrac{(a + 3)(a + 2)}{a(a + 2)}$

$$= \dfrac{a^2 + a}{a(a + 2)} - \dfrac{a^2 + 5a + 6}{a(a + 2)} = \dfrac{-4a - 6}{a(a + 2)}$$

$$= \dfrac{-2(2a + 3)}{a(a + 2)}$$

65. $\dfrac{3x}{2x^2 - 2} - \dfrac{2}{4x^2 - 5x + 1} = \dfrac{3x}{2(x + 1)(x - 1)} - \dfrac{2}{(4x - 1)(x - 1)};$

L.C.D. $= 2(x + 1)(x - 1)(4x - 1)$

$$\dfrac{3x}{2(x - 1)(x - 1)} \times \dfrac{(4x - 1)}{(4x - 1)} - \dfrac{2}{(4x - 1)(x - 1)} \times \dfrac{(2)(x + 1)}{(2)(x + 1)}$$

$$= \dfrac{12x^2 - 3x - 4(x + 1)}{2(x + 1)(x - 1)(4x - 1)} = \dfrac{12x^2 - 3x - 4x - 4}{2(x - 1)(x - 1)(4x - 1)}$$

$$= \dfrac{12x^2 - 7x - 4}{2(x - 1)(x + 1)(4x - 1)}$$

69. $\dfrac{x}{2} - 3 = \dfrac{x - 10}{4}$ L.C.D. is 4; multiply all terms by the L.C.D.

$$\dfrac{4(x)}{2} - 4(3) = \dfrac{4(x - 10)}{4}$$

$$2x - 12 = x - 10$$
$$x = 2$$

73. $\dfrac{2}{t} - \dfrac{1}{at} = 2 + \dfrac{a}{t}$ L.C.D. is $a(t)$; multiply all terms by L.C.D.

$$\dfrac{at(2)}{t} - \dfrac{at(1)}{at} = at(2) + \dfrac{at(a)}{t}$$

$$2a - 1 = 2at + a^2$$

$$-a^2 + 2a - 1 = 2at$$

$$\dfrac{-a^2 + 2a - 1}{2a} = t$$

$$\dfrac{-(a^2 - 2a + 1)}{2a} = t$$

$$-\dfrac{(a - 1)^2}{2a} = t$$

77. $xy = \dfrac{1}{4}[x^2 + 2xy + y^2 - (x^2 - 2xy + y^2)]$

$xy = \dfrac{1}{4}[x^2 + 2xy + y^2 - x^2 + 2xy - y^2]$

$xy = \dfrac{1}{4}(4xy) = xy$

81. $\pi r_1^2 \ell - \pi r_2^2 \ell = \pi \ell (r_1^2 - r_2^2) = \pi \ell (r_1 + r_2)(r_1 - r_2)$

85. $(W^2 - 2L^2)^2 + 4L^2(W^2 + k^2 - 2L^2) = W^4 - 4W^2L^2 + 4L^4 + 4W^2L^2 + 4k^2L^2 - 8L^4$
$$= W^4 + 4k^2L^2 - 4L^4$$

89. $\left(\dfrac{2wtv^2}{Dg}\right)\left(\dfrac{b\pi^2D^2}{n^2}\right)\left(\dfrac{6}{bt^2}\right) = \dfrac{12wtv^2b\pi^2D^2}{Dgn^2bt^2} = \dfrac{12wv^2\pi^2D}{gn^2t}$

93. $1 - \dfrac{d^2}{2} + \dfrac{d^4}{24} - \dfrac{d^6}{120} = \dfrac{120}{120} - \dfrac{60d^2}{120} + \dfrac{5d^4}{120} - \dfrac{d^6}{120} = \dfrac{120 - 60d^2 + 5d^4 - d^6}{120}$

97. $1 - \dfrac{3a}{4r} - \dfrac{a^3}{4r^3} = \dfrac{4r^3}{4r^3} - \dfrac{3ar^2}{4r^3} - \dfrac{a^3}{4r^3} = \dfrac{4r^3 - 3ar^2 - a^3}{4r^3}$

101. $R = \dfrac{wL}{H(w + L)}$

$H(w + L)R = wL$

$HwR + HLR = wL$

$\qquad HwR = wL - HLR$

$\qquad HwR = (w - HR)L$

$\dfrac{HwR}{w - HR} = L$

105. $1 = \left(\dfrac{1}{4} + \dfrac{1}{24}\right)t$

$1 = \dfrac{7}{24}\,t; \qquad t = \dfrac{24}{7} = 3.4h$

109. $d = \dfrac{w_\alpha}{w_\alpha - w_\omega} = \dfrac{1.097w_\omega}{1.097w_\omega - w_\omega} = \dfrac{1.097w_\omega}{w_\omega(1.097 - 1)} = \dfrac{1.097}{0.097} = 11.3$

CHAPTER 6

Quadratic Equations

1. $x^2 + 5 = 8x$; $x^2 - 8x + 5 = 0$; $a = 1$, $b = -8$, $c = 5$

5. $x^2 = (x + 2)^2$; $x^2 = x^2 + 4x + 4$; $4x + 4 = 0$
    Not quadratic; there is no x^2 term.

9. $x^2 - 4 = 0$; $(x + 2)(x - 2) = 0$;
    $x + 2 = 0$ or $x = -2$; $x - 2 = 0$ or $x = 2$

13. $x^2 - 8x - 9 = 0$; $(x - 9)(x + 1) = 0$;
    $x - 9 = 0$ or $x = 9$; $x + 1 = 0$ or $x = -1$

17. $x^2 = -2x$; $x^2 + 2x = 0$;
    $x(x + 2) = 0$; $x = 0$;
    $x + 2 = 0$ or $x = -2$

21. $3x^2 - 13x + 4 = 0$; $(3x - 1)(x - 4) = 0$;
    $3x - 1 = 0$ or $3x = 1$;
    $x = \frac{1}{3}$; $x - 4 = 0$ or $x = 4$

25. $6x^2 = 13x - 6$; $6x^2 - 13x + 6 = 0$;
    $(3x - 2)(2x - 3) = 0$; $3x - 2 = 0$ or $3x = 2$, $x = \frac{2}{3}$;
    $2x - 3 = 0$ or $2x = 3$, $x = \frac{3}{2}$

29. $x^2 - x - 1 = 1$; $x^2 - x - 2 = 0$;
    $(x - 2)(x + 1) = 0$; $x - 2 = 0$ or $x = 2$;
    $x + 1 = 0$ or $x = -1$

33. $40x - 16x^2 = 0$; $16x^2 - 40x = 0$;
    $8x(2x - 5) = 0$; $8x = 0$ or $x = 0$;
    $2x - 5 = 0$ or $2x = 5$, $x = \frac{5}{2}$

70

37. $(x + 2)^3 = x^3 + 8$; $x^3 + 6x^2 + 12x + 8 = x^3 + 8$;
$6x^2 + 12x = 0$; $6x(x + 2) = 0$; $6x = 0$ or $x = 0$;
$x + 2 = 0$ or $x = -2$

41. $P^2 - 3P = 70$
$P^2 - 3P - 70 = 0$
$(P + 7)(P - 10) = 0$
                $P + 7 = 0$, or $P = -7$
                $P - 10 = 0$, or $P = 10$ Pa

45. $\dfrac{1}{(x - 3)} + \dfrac{4}{x} = 2$

$\dfrac{(x - 3)(x)}{(x - 3)} + \dfrac{4(x - 3)(x)}{x} = 2(x - 3)(x)$

$x + 4x - 12 = 2x^2 - 6x$
$2x^2 - 11x + 12 = 0$
$(2x - 3)(x - 4) = 0$
$2x - 3 = 0$ or $x = 3/2$
$x - 4 = 0$ or $x = 4$

49.   If $k_c = 2$, $\dfrac{1}{k_c} = \dfrac{1}{2}$

$\dfrac{1}{k_c} = \dfrac{1}{k} + \dfrac{1}{k + 3}$

$\dfrac{1}{2} = \dfrac{1}{k} + \dfrac{1}{k + 3}$

$\dfrac{2(k)(k + 3)}{2} = \dfrac{2k(k + 3)}{k} + \dfrac{(2k)(k + 3)}{(k + 3)}$

$k^2 + 3k = 2k + 6 + 2k$
$k^2 - k - 6 = 0$
$(k + 2)(k - 3) = 0$
        $k + 2 = 0$, or $k = -2$
        $k - 3 = 0$, or $k = 3$ and $k + 3 = 6$

Exercises 6-2, page 193

1. $x^2 = 25$; $\sqrt{x^2} = \pm\sqrt{25}$; $x = 5$ or $x = -5$

5. $(x - 2)^2 = 25$; $\sqrt{(x - 2)^2} = \pm\sqrt{25}$;
    $x - 2 = 5$ or $x = 7$; $x - 2 = -5$ or $x = -3$

9. $x^2 + 2x - 8 = 0$; $x^2 + 2x = 8$; $x^2 + 2x + 1 = 8 + 1$;
    $(x + 1)^2 = 9$; $x + 1 = 3$ or $x = 2$;
    $x + 1 = -3$ or $x = -4$

13. $x^2 - 4x + 2 = 0$; $x^2 - 4x = -2$; $x^2 - 4x + 4 = -2 + 4$;
    $(x - 2)^2 = 2$; $x - 2 = \pm\sqrt{2}$ or $x = 2 \pm\sqrt{2}$

17. $2s^2 + 5s = 3$; $2s^2 + 5s - 3 = 0$; $s^2 + \dfrac{5}{2}s = \dfrac{3}{2}$;

$s + \dfrac{5}{2}s + \dfrac{25}{16} = \dfrac{3}{2} + \dfrac{25}{16}$; $(s + \dfrac{5}{4})^2 = \dfrac{49}{16}$;

$s + \dfrac{5}{4} = \dfrac{7}{4}$ or $s = \dfrac{1}{2}$; $s + \dfrac{5}{4} = -\dfrac{7}{4}$; $s = -3$

21. $2y^2 - y - 2 = 0$; $y^2 - \frac{1}{2}y - 1 = 0$; $y^2 - \frac{1}{2}y = 1$;

$y^2 - \frac{1}{2}y + \frac{1}{16} = 1 + \frac{1}{16}$; $(y - \frac{1}{4})^2 = \frac{17}{16}$; $y - \frac{1}{4} = \pm\frac{\sqrt{17}}{4}$;

$y = \frac{1 \pm\sqrt{17}}{4} = \frac{1}{4}(1 \pm\sqrt{17})$

Exercises 6-3, page 197

1. $x^2 + 2x - 8 = 0$; $a = 1$, $b = 2$, $c = -8$

$x = \frac{-2 \pm\sqrt{4 - 4(1)(-8)}}{2} = \frac{-2 \pm\sqrt{36}}{2}$

$= \frac{-2 \pm 6}{2} = 2, -4$

5. $x^2 - 4x + 2 = 0$; $a = 1$, $b = -4$, $c = 2$

$x = \frac{4 \pm\sqrt{16 - 4(1)(2)}}{2} = \frac{4 \pm\sqrt{8}}{2}$;

$= \frac{4 \pm2\sqrt{2}}{2} = 2 \pm\sqrt{2}$

9. $2s^2 + 5s = 3$; $2s^2 + 5s - 3 = 0$; $a = 2$, $b = 5$, $c = -3$

$s = \frac{-5 \pm\sqrt{25 - 4(2)(-3)}}{4} = \frac{-5 \pm\sqrt{49}}{4}$

$= \frac{-5 \pm 7}{4} = -3, \frac{1}{2}$

13. $2y^2 - y - 2 = 0$; $a = 2$, $b = -1$, $c = -2$

$y = \frac{1 \pm\sqrt{1 - 4(2)(-2)}}{4} = \frac{1 \pm\sqrt{17}}{4}$

$= \frac{1}{4}(1 \pm\sqrt{17})$

17. $2t^2 + 10t = -15$; $2t^2 + 10t + 15 = 0$; $a = 2$, $b = 10$, $c = 15$

$t = \frac{-10 \pm\sqrt{100 - 4(2)(15)}}{4} = \frac{-10 \pm\sqrt{-20}}{4}$

$= \frac{-10 \pm2\sqrt{-5}}{4} = \frac{-5 \pm\sqrt{-5}}{2}$

$= \frac{1}{2}(-5 \pm\sqrt{-5})$

21. $4x^2 = 9$; $4x^2 - 9 = 0$; $a = 4$, $b = 0$, $c = -9$

$$x = \frac{0 \pm \sqrt{0 - 4(4)(-9)}}{8} = \frac{\pm\sqrt{144}}{8}$$

$$= \frac{\pm 12}{8} = \frac{3}{2}, -\frac{3}{2}$$

25. $x^2 - 0.20x - 0.40 = 0$; $a = 1$, $b = 0.20$, $c = -0.40$

$$x = \frac{0.20 \pm \sqrt{(0.20)^2 - 4(1)(-0.40)}}{2(1)} = \frac{0.20 \pm \sqrt{1.64}}{2}$$

$$= \frac{0.20 \pm 1.28}{2} = 0.10 \pm 0.64 = -0.54,\ 0.74$$

29. $x^2 + 2cx - 1 = 0$; $a = 1$, $b = 2c$, $c = -1$

$$x = \frac{-2c \pm \sqrt{4c^2 - 4(1)(-1)}}{2} = \frac{-2c \pm \sqrt{4c^2 + 4}}{2}$$

$$= \frac{-2c \pm \sqrt{4(c^2 + 1)}}{2} = \frac{-2c \pm 2\sqrt{c^2 + 1}}{2}$$

$$= -c \pm \sqrt{c^2 + 1}$$

33. $D_0^2 - DD_0 - 0.25D^2 = 0$; $D = 3.625$

$$D_0^2 - 3.625D_0 - 0.25(3.625)^2 = 0$$

$a = 1$, $b = -3.625$, $c = -0.25(3.625)^2$
Use the quadratic formula

$$D_0 = \frac{-b \pm \sqrt{b^2 - 4ac}}{2a}$$

$$= \frac{-(-3.625) \pm \sqrt{(-3.625)^2 - 4(1)(-0.25)(3.625)^2}}{2(1)}$$

$$= \frac{3.625 \pm \sqrt{3.625^2 + 1(3.625)^2}}{2}$$

$$= \frac{3.625 \pm \sqrt{2(3.625)^2}}{2} = \frac{3.625 \pm 3.625\sqrt{2}}{2} = \frac{3.625(1 \pm \sqrt{2})}{2}$$

$$= \frac{3.625(2.414)}{2} \text{ or } \frac{3.625(-0414)}{2} \text{ (not valid)}$$

$$= 4.376$$

37. Let w = width of the door; $w + 2.0$ = the height of the door.
     $A = w \times h$; $29 = w(w + 2.0)$; $w^2 + 2w = 29$; $w^2 + 2w - 29 = 0$
     Using the quadratic formula, $a = 1$, $b = 2$, $c = -29$,

$$w = \frac{-2 \pm \sqrt{2^2 - 4(1)(-29)}}{2(1)} = \frac{-2 \pm \sqrt{4 + 116}}{2} = \frac{-2 \pm \sqrt{120}}{2}$$

$$w = \frac{-2 + 10.95}{2} = 4.48 \text{ ft or } w = \frac{-2 - 11.0}{2} = -6.48 \text{ ft}$$

   $h = 4.48 + 2 = 6.48$ ft   The negative solution is meaningless.

Exercises 6-4, page 202

1. $y = x^2 - 6x + 5$; $a = 1$, $b = -6$.   This
    means that the x-coordinate of the
    extreme is $\frac{-b}{2a} = \frac{-(-6)}{2(1)} = \frac{6}{2} = 3$, and the
    y-coordinate is $y = 3^2 - 6(3) + 5 =$
    $9 - 18 + 5 = -4$. Thus the extreme point is
    $(3,-4)$. Since $a > 0$, it is a minimum point.
    Since $c = 5$, the y-intercept is $(0,5)$.
       Use the minimum point $(3,-4)$ and the
    y-intercept $(0,5)$, and the fact that the
    graph is a parabola, to get an approxi-
    mate sketch of the graph.

5. $y = x^2 - 4x$; $a = 1$, $b = -4$.   This
    means that the x-coordinate of the
    extreme is $\frac{-b}{2a} = \frac{-(-4)}{2(1)} = \frac{4}{2} = 2$, and
    the y-coordinate is $y = 2^2 - 4(2) = -4$.
    Thus the extreme point is $(2,-4)$. Since
    $a > 0$, it is a minimum point. Since
    $c = 0$, the y-intercept is $(0,0)$.
       Use the minimum point $(2,-4)$ and the
    y-intercept $(0,0)$, and the fact that
    the graph is a parabola, to get an
    approximate sketch of the graph.

9. $y = x^2 - 4$; $a = 1, b = 0, c = -4$.   This
    means that the x-coordinate of the extreme
    is $\frac{-b}{2a} = \frac{-0}{2(1)} = 0$, and the y-coordinate is
    $y = 0^2 - 4 = -4$.   Thus the extreme point
    is $(0,-4)$.   Since $a > 0$, it is a minimum
    point.   This is also the y-intercept.
    The x-intercepts are found by setting
    $y = 0$; $0 = x^2 - 4$; $(x + 2)(x - 2) = 0$;
    $x = -2$ or $x = 2$.   Therefore, the x-inter-
    cepts are $(-2,0)$ and $(2,0)$.
        Use the minimum point $(0,-4)$ and the
    x-intercepts $(-2,0)$ and $(2,0)$ to make an
    approximate sketch of the graph.

13. $y = 2x^2 + 3$; $a = 2, b = 0, c = 3$.   This
    means that the x-coordinate of the extreme
    is $\frac{-b}{2a} = \frac{0}{2(2)} = 0$, and the y-coordinate is
    $y = 2(0^2) + 3 = 3$.   Thus, the extreme
    point is $(0,3)$.   Since $a > 0$, this is a
    minimum point.   Two other points which
    may be used are: If $x = 1$, $y = 2(1^2) + 3 = 5$.
    Therefore $(1,5)$ is a point on the
    parabola.   If $x = -1$, $y = 2(-1)^2 + 3 = 5$.
    Therefore $(-1,5)$ is a point on the para-
    bola.   Sketch the graph using these
    points.

17. $2x^2 - 3 = 0$;   let $y = 2x^2 - 3$; $a = 2$,
    $b = 0, c = -3$.   Use the minimum point
    $(0,-3)$ and the points $(2,5)$ and $(-2,5)$ to
    sketch the graph.   The points $(1,-1)$ and
    $(-1,-1)$ were added for accuracy.   The
    x-intercepts are approximately $(1.2,0)$
    and $(-1.2,0)$.   These are also the solu-
    tions to the equation.

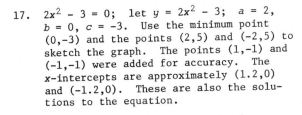
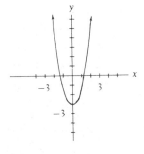

21. $x(2x - 1) = -3;$ $2x^2 - x + 3 = 0;$
the minimum point is $(\frac{1}{4}, 2\frac{7}{8})$. Therefore, there are no real roots, since
the graph cannot cross the axis.

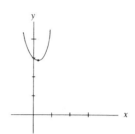

25. $A = w(7 - 2w);$ $7w - 2w^2 = 0;$
$-2w^2 + 7w = 0.$ Since $a < 0,$
$(\frac{7}{4}, 6\frac{1}{8})$ is a maximum point. The
points $(0,0)$, $(3,3)$, and $(\frac{7}{2}, 0)$ were
also used to sketch the graph.

29. $s = 150 + 250t - 16t^2$

The extreme point has x-coordinate of

$\frac{-b}{2a} = \frac{-250}{-32} \approx 7.8,$ and y-coordinate

of $150 + 250(7.8) - 16(7.8)^2 \approx 1130.$

a) Missile hits ground after approx-
    imately 16 seconds.
b) Maximum altitude is 1130 ft.
c) 800 ft is attained after approx-
    imately 3.3 s and 12.3 s.

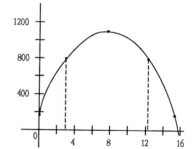

29M. $s = 50 + 90t - 4.9t^2$
The extreme point has x-coordinate of

$\frac{-b}{2a} = \frac{-90}{-9.8} \approx 9.2,$ and y-coordinate

of $50 + 90(9.2) - 4.9(9.2)^2 \approx 460.$

a) Missile hits the ground after approx-
    imately 19 seconds.
b) Maximum altitude is 460 meters.
c) 250 m is attained after approx-
    imately 2.6 s and 15.8 s.

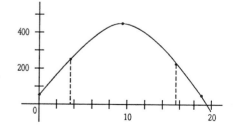

Review Exercises for Chapter 6, page 202

1. $x^2 + 3x - 4 = 0$; $(x + 4)(x - 1) = 0$
$x + 4 = 0$ or $x = -4$; $x - 1 = 0$ or $x = 1$

5. $3x^2 + 11x = 4$; $3x^2 + 11x - 4 = 0$
$(3x - 1)(x + 4) = 0$; $3x - 1 = 0$ or $x = \dfrac{1}{3}$
$x + 4 = 0$ or $x = -4$

9. $6s^2 = 25s$; $6s^2 - 25s = 0$
$s(6s - 25) = 0$; $s = 0$
$6s - 25 = 0$ or $s = \dfrac{25}{6}$

13. $x^2 - x - 110 = 0$; $a = 1$, $b = -1$, $c = -110$
$$x = \frac{1 \pm \sqrt{1 - 4(1)(-110)}}{2} = \frac{1 \pm \sqrt{441}}{2} = \frac{1 \pm 21}{2} = -10,\ 11$$

17. $2x^2 - x = 36$; $2x^2 - x - 36 = 0$; $a = 2$, $b = -1$, $c = -36$
$$x = \frac{1 \pm \sqrt{1 - 4(2)(-36)}}{4} = \frac{1 \pm \sqrt{289}}{4} = \frac{1 \pm 17}{4} = -4,\ \frac{9}{2}$$

21. $2.1x^2 + 2.3x + 5.5 = 0$ or $21x^2 + 23x + 55 = 0$; $a = 21$, $b = 23$, $c = 55$
$$x = \frac{-23 \pm \sqrt{23^2 - 4(21)(55)}}{2(21)} = \frac{-23 \pm \sqrt{529 - 4620}}{42} = \frac{-23 \pm \sqrt{-4091}}{42}$$

25. $x^2 + 4x - 4 = 0$; $a = 1$, $b = 4$, $c = -4$
$$x = \frac{-4 \pm \sqrt{16 - 4(1)(-4)}}{2} = \frac{-4 \pm \sqrt{32}}{2} = \frac{-4 \pm 4\sqrt{2}}{2} = -2 \pm 2\sqrt{2}$$

29. $4v^2 = v + 5$; $4v^2 - v - 5 = 0$
$(4v - 5)(v + 1) = 0$; $4v - 5 = 0$ or $v = \dfrac{5}{4}$
$v + 1 = 0$ or $v = -1$

33. $a^2x^2 + 2ax + 2 = 0$; $a = a^2$, $b = 2a$, $c = 2$
$$x = \frac{-2a \pm \sqrt{4a^2 - 4(a^2)(2)}}{2a^2} = \frac{-2a \pm \sqrt{-4a^2}}{2a^2}$$
$$= \frac{-2a \pm 2a\sqrt{-1}}{2a^2} = \frac{-1 \pm \sqrt{-1}}{a}$$

37. $x^2 - x - 30 = 0$; $x^2 - x = 30$; $x^2 - x + \dfrac{1}{4} = 30 + \dfrac{1}{4}$
$\left(x - \dfrac{1}{2}\right)^2 = \dfrac{121}{4}$; $x - \dfrac{1}{2} = \pm\dfrac{11}{2}$; $x = 6,\ -5$

41. $\dfrac{x-4}{x-1} = \dfrac{2}{x}$;   Multiply by L.C.D. $x(x-1)$; $x^2 - 4x = 2x - 2$;

$x^2 - 6x + 2 = 0$

$a = 1$, $b = -6$, $c = 2$

$x = \dfrac{6 \pm \sqrt{36 - 4(1)(2)}}{2} = \dfrac{6 \pm \sqrt{28}}{2} = \dfrac{6 \pm 2\sqrt{7}}{2} = 3 \pm \sqrt{7}$

45. $y = 2x^2 - x - 1$; $a = 2$, $b = -1$, $c = -1$,

$\dfrac{-b}{2a} = \dfrac{-(-1)}{2(2)} = \dfrac{1}{4}$. $y = 2(\tfrac{1}{4})^2 - \dfrac{1}{4} - 1 = -\dfrac{9}{8}$.

Since $a > 0$, $(\tfrac{1}{4}, -\tfrac{9}{8})$ is a minimum point.

The y-intercept is (-1).   Therefore $(0,-1)$ is a point on the graph. $(1,0)$ and $(-1,2)$ are additional points on the parabola.

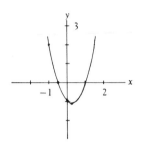

49. $2x^2 + x - 4 = 0$;   let $y = 2x^2 + x - 4$.

Use the minimum point $(-\tfrac{1}{4}, -4\tfrac{1}{8})$ and the points $(2,6)$, $(-2,2)$, $(0,-4)$, and $(1,-1)$ to sketch the graph.   The x-intercepts are approximately $(1.2,0)$ and $(-1.7,0)$. These are also the solutions to the equation.

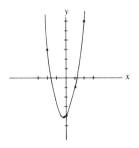

53. $v = 5.2x - x^2$; $v = 4.8$
      $4.8 = 5.2x - x^2$; $x^2 - 5.2x + 4.8 = 0$

$x = \dfrac{-(-5.2) \pm \sqrt{(-5.2)^2 - 4(4.8)}}{2}$

$= \dfrac{5.2 \pm \sqrt{27.04 - 19.2}}{2}$

$= \dfrac{5.2 \pm \sqrt{7.84}}{2}$

$x = \dfrac{8.0}{2} = 4$ or $x = \dfrac{2.4}{2} = 1.2$

57. $\dfrac{n^2}{500\ 000} = 144 - \dfrac{n}{500}$

$n^2 = 500\ 000(144) - 1000n$
$n^2 + 1000n - 72\ 000\ 000 = 0$
$(n + 9000)(n - 8000) = 0$
$n + 9000 = 0$ or $n = -9000$
$n - 8000 = 0$ or $n = 8000$

61. $p = 0.090t - 0.015t^2$

$\dfrac{-b}{2a} = \dfrac{-0.090}{-0.030} = 3;$

$p = 0.090(3) - 0.015(9)$
$\quad = 0.270 - 0.135$
$\quad = 0.135$

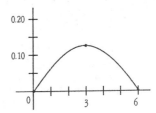

65. Let x = the original edge length, and $x + 0.20$ = expanded length
$(x + 0.20)^3 - x^3 = 6.00$
$x^3 + 0.60x^2 + 0.12x + 0.008 - x^3 = 6.00$
$0.60x^2 + 0.12x - 5.992 = 0$

$$x = \dfrac{-0.12 \pm \sqrt{(0.12)^2 - 4(0.60)(-5.992)}}{2(0.60)} = \dfrac{-0.12 \pm \sqrt{14.3952}}{1.20}$$

$$= \dfrac{-0.12 \pm 3.79}{1.2} = 3.06$$

69. The number of poles per kilometre is equal to the number of spaces between poles. Let n be the number of poles and spaces, and let x be the length of each space. Since 1 km = 1000 m, $n = 1000 \div x$. When the number of poles decreases to $n - 5$, the length increases to $x + 10$, so
$n - 5 = \dfrac{1000}{x + 10};$ $\dfrac{1000}{x + 10} + 5 = n.$ Since $n = \dfrac{1000}{x}$ also,
then $\dfrac{1000}{x + 10} + 5 = \dfrac{1000}{x}.$
Multiply by L.C.D. $(x + 10)(x)$ and simplify, getting $5x^2 + 50x - 10,000$
$= 0$ or $x^2 + 10x - 2000 = 0;$ $(x + 50)(x - 40) = 0;$ $x + 50 = 0$ or
$x = -50;$ $x - 40 = 0$ or $x = 40.$
$n = \dfrac{1000}{x} = \dfrac{1000}{40} = 25$

The solution is $x = 40$ since the distance cannot be negative, and the number of poles is 25 per kilometre.

CHAPTER 7

Trigonometric Functions of Any Angle

Exercises 7-1, page 207

1. sin 60° is positive since 60° is in Quad I, where all functions are positive.
    cos 120° is negative since 120° is in Quad II, where cos θ is negative.
    tan 320° is negative since 320° is in Quad IV, where tan θ is negative.

5. cot 186° is positive since 186° is in Quad III, where cot θ is positive.
    sec 280° is positive since 280° is in Quad IV, where sec θ is positive.
    sin 470° = sin(470° − 360°) = sin 110°, which is positive since 110°
    is in Quad II, where sin θ is positive.

9. (2,1); $x = 2$, $y = 1$, $r = \sqrt{x^2 + y^2} = \sqrt{5}$

$$\sin \theta = \frac{y}{r} = \frac{1}{\sqrt{5}} \qquad \cos \theta = \frac{x}{r} = \frac{2}{\sqrt{5}} \qquad \tan \theta = \frac{y}{x} = \frac{1}{2}$$

$$\csc \theta = \frac{r}{y} = \frac{\sqrt{5}}{1} = \sqrt{5} \qquad \sec \theta = \frac{r}{x} = \frac{\sqrt{5}}{2} = \frac{1}{2}\sqrt{5} \qquad \cot \theta = \frac{x}{y} = \frac{2}{1} = 2$$

13. (−5,12); $x = -5$, $y = 12$, $r = \sqrt{x^2 + y^2} = \sqrt{169} = 13$

$$\sin \theta = \frac{y}{r} = \frac{12}{13} \qquad \cos \theta = \frac{x}{r} = \frac{-5}{13} = -\frac{5}{13} \qquad \tan \theta = \frac{y}{x} = \frac{12}{-5} = -\frac{12}{5}$$

$$\csc \theta = \frac{r}{y} = \frac{13}{12} \qquad \sec \theta = \frac{r}{x} = \frac{13}{-5} = -\frac{13}{5} \qquad \cot \theta = \frac{x}{y} = \frac{-5}{12} = -\frac{5}{12}$$

17. sin θ positive and cos θ negative

    sin θ is positive only in Quad I and Quad II.
    cos θ is negative only in Quad II and Quad III.
    The terminal side of θ must lie in Quad II to meet both conditions.

21. csc θ negative and tan θ negative

csc θ is negative only in Quad III and Quad IV.
tan θ is negative only in Quad II and Quad IV.
The terminal side of θ must lie in Quad IV to meet both conditions.

Exercises 7-2, page 214

1. sin 160° = +sin(180° − 160°) = sin 20° (Eq. 7-3, since 160° is in
    Quad II)
    cos 220° = −cos(220° − 180°) = −cos 40° (Eq. 7-4, since 200° is in
    Quad III)

5. sin(−123°) = sin(360° − 123°)
                = sin 237° = −sin(237° − 180°)
                = −sin 57°
    cot 174° = −cot(180° − 174°)
             = −cot 6°

9. sin 195° = −sin(195° − 180°)
              = −sin 15° = −0.2588

13. tan 219.15° = tan(219.15° − 180°)
                 = tan 39.15° = 0.8141

17. tan 152.4°;   152.4, $\boxed{\text{TAN}}$, $\boxed{=}$ −0.5228

21. cos 110° ; 110 , $\boxed{\text{COS}}$, $\boxed{=}$ −0.3420

25. sin θ = −0.8480; 0.8480, $\boxed{\text{INV}}$, $\boxed{\text{SIN}}$ 58.00°.
     θ is a third or fourth quadrant angle since the sine is negative.
     Therefore, θ = 180° + 58° = 237.99° or θ = 360° − 58° = 302.01°.

     Depending on your calculator, you may use $\sin^{-1}$ or $\boxed{\text{ARCSIN}}$ instead
     of $\boxed{\text{INV}}$, $\boxed{\text{SIN}}$.

29. tan θ = 0.283 ; 0.283 , $\boxed{\text{INV}}$, $\boxed{\text{TAN}}$ 15.82.
     θ is a first or third quadrant angle since the tangent is positive.
     Therefore, θ = 15.82° or θ = 180° + 15.82° = 195.82°.

33. sin θ = 0.870 , cos θ < 0; 0.870 , $\boxed{\text{INV}}$, $\boxed{\text{SIN}}$ 60.46. θ is a second
     quadrant angle since sin θ > 0 and cos θ < 0. Therefore,
     θ = 180° − 60.5.° = 119.5°.

37. tan θ = −1.366, cos θ > 0; 1.366, $\boxed{\text{INV}}$, $\boxed{\text{TAN}}$ 53.79.
     θ is a fourth quadrant angle since tan θ < 0, and cos θ > 0. Therefore,
     θ = 360° − 53.79° = 306.21°.

41. sin θ is negative, so θ is in Quad III or Quad IV.
     cos θ is positive, so θ is in Quad I or Quad IV.
     θ is in Quad IV.

     sin 35.0° = 0.5736; -sin 35.0° = -sin(360° - θ)
     35.0° = 360° - θ or θ = 360° - 35.0° = 325.0°
     tan 325.0° = -tan(360° - 325°)
                = -tan 35.0° = -0.7003

45. sin 90° = 1.000; sin 45° = 0.7071
     2 sin 45° = 2(0.7071) = 1.4142, so sin 90° < 2 sin 45°

49. $i = i_m$ sin θ; i_m = 0.0259, θ = 495.2°

     i = 0.0259(sin 495.2°) = 0.0259(sin 135.2°) = 0.0259(sin 44.8°)
       = 0.0259(0.7046) = 0.0183

53. $\cos θ = \frac{x}{r}$ $\cos(-θ) = \frac{x}{r}$; $\cos(-θ) = \cos θ$

     $\tan θ = \frac{y}{x}$ $\tan(-θ) = \frac{-y}{x}$; $\tan(-θ) = -\tan θ$

     $\cot θ = \frac{x}{y}$ $\cot(-θ) = \frac{x}{-y}$; $\cot(-θ) = -\cot θ$

     $\sec θ = \frac{r}{x}$ $\sec(-θ) = \frac{r}{x}$; $\sec(-θ) = \sec(θ)$

     $\csc θ = \frac{r}{y}$ $\csc(-θ) = \frac{r}{-y}$; $\csc(-θ) = -\csc(θ)$

Exercises 7-3, page 219

1. $15° = \frac{π}{180}(15)$ $150° = \frac{π}{180}(150)$

    $= \frac{15π}{180} = \frac{π}{12}$ $= \frac{150π}{180} = \frac{5π}{6}$

5. $210° = \frac{π}{180}(210)$ $270° = \frac{π}{180}(270)$

    $= \frac{210π}{180} = \frac{7π}{6}$ $= \frac{270π}{180} = \frac{3π}{2}$

9. $\frac{2π}{5} = \frac{180°}{π}(\frac{2π}{5})$ $\frac{3π}{2} = \frac{180°}{π}(\frac{3π}{2})$

    $= \frac{360π°}{5π} = 72°$ $= \frac{540π°}{2π} = 270°$

13. $\dfrac{17\pi}{18} = \dfrac{180°}{\pi}(\dfrac{17\pi}{18})$

$= \dfrac{10(17\pi)°}{\pi} = 170°$

$\dfrac{5\pi}{3} = \dfrac{180°}{\pi}(\dfrac{5\pi}{3})$

$= \dfrac{60(5\pi)°}{\pi} = 300°$

17. $23.0° = \dfrac{\pi}{180}(23.0°)$

$= \dfrac{23.0\pi}{180} - \dfrac{72.22}{180}$

$= 0.401$

21. $= 333.5° = \dfrac{\pi}{180}$ (333.5)

$= \dfrac{333.5\pi}{180} = \dfrac{1048}{180}$

$= 5.821$

25. $0.750 = \dfrac{180°}{3.14}(0.750)$

$= \dfrac{135°}{3.14} = 43.0°$

29. $2.45 = \dfrac{180°}{3.14}(2.45)$

$= \dfrac{441°}{3.14} = 140°$

33. $\sin\dfrac{\pi}{4} = \sin(\dfrac{\pi}{4})(\dfrac{180}{\pi})$

$= \sin 45° = 0.7071$

37. $\cos\dfrac{5\pi}{6} = \cos(\dfrac{5\pi}{6})(\dfrac{180}{\pi})$

$= \cos 150°$
$= -\cos(180° - 150°)$
$= -\cos 30° = -0.8660$

41. tan 0.7359. Set calculator on radian mode. 0.7359, $\boxed{\text{TAN}}$ 0.9056

45. cos 2.07. Set calculator on radian mode. 2.07, $\boxed{\text{COS}}$ -0.48

49. $\sin\theta = 0.3090;\quad \theta = 0.3142$
Also, $\sin(\pi - \theta) = 0.3090;\quad \pi - \theta = 0.3141$
$\theta = \pi - 0.3141 = 2.827$

53. $\cos\theta = 0.6742;\quad \theta = 0.8309$
Also, $\cos(2\pi - \theta) = 0.6742;\quad 2\pi - \theta = 0.8309$
$\theta = 6.283 - 0.8309 = 5.452$

57. $V = \dfrac{1}{2}Wb\theta^2;\quad W = 8.75,\ b = 0.75;\ \theta = 5.5° = \dfrac{5.5}{1}\left(\dfrac{\pi}{180°}\right) = 0.0960$

$V = \dfrac{1}{2}(8.75)(0.75)(0.0960)^2 = 0.030$

Exercises 7-4, page 224

1. $s = \theta r;\ r = 320,\ \theta = 62.0° = \dfrac{62.0°}{1}\left(\dfrac{\theta}{180°}\right) = 1.08$

$s = 1.08(320) = 346$

5. $s = \theta r; \quad \theta = \dfrac{s}{r} = \dfrac{3.25}{8.50}$

$= 0.382$ or $21.9°$

9. In making a U-turn, a car will travel the distance which is $\dfrac{1}{2}$ the circumference of a circle. Therefore,

$s = \dfrac{2\pi r}{2} = \pi r$

$s = vt; \quad \pi r = v(6); \quad v = \dfrac{\pi r}{6}$

By Eq. (7-12), $v = wr; \quad \dfrac{\pi r}{6} = wr$

$w = \dfrac{\pi}{6}$ rad/s $= 0.52$ rad/s

13. $s = s_0 - s_1$ where $s_0 = \theta(93.67 + 4.71)$ and $s_1 = \theta(93.67)$

$28.0° = \dfrac{28°}{1}\left(\dfrac{\pi}{180°}\right)$ $s_0 = 0.4887(98.38)$ $s_1 = 0.4887(93.67)$

$= 0.4887$ $= 48.08$ $= 45.78$

$s = 2.30$

13M. $s = s_0 - s_1$ where $s_0 = \theta(28.55 + 1.44)$ and $s_1 = \theta(28.55)$

$28.0° = \dfrac{28°}{1}\left(\dfrac{\pi}{180°}\right)$ $= 0.4887(29.99)$ $= 0.4887(28.55)$

$= 0.4887$ $= 14.66$ $= 13.95$

$s = 0.7100$

17. $A = A_0 - A_1$ where $A_0 = \dfrac{1}{2}\theta(28.50 + 15.2)^2$ and $A_1 = \dfrac{1}{2}\theta(285.0)^2$

$\theta = 15.6° = \dfrac{15.6°}{1}\left(\dfrac{\pi}{180°}\right)$ $A_0 = \dfrac{1}{2}(0.272)(300.2)^2 = 12256$

$= 0.272$ $A_1 = \dfrac{1}{2}(0.272)(285.0)^2 = 11047$

$A = 1209, \quad V = 1209(0.305) = 369$

21. $v = \omega r; \quad v = 3.5$ mi/hr $= 308$ ft/min; $\quad r = \dfrac{1}{2}(12.0) = 6.0$ ft

$\omega = \dfrac{v}{r} = \dfrac{308}{6.0} = 51.3$ rad/min $= 8.17$ r/min

21M. $v = \omega r; \quad v = 5.6$ km/hr $= 93.3$ m/s; $\quad r = \dfrac{1}{2}(3.66) = 1.83$ m

$\omega = \dfrac{v}{r} = \dfrac{93.3}{1.83} = 50.9$ rad/min $= 8.11$ r/min

25.

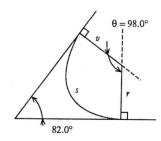

$\theta = 98.0°$

$s = \theta r; \; \theta = 98.0° = 1.71 \text{ rad}, \; r = 115.0 \text{ ft}$

$s = 1.71(15.0) = 25.7 \text{ ft}$

82.0°

25M. $s = \theta r = 5.5(98.0)\left(\dfrac{\pi}{180}\right) = 9.41$

29. $v = \dfrac{1}{4}(6.50) = 1.625; \; r = 3.75$

   $\omega = \dfrac{v}{r} = \dfrac{1.625}{3.75} = 0.433 \text{ rad/s}$

33. $2000 \text{ r/min} = 33.3 \text{ r/s} = 33.3(2\pi) \text{ rad/s} = 209 \text{ rad/s}$

37. $\sin \theta = \tan \theta = \theta$ for small radian angles.

   $1'' = \dfrac{1}{3600}^{°} = \dfrac{1}{3600}(\dfrac{\pi}{180}) \text{ rad} = \dfrac{\pi}{6.48 \times 10^{5}} = 0.4848 \times 10^{-5} \text{ rad} = \theta$

   $\sin \theta = \theta$
   $\sin 0.4848 \times 10^{-5} = 0.4848 \times 10^{-5} = 4.848 \times 10^{-6}$

Review   Exercises for Chapter 7, page 227

   1. $(6,8); \; x = 6, \; y = 8, \; r = \sqrt{6^2 + 8^2} = 10$

   $\sin \theta = \dfrac{y}{r} = \dfrac{8}{10} = \dfrac{4}{5}$ $\cos \theta = \dfrac{x}{r} = \dfrac{6}{10} = \dfrac{3}{5}$ $\tan \theta = \dfrac{y}{x} = \dfrac{8}{6} = \dfrac{4}{3}$

   $\csc \theta = \dfrac{r}{y} = \dfrac{5}{4}$ $\sec \theta = \dfrac{r}{x} = \dfrac{5}{3}$ $\cot \theta = \dfrac{x}{y} = \dfrac{3}{4}$

   5. $\cos 132° = -\cos(180° - 132°) = -\cos 48°$ (Eq. 7-3)
       $\tan 194° = \tan(194° - 180°) = \tan 14°$ (Eq. 7-4)

   9. $40° = 40°(\dfrac{\pi}{180°}) = \dfrac{2\pi}{9}; \quad 153° = 153°(\dfrac{\pi}{180°}) = \dfrac{17\pi}{20}$

   13. $\dfrac{7\pi}{5} = \dfrac{7\pi}{5}(\dfrac{180°}{\pi}) = 252°; \quad \dfrac{13\pi}{18} = \dfrac{13\pi}{18}(\dfrac{180°}{\pi}) = 130°$

   17. $0.560 = 0.560(\dfrac{180°}{\pi}) = \dfrac{100.8°}{\pi} = 32.1°$

21. $102° = 102\left(\dfrac{\pi}{180}\right) = \dfrac{320}{180} = 1.78$

25. $262.05° = 262.05 \times \dfrac{\pi}{180°} = 4.574$

29. cos 245.5°;   245.5, $\boxed{\text{COS}}$ −0.415

33. csc 247.82°;   247.82, $\boxed{\text{SIN}}$, $\boxed{\dfrac{1}{x}}$ −1.080

37. tan 301.4°;   301.4, $\boxed{\text{TAN}}$ −1.64

41. $\sin \dfrac{9\pi}{5}$;   9, $\boxed{\times}$, 3.14159, $\boxed{\div}$, 5, $\boxed{=}$, $\boxed{\text{DRG}}$, $\boxed{\text{SIN}}$ −0.5878

45. sin 0.5906;   0.5906, $\boxed{\text{DRG}}$, $\boxed{\text{SIN}}$ 0.5569

49. tan θ = 0.1817, 0° ≤ θ < 360° (tan θ > 0 in Q I and Q III)
     θ = 10.30°   (Q I)
     θ = 180° + 10.30° = 190.30°   (Q III)

53. cos θ = 0.8387, 0 ≤ θ < 2π (cos θ > 0 in Q I and Q IV)
     θ = 0.5759   (Q I)
     θ = 2π − 0.5759 = 5.707   (Q IV)

57. cos θ = −0.7222, sin θ < 0. For cos θ < 0, θ must be in the 2nd
     quadrant or the 3rd quadrant.

     For sin θ < 0, θ must be in the 3rd quadrant or the 4th quadrant.
     cos < 0 and sin θ < 0 in the 3rd quadrant.

     The reference angle is $\boxed{\text{INV}}$, 0.7222, 43.76°.
     Therefore, θ = 180° + 43.76° = 223.76°.

61. $p = p_m \sin^2 377t$; $p_m = 0.120$ W; $t = 2.00$ ms = 0.002 s
     $p = 0.120[\sin^2 (377)(0.002)] = 0.120[\sin (377)(0.002)]^2$
     $= 0.120[\sin 0.754]^2 = 0.120[0.6846]^2 = 0.120[0.4687]$
     $= 0.0562$

65. If θ = 55.25° arc for 40.00°C temperature change, 1°C temperature change
     moves the needle through $\dfrac{55.25}{40.00}$ or 1.38125° per °C. To move through
     150°C, the needle must move 150(1.38125) or 207.1875° or 1.151π rad.
     If the needle is 5.250 cm long, it moves through a distance s of $s = r\theta$
     $= 5.25(1.151\pi) = 18.98$ cm.

69. $d = 25.0$ cm; $r = 12.5$ cm; 60.0 r/s = 60.0(2π) = 120π rad/s
     $v = \omega r = 120\pi(12.5) = 1500\pi = 4710$ cm/s

73. $s = \theta r$; $\theta = 26.0° = 26.0 \left(\dfrac{\pi}{180}\right) = 0.454$ rad; $r = 24.0$ ft

$s = 0.454(24.0) = 10.9$ ft; 6 lines (20.0 ft) = 120.0 ft
4 lines (10.0 ft) = 40.0 ft
Total lines = 170.9 ft

73M. $s = \theta r$; $\theta = 26.0° = 26.0 \left(\dfrac{\pi}{180}\right) = 0.454$ rad; $r = 7.20$ m

$s = 0.454(7.20) = 3.27$ m
6 lines (6.00 m) = 36.0 m
4 lines (3.00 m) = 12.0 m
Total lines = 51.27 m

CHAPTER 8

Vectors and Oblique Triangles

Exercises 8-1, page 234

1. (a) 300 km southwest is a vector; it has magnitude and direction
   (b) 300 km is scalar; it has magnitude but not direction

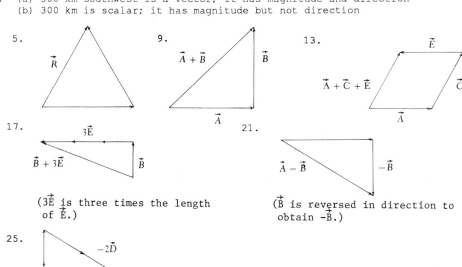

5.

9.

13.

17.

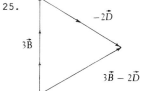

21.

(3$\vec{E}$ is three times the length of $\vec{E}$.)

($\vec{B}$ is reversed in direction to obtain $-\vec{B}$.)

25.

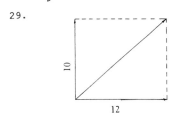

($\vec{D}$ is reversed in direction for $-\vec{D}$.)

29.

From the scale drawing, we find that the resultant force is about 16, and acts at an angle of about 40°.

33.

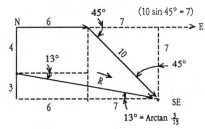

$$R = \sqrt{3^2 + 13^2} = 13$$

Exercises 8-2, page 237

1. $V_x = 750 \cos 28.0° = 662$
    $V_y = 750 \sin 28.0° = 352$

5. Let $V = 8.60$
    $V_x = V \cos 68.0° = 8.60(0.3746)$
    $\quad = 3.22$
    $V_y = V \sin 68.0° = 8.60(0.9272)$
    $\quad = 7.97$

9. Let $V = 9.04$
    $V_x = V \cos 283.3° = 9.04(0.2300)$
    $\quad = 2.08$
    $V_y = V \sin 283.3° = 9.04(-0.9732)$
    $\quad = -8.80$

13.

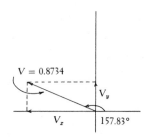

$V_x = V \cos 157.83° = -V \cos 22.17°$
$V_y = V \sin 157.83° = V \sin 22.17°$

$V_x = -0.8734(0.9261) = -0.8088$
$V_y = 0.8734(0.3774) = 0.3296$

17.

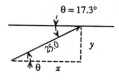

$x = 25.0 \cos 17.3° = 23.9$ km/h

$y = 25.0 \sin 17.3° = 7.43$ km/h

21. The magnitude of the vector is 145, and $\theta = 37.5°$.
     The component directed toward the east is $V_x = 145 \cos 37.5°$
       $= 145(0.7933) = 115$ km.
     The component directed toward the north is $V_y = 145 \sin 37.5°$
       $= 145(0.6088) = 88.3$ km.

Exercises 8-3, page 242

1. $R = \sqrt{A^2 + B^2}$; $A = 14.7$, $B = 19.2$

     $R = \sqrt{14.2^2 + 19.2^2} = \sqrt{585} = 24.2$

     $\tan \theta = \dfrac{B}{A} = \dfrac{19.2}{14.7}$

     $= 1.306$

     $\theta = 52.6°$

     (with $\vec{A}$)

5. $R_x = 5.18$, $R_y = 8.56$; $R = \sqrt{R_x{}^2 + R_y{}^2}$

     $R = \sqrt{5.18^2 + 8.56^2} = \sqrt{100} = 10.0$

     $\tan \theta = \dfrac{8.56}{5.18} = 1.653$; $\theta = 58.8°$

9. $R_x = -646$, $R_y = 2030$; $R = \sqrt{R_x{}^2 + R_y{}^2}$

     $R = \sqrt{(-646)^2 + 2030^2} = \sqrt{4,538,000} = 2130$

     $\tan \theta = \dfrac{2030}{-646} = -3.142$; $\tan 72.3° = +3.142$, so $\theta = 180° - 72.3° = 107.7°$

     (θ is in Q II since R_x is negative and R_y is positive.)

13. $A = 18.0$, $\theta_A = 0.0°$; $A_x = 18.0 \cos 0.0° = 18.0$; $A_y = 18.0 \sin 0.0° = 0.0$

     $B = 12.0$, $\theta_B = 27.0°$; $B_x = 12.0 \cos 27.0° = 10.7$; $B_y = 12.0 \sin 27.0°$
       $= 5.45$

     $R_x = A_x + B_x = 28.7$, $R_y = A_y + B_y = 5.45$

     $R = \sqrt{28.7^2 + 5.45^2} = 29.2$

     $\tan \theta = \dfrac{R_y}{R_x} = \dfrac{5.45}{28.7} = 0.1899$

     $\theta = 10.8°$

17. $A = 9.821$, $\theta_A = 34.27°$; $A_x = 9.821 \cos 34.27° = 8.116$
     $A_y = 9.821 \sin 34.27° = 5.530$

     $B = 17.45$, $\theta_B = 752.50°$ with the same cosine as $32.50°$;
     $B_x = 17.45 \cos 32.50° = 14.72$
     $B_y = 17.45 \sin 32.50° = 9.376$

     $R_x = A_x + B_x = 8.116 + 14.72 = 22.84$; $R_y = A_y + B_y = 5.530 + 9.376 = 14.91$

     $\tan \theta = \dfrac{R_y}{R_x} = \dfrac{14.91}{22.84} = 0.6528$; $\theta = 33.14°$

     $R = \sqrt{R_x{}^2 + R_y{}^2} = 27.27$

21. $A = 21.9$, $\theta_A = 236.2°$; $A_x = 21.9 \cos 236.2° = -12.2$; $A_y = 21.9 \sin 236.2°$
     $= -18.2$

   $B = 96.7$, $\theta_B = 11.5°$; $B_x = 96.7 \cos 11.5° = 94.8$; $B_y = 96.7 \sin 11.5°$
   $= 19.3$

   $C = 62.9$, $\theta_C = 143.4°$; $C_x = 62.9 \cos 143.4° = -50.5$; $C_y = 62.9 \sin 143.4°$
   $= 37.5$

   $R_x = A_x + B_x + C_x = 32.1$, $R_y = A_y + B_y + C_y = 38.6$
   $R = \sqrt{R_x^2 + R_y^2} = 50.1$

   $\tan \theta = \dfrac{R_y}{R_x} = \dfrac{38.6}{32.1} = 1.202$; $\theta = 50.3°$

Exercises 8-4, page 245

1. $F_x = 5.75$, $F_y = 3.25$, $F = \sqrt{(5.75)^2 + (3.25)^2} = 6.60$

   $\tan x = \dfrac{F_y}{F_x} = \dfrac{3.25}{5.75} = 0.565$; $\theta = 29.5°$ from F_x

5.

   $R_x = 1580 + 1640 \cos 35° = 2920$
   $R_y = 1640 \sin 35° = 941$
   $R = \sqrt{941^2 + 2920^2} = 3070$
   $\tan \theta = -\dfrac{941}{2920} = 0.322$; $\theta = 17.8°$

9. $F_x = 22.0$, $F_y = 12.5$; $F = \sqrt{22.0^2 + 12.5^2} = 25.3$

   $\tan \theta = \dfrac{12.5}{22.0} = 0.568$; $\theta = 29.6°$ S of E

13.

   $F_x = 60 \sin 12.0° = 12.5$
   $F_y = 60 \cos 12.0° = 58.7$
   $R = \sqrt{(550 - 12.5)^2 + 58.7^2} = 540$
   $\tan \theta = \dfrac{58.7}{537.5} = 0.109$; $\theta = 6.2°$

17. $d = 8.20$; $r = 4.10$; $\omega = 210$; $\alpha = 320$
   $a_T = r\alpha = 4.10(320) = 1312$
   $a_R = r\omega^2 = 4.10(210)^2 = 180\ 800$
   $a = \sqrt{(1312)^2 + (180\ 800)^2} = 181\ 000$; $\tan \emptyset = \dfrac{180\ 800}{1312} = 137.8$;
   $\emptyset = 89.6°$

21. $V_v = 9.80t = 9.80(2.00) = 19.6$ m/s; $F_y = 19.6 + 15.0 = 34.6$ m/s

$V = \sqrt{(75.0)^2 + (34.6)^2} = 82.6$; $\tan\theta = \dfrac{34.6}{75.0} = 0.461$; $\theta = 24.7°$

Exercises 8-5, page 252

1. $a = 45.7$, $A = 65.0°$, $B = 49.0°$

$\dfrac{a}{\sin A} = \dfrac{b}{\sin B}$; $\dfrac{45.7}{\sin 65.0°} = \dfrac{b}{\sin 49.0°}$

$b = \dfrac{45.7 \sin 49.0°}{\sin 65.0°} = \dfrac{45.7(0.7547)}{0.9063} = 38.1$

$C = 180° - A - B = 66.0°$

$\dfrac{a}{\sin A} = \dfrac{c}{\sin C}$; $\dfrac{45.7}{\sin 65.0°} = \dfrac{c}{\sin 66.0°}$

$c = \dfrac{45.7 \sin 66.0°}{\sin 65.0°} = \dfrac{45.7(0.9135)}{0.9063} = 46.1$

5. $a = 4.601$, $b = 3.107$, $A = 18.23°$

$\dfrac{a}{\sin A} = \dfrac{b}{\sin B}$; $\dfrac{4.601}{\sin 18.23°} = \dfrac{3.107}{\sin B}$

$\sin B = \dfrac{3.107 \sin 18.23°}{4.601} = \dfrac{3.107(0.3128)}{4.601} = 0.2112$

$B = 12.20°$

$C = 180° - 18.23° - 12.19° = 149.57°$

$\dfrac{a}{\sin A} = \dfrac{c}{\sin C}$; $\dfrac{4.601}{\sin 18.23°} = \dfrac{c}{\sin 149.57°}$

$c = \dfrac{4.601 \sin 149.57°}{\sin 18.23°} = \dfrac{4.601(0.5063)}{0.3128} = 7.448$

9. $b = 0.0742$, $B = 51.0°$, $C = 3.36°$

$\dfrac{b}{\sin B} = \dfrac{c}{\sin C}$; $\dfrac{0.0742}{\sin 51.0°} = \dfrac{c}{\sin 3.36°}$

$c = \dfrac{0.0742 \sin 3.36°}{\sin 51.0°} = \dfrac{0.0742(0.0586)}{0.7771} = 0.00560$

$A = 180° - 51.0° - 3.36° = 125.64°$

$\dfrac{b}{\sin B} = \dfrac{a}{\sin A}$; $\dfrac{0.0742}{\sin 51.0°} = \dfrac{a}{\sin 125.64°}$

$a = \dfrac{0.0742 \sin 125.64°}{\sin 51.0°} = \dfrac{0.0742(0.8127)}{0.7771} = 0.0776$

13. $b = 4384$; $B = 47.43°$; $C = 64.56°$

$A = 180° - 47.43° - 64.56° = 68.01°$

$$\frac{a}{\sin A} = \frac{b}{\sin B}; \quad \frac{a}{\sin 68.01°} = \frac{4384}{\sin 47.43°}$$

$$a = \frac{4384 \sin 68.01°}{\sin 47.43°} = \frac{4384(0.9272)}{0.7365} = 5520$$

$$\frac{c}{\sin C} = \frac{b}{\sin B}; \quad \frac{c}{\sin 64.56°} = \frac{4384}{\sin 47.43°}$$

$$c = \frac{4384 \sin 64.56°}{\sin 47.43°} = \frac{4384(0.9030)}{0.7365} = 5376$$

17. $b = 2880$, $c = 3650$, $B = 31.4°$

$$\frac{c}{\sin C} = \frac{b}{\sin B}; \quad \frac{3650}{\sin C} = \frac{2880}{\sin 31.4°}$$

$$\sin C = \frac{3650 \sin 31.4°}{2880} = \frac{3650(0.5210)}{2880} = 0.6603$$

$C = 41.3°$ or $C = 180° - 41.3° = 138.7°$

Let $C_1 = 41.3°$ and let $C_2 = 138.7°$.

Then $A_1 = 180° - B - C_1$ and $A_2 = 180° - B - C_2$.

$A_1 = 107.3°$ $A_2 = 9.9°$

$$\frac{a_1}{\sin A_1} = \frac{b}{\sin B} \qquad\qquad \frac{a_2}{\sin A_2} = \frac{b}{\sin B}$$

$$\frac{a_1}{\sin 107.3°} = \frac{2880}{\sin 31.4°} \qquad \frac{a_2}{\sin 9.9°} = \frac{2880}{\sin 31.4°}$$

$$a_1 = \frac{2880 \sin 107.3°}{\sin 31.4°} \qquad a_2 = \frac{2880 \sin 9.9°}{\sin 31.4°}$$

$$a_1 = \frac{2880(0.9548)}{0.5210} \qquad\quad a_2 = \frac{2880(0.1719)}{0.5210}$$

$a_1 = 5280$ $a_2 = 950$

21.

$\theta = 360° ÷ 5 = 72°$

$\emptyset = \dfrac{1}{2}(180° - 72°) = 54°$

$$\frac{x}{\sin 90°} = \frac{2.00}{\sin 54°}; \quad x = \frac{2.00 \sin 90°}{\sin 54°} = 2.47 \text{ ft}$$

21M. $$\frac{x}{\sin 90°} = \frac{1.30}{\sin 54°}; \quad x = \frac{1.30 \sin 90°}{\sin 54°} = 1.61 \text{ m}$$

25.

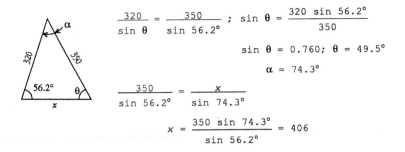

$$\frac{320}{\sin \theta} = \frac{350}{\sin 56.2°} \; ; \; \sin \theta = \frac{320 \sin 56.2°}{350}$$

$$\sin \theta = 0.760; \; \theta = 49.5°$$

$$\alpha = 74.3°$$

$$\frac{350}{\sin 56.2°} = \frac{x}{\sin 74.3°}$$

$$x = \frac{350 \sin 74.3°}{\sin 56.2°} = 406$$

29. $A = 180° - 89.2° = 90.8°$
      $S = 180° - 90.8° = 86.5° = 2.7°$

$$\frac{b}{\sin 86.5°} = \frac{1290}{\sin 2.7°}$$

$$b = \frac{1290 \sin 86.5°}{\sin 2.7°} = \frac{1290(0.9981)}{0.0471}$$

$$= 27,300 \text{ km}$$

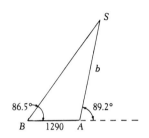

Exercises 8-6, page 257

1. $a = 6.00$, $b = 7.56$, $c = 54.0°$

$$c = \sqrt{6.00^2 + 7.56^2 - 2(6.00)(7.56)(\cos 54.0°)} = \sqrt{39.8} = 6.31$$

$$\frac{a}{\sin A} = \frac{c}{\sin C}; \quad \frac{6.00}{\sin A} = \frac{6.31}{\sin 54.0°}$$

$$\sin A = \frac{6.00 \sin 54.0°}{6.31} = \frac{6.00(0.8090)}{6.31}$$

$$\sin A = 0.7693; \quad A = 50.3°$$

$$\frac{b}{\sin B} = \frac{c}{\sin C}; \quad \frac{7.56}{\sin B} = \frac{6.31}{\sin 54.0°}$$

$$\sin B = \frac{7.56 \sin 54.0°}{6.31} = \frac{7.56(0.8090)}{6.31}$$

$$\sin B = 0.9693; \quad B = 75.7°; \quad \text{or,} \quad B = 180° - A - C = 75.7°$$

5. $a = 39.53$, $b = 45.22$, $c = 67.15$

$$\cos A = \frac{b^2 + c^2 - a^2}{2bc} = \frac{45.22^2 + 67.15^2 - 39.53^2}{2(45.22)(67.15)} = 0.8219$$

$A = 34.72°$

$$\cos B = \frac{a^2 + c^2 - b^2}{2ac} = \frac{39.53^2 + 67.15^2 - 45.22^2}{2(39.53)(67.15)} = 0.7585$$

$B = 40.67°$

$$\cos C = \frac{a^2 + b^2 - c^2}{2ab} = \frac{39.53^2 + 45.22^2 - 67.15^2}{2(45.22)(39.53)} = -0.2522$$

$C = 104.61°$

9. $a = 320$, $b = 847$, $c = 158.0°$

$$c = \sqrt{a^2 + b^2 - 2ab(\cos C)} = \sqrt{1,322,000} = 1150$$

$$\frac{a}{\sin A} = \frac{c}{\sin C}; \quad \frac{320}{\sin A} = \frac{1150}{\sin 158.0°}$$

$$\sin A = \frac{320 \sin 158.0°}{1150} = \frac{320(0.3746)}{1150} = 0.1042$$

$A = 6.0°$

$$\frac{b}{\sin B} = \frac{c}{\sin C}; \quad \frac{847}{\sin B} = \frac{1150}{\sin 158.0°}$$

$$\sin B = \frac{847 \sin 158.0°}{1150} = \frac{847(0.3746)}{1150}$$

$\sin B = 0.2759$; $B = 16.0°$;   or, $B = 180° - A - C = 16.0°$

13. $b = 103.7$, $c = 159.1$, $C = 104.67°$

Use the law of sines first:

$$\frac{b}{\sin B} = \frac{c}{\sin C}; \quad \frac{103.7}{\sin B} = \frac{159.1}{\sin 104.67°}$$

$$\sin B = \frac{103.7 \sin 104.67°}{159.1} = \frac{103.7(0.9674)}{159.1} = 0.6305$$

$B = 39.09°$

$A = 180° - 39.09° - 104.67° = 36.24°$

$$a = \sqrt{a^2 + c^2 - 2ac(\cos B)} = \sqrt{103.7^2 + 159.1^2 - 2(103.7)(159.1)\cos 36.24°}$$

$a = 97.22$

17. $a = 723$, $b = 598$, $c = 158$

$$\cos A = \frac{b^2 + c^2 - a^2}{2bc} = \frac{-140,200}{189,000} = -0.7418$$

$A = 137.9°$

$$\cos B = \frac{a^2 + c^2 - b^2}{2ac} = \frac{190,000}{228,500} = 0.8315$$

$B = 33.7°$

$$\cos C = \frac{a^2 + b^2 - c^2}{2ab} = \frac{855,400}{864,700} = 0.9892$$

$C = 8.4°$

21.

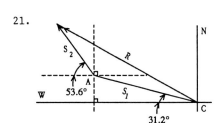

$s_1 = vt_1 = 23.5(2) = 47.0$ at $32.0°$ N of W

$s_2 = vt_2 = 23.5(1) = 23.5$ at $32.1° + 21.5°$

or $53.6°$ N or W

$\angle A = 32.1° + 90° + (90° - 53.6°) = 158.5°$

$R^2 = 47.02^2 + 23.5^2 - 2(47.0)(23.5)(\cos 158.5°)$;

$R^2 = 4816$; $R = 69.4$

25.

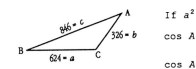

If $a^2 = b^2 + c^2 - 2bc \cos A_2$ then

$$\cos A = \frac{a^2 - b^2 - c^2}{2bc} = \frac{624^2 - 326^2 - 846^2}{2(326)(846)}$$

$\cos A = -0.784$, and $A = 141.7°$

Also, $\cos C = \dfrac{c^2 - a^2 - b^2}{2ab} = \dfrac{846^2 - 624^2 - 326^2}{2(624)(326)} = 0.541$; $C = 57.3°$

29.

$c^2 = 12.7^2 + 11.5^2 - 2(12.7)(11.5)\cos 23.6°$

$c^2 = 25.87$

$c = 5.09$

Review   Exercises for Chapter 8, page 258

1. $A = 65.0$, $\theta_A = 28.0°$; $A_x = 65.0 \cos 28.0° = 57.4$, $A_y = 65.0 \sin 28.0°$
                                                                $= 30.5$

5. $A = 327$, $B = 505$; $R = \sqrt{327^2 + 505^2} = \sqrt{362,000} = 602$

$\tan \theta = \dfrac{327}{505} = 0.6475$, $\theta = 32.9°$ with B, $57.1°$ with A

9. $A = 780$, $\theta_A = 28.0°$ $B = 346$, $\theta_B = 320.0°$

$A_x = 780 \cos 28.0° = 689$ $B_x = 346 \cos 320.0° = 265$

$A_y = 780 \sin 28.0° = 366$ $B_y = 346 \sin 320.0° = -222$

$R_x = 689 + 265 = 954$; $R_y = 366 - 222 = 144$

$R = \sqrt{954^2 + 144^2} = \sqrt{930,700} = 965$

$\tan \theta = \dfrac{R_y}{R_x} = \dfrac{144}{954} = 0.1509$; $\theta = 8.6°$

13. $A = 51.33$, $\theta_A = 12.25°$ $B = 42.61$, $\theta_B = 291.77°$

$A_x = 51.33 \cos 12.25° = 50.16$ $B_x = 42.61 \cos 291.77° = 15.80$

$A_y = 51.33 \sin 12.25° = 10.89$ $B_y = 42.61 \sin 291.77° = -39.57$

$R_x = 50.16 + 15.80 = 65.96$; $R_y = 10.89 - 39.57 = -28.68$

$R = \sqrt{65.96^2 + (-28.68)^2} = 71.93$

$\tan \theta = \dfrac{R_y}{R_x} = \dfrac{-28.68}{65.96} = -0.4348$

$\theta_{ref} = 23.50°$ (since y is negative and x is positive, Quad IV)

$\theta = 336.50°$

17. $A = 48.0°$, $B = 68.0°$, $a = 14.5$; $C = 180° - 48.0° - 68.0° = 64.0°$

$\dfrac{a}{\sin A} = \dfrac{c}{\sin C}$; $\dfrac{14.5}{\sin 48.0°} = \dfrac{c}{\sin 64.0°}$

$c = \dfrac{14.5 \sin 64.0°}{\sin 48.0°} = \dfrac{14.5(0.8988)}{0.7431}$

$c = 17.5$

$\dfrac{b}{\sin B} = \dfrac{a}{\sin A}$; $\dfrac{b}{\sin 68.0°} = \dfrac{14.5}{\sin 48.0°}$

$b = \dfrac{14.5 \sin 68.0°}{\sin 48.0°} = \dfrac{14.5(0.9272)}{0.7431}$

$b = 18.1$

21. $A = 17.85°$, $B = 154.16°$, $c = 7863$

$C = 180° - 17.85° - 154.16° = 7.99°$

$$\frac{a}{\sin 17.85°} = \frac{7863}{\sin 7.99°};$$

$$a = \frac{7863 \sin 17.85°}{\sin 7.99°} = 17,340$$

$$\frac{b}{\sin 154.16°} = \frac{7863}{\sin 7.99°};$$

$$b = \frac{7863 \sin 154.16°}{\sin 7.99°} = 24,660$$

21M. $a = 17\ 340$; $b = 24\ 660$; $C = 7.99°$

25. The angle opposite b may be either an acute or an obtuse angle. See Example E. $b = 14.5$, $c = 13.0$, $C = 56.6°$

Case I: $\dfrac{13.0}{\sin 56.6°} = \dfrac{14.5}{\sin B_1}$; $\sin B_1 = \dfrac{14.5 \sin 56.6°}{13.0} = 0.9312$

$B_1 = 68.6°$; $A_1 = 180° - 68.6° - 56.6° = 54.8°$

$\dfrac{a_1}{\sin 54.8°} = \dfrac{13.0}{\sin 56.6°}$; $a_1 = \dfrac{13.0 \sin 54.8°}{\sin 56.6°} = 12.7$

Case II: $B_2 = 180° - 68.6° = 111.4°$

$\qquad\qquad A_2 = 180° - 56.6° - 111.4° = 12.0°$

$\dfrac{a_2}{\sin 12.0°} = \dfrac{13.0}{\sin 56.6°}$; $a_2 = \dfrac{13.0 \sin 12.0°}{\sin 56.6°} = 3.24$

27. $a = 186$, $B = 130.0°$, $c = 106$

$b = \sqrt{a^2 + c^2 - 2ac(\cos B)}$

$\quad = \sqrt{186^2 + 106^2 - 2(186)(106)\cos 130°}\ = \sqrt{71,180}$

$\quad = 267$

$\dfrac{c}{\sin C} = \dfrac{b}{\sin B}$; $\dfrac{106}{\sin C} = \dfrac{267}{\sin 130.0°}$; $\sin C = \dfrac{106 \sin 130.0°}{267} = \dfrac{106(0.7660)}{267}$

$\sin C = 0.3041$; $C = 17.7°$

$A = 180° - 130° - 17.7° = 32.3°$

29. $a = 7.86$, $b = 2.45$, $C = 22.0°$

$c = \sqrt{a^2 + b^2 - 2ab\cos C} = \sqrt{7.86^2 + 2.45^2 - 2(7.86)(2.45)\cos 22.0°} = 5.66$

$\dfrac{5.66}{\sin 22.0°} = \dfrac{2.45}{\sin B}$; $\sin B = \dfrac{2.45 \sin 22.0°}{5.66} = 0.1622$

$B = 9.3°$

$A = 180° - 22.0° - 9.3° = 148.7°$

33. $a = 17$, $b = 12$, $c = 25$

$$\cos A = \frac{b^2 + c^2 - a^2}{2bc} = \frac{12^2 + 25^2 - 17^2}{2(12)(25)}$$

$$= \frac{480}{600} = 0.8$$

$$A = 37°$$

$$\cos C = \frac{a^2 + b^2 - c^2}{2ab} = \frac{17^2 + 12^2 - 25^2}{2(17)(12)}$$

$$= \frac{-192}{408} = -0.4706$$

$$C = 118°$$

$$B = 180° - 37° - 118° = 25°$$

37. $V_x = 175.616 \cos 152.48° = -155.7$ lb

$V_y = 175.616 \sin 152.48° = 81.14$ lb

41. $F = 15.0$, $\theta = 6.0°$; $F_v = 15.0 \cos 6.0° = 14.9$ mN

45. $1.25^2 = 2.70^2 + x^2 - 2(2.70)(x) \cos 27.5°$

$1.56 = 7.29 + x^2 - 4.79x$

$$x^2 - 4.79x + 5.73 = 0; \quad x = \frac{4.79 \pm \sqrt{(-4.79)^2 - 4(5.73)}}{2} = \frac{4.79 \pm 0.155}{2}$$

$$x = 2.47 \text{ or } 2.32$$

Answers may vary according to rounding.

49.

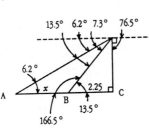

$$\frac{y}{\sin 90} = \frac{2.25}{\sin 76.5} ;$$

$$y = \frac{2.25(1)}{0.972} = 2.31$$

$$\frac{x}{\sin 7.3°} = \frac{y}{\sin 6.2°} = \frac{2.31}{\sin 6.2°}$$

$$x = \frac{2.32 \sin 7.3°}{\sin 6.2°} = \frac{0.2935}{0.1080} = 2.72 \text{ km}$$

Answers may vary according to rounding.

53.

$$F = \sqrt{480^2 + 650^2} = 808 \ N$$

$$\tan \theta = \frac{650}{480} = 1.354; \ \theta = 53.6° \text{ E of N}$$

CHAPTER 9

Graphs of the Trigonometric Functions

Exercises 9-1, page 264

1. $y = \sin x$

x	$-\pi$	$-\dfrac{3\pi}{4}$	$-\dfrac{\pi}{2}$	$-\dfrac{\pi}{4}$	0	$\dfrac{\pi}{4}$	$\dfrac{\pi}{2}$	$\dfrac{3\pi}{4}$	π	$\dfrac{5\pi}{4}$	$\dfrac{3\pi}{2}$
y	0	-0.7	-1	-0.7	0	0.7	1	0.7	0	-0.7	-1

x	$\dfrac{7\pi}{4}$	2π	$\dfrac{9\pi}{4}$	$\dfrac{5\pi}{2}$	$\dfrac{11\pi}{4}$	3π
y	-0.7	0	0.7	1	0.7	0

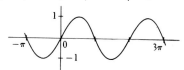

5. $y = 3 \sin x$; $\sin x$ has its amplitude value at $x = \dfrac{\pi}{2}$ and $x = \dfrac{3\pi}{2}$, and has intercepts at $x = 0$, $x = \pi$, and $x = 2\pi$; the graph can be sketched with these values.

x	0	$\dfrac{\pi}{2}$	π	$\dfrac{3\pi}{2}$	2π
$\sin x$	0	1	0	-1	0
$3 \sin x$	0	3	0	-3	0

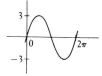

9. $y = 2 \cos x$; $\cos x$ has its amplitude value at $x = 0$, $x = \pi$, and $x = 2\pi$, and has intercepts at $x = \dfrac{\pi}{2}$ and $x = \dfrac{3\pi}{2}$. The graph can be sketched with these values.

x	0	$\dfrac{\pi}{2}$	π	$\dfrac{3\pi}{2}$	2π
$\cos x$	1	0	-1	0	1
$2 \cos x$	2	0	-2	0	2

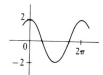

13. $y = -\sin x = -1(\sin x)$; see exercise 5 above. (The negative sign will invert the graph values.)

x	0	$\frac{\pi}{2}$	π	$\frac{3\pi}{2}$	2π
$\sin x$	0	1	0	-1	0
$-\sin x$	0	-1	0	1	0

17. $y = -\cos x = -1(\cos x)$; see exercise 9 above. (The negative sign will invert the graph values.)

x	0	$\frac{\pi}{2}$	π	$\frac{3\pi}{2}$	2π
$\cos x$	1	0	-1	0	1
$-\cos x$	-1	0	1	0	-1

21. Sketch $y = \sin x$ for $x = 0, 1, 2, 3, 4, 5, 6, 7$.

x	0	1	2	3	4	5	6	7
$\sin x$	0.0	0.84	0.91	0.14	-0.76	-0.96	-0.28	0.66

Exercises 9-2, page 268

1. Since $\sin bx$ has period $\frac{2\pi}{b}$, $y = 2 \sin 6x$ has a period of $\frac{2\pi}{6}$, or $\frac{\pi}{3}$.

5. $y = -2 \sin 12x$ has a period of $\frac{2\pi}{12}$, or $\frac{\pi}{6}$. See exercise 1.

9. $y = 5 \sin 2\pi x$ has a period of $\frac{2\pi}{2\pi}$, or 1.

13. $y = 3 \sin \frac{1}{3}x$ has a period of $\frac{2\pi}{\frac{1}{3}} = \frac{2\pi}{1} \times \frac{3}{1} = 6\pi$.

17. $y = 0.4 \sin \frac{2\pi x}{3}$ has a period of $\frac{2\pi}{1} \div \frac{2\pi}{3} = \frac{2\pi}{1} \times \frac{3}{2\pi} = 3$.

21. $y = 2 \sin 6x$ has amplitude of 2 and period of $\frac{\pi}{3}$. (See exercise 1.)

x	0	$\frac{\pi}{12}$	$\frac{\pi}{6}$	$\frac{\pi}{4}$	$\frac{\pi}{3}$
$6x$	0	$\frac{\pi}{2}$	π	$\frac{3\pi}{2}$	2π
$\sin 6x$	0	1	0	-1	0
$2 \sin 6x$	0	2	0	-2	0

25. $y = -2 \sin 12x$ has amplitude $|-2| = 2$, period of $\frac{\pi}{6}$. (See exercise 5.)

x	0	$\frac{\pi}{24}$	$\frac{\pi}{12}$	$\frac{\pi}{8}$	$\frac{\pi}{6}$
$12x$	0	$\frac{\pi}{2}$	π	$\frac{3\pi}{2}$	2π
$\sin 12x$	0	1	0	-1	0
$-2 \sin 12x$	0	-2	0	2	0

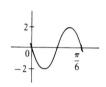

29. $y = 5 \sin 2\pi x$; amplitude is 5, period is 1. (See exercise 9.)

x	0	$\frac{1}{4}$	$\frac{1}{2}$	$\frac{3}{4}$	1
$2\pi x$	0	$\frac{\pi}{2}$	π	$\frac{3\pi}{2}$	2π
$\sin 2\pi x$	0	1	0	-1	0
$5 \sin 2\pi x$	0	5	0	-5	0

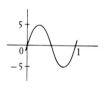

33. $y = 3 \sin \frac{1}{3}x$; amplitude is 3, period is 6π. (See exercise 13.)

x	0	$\frac{3\pi}{2}$	3π	$\frac{9\pi}{2}$	6π
$\frac{1}{3}x$	0	$\frac{\pi}{2}$	π	$\frac{3\pi}{2}$	2π
$\sin \frac{1}{3}x$	0	1	0	-1	0
$3 \sin \frac{1}{3}x$	0	3	0	-3	0

37. $y = 0.4 \sin \frac{2\pi}{3}x$; amplitude is 0.4, period is 3. (See Exercise 17.)

x	0	$\frac{3}{4}$	$\frac{3}{2}$	$\frac{9}{4}$	3
$\frac{2\pi}{3}x$	0	$\frac{\pi}{2}$	π	$\frac{3\pi}{2}$	2π
$0.4 \sin \frac{2\pi}{3}x$	0	0.4	0	-0.4	0

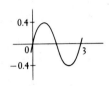

41. $B = \frac{2\pi}{\pi/3} = 6$; $y = \sin 6x$

45. $v = 170 \sin 120\pi t$; amplitude is 170; period is $\frac{2\pi}{120\pi} = \frac{1}{60}$, or 60

cycles from $t = 0$ to $t = 1$ s, and 3 cycles from $t = 0$ to $t = 0.05$ s.

t	0	$\frac{1}{60}$	$\frac{2}{60}$	$\frac{3}{60}$
$120\pi t$	0	2π	4π	6π
$\sin 120\pi t$	0	0	0	0
$170 \sin 120\pi t$	0	0	0	0

t	0	$\frac{1}{240}$	$\frac{1}{120}$	$\frac{3}{240}$	$\frac{1}{60}$
$120\pi t$	0	$\frac{\pi}{2}$	π	$\frac{3\pi}{2}$	2π
$\sin 120\pi t$	0	1	0	-1	0
$170 \sin 120\pi t$	0	170	0	-170	0

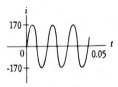

one cycle with all intercepts
and maximum value

Exercises 9-3, page 271

1. $y = \sin(x - \frac{\pi}{6})$, $a = 1$, $b = 1$, $c = -\frac{\pi}{6}$

amplitude is $|a| = 1$, period is $\frac{2\pi}{b} = \frac{2\pi}{1} = 2\pi$

displacement is $-\frac{c}{b} = -(-\frac{\pi}{6}) = \frac{\pi}{6}$

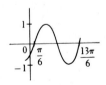

5. $y = 2 \sin(2x + \frac{\pi}{2})$, $a = 2$, $b = 2$, $c = \frac{\pi}{2}$

amplitude $|a| = 2$, period is $\frac{2\pi}{b} = \frac{2\pi}{2} = \pi$

displacement is $-\frac{c}{b} = -\frac{\frac{\pi}{2}}{2} = -\frac{\pi}{2} \times \frac{1}{2} = -\frac{\pi}{4}$

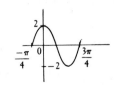

9. $y = \frac{1}{2} \sin(\frac{1}{2}x - \frac{\pi}{4})$, $a = \frac{1}{2}$, $b = \frac{1}{2}$, $c = -\frac{\pi}{4}$

amplitude $|a| = \frac{1}{2}$, period $\frac{2\pi}{b} = \frac{2\pi}{\frac{1}{2}} = \frac{2\pi}{1} \times \frac{2}{1} = 4\pi$

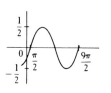

displacement $-\frac{c}{b} = -(-\frac{\pi}{4}) \div \frac{1}{2} = \frac{\pi}{4} \times \frac{2}{1} = \frac{\pi}{2}$

13. $y = \sin(\pi x + \frac{\pi}{8})$, $a = 1$, $b = \pi$, $c = \frac{\pi}{8}$

amplitude $|a| = 1$, period $\frac{2\pi}{b} = \frac{2\pi}{\pi} = 2$

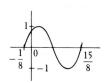

displacement $-\frac{c}{b} = -\frac{\pi}{8} \div \frac{\pi}{1} = -\frac{1}{8}$

17. $y = -0.6 \sin(2\pi x - 1)$, $a = -0.6$, $b = 2\pi$, $c = -1$

amplitude $|a| = |-0.6| = 0.6$

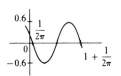

period $\frac{2\pi}{b} = \frac{2\pi}{2\pi} = 1$, displacement $-\frac{c}{b} = -\frac{(-1)}{2\pi} = \frac{1}{2\pi}$

21. $y = \sin(\pi^2 x - \pi)$, $a = 1$, $b = \pi^2$, $c = -\pi$

amplitude $|a| = 1$, period $\frac{2\pi}{b} = \frac{2\pi}{\pi^2} = \frac{2}{\pi}$

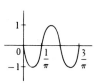

displacement $-\frac{c}{b} = -\frac{(-\pi)}{\pi^2} = \frac{1}{\pi}$

25. $y = A \sin 2\pi\left(\frac{t}{T} - \frac{x}{\lambda}\right)$, $A = 2.00$ cm, $T = 0.100$ s, $\lambda = 20.0$ cm, $x = 5.00$ cm

$y = 2.00 \sin\left(\frac{2\pi t}{0.100} - \frac{10.00\pi}{20.0}\right) = 2.00 \sin(20.0\pi t - 0.500\pi)$

amplitude $|2.00| = 2.00$, period $\frac{2\pi}{20.0} = 0.314$

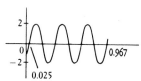

displacement $\frac{0.500\pi}{20.0\pi} = \frac{1}{40} = 0.025$

<u>Exercises 9-4</u>, page 275

1.

x	$-\frac{\pi}{2}$	$-\frac{\pi}{3}$	$-\frac{\pi}{4}$	$-\frac{\pi}{6}$	0	$\frac{\pi}{6}$	$\frac{\pi}{4}$	$\frac{\pi}{3}$	$\frac{\pi}{2}$	$\frac{2\pi}{3}$	$\frac{3\pi}{4}$	$\frac{5\pi}{6}$	2π
tan x	*	-1.7	-1	-0.58	0	0.58	1	1.7	*	-1.7	-1	-0.58	0

y = TAN x

handwritten: -3.4 -2 -1.16 0 1.16 2 3.4 - -3.4 -2 -1.16

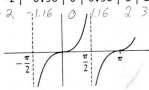

5. For y = 2 tan x, first sketch the graph of y = tan x, then multiply the y-values of the tangent function by 2 and graph.

9. y = -2 cot x; sketch the graph of y = cot x, multiply the y-values by -2 and resketch the function. It will be inverted.

13.

x	0	$\frac{\pi}{8}$	$\frac{\pi}{4}$	$\frac{3\pi}{8}$	$\frac{\pi}{2}$	$\frac{5\pi}{8}$	$\frac{3\pi}{4}$	$\frac{7\pi}{8}$	π
2x	0	$\frac{\pi}{4}$	$\frac{\pi}{2}$	$\frac{3\pi}{4}$	π	$\frac{5\pi}{4}$	$\frac{3\pi}{2}$	$\frac{7\pi}{4}$	2π
tan 2x	0	1	*	-1	0	1	*	-1	0

17.

x	$\frac{\pi}{12}$	$\frac{\pi}{6}$	$\frac{\pi}{4}$	$\frac{\pi}{3}$	$\frac{5\pi}{12}$	$\frac{\pi}{2}$	$\frac{7\pi}{12}$	$\frac{2\pi}{3}$	$\frac{3\pi}{4}$	$\frac{5\pi}{6}$	$\frac{11\pi}{12}$	π
$2x + \frac{\pi}{6}$	$\frac{\pi}{3}$	$\frac{\pi}{2}$	$\frac{2\pi}{3}$	$\frac{5\pi}{6}$	π	$\frac{7\pi}{6}$	$\frac{4\pi}{3}$	$\frac{3\pi}{2}$	$\frac{5\pi}{3}$	$\frac{11\pi}{6}$	2π	$\frac{13\pi}{6}$
$\cot(2x + \frac{\pi}{6})$	0.6	0	-0.6	-1.7	*	1.7	0.6	0	-0.6	-1.7	*	1.7
$2\cot(2x + \frac{\pi}{6})$	1.2	0	-1.2	-3.4	*	3.4	1.2	0	-1.2	-3.4	*	3.4

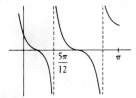

21. $y = \sec x$ has a reciprocal of
     $y = \cos x$. Sketch the graph of
     $y = \cos x$; the reciprocal of
     $\cos x = 1$ is also 1. Where
     $\cos x = 1$, $\sec x = 1$. When
     $\cos x$ is 0, $\sec x$ would be
     undefined. As $\cos x$ becomes
     smaller, $\sec x$ becomes larger.

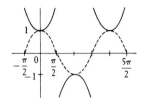

25. $d = a \sec \theta$; $a = 3.00$, $\theta > 0$, $d > 0$
     $d = 3.00 \sec \theta$

θ	0	$\pi/2$
$\sec \theta$	1	$\star$
$3.00 \sec \theta$	3	$\star$

     $\star$ = undefined

Exercises 9-5, page 279

1. $y = R \sin \omega t = 2.40 \sin 2.00t$
    amplitude is 2.40
    period is 1.00π, 2.00π for 2 cycles
    displacement is 0

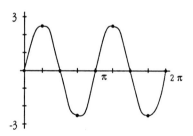

5. $D = A \sin (\omega t + \alpha) = 500 \sin (3.60t + 0)$

    $= 500 \sin 3.60t$

    amplitude is 500
    period is $2\pi/3.60 = 1.75$,
            3.49 for 2 cycles
    displacement is 0

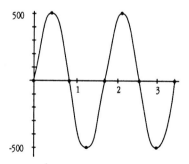

9. $y = A \sin 2\pi\left(\dfrac{t}{T} - \dfrac{x}{\lambda}\right)$

$= 3.20 \sin 2\pi\left(\dfrac{t}{0.050} - \dfrac{5.00}{40.0}\right)$

$= 3.20 \sin\left(\dfrac{2\pi t}{0.050} - \dfrac{10.0\pi}{40.0}\right)$

$= 3.20 \sin\left(40.0\ \pi t - \dfrac{\pi}{4.0}\right)$

amplitude is 3.20
period is $2\pi/40.0\pi$

    $= 0.050$ for 1 cycle,
    0.100 for 2 cycles

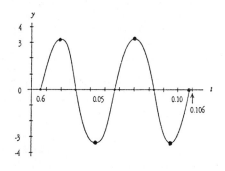

13. $y = 150 \cos 200\pi t$
     amplitude is 1.50
     displacement is 0
     period is $2\pi/200\pi = 1/100$

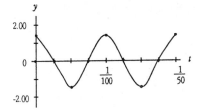

Exercises 9-6, page 282

1. $y = 1 + \sin x$

x	0	$\pi/2$	π	$3\pi/2$	2π
$\sin x$	0	1	0	-1	0
1	1	1	1	1	1
$1 + \sin x$	1	2	1	0	1

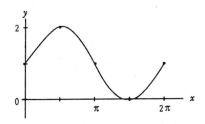

5. $y = \frac{1}{3}x + \sin 2x$

x	$-\pi/2$ -1.57	$-\pi/4$ -0.785	0 0	$\pi/4$ 0.785	$\pi/2$ 1.57	$3\pi/4$ 2.36	π 3.14
$\sin 2x$	0	−1	0	1	0	−1	0
$1/3x$	−0.52	−0.26	0	0.26	0.52	0.79	1.05
$\frac{1}{3}x + \sin 2$	−0.52	−1.26	0	1.26	0.52	−0.21	1.05

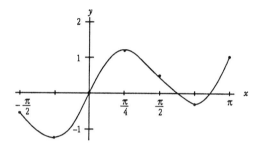

9.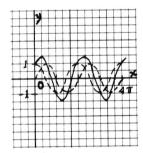

13.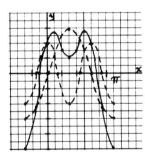

17.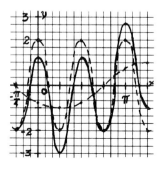

21.

t	0	$\frac{\pi}{4}$	$\frac{\pi}{2}$	$\frac{3\pi}{4}$	π	$\frac{5\pi}{4}$	$\frac{3\pi}{2}$	$\frac{7\pi}{4}$	2π
x	0.0	0.7	1.0	0.7	0.0	−0.7	−1.0	−0.7	0.0
y	0.0	0.7	1.0	0.7	0.0	−0.7	−1.0	−0.7	0.0
Pt	1	2	3	4	5	6	7	8	9

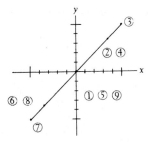

25.

t	0	$\frac{1}{6}$	$\frac{1}{3}$	$\frac{1}{2}$	$\frac{2}{3}$	$\frac{5}{6}$	1	$\frac{7}{6}$	$\frac{4}{3}$	$\frac{3}{2}$	$\frac{5}{3}$	$\frac{11}{6}$	2
x	0.9	0.5	0	−0.5	−0.9	−1	−0.9	−0.5	0	0.5	0.9	1	0.9
y	0	1	1.7	2	1.7	1	0	−1	−1.7	−2	−1.7	−1	0
Pt	1	2	3	4	5	6	7	8	9	10	11	12	13

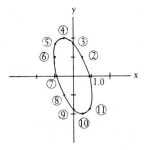

29. $T = 18 - 10 \cos 2\pi (t - 0.10)$; 0 t 2

t	0.10	0.35	0.60	0.85	1.10
$2\pi(t - 0.10)$	0	$\pi/2$	π	$3\pi/2$	2π
$\cos 2\pi(t - 0.10)$	1	0	-1	0	1
$18 - 10\cos 2\pi(t - 0.10)$	8	18	28	18	8

continued

1.35	1.6	1.85	2.2
$5\pi/2$	3π	$7\pi/2$	4π
0	-1	0	1
18	28	18	8

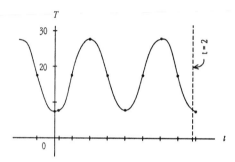

33. $i = 0.32 + 0.50 \sin t - 0.20 \cos 2t$

t	0	$\pi/4$	$\pi/2$	$3\pi/4$	π	$5\pi/4$	$3\pi/2$	$7\pi/4$	2π
$2t$	0	$\pi/2$	π	$3\pi/2$	2π	$5\pi/2$	3π	$7\pi/2$	4π
$0.50 \sin t$	0	0.35	0.50	0.35	0	-0.35	-0.50	-0.35	0
$-0.20 \cos 2t$	-0.20	0	0.20	0	-0.20	0	0.20	0	-0.20
$+0.32$	0.32	0.32	0.32	0.32	0.32	0.32	0.32	0.32	0.32
i	0.12	0.67	1.02	0.67	0.12	-0.03	0.02	-0.03	0.12

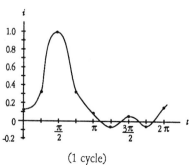

(1 cycle)

Review Exercises for Chapter 9, page 286

1. amplitude is $\frac{2}{3}$

    period is 2π

    displacement is 0

5. amplitude is 2

    period is $\frac{2\pi}{3}$

    displacement is 0

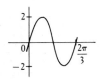

9. amplitude is 3

    period is $\frac{2\pi}{\frac{1}{3}} = 6\pi$

    displacement is 0

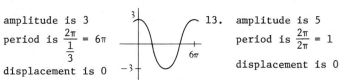

13. amplitude is 5

    period is $\frac{2\pi}{2\pi} = 1$

    displacement is 0

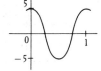

17. amplitude is 2

    period is $\frac{2\pi}{3}$

    displacement is $-(-\frac{\pi}{2}) \div 3 = \frac{\pi}{6}$

21. amplitude is 1, but inverted from
     sine function

     period is $\frac{2\pi}{\pi} = 2$

     displacement is $-\frac{\pi}{6} \div \pi = -\frac{1}{6}$

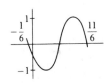

25. Graph is the tangent function with
     y-values multiplied by 3.

29. $y = 2 + \frac{1}{2} \sin 2x$

x	0	$\pi/4$	p/2	$3\pi/4$	π	$5\pi/4$
$2x$	0	$\pi/2$	π	$3\pi/2$	2π	$5\pi/2$
1/2 sin 2x	0	0.5	0	-0.5	0	0.5
2	2	2	2	2	2	2
y	2	2.5	2	1.5	2	2.5

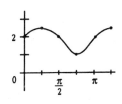

33.

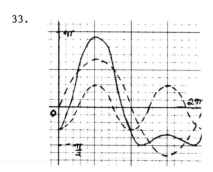

37.   amplitude is 2, so $a = 2$
period is $\pi = 2\pi/b$, so $b = 2$
displacement is $-\pi/4 = -c/b$
$= -c/2 = -2c/4$; $2c = \pi$ so $c = \pi/2$
$$y = 2 \sin \left(2x + \frac{\pi}{2}\right)$$

41.

t	-1	$-\dfrac{3}{4}$	$-\dfrac{1}{2}$	$-\dfrac{1}{4}$	0	$\dfrac{1}{4}$	$\dfrac{1}{2}$	$\dfrac{3}{4}$	1
x	-1	0	1	0	-1	0	1	0	-1
y	0	-1.4	-2	-1.4	0	1.4	2	1.4	0

Negative t's do not have any physical significance.

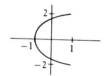

45. $$R = \frac{v_0^2 \sin 2\theta}{g} = \frac{(1000)^2 \sin 2\theta}{9.8} = \frac{10^6 \sin 2\theta}{9.8} = 10^5 \sin 2\theta$$

amplitude is 10^5
displacement is 0
period is $2\pi/2 = \pi$

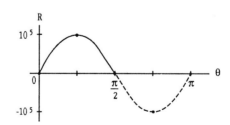

49. $y = A \sin (6t + 0.5) = 5.2 \sin(6t + 0.5)$

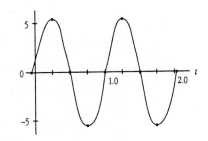

amplitude is 5.2
period is $2\pi/6 = \pi/3$ for 1 cycle,
              $2\pi/3$ for 2 cycles
displacement is $-0.5 + 6 = -0.08$

53. $y = 4 \sin 2t - 2 \cos 2t$

t	0	$\pi/4$	$\pi/2$	$3\pi/4$	π	$5\pi/4$	$3\pi/2$	$7\pi/4$	2π
$2t$	0	$\pi/2$	π	$3\pi/2$	2π	$5\pi/2$	3π	$7\pi/2$	4π
$-2 \cos 2t$	-2	0	2	0	-2	0	2	0	-2
$4 \sin 2t$	0	4	0	-4	0	4	0	-4	0
y	-2	4	2	-4	-2	4	2	-4	-2

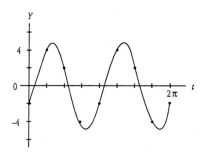

57. $Z = R \sec \theta, \ -\pi/2 < \theta < \pi/2$

The graph is the secant function
with values multiplied by R.

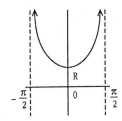

CHAPTER 10

Exponents and Radicals

Exercises 10-1, page 292

1. $x^7 \times x^{-4} = x^{7+(-4)}$

    $= x^3$

9. $(2ax^{-1})^2 = \left(\dfrac{2a}{x}\right)^2 = \dfrac{4a^2}{x^2}$

13. $(-4)^0 = 1$ by Eq. (10-5)

21. $\left(\dfrac{2}{n^3}\right)^{-1} = \dfrac{2^{-1}}{n^{-3}}$

    $= \dfrac{n^3}{2}$

25. $(a + b)^{-1} = \dfrac{1}{(a + b)^1}$

    $= \dfrac{1}{a + b}$

33. $\left(\dfrac{3a^2}{4b}\right)^{-3}\left(\dfrac{4}{a}\right)^{-5} = \left(\dfrac{3^{-3}a^{-6}}{4^{-3}b^{-3}}\right)\left(\dfrac{4^{-5}}{a^{-5}}\right)$

    $= \left(\dfrac{4^3 b^3}{3^3 a^6}\right)\left(\dfrac{a^5}{4^5}\right) = \dfrac{4^3 a^5 (b^3)}{4^3 a^5 (432a)}$

    $= \dfrac{b^3}{432a}$

37. $(x^2 y^{-1})^2 - x^{-4} = x^4 y^{-2} - x^{-4}$

    $= \dfrac{x^4}{y^2} - \dfrac{1}{x^4} = \dfrac{x^4(x^4)}{y^2(x^4)} - \dfrac{1(y^2)}{x^4(y^2)}$

    $= \dfrac{x^8 - y^2}{x^4 y^2}$

5. $5 \times 5^{-3} = 5^{1+(-3)}$

    $= 5^{-2} = \dfrac{1}{5^2}$

    $= \dfrac{1}{25}$

17. $3x^{-2} = 3\left(\dfrac{1}{x^2}\right)$

    $= \dfrac{3}{x^2}$

29. $(2 \times 3^{-2})^2 \left(\dfrac{3}{2}\right)^{-1} = \left(2 \times \dfrac{1}{3^2}\right)^2 \left(\dfrac{2}{3}\right)$

    $= \left(\dfrac{2}{9}\right)^2 \left(\dfrac{2}{3}\right) = \left(\dfrac{4}{81}\right)\left(\dfrac{2}{3}\right)$

    $= \dfrac{8}{243} = \dfrac{2^3}{3^5}$

41. $2 \times 3^{-1} + 4 \times 3^{-2} = 2 \times \frac{1}{3} + 4 \times \frac{1}{3^2}$

$$= \frac{2}{3} + \frac{4}{9} = \frac{6}{9} + \frac{4}{9}$$

$$= \frac{10}{9}$$

45. $(n^{-2} - 2n^{-1})^2 = (\frac{1}{n^2} - \frac{2}{n})^2$

$$= \frac{1}{n^4} - \frac{4}{n^3} + \frac{4}{n^2} = \frac{1 - 4n + 4n^2}{n^4}$$

$$= \frac{4n^2 - 4n + 1}{n^4}$$

49. $\dfrac{x - y^{-1}}{x^{-1} - y} = \dfrac{x - \frac{1}{y}}{\frac{1}{x} - y}$

$$= \frac{\frac{x(y)}{1(y)} - \frac{1}{y}}{\frac{1}{x} - \frac{y(x)}{1(x)}} = \frac{\frac{xy - 1}{y}}{\frac{1 - xy}{x}}$$

$$= \frac{xy - 1}{y} \times \frac{x}{1 - xy} = \frac{-1(1 - xy)}{y} \times \frac{x}{(1 - xy)}$$

$$= -\frac{x}{y}$$

53. $2t^{-2} + t^{-1}(t + 1) = \frac{2}{t^2} + \frac{1}{t}(t + 1)$

$$= \frac{2}{t^2} + \frac{(t + 1)(t)}{t(t)} = \frac{2 + t^2 + t}{t^2}$$

$$= \frac{t^2 + t + 2}{t^2}$$

57. $4^2 \times 64 = 4^2 \times 4^3 = 4^5$

$4^2 \times 64 = (2^2)^2 \times 2^6 = 2^4 \times 2^6 = 2^{10}$ 61. $\dfrac{Pa \times m^6 \times mol^{-2}}{m^3 \times mol^{-1}} = \dfrac{Pa \times m^6}{m^3 \times mol^2} = \dfrac{Pa \times m^3}{mol^2}$

Exercises 10-2, page 297

1. $(25)^{1/2} = \sqrt{25}$

$\qquad = 5$

5. $8^{4/3} = \sqrt[3]{8^4}$

$\qquad = (\sqrt[3]{8})^4 = 2^4$

$\qquad = 16$

9. $8^{-1/3} = (\frac{1}{8})^{1/3} = \sqrt[3]{\frac{1}{8}}$

$\qquad = \frac{1}{2}$

13. $5^{1/2} 5^{3/2} = 5^{1/2 + 3/2}$

$\qquad = 5^2 = 25$

17. $\dfrac{121^{-1/2}}{100^{1/2}} = \dfrac{1}{(100^{1/2})(121^{1/2})}$

$= \dfrac{1}{\sqrt{100}\sqrt{121}} = \dfrac{1}{10(11)}$

$= \dfrac{1}{110}$

21. $\dfrac{(-27)^{1/3}}{6} = \dfrac{\sqrt[3]{(-27)^1}}{6}$

$= \dfrac{\sqrt[3]{-27}}{6} = \dfrac{-3}{6}$

$= -\dfrac{1}{2}$

25. $(125)^{-2/3} - (100)^{-3/2} = (\sqrt[3]{125})^{-2} - (\sqrt{100})^{-3}$

$= (5)^{-2} - (10)^{-3} = \dfrac{1}{5^2} - \dfrac{1}{10^3}$

$= \dfrac{1}{25} - \dfrac{1}{1000} = \dfrac{1(40)}{25(40)} - \dfrac{1}{1000}$

$= \dfrac{39}{1000}$

29. $(17.98)^{1/4} = 17.98,\ \boxed{y^x}\ ,\ 0.25,\ \boxed{=}\ 2.059$

33. $a^{2/3}a^{1/2} = a^{2/3+1/2}$

$= a^{4/6+3/6} = a^{7/6}$

37. $\dfrac{s^{1/4}s^{2/3}}{s^{-1}} = \dfrac{s^{1/4+2/3}}{s^{-1}}$

$= \dfrac{s^{3/12+8/12}}{s^{-1}} = \dfrac{s^{11/12}}{\dfrac{1}{s}}$

$= s^{11/12} \times s = s^{11/12+1}$

$= s^{23/12}$

41. $(8a^3b^6)^{1/3} = 8^{1/3}a^{3(1/3)}b^{6(1/3)}$

$= \sqrt[3]{8}ab^2 = 2ab^2$

45. $\dfrac{1}{2}(4x^2 + 1)^{-1/2}(8x) = 4x(4x^2 + 1)^{-1/2}$

$= \dfrac{4x}{(4x^2 + 1)^{1/2}}$

49. $\left(\dfrac{4a^{5/6}b^{-1/5}}{a^{2/3}b^2}\right)^{-1/2} = (4a^{5/6-4/6}b^{-1/5-10/5})^{-1/2}$

$= (4a^{1/6}b^{-11/5})^{-1/2} = 4^{-1/2}a^{(1/6)(-1/2)}b^{(-11/5)(-1/2)}$

$= \dfrac{1}{4^{1/2}}a^{-1/12}b^{11/10} = \dfrac{1}{2} \times \dfrac{1}{a^{1/12}} \times b^{11/10}$

$= \dfrac{b^{11/10}}{2a^{1/12}}$

53. $(x^{-1} + 2x^{-2})^{-1/2} = (\frac{1}{x} + \frac{2}{x^2})^{-1/2}$

$= (\frac{x + 2}{x^2})^{-1/2} = (\frac{x^2}{x + 2})^{1/2}$

$= \sqrt{\frac{x^2}{x + 2}} = \frac{\sqrt{x^2}}{\sqrt{x + 2}}$

$= \frac{x}{(x + 2)^{1/2}}$

57. $[(a^{1/2} - a^{-1/2})^2 + 4]^{1/2} = [(a^{1/2} - \frac{1}{a^{1/2}})^2 + 4]^{1/2}$

$= [(\frac{a - 1}{a^{1/2}})^2 + 4]^{1/2} = (\frac{a^2 - 2a + 1}{a} + 4)^{1/2}$

$= (\frac{a^2 - 2a + 1 + 4a}{a})^{1/2} = (\frac{a^2 + 2a + 1}{a})^{1/2}$

$= [\frac{(a + 1)^2}{a}]^{1/2} = \frac{\sqrt{(a + 1)^2}}{a^{1/2}}$

$= \frac{a + 1}{a^{1/2}}$

61. $f(x) = 3x^{1/2}$

x	0	1	2	3	4	...9	...16
y	0	3	4.2	5.2	6	9	12

65. $(A/s)^{-1/4} = (2s/s)^{-1/4} = (2)^{-1/4} = (1/2)^{1/4} = \sqrt[4]{0.5} = 0.84$

Exercises 10-3, page 301

1. $\sqrt{24} = \sqrt{4(6)}$

$= \sqrt{4}\sqrt{6} = 2\sqrt{6}$

5. $\sqrt{x^2 y^5} = \sqrt{x^2 y^4 (y^1)}$

$= \sqrt{x^2 y^4}\sqrt{y} = xy^2\sqrt{y}$

9. $\sqrt{5x^2} = \sqrt{x^2}\sqrt{5}$

$= x\sqrt{5}$

13. $\sqrt[3]{16} = \sqrt[3]{2^4}$

$= \sqrt[3]{2^3 (2)} = 2\sqrt[3]{2}$

17. $\sqrt[3]{8a^2} = \sqrt[3]{2^3 a^2}$

$= 2\sqrt[3]{a^2}$

21. $\sqrt[5]{8}\ \sqrt[5]{4} = \sqrt[5]{8(4)}$

$= \sqrt[5]{32} = \sqrt[5]{2^5}$

$= 2$

25. $\sqrt{\dfrac{3}{2}} = \sqrt{\dfrac{3}{2} \times \dfrac{2}{2}}$

$\qquad = \sqrt{\dfrac{6}{4}} = \dfrac{\sqrt{6}}{\sqrt{4}}$

$\qquad = \dfrac{\sqrt{6}}{2} = \dfrac{1}{2}\sqrt{6}$

29. $\sqrt[3]{\dfrac{3}{4}} = \sqrt[3]{\dfrac{3}{2(2)} \times \dfrac{2}{2}}$

$\qquad = \sqrt[3]{\dfrac{6}{2^3}} = \dfrac{\sqrt[3]{6}}{\sqrt[3]{2^3}}$

$\qquad = \dfrac{\sqrt[3]{6}}{2} = \dfrac{1}{2}\sqrt[3]{6}$

33. $\sqrt[4]{400} = \sqrt[4]{16(25)}$

$\qquad = 2\,\sqrt[4]{5^2} = 2(5^{2/4})$

$\qquad = 2(5^{1/2}) = 2\sqrt{5}$

37. $\sqrt{4 \times 10^4} = 10^2\sqrt{4}$

$\qquad = 10^2(2) = 200$

41. $\sqrt[4]{4a^2} = \sqrt[4]{2^2a^2}$

$\qquad = 2^{2/4}a^{2/4} = 2^{1/2}a^{1/2}$

$\qquad = (2a)^{1/2} = \sqrt{2a}$

45. $\sqrt[4]{\sqrt[3]{16}} = \sqrt[12]{16}$

$\qquad = \sqrt[12]{2^4} = 2^{4/12}$

$\qquad = 2^{1/3} = \sqrt[3]{2}$

49. $\sqrt{\dfrac{1}{2} - \dfrac{1}{3}} = \sqrt{\dfrac{1(3)}{2(3)} - \dfrac{1(2)}{3(2)}}$

$\qquad = \sqrt{\dfrac{1}{6}} = \sqrt{\dfrac{1 \times 6}{6 \times 6}}$

$\qquad = \sqrt{\dfrac{6}{36}} = \dfrac{\sqrt{6}}{6}$

$\qquad = \dfrac{1}{6}\sqrt{6}$

55. $\sqrt{a^2 + 2ab + b^2} = \sqrt{(a + b)^2}$

$\qquad = a + b$

53. $\sqrt{\dfrac{x}{2x + 1}} = \sqrt{\dfrac{x(2x + 1)}{(2x + 1)^2}} = \dfrac{\sqrt{2x^2 + x}}{2x + 1}$

57. $\sqrt{4x^2 + 1}$ cannot be simplified any further

61. $E = 100\left(1 - \dfrac{1}{\sqrt[5]{R^2}}\right) = 100\left(1 - \dfrac{1}{R^{2/5}}\right) = 100(1 - R^{-2/5})$

$E = 100\,[1 - (7.35)^{-2/5}] = 100\,(1 - 0.45) = 55\%$

Exercises 10-4, page 304

1. $2\sqrt{3} + 5\sqrt{3} = (2 + 5)\sqrt{3}$

$\qquad = 7\sqrt{3}$

5. $\sqrt{5} + \sqrt{20} = \sqrt{5} + \sqrt{4(5)}$

$\qquad = \sqrt{5} + 2\sqrt{5} = (1 + 2)\sqrt{5}$

$\qquad = 3\sqrt{5}$

9. $\sqrt{8a} - \sqrt{32a} = \sqrt{4(2a)} - \sqrt{16(2a)} = 2\sqrt{2a} - 4\sqrt{2a} = -2\sqrt{2a}$

13. $2\sqrt{20} - \sqrt{125} - \sqrt{45} = 2\sqrt{4(5)} - \sqrt{25(5)} - \sqrt{9(5)}$

$= 2(2)\sqrt{5} - 5\sqrt{5} - 3\sqrt{5} = (4 - 8)\sqrt{5}$

$= -4\sqrt{5}$

17. $\sqrt{60} + \sqrt{\dfrac{5}{3}} = \sqrt{4(15)} + \sqrt{\dfrac{5(3)}{3(3)}}$

$= 2\sqrt{15} + \sqrt{\dfrac{15}{9}} = 2\sqrt{15} + \dfrac{1}{3}\sqrt{15}$

$= \dfrac{7}{3}\sqrt{15}$

21. $\sqrt[3]{81} + \sqrt[3]{3000} = \sqrt[3]{27(3)} + \sqrt[3]{1000(3)}$

$= 3\sqrt[3]{3} + 10\sqrt[3]{3} = 13\sqrt[3]{3}$

25. $\sqrt{a^3b} - \sqrt{4ab^5} = a\sqrt{ab} - 2b^2\sqrt{ab}$

$= (a - 2b^2)\sqrt{ab}$

29. $\sqrt[3]{24a^2b^4} - \sqrt[3]{3a^5b} = \sqrt[3]{8(3)a^2b^3(b)} - \sqrt[3]{3a^3(a^2)b}$

$= 2b\sqrt[3]{3a^2b} - a\sqrt[3]{3a^2b} = (2b - a)\sqrt[3]{3a^2b}$

33. $\sqrt[3]{\dfrac{a}{b}} - \sqrt[3]{\dfrac{8b^2}{a^2}} = \sqrt[3]{\dfrac{a(b^2)}{b(b^2)}} - \sqrt[3]{\dfrac{8b^2(a)}{a^2(a)}}$

$= \dfrac{1}{b}\sqrt[3]{ab^2} - \dfrac{2}{a}\sqrt[3]{ab^2} = (\dfrac{1}{b} - \dfrac{2}{a})\sqrt[3]{ab^2}$

$= \dfrac{(a - 2b)\ \sqrt[3]{ab^2}}{ab}$

37. $3\sqrt{45} + 3\sqrt{75} - 2\sqrt{500} = 3\sqrt{9(5)} + 3\sqrt{25(3)} - 2\sqrt{100(5)}$

$= 9\sqrt{5} + 15\sqrt{3} - 20\sqrt{5} = 15\sqrt{3} - 11\sqrt{5}$

$= 15(1.732) - 11(2.236) = 25.980 - 24.596$

$= 1.384$

Using calculator: 3, $\boxtimes$, 45, $\boxed{\text{INV}}$, $\boxed{x^2}$, $\boxplus$, 3, $\boxtimes$, 75, $\boxed{\text{INV}}$, $\boxed{x^2}$, $\boxminus$, 2,

$\boxtimes$, 500, $\boxed{\text{INV}}$, $\boxed{x^2}$, $\boxed{=}$ 1.3840144

41. $x - 2x - 2 = 0$ $x + 2x - 11 = 0$

$x = \dfrac{2 \pm \sqrt{4 + 8}}{2}$ $x = \dfrac{-2 \pm \sqrt{4 + 44}}{2}$

$= 1 \pm \dfrac{1}{2}\sqrt{12}$ $= -1 \pm \dfrac{1}{2}\sqrt{48}$

$\left(1 + \dfrac{1}{2}\ \sqrt{12}\right) + \left(-1 + \dfrac{1}{2}\sqrt{48}\right) = \left(1 + \dfrac{1}{2}\sqrt{4 \times 3}\right) + \left(-1 + \dfrac{1}{2}\sqrt{16 \times 3}\right)$

$= (1 + \sqrt{3}) + (-1 + 2\sqrt{3}) = 3\sqrt{3}$

<u>Exercises 10-5,</u> page 307

1. $\sqrt{3}\sqrt{10} = \sqrt{3(10)}$
    $= \sqrt{30}$

5. $\sqrt[3]{4}\ \sqrt[3]{2} = \sqrt[3]{4(2)}$
    $= \sqrt[3]{8} = 2$

9. $(5\sqrt{2})^2 = 25\sqrt{2}\sqrt{2}$
    $= 25(2) = 50$

13. $\sqrt{\frac{2}{3}\sqrt{5}} = \sqrt{\frac{2}{3}(\frac{5}{1})}$
    $= \sqrt{\frac{10}{3}} = \sqrt{\frac{10(3)}{3(3)}}$
    $= \frac{1}{3}\sqrt{30}$

17. $\sqrt{3}(\sqrt{2} - \sqrt{5}) = \sqrt{3(2)} - \sqrt{3(5)}$
    $= \sqrt{6} - \sqrt{15}$

21. $(2 - \sqrt{5})(2 + \sqrt{5}) = 4 - \sqrt{5}\sqrt{5}$
    $= 4 - 5 = -1$

25. $(3\sqrt{5} - 2\sqrt{3})(6\sqrt{5} + 7\sqrt{3}) = 18\sqrt{5}\sqrt{5} + 21\sqrt{5}\sqrt{3} - 12\sqrt{5}\sqrt{3} - 14\sqrt{3}\sqrt{3}$
    $= 18(5) + 21\sqrt{15} - 12\sqrt{15} - 14(3)$
    $= 90 + 9\sqrt{15} - 42$
    $= 48 + 9\sqrt{15}$

29. $\sqrt{a}\ (\sqrt{ab} + \sqrt{c^3}) = \sqrt{a}\sqrt{ab} + \sqrt{a}\sqrt{c^3}$
    $= \sqrt{a^2b} + \sqrt{ac^3} = a\sqrt{b} + c\sqrt{ac}$

33. $(\sqrt{2}a - \sqrt{b})(\sqrt{2}a + 3\sqrt{b}) = \sqrt{2}a\sqrt{2}a + 3\sqrt{2}a\sqrt{b} - \sqrt{b}\sqrt{2}a - 3\sqrt{b}\sqrt{b}$
    $= 2a + 3\sqrt{2ab} - \sqrt{2ab} - 3b = 2a - 3b + 2\sqrt{2ab}$

37. $\sqrt{2}\ \sqrt[3]{3} = 2^{1/2}3^{1/3}$
    $= 2^{3/6}3^{2/6} = \sqrt[6]{2^3 3^2}$
    $= \sqrt[6]{8(9)} = \sqrt[6]{72}$

41. $(\sqrt[5]{\sqrt{6}} - \sqrt{5})(\sqrt[5]{\sqrt{6}} + \sqrt{5}) = \sqrt[5]{(\sqrt{6} - \sqrt{5})(\sqrt{6} + \sqrt{5})}$
    $= \sqrt[5]{\sqrt{6}\sqrt{6} - \sqrt{5}\sqrt{5}} = \sqrt[5]{6 - 5}$
    $= \sqrt[5]{1} = 1$

45. $\left(\sqrt{\frac{2}{a}} + \sqrt{\frac{a}{2}}\right)\left(\sqrt{\frac{2}{a}} - 2\sqrt{\frac{a}{2}}\right) = \sqrt{\frac{2}{a}}\sqrt{\frac{2}{a}} - 2\sqrt{\frac{a}{2}}\sqrt{\frac{2}{a}} + \sqrt{\frac{a}{2}}\sqrt{\frac{2}{a}} - 2\sqrt{\frac{a}{2}}\sqrt{\frac{a}{2}}$
    $= \frac{2}{a} - 2\sqrt{\frac{2a}{2a}} + \sqrt{\frac{2a}{2a}} - 2(\frac{a}{2}) = \frac{2}{a} - 2(1) + 1 - 2(\frac{a}{2})$
    $= \frac{2}{a} - 1 - 2(\frac{a}{2}) = \frac{2(2)}{a(2)} - \frac{1(2a)}{1(2a)} - \frac{2a(a)}{2(a)}$
    $= \frac{4 - 2a - 2a^2}{2a} = \frac{2(2 - a - a^2)}{2(a)}$
    $= \frac{2 - a - a^2}{a}$

49. $(\sqrt{11} + \sqrt{6})(\sqrt{11} - 2\sqrt{6}) = 11 - \sqrt{66} - 12$

$$= -1 - \sqrt{66} = -1 - 8.12 = -9.1240384$$

Using a calculator: 11, $\boxed{\text{INV}}$, $\boxed{x^2}$, $\boxed{+}$, 6, $\boxed{\text{INV}}$, $\boxed{x^2}$, $\boxed{=}$ $\boxed{\times}$, $\boxed{(}$,

11, $\boxed{\text{INV}}$, $\boxed{x^2}$, -2, $\boxed{\times}$, 6, $\boxed{\text{INV}}$, $\boxed{x^2}$, $\boxed{)}$, $\boxed{=}$ -9.1240384

53. $2\sqrt{x} + \dfrac{1}{\sqrt{x}} = \dfrac{2\sqrt{x}}{1} + \dfrac{1}{\sqrt{x}} = \dfrac{2\sqrt{x}\ \sqrt{x}}{\sqrt{x}} + \dfrac{1}{\sqrt{x}} = = \dfrac{2x + 1}{\sqrt{x}}$

57. $(1 - \sqrt{2})^2 - 2(1 - \sqrt{2}) - 1 = (1 - 2\sqrt{2} + 2) - 2 + 2\sqrt{2} - 1$

$$= 3 - 3 - 2\sqrt{2} + 2\sqrt{2} = 0$$

Exercises 10-6, page 310

1. $\dfrac{\sqrt{21}}{\sqrt{3}} = \dfrac{\sqrt{21}\sqrt{3}}{\sqrt{3}\sqrt{3}}$

$= \dfrac{\sqrt{63}}{3} = \dfrac{\sqrt{9(7)}}{3}$

$= \dfrac{3\sqrt{7}}{3} = \sqrt{7}$

alternate method: $\dfrac{\sqrt{21}}{\sqrt{3}} = \sqrt{\dfrac{21}{3}}$

$$= \sqrt{7}$$

5. $\dfrac{\sqrt[3]{x^2}}{\sqrt[3]{24}} = \dfrac{\sqrt[3]{x^2}}{\sqrt[3]{8(3)}}$

$= \dfrac{\sqrt[3]{x^2}}{2\ \sqrt[3]{3}} = \dfrac{\sqrt[3]{x^2}}{2\ \sqrt[3]{3}}\ \dfrac{\sqrt[3]{9}}{\sqrt[3]{9}}$

$= \dfrac{\sqrt[3]{9x^2}}{2\ \sqrt[3]{27}} = \dfrac{\sqrt[3]{9x^2}}{2(3)}$

$= \dfrac{1}{6}\ \sqrt[3]{9x^2}$

9. $\dfrac{\sqrt{a}}{\sqrt[3]{4}} = \dfrac{\sqrt{a}}{\sqrt[3]{4}}\ \dfrac{(\sqrt[3]{4})^2}{(\sqrt[3]{4})^2} = \dfrac{\sqrt{a}\ \sqrt[3]{16}}{4} = \dfrac{\sqrt{a}\ 2\ \sqrt[3]{2}}{4}$

$= \dfrac{1}{2}\sqrt{a}\ \sqrt[3]{2} = \dfrac{1}{2}a^{1/2}2^{1/3}$

$= \dfrac{1}{2}a^{3/6}2^{2/6} = \dfrac{1}{2}\sqrt[6]{4a^3}$

13. $\dfrac{\sqrt{2a} - b}{\sqrt{a}} = \dfrac{(\sqrt{2a} - b)(\sqrt{a})}{\sqrt{a}\sqrt{a}}$

$= \dfrac{\sqrt{2a}\sqrt{a} - b\sqrt{a}}{a} = \dfrac{\sqrt{2a^2} - b\sqrt{a}}{a}$

$= \dfrac{a\sqrt{2} - b\sqrt{a}}{a}$

17. $\dfrac{1}{\sqrt{7} + \sqrt{3}} = \dfrac{1(\sqrt{7} - \sqrt{3})}{(\sqrt{7} + \sqrt{3})(\sqrt{7} - \sqrt{3})}$

$= \dfrac{\sqrt{7} - \sqrt{3}}{(\sqrt{7})^2 - (\sqrt{3})^2} = \dfrac{\sqrt{7} - \sqrt{3}}{7 - 3}$

$= \dfrac{\sqrt{7} - \sqrt{3}}{4} = \dfrac{1}{4}(\sqrt{7} - \sqrt{3})$

21. $\dfrac{3}{2\sqrt{5} - 6} = \dfrac{3(2\sqrt{5} + 6)}{(2\sqrt{5} - 6)(2\sqrt{5} + 6)}$

$= \dfrac{6\sqrt{5} + 18}{(2\sqrt{5})^2 - 6^2} = \dfrac{6(\sqrt{5} + 3)}{4(5) - 36}$

$= \dfrac{6(\sqrt{5} + 3)}{-16} = -\dfrac{3}{8}(\sqrt{5} + 3)$

25. $\dfrac{\sqrt{2} - 1}{\sqrt{7} - 3\sqrt{2}} = \dfrac{(\sqrt{2} - 1)(\sqrt{7} + 3\sqrt{2})}{(\sqrt{7} - 3\sqrt{2})(\sqrt{7} + 3\sqrt{2})}$

$= \dfrac{\sqrt{2}\sqrt{7} + 3\sqrt{2}\sqrt{2} - \sqrt{7} - 3\sqrt{2}}{(\sqrt{7})^2 - (3\sqrt{2})^2} = \dfrac{\sqrt{14} + 3(2) - \sqrt{7} - 3\sqrt{2}}{7 - 18}$

$= \dfrac{-1(-\sqrt{14} - 6 + \sqrt{7} + 3\sqrt{2})}{-11} = \dfrac{1}{11}(\sqrt{7} + 3\sqrt{2} - 6 - \sqrt{14})$

29. $\dfrac{2\sqrt{3} - 5\sqrt{5}}{\sqrt{3} + 2\sqrt{5}} = \dfrac{(2\sqrt{3} - 5\sqrt{5})(\sqrt{3} - 2\sqrt{5})}{(\sqrt{3} + 2\sqrt{5})(\sqrt{3} - 2\sqrt{5})}$

$= \dfrac{2\sqrt{3}\sqrt{3} - 4\sqrt{3}\sqrt{5} - 5\sqrt{5}\sqrt{3} + 10\sqrt{5}\sqrt{5}}{(\sqrt{3})^2 - (2\sqrt{5})^2} = \dfrac{2(3) - 4\sqrt{15} - 5\sqrt{15} + 10(5)}{3 - 4(5)}$

$= \dfrac{56 - 9\sqrt{15}}{-17} = \dfrac{1}{17}(-56 + 9\sqrt{15})$

33. $\dfrac{2\sqrt{x}}{\sqrt{x} - \sqrt{y}} = \dfrac{(2\sqrt{x})}{(\sqrt{x} - \sqrt{y})} \times \dfrac{(\sqrt{x} + \sqrt{y})}{(\sqrt{x} + \sqrt{y})}$

$= \dfrac{2x + 2\sqrt{xy}}{x - y}$

37. $\dfrac{\sqrt{2c} + 3d}{\sqrt{2c} - d} = \dfrac{(\sqrt{2c} + 3d)}{(\sqrt{2c} - d)} \times \dfrac{(\sqrt{2c} + d)}{(\sqrt{2c} + d)} = \dfrac{2c + 3d\sqrt{2c} + d\sqrt{2c} + 3d^2}{2c - d^2}$

$= \dfrac{2c + 4d\sqrt{2c} + 3d^2}{2c - d^2}$

41. $\dfrac{2\sqrt{7}}{\sqrt{7} + \sqrt{6}} = \dfrac{2\sqrt{7}}{(\sqrt{7} + \sqrt{6})} \times \dfrac{(\sqrt{7} - \sqrt{6})}{(\sqrt{7} - \sqrt{6})}$

$= \dfrac{14 - 2\sqrt{42}}{7 - 6} = \dfrac{14 - 2\sqrt{42}}{1}$

$= 14 - 12.961481 = 1.0385186$

Using calculator: 2, $\boxed{\times}$, 7, $\boxed{\text{INV}}$, $\boxed{x^2}$, $\boxed{=}$, $\boxed{(}$, $\boxed{\div}$, 7, $\boxed{\text{INV}}$, $\boxed{x^2}$, $\boxed{+}$,

6, $\boxed{\text{INV}}$, $\boxed{x^2}$, $\boxed{)}$, $\boxed{=}$ 1.0385186

45. $\dfrac{\sqrt{5} + \sqrt{2}}{3\sqrt{6}} = \dfrac{(\sqrt{5} + \sqrt{2})(\sqrt{5} - \sqrt{2})}{3\sqrt{6}(\sqrt{5} - \sqrt{2})} = \dfrac{5 - 2}{3\sqrt{30} - 3\sqrt{12}} = \dfrac{3}{3\sqrt{30} - 6\sqrt{3}} = \dfrac{1}{\sqrt{30} - 2\sqrt{3}}$

49. $v = \dfrac{\sqrt{gs}}{\sqrt{12w}} = \dfrac{(\sqrt{gs})(\sqrt{12w})}{(\sqrt{12w})(\sqrt{12w})} = \dfrac{\sqrt{12gsw}}{12w} = \dfrac{2\sqrt{3gsw}}{12w} = \dfrac{\sqrt{3gsw}}{6w}$

Review Exercises for Chapter 10, page 312

1. $2a^{-2}b^0 = 2a^{-2}(1)$

$\quad = 2a^{-2} = \dfrac{2}{a^2}$

5. $3(25)^{3/2} = 3(\sqrt{25})^3$

$\quad = 3(5)^3 = 3(125) = 375$

9. $(\dfrac{3}{t^2})^{-2} = \dfrac{3^{-2}}{t^{-4}}$

$\quad = \dfrac{t^4}{3^2} = \dfrac{t^4}{9}$

13. $(2a^{1/3}b^{5/6})^6 = 2^6 a^{(1/3)(6)} b^{(5/6)(6)}$

$\quad = 64a^2 b^5$

17. $2x^{-2} - y^{-1} = \dfrac{2}{x^2} - \dfrac{1}{y}$

$\quad = \dfrac{2(y)}{x^2(y)} - \dfrac{1(x^2)}{y(x^2)} = \dfrac{2y - x^2}{x^2 y}$

21. $(a - 3b^{-1})^{-1} = (a - \dfrac{3}{b})^{-1}$

$\quad = \left(\dfrac{a(b)}{1(b)} - \dfrac{3}{b}\right)^{-1} = \left(\dfrac{ab - 3}{b}\right)^{-1}$

$\quad = \dfrac{(ab - 3)^{-1}}{b^{-1}} = \dfrac{b}{ab - 3}$

25. $(8a^3)^{2/3}(4a^{-2} + 1)^{1/2} = (8^{2/3}a^{3(2/3)})(4a^{-2} + 1)^{1/2}$

$\quad = (\sqrt[3]{8})^2 a^2 (\dfrac{4}{a^2} + \dfrac{a^2}{a^2})^{1/2} = 4a^2(\dfrac{4 + a^2}{a^2})^{1/2}$

$\quad = 4a^2 \dfrac{(4 + a^2)^{1/2}}{a^2(1/2)} = 4a^2 \dfrac{(4 + a^2)^{1/2}}{a}$

$\quad = 4a(a^2 + 4)^{1/2}$

29. $\sqrt{68} = \sqrt{4(17)}$

$\quad = 2\sqrt{17}$

33. $\sqrt{9a^3 b^4} = \sqrt{9(a^2)(a)(b^2)(b^2)}$

$\quad = 3ab^2\sqrt{a}$

37. $\dfrac{5}{\sqrt{2s}} = \dfrac{5\sqrt{2s}}{\sqrt{2s}\sqrt{2s}}$

$\quad = \dfrac{5\sqrt{2s}}{2s}$

41. $\sqrt[4]{8m^6 n^9} = \sqrt[4]{8m^4 (m^2)(n^8)(n^1)}$

$\quad = mn^2 \sqrt[4]{8m^2 n}$

45. $\sqrt{200} + \sqrt{32} = \sqrt{100(2)} + \sqrt{16(2)}$

$\quad = 10\sqrt{2} + 4\sqrt{2} = 14\sqrt{2}$

49. $a\sqrt{2x^3} + \sqrt{8a^2 x^3} = ax\sqrt{2x} + 2ax\sqrt{2x}$

$\quad = 3ax\sqrt{2x}$

53. $\sqrt{5}(2\sqrt{5} - \sqrt{11}) = 2\sqrt{5}\sqrt{5} - \sqrt{5}\sqrt{11}$

$\quad = 2(5) - \sqrt{55} = 10 - \sqrt{55}$

57. $(2 - 3\sqrt{17})(3 + \sqrt{17}) = 2(3) + 2\sqrt{17} - 3(3\sqrt{17}) - 3\sqrt{17}\sqrt{17}$

$\quad = 6 + 2\sqrt{17} - 9\sqrt{17} - 3(17) = -45 - 7\sqrt{17}$

61. $\dfrac{\sqrt{3x}}{2\sqrt{3x} - \sqrt{y}} = \dfrac{\sqrt{3x}}{(2\sqrt{3x} - \sqrt{y})} \times \dfrac{(2\sqrt{3x} + \sqrt{y})}{(2\sqrt{3x} + \sqrt{y})}$

$= \dfrac{2(3x) + \sqrt{3xy}}{4(3x) - y} = \dfrac{6x + \sqrt{3xy}}{12x - y}$

65. $\dfrac{\sqrt{7} - \sqrt{5}}{\sqrt{5} + 3\sqrt{7}} = \dfrac{(\sqrt{7} - \sqrt{5})}{(\sqrt{5} + 3\sqrt{7})} \times \dfrac{(\sqrt{5} - 3\sqrt{7})}{(\sqrt{5} - 3\sqrt{7})}$

$= \dfrac{\sqrt{35} - 21 - 5 + 3\sqrt{35}}{5 - 63} = \dfrac{4\sqrt{35} - 26}{-58}$

$= \dfrac{2(2\sqrt{35} - 13)}{-58} = \dfrac{2\sqrt{35} - 13}{-29}$

$= \dfrac{13 - 2\sqrt{35}}{29}$

69. $\sqrt{4b^2 + 1}$ is in simplest form.

73. $\sqrt{52} + 4\sqrt{24} - \sqrt{54} = \sqrt{4(13)} + 4\sqrt{4(6)} - \sqrt{9(6)}$

$= 2\sqrt{13} + 8\sqrt{6} - 3\sqrt{6} = 2\sqrt{13} + 5\sqrt{6}$

$= 2(3.606) + 5(2.449) = 7.212 + 12.245$

$= 19.457$

Using calculator: 52, $\boxed{\text{INV}}$, $\boxed{x^2}$, $\boxed{+}$, 4, $\boxed{\times}$, 24, $\boxed{\text{INV}}$, $\boxed{x^2}$, $\boxed{-}$, 54,

$\boxed{\text{INV}}$, $\boxed{x^2}$, $\boxed{=}$ 19.458551

77. $i = 100[(c_2/c_1)^{1/n} - 1] = 100[(442.3 + 247.0)^{1/10} - 1]$

$= 100[1.79^{1/10} - 1] = 100(1.06 - 1)$

$= 6\%$

81. $\dfrac{v}{n_2^{-2} - n_1^{-2}} = \dfrac{v}{\dfrac{1}{n_2^2} - \dfrac{1}{n_1^2}} = \dfrac{v}{\dfrac{n_1^2}{n_2^2 \, n_1^2} - \dfrac{n_2^2}{n_2^2 \, n_1^2}} = \dfrac{v}{\dfrac{n_1^2 - n_2^2}{n_2^2 \, n_1^2}}$

$= \dfrac{v}{1} \times \dfrac{n_2^2 \, n_1^2}{n_1^2 - n_2^2} = \dfrac{vn_2^2 \, n_1^2}{n_1^2 - n_2^2}$

85. $\sqrt{3^2 + 3^2} + \sqrt{2^2 + 2^2} + \sqrt{1^2 + 1^2} = \sqrt{18} + \sqrt{8} + \sqrt{2} = 3\sqrt{2} + 2\sqrt{2} + \sqrt{2}$

$= 6\sqrt{2}$

CHAPTER 11

Complex Numbers

Exercises 11-1, page 319

1. $\sqrt{-81} = \sqrt{81(-1)}$
   $= \sqrt{81}\sqrt{-1} = 9j$

5. $\sqrt{-0.36} = \sqrt{0.36(-1)}$
   $= \sqrt{0.36}\sqrt{-1} = 0.6j$

9. $\sqrt{\dfrac{-7}{4}} = \dfrac{\sqrt{-7}}{2}$

   $= \dfrac{\sqrt{7(-1)}}{2} = \dfrac{j\sqrt{7}}{2} = \dfrac{1}{2}j\sqrt{7}$

13. $(\sqrt{-7})^2 = \sqrt{-7}\sqrt{-7}$
    $= (\sqrt{7}j)(\sqrt{7}j) = 7j^2$
    $= 7(-1) = -7$
    $\sqrt{(-7)^2} = \sqrt{49}$
    $= 7$

17. $j^7 = j^4 j^3$
    $= (1)j^2 j = (1)(-1)j$
    $= -j$

21. $j^2 - j^6 = j^2(1 - j^4)$
    $= j^2(1 - 1) = j^2(0)$
    $= 0$

25. $2 + \sqrt{-9} = 2 + \sqrt{3^2(-1)}$
    $= 2 + \sqrt{3^2}\sqrt{-1}$
    $= 2 + 3j$

29. $8 - \sqrt{4} + \sqrt{-4} = 8 - 2 + \sqrt{4(-1)} = 8 - 2 + 2\sqrt{-1} = 6 + 2j$

33. $\sqrt{18} - \sqrt{-8} = \sqrt{9(2)} - \sqrt{4(2)(-1)}$
    $= 3\sqrt{2} - 2\sqrt{2}\sqrt{-1}$
    $= 3\sqrt{2} - 2\sqrt{2}j$

37. $6 - 7j$ has $6 + 7j$ as conjugate.

41. $7x - 2yj = 14 + 4j; \quad 7x = 14 \quad$ and $\quad -2yj = 4j$
    $\qquad\qquad\qquad\qquad\quad x = 2$
    $\qquad\qquad\qquad\qquad\qquad\qquad\qquad\quad y = \dfrac{4j}{-2j} = -2$

45. $x - y = 1 - xj - yj - j$
    $x - y + xj + yj = 1 - j$
    $(x - y) + (x + y)j = 1 - j; \quad x - y = 1 \quad$ and $\quad (x + y)j = -j$
    $\qquad\qquad\qquad\qquad\qquad\qquad\qquad\qquad\qquad\qquad\qquad\qquad x + y = -1$
    Solving the system $\begin{cases} x - y = 1 \\ x + y = -1 \end{cases}$ gives $x = 0$ and $y = -1$.

49. Yes. $x^2 + 64 = 0$. If $8j$ is a solution, $(8j)^2 + 64 = 0$; $64j^2 + 64 = 0$;
     $64(-1) + 64 = 0$; $-64 + 64 = 0$; $0 = 0$. If $-8j$ is a solution,
     $(-8j)^2 + 64 = 0$; $64j^2 + 64 = 0$; $0 = 0$.

Exercises 11-2, page 322

1. $(3 - 7j) + (2 - j) = (3 + 2) + (-7 - 1)j$
    $$= 5 - 8j$$

5. $(4 + \sqrt{-16}) + (3 - \sqrt{-81}) = (4 + 4j) + (3 - 9j)$
    $$= (4 + 3) + (4 - 9)j = 7 - 5j$$

9. $j - (j - 7) - 8 = j - j + 7 - 8$
    $$= 0j - 1 = -1$$

13. $(7 - j)(7j) = 49j - 7j^2$
     $$= 49j - 7(-1) = 49j + 7$$
     $$= 7 + 49j$$

17. $(4 - j)(5 + 2j) = 20 + 8j - 5j - 2j^2$
     $$= 20 + 3j - 2(-1) = 22 + 3j$$

21. $(\sqrt{-18} \; \sqrt{-4})(3j) = (3\sqrt{2}j)(2j)(3j)$
     $$= 18\sqrt{2}j^3 = 18\sqrt{2}j^2j$$
     $$= 18\sqrt{2}(-1)j = -18\sqrt{2}j$$

25. $\sqrt{-108} - \sqrt{-27} = \sqrt{36(3)(-1)} - \sqrt{9(3)(-1)}$
     $$= 6\sqrt{3}j - 3\sqrt{3}j = 3\sqrt{3}j$$

29. $7j^3 - 7\sqrt{-9} = 7j^3 - 7\sqrt{9(-1)}$
     $$= 7j^2j - 7(3)j = 7(-1)j - 21j$$
     $$= -7j - 21j = -28j$$

33. $(3 - 7j)^2 = (3 - 7j)(3 - 7j)$
     $$= (9 + 49j^2) + (-21 - 21)j = 9 + 49(-1) + (-42)j$$
     $$= -40 - 42j$$

37. $\dfrac{6j}{2 - 5j} = \dfrac{6j(2 + 5j)}{(2 - 5j)(2 + 5j)}$

     $$= \frac{12j + 30j^2}{2^2 - 25j^2} = \frac{12j + 30(-1)}{4 - 25(-1)}$$

     $$= \frac{12j - 30}{4 + 25} = \frac{-30 + 12j}{29}$$

     $$= \frac{1}{29}(-30 + 12j)$$

41. $\dfrac{1 - j}{3j} \times \dfrac{3j}{3j} = \dfrac{3j - 3j^2}{9(j^2)} = \dfrac{3j - 3(-1)}{9(-1)} = \dfrac{3j + 3}{-9} = \dfrac{-1}{3}(1 + j)$

45. $\dfrac{\sqrt{-16} - \sqrt{2}}{\sqrt{2} + j} = \dfrac{4j - \sqrt{2}}{\sqrt{2} + j}$

$\qquad = \dfrac{(-\sqrt{2} + 4j)(\sqrt{2} - j)}{(\sqrt{2} + j)(\sqrt{2} - j)} = \dfrac{-\sqrt{2}^2 - 4j^2 + (4\sqrt{2} + \sqrt{2})j}{2 - j^2}$

$\qquad = \dfrac{-2 + 4 + 5\sqrt{2}j}{2 - (-1)} = \dfrac{2 + 5\sqrt{2}j}{3} = \dfrac{1}{3}(2 + 5\sqrt{2}j)$

49. $(-1 + j)^2 + 2(-1 + j) + 2 = (1 - 2j + j^2) - 2 + 2j + 2$

$\qquad = 1 + j^2 = 1 - 1 = 0$

53. $\dfrac{2 - 3j}{2 + 3j} \times \dfrac{2 - 3j}{2 - 3j} = \dfrac{(2 - 3j)^2}{2^2 - (3j)^2} = \dfrac{2^2 - 12j + 9j^2}{4 - 9j^2} = \dfrac{4 - 9 - 12j}{4 + 9}$

$\qquad = \dfrac{-5 - 12j}{13} = \dfrac{1}{13}(5 + 12j)$

57. $(x + yj) + (x - yj) = 2x + 0j = 2x$ where x and y are real numbers.

Exercises 11-3, page 324

1.

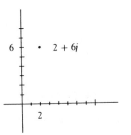

6 $\cdot$ 2 + 6j

2

5.

3 + 4j   5 + 4j

2

$2 + (3 + 4j)$
$= 5 + 4j$

9. Subtracting $1 - 4j$ is equivalent to adding $-1 + 4j$: $5 + (-1 + 4j)$.

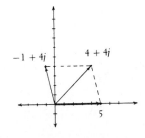

$-1 + 4j$ $4 + 4j$

5

$5 - (1 - 4j) = 5 - 1 + 4j$
$\qquad\qquad\quad = 4 + 4j$

13. $-(4 - 6j)$

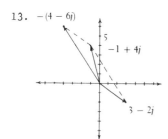

$$
\begin{aligned}
(3 - 2j) &- (4 - 6j) \\
&= (3 - 4) + (-2 + 6)j \\
&= -1 + 4j
\end{aligned}
$$

17.

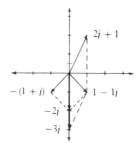

$$
\begin{aligned}
(3.0 + 2.5j) &+ (1.5 - 0.5j) \\
&= (3.0 + 1.5) + (2.5 - 0.5)j \\
&= 4.5 + 2.0j
\end{aligned}
$$

21.

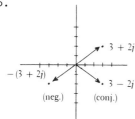

$$
\begin{aligned}
(2j + 1) &- 3j - (j + 1) \\
&= 2j + 1 - 3j - j - 1 \\
&= 2j - 4j + 1 - 1 = -2j
\end{aligned}
$$

25.

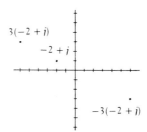

29.

$$
\begin{aligned}
&-2 + j \\
&3(-2 + j) = -6 + 3j \\
&-3(-2 + j) = 6 - 3j
\end{aligned}
$$

Exercises 11-4, page 328

1.

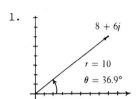

$8 + 6j$

$r = \sqrt{8^2 + 6^2} = \sqrt{100} = 10$

$\tan \theta = \dfrac{6}{8} = 0.7500$

$\theta = 36.9°$

$10(\cos 36.9° + j \sin 36.9°)$

5.

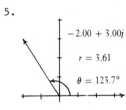

$-2.00 + 3.00j$

$r = \sqrt{(-2.00)^2 + (3.00)^2} = \sqrt{13.0} = 3.61$

$\tan \theta = \dfrac{3.00}{-2.00} = -1.500$ in Q II

$\theta = 123.7°$

$3.61(\cos 123.7° + j \sin 123.7°)$

9.

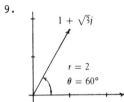

$1 + j\sqrt{3} = 1 + \sqrt{3}\,j$

$r = \sqrt{1^2 + (\sqrt{3})^2} = \sqrt{4} = 2$

$\tan \theta = \dfrac{\sqrt{3}}{1} = \sqrt{3}.$ Using $\sqrt{3} = 1.732$, then

$\theta = 60.0°.$

$2(\cos 60° + j \sin 60°)$

13.

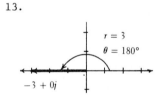

$-3 = -3 + 0j$

$r = \sqrt{(-3)^2 + 0^2} = 3$

$\tan \theta = \dfrac{0}{-3} = 0$

$\theta = 180°$ since "x" is neg.

$3(\cos 180° + j \sin 180°)$

17.

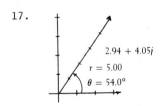

$5.00(\cos 54.0° + j \sin 54.0°)$
$x = 5.00 \cos 54.0° = 5.00(0.5878) = 2.94$
$y = 5.00 \sin 54.0° = 5.00(0.8090) = 4.05$

$2.95 + 4.05j$

21.

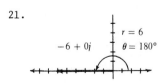

$6(\cos 180° + j \sin 180°)$
$x = 6 \cos 180° = 6(-1) = -6$
$y = 6 \sin 180° = 6(0) = 0$

$-6 + 0j = -6$

25.

$r = 12.36$
$\theta = 345.56°$

$11.97 - 3.08j$

12.36(cos 345.56° + j sin 345.56°)
x = 12.36 cos 345.56° = 12.36(0.9684)
                                   = 11.97
y = 12.36 sin 345.56° = 12.36(-0.2494)
                                   = -3.08

11.97 - 3.082j

29.

$r = 4.75$
$\theta = 172.8°$

$-4.71 + 0.60j$

4.75∠172.8°
x = 4.75 cos 172.8° = -4.71
y = 4.75 sin 172.8° = 0.595

-4.71 + 0.595j

33.

$0 + 7.32j$

$r = 7.32$
$\theta = -270° = 90°$

7.32∠270 = 7.32 ∠90°
x = 7.32 cos 90° = 7.32(0) = 0
y = 7.32 sin 90° = 7.32(1) = 7.32

0 + 7.32j = 7.32j

37. 25.6 - 34.2j

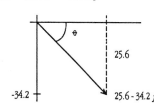

$R = \sqrt{(25.6)^2 + (34.2)^2} = \sqrt{1825} = 42.7$

$\tan \theta = \dfrac{-34.2}{25.6} = -1.3359$

$\theta = -53.2° = 306.8°$

Exercises 11-5, page 332

1. 3.00(cos 60.0° + j sin 60.0°); r = 3.00, θ = 60.0° = $\dfrac{60.0\pi}{180}$ = 1.05 rad;
    3.00 $e^{1.05j}$

5. 375.5(cos 95.46° + j sin 95.46°); r = 375.5, θ = 95.46° = $\dfrac{95.46\pi}{180}$
    = 1.666 rad; 375.5$e^{1.666j}$

9. 4.06 ∠-61.4°; r = 4.06, θ = -61.4° = $\dfrac{-61.4\pi}{180}$ = -1.07
    = 4.06$e^{-1.07j}$

13. $3 - 4j$; $r = \sqrt{3^2 + (-4)^2} = \sqrt{25} = 5$; $\tan \theta = \dfrac{-4}{3} = -1.333$; $\theta = 306.9°$
= 5.36 rad; $5.00\, e^{5.36j}$

17. $5.90 + 2.40j$; $r = \sqrt{(5.90)^2 + (2.40)^2} = \sqrt{40.6} = 6.37$; $\tan \theta = \dfrac{2.40}{5.90}$
= 0.4068; $\theta = 22.1° = \dfrac{22.1\pi}{180} = 0.386$ rad; $6.37\, e^{0.386j}$

21. $3.00e^{0.500j}$; $r = 3.00$, $\theta = 0.500$ rad $= \dfrac{0.500(180)}{\pi} = 28.6°$
$3.00(\cos 28.6° + j \sin 28.6°)$; $x = 3.00 \cos 28.6° = 3.00(0.8780) = 2.63$
 $y = 3.00 \sin 28.6° = 3.00(0.4787) = 1.44$
$2.63 + 1.44j$

25. $3.20e^{5.41j}$ $r = 3.20$, $\theta = 5.41$ rad $= \dfrac{5.41(180)}{\pi} = 310.0°$
$3.20(\cos 310.0° + j \sin 310.0°)$; $x = 3.20 \cos 310.0° = 3.20(0.6428)$
 $= 2.06$
 $y = 3.20 \sin 310.0° = 3.20(-0.7660)$
 $= -2.45$
 $2.06 - 2.45j$

29.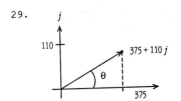

$\sqrt{375^2 + 110^2} = 391$; $391\ \Omega$

$\tan \theta = \dfrac{110}{375} = 0.2933$

$\theta = 16.3° = \dfrac{16.3\pi}{180} = 0.285$

$375 + 110j = 391\, e^{0.285j}$ ohms; $391\ \Omega$

Exercises 11-6, page 338

1. $[4(\cos 60° + j \sin 60°)][2(\cos 20° + j \sin 20°)]$
= $4(2)[\cos(60° + 20°) + j \sin(60° + 20°)] = 8(\cos 80° + j \sin 80°)$

5. $\dfrac{8(\cos 100° + j \sin 100°)}{4(\cos 65° + j \sin 65°)} = \dfrac{8}{4}[\cos(100° - 65°) + j \sin(100° - 65°)]$
$= 2(\cos 35° + j \sin 35°)$

9. $[2(\cos 35° + j \sin 35°)]^3 = 2^3[\cos 3(35°) + j \sin 3(35°)]$
$= 8(\cos 105° + j \sin 105°)$

13. $2.78\underline{/56.8°} + 1.37\underline{/207.3°}$
$2.78(\cos 56.8° + j \sin 56.8°) + 1.37(\cos 207.3° + j \sin 207.3°)$
$1.5222 + 2.3262j - 1.2174 - 0.6283j = 0.3048 + 1.6979j$
Change this to polar form: $r = \sqrt{(0.3048)^2 + (1.6979)^2} = 1.73$
$\tan \theta = \dfrac{1.6979}{0.3048} = 5.5705$; $\theta = 79.8°$

Therefore, $2.78\underline{/56.8°} + 1.37\underline{/207.3°} = 1.73\underline{/79.8°}$

17. $(3 + 4j)$; $r_1 = \sqrt{9 + 16} = 5$, $\tan \theta_1 = 53.1°$
$(5 - 12j)$; $r_2 = \sqrt{25 + 144} = 13$, $\tan \theta_2 = 292.6°$
$5(\cos 53.1° + j \sin 53.1°)(13)(\cos 292.6° + j \sin 292.6°)$
$= 65.0[\cos(53.1° + 292.6°) + j \sin(53.1° + 292.6°)]$
$= 65.0[\cos 345.7° + j \sin 345.7°]$; $x = 65 \cos 345.7° = 65(0.9690)$
$= 63.0$
$y = 65 \sin 345.7° = 65(-0.2470)$
$= -16.1$

$63.0 - 16.1j$

Rectangular form: $(3 + 4j)(5 - 12j) = 15 + 3(-12j) + 5(+4j) + 4j(-12j)$
$= 15 - 36j + 20j - 48j^2$
$= 15 - 16j - 48(-1)$
$= 63 - 16j$

21. $\dfrac{7}{1 - 3j} = \dfrac{7 + 0j}{1 - 3j}$; $r_1 = \sqrt{7^2 + 0^2} = 7$, $\tan \theta_1 = \dfrac{0}{7} = 0$, $\theta_1 = 0°$

$r_2 = \sqrt{1^2 + (-3)^2} = 3.16$, $\tan \theta_2 = \dfrac{-3}{1} = -3$; $\theta_2 = 288.4°$

$\dfrac{7(\cos 0° + j \sin 0°)}{3.16(\cos 288.4° + j \sin 288.4°)} = \dfrac{7}{3.16}[\cos(0° - 288.4°)$

$+ j \sin(0° - 288.4°)]$

$= 2.21[\cos(-288.4° + j \sin(-288.4°)]$

$= 2.21(\cos 71.6° + j \sin 71.6°)$

$x = 2.21 \cos 71.6° = 2.21(0.3156) = 0.697$
$y = 2.21 \sin 71.6° = 2.21(0.9489) = 2.10$

$0.697 + 2.10j$

Rectangular form: $\dfrac{7}{1 - 3j} = \dfrac{7(1 + 3j)}{(1 - 3j)(1 + 3j)} = \dfrac{7 + 21j}{1 - 9j^2} = \dfrac{7 + 21j}{1 - 9(-1)}$
$= \dfrac{7 + 21j}{10} = \dfrac{7}{10} + \dfrac{21}{10}j$

25. $(3 + 4j)^4$; $r = \sqrt{3^2 + 4^2} = 5$; $\tan \theta = \dfrac{4}{3}$, $\theta = 53.1°$

$[5(\cos 53.1° + j \sin 53.1°)]^4 = 5^4[\cos 4(53.1°) + j \sin 4(53.1°)]$
$= 625(\cos 212.4° + j \sin 212.4°)$

$x = 625 \cos 212.4° = 625[-\cos(212.4° - 180°)] = -625 \cos 32.4° = -528$
$y = 625 \sin 212.4° = 625[-\sin(212.4° - 180°)] = -625 \cos 32.4° = -335$

$-528 - 335j$

Rectangular form: $(3 + 4j)^4 = [(3 + 4j)^2]^2 = [9 + 24j + 16j^2]^2$
$= [9 + 24j + 16(-1)]^2$
$= [9 + 24j - 16]^2$
$= [-7 + 24j]^2$
$[-7 + 24j]^2 = 49 - 336j + 576j^2 = 49 - 336j + 576(-1) = -527 - 336j$

29. $\sqrt{4(\cos\ 60° + j\ \sin\ 60°)} = [4(\cos\ 60° + j\ \sin\ 60°)]^{1/2}$

First root: $4^{1/2}[\cos\ \frac{1}{2}(60°) + j\ \sin\ \frac{1}{2}(60°)] = 2(\cos\ 30° + j\ \sin\ 30°)$

Second root: $4^{1/2}[\cos\ \frac{1}{2}(60° + 360°) + j\ \sin\ \frac{1}{2}(60° + 360°)]$

$$= 2[\cos\ \frac{1}{2}(420°) + j\ \sin\ \frac{1}{2}(420°)]$$

$$= 2(\cos\ 210° + j\ \sin\ 210°)$$

33. $\sqrt[4]{I} = \sqrt[4]{\cos\ 0° + j\ \sin\ 0°} = (\cos\ 0° + j\ \sin\ 0°)^{1/4}$

First root: $1^{1/4}[\cos\ \frac{1}{4}(0°) + j\ \sin\ \frac{1}{4}(0°)] = \cos\ 0° + j\ \sin\ 0°$

$$= 1 + 0j = 1$$

Second root: $\cos\ \frac{1}{4}(0° + 360°) + j\ \sin\ \frac{1}{4}(0° + 360°) = \cos\ 90° + j\ \sin\ 90°$

$$= 0 + j = j$$

Third root: $\cos\ \frac{1}{4}(0° + 720°) + j\ \sin\ \frac{1}{4}(0° + 720°) = \cos\ 180° + j\ \sin\ 180°$

$$= -1 + 0j = -1$$

Fourth root: $\cos\ \frac{1}{4}(0° + 1080°) + j\ \sin\ \frac{1}{4}(0° + 1080°)$

$$= \cos\ 270° + j\ \sin\ 270°$$

$$= 0 + j(-1) = -j$$

37. $[\frac{1}{2}(1 - \sqrt{3}j)]^3 = [\frac{1}{2}(1 - \sqrt{3}j)]^2[\frac{1}{2}(1 - \sqrt{3}j)] = \frac{1}{4}[1 - 2\sqrt{3}j + 3j^2][\frac{1}{2}(1 - \sqrt{3}j)]$

$$= \frac{1}{4} \times \frac{1}{2}[1 - 2\sqrt{3}j + 3(-1)][1 - \sqrt{3}j] = \frac{1}{8}[-2 - 2\sqrt{3}j][1 - \sqrt{3}j]$$

$$= \frac{1}{8}[-2 - 2(-\sqrt{3}j) - 2\sqrt{3}j + 2(\sqrt{3})^2j^2]$$

$$= \frac{1}{8}[-2 + 2\sqrt{3}j - 2\sqrt{3}j + 2(3)(-1)] = \frac{1}{8}(-8) = -1$$

Exercises 11-7, page 344

1. $I = 5.75 \cdot mA = 0.00575\ A;\ V_R = IX_R = 0.00575(2250) = 12.9\ V$

5. $X_L = 2\pi fL = 2\pi(60)(0.0429) = 16.2$

$X_c = \frac{1}{2\pi fC} = \frac{1}{2\pi(60)(0.0000862)} = 30.8$

$Z = 0 + (X_L - X_c)j = 0 + (16.2 - 30.8)j = -14.6j$

$|Z| = 14.6j$ $\tan\ \theta$ is undefined, so $\theta = -90°$

$Z = 14.6\ \Omega$

9. $i = 3.90 - 6.04\ mA,\ Z = (5.16 + 1.14)\ k\Omega$

$V = (3.90 - 6.04j) \times 10^{-3}\ (5.16 + 1.14j) \times 10^3$

$$= 10.1 - 31.2j + 4.4j - 6.89j^2 = 27.0 - 26.8j$$

$|V| = \sqrt{(27.0)^2 + (-26.8)^2} = \sqrt{729 + 718} = \sqrt{1447} = 38.0V$

13. $X = 2\pi f L$; $1200 = 2\pi(280)(L)$; $L = 1200 \div 2\pi(280) = 0.682\ H$

17. $C = \dfrac{1}{(2\pi f)^2 L} = \dfrac{1}{[2\pi(680 \times 10^3)^2](4.20 \times 10^{-3})}$

$= \dfrac{1}{4\pi^2(4.62 \times 10^5 \times 10^6)(4.20 \times 10^{-3})}$

$= \dfrac{1}{766 \times 10^8} = 1.30 \times 10^{-11} = 13pf$

Review Exercises for Chapter 11, page 346

1. $(6 - 2j) + (4 + j) = 6 + 4 + (-2 + 1)j$
$= 10 - j$

5. $(2 + j)(4 - j) = 2(4) + 2j - j^2$
$= 8 + 2j - (-1) = 9 + 2j$

9. $\dfrac{3}{7 - 6j} = \dfrac{3(7 + 6j)}{(7 - 6j)(7 + 6j)} = \dfrac{21 + 18j}{49 - 36j^2}$

$= \dfrac{21 + 18j}{49 + 36} = \dfrac{21 + 18j}{85}$

$= \dfrac{1}{85}(21 + 18j)$

13. $\dfrac{5j - (3 - j)}{4 - 2j} = \dfrac{6j - 3}{4 - 2j} = \dfrac{(-3 + 6j)(4 + 2j)}{(4 - 2j)(4 + 2j)}$

$= \dfrac{-12 + 18j + 12j^2}{16 - 4j^2} = \dfrac{-12 + 18j + 12(-1)}{16 - 4(-1)}$

$= \dfrac{-24 + 18j}{20} = \dfrac{-12 + 9j}{10} = \dfrac{1}{10}(-12 + 9j)$

17. $3x - 2j = yj - 2$; $3x = -2$ and $-2j = yj$

$x = -\dfrac{2}{3}$ $y = -2$

21.

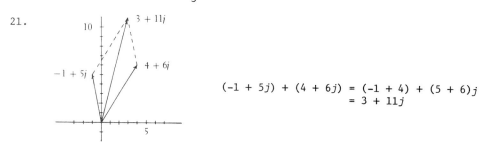

$(-1 + 5j) + (4 + 6j) = (-1 + 4) + (5 + 6)j$
$= 3 + 11j$

25. $1 - j = r(\cos\theta + j\sin\theta)$ where $r = \sqrt{1^2 + (-1)^2} = 1.41$

   $\tan\theta = \dfrac{-1}{1} = -1;\quad \theta = 315°$

   $1.41(\cos 315° + j\sin 315°) = re^{j\theta}$ where $r = 1.41$ and $\theta = \dfrac{315\pi}{180}$

   $\quad = 1.75\pi = 5.50$

   $1 - j = 1.41(\cos 315° + j\sin 315°)$

   $\quad = 1.41e^{5.50j}$

29. $1.07 + 4.55j = 4(\cos\theta + j\sin\theta)$ where $r = \sqrt{1.07^2 + 4.55^2} = 4.67$

   $\tan\theta = \dfrac{4.55}{1.07} = 4.252,\ \theta = 76.7°$

   $4.67(\cos 76.7° + j\sin 76.7°) = 4.67e^{j\theta}$ where $\theta = \dfrac{76.7\pi}{180} = 1.34$

   $1.07 + 4.55j = 4.67(\cos 76.7° + j\sin 76.7°)$

   $\quad = 4.67e^{1.34j}$

33. $2(\cos 225° + j\sin 225°);\quad x = 2\cos 225° = 2(-0.7071) = -1.41$
   $\qquad\qquad\qquad\qquad\qquad\quad\ y = 2\sin 225° = 2(-0.7071) = -1.41$

   $2(\cos 225° + j\sin 225°) = -1.41 - 1.41j$

37. $0.62\underline{/-72°} = 0.62[\cos(-72°) + j\sin(-72°)] = 0.19 - 0.59j$

41. $2.00e^{0.25j};\quad \theta = \dfrac{0.25(180)}{\pi} = 14.3°,\ r = 2.00$

   $2.00(\cos 14.3° + j\sin 14.3°) = 2.00[0.9690 + j(0.2470)]$
   $\qquad\qquad\qquad\qquad\qquad\quad = 1.94 + 0.495j$

45. $[3(\cos 32° + j\sin 32°)][5(\cos 52° + j\sin 52°)]$
   $= 15[\cos(32° + 52°) + j\sin(32° + 52°)]$
   $= 15(\cos 84° + j\sin 84°)$

49. $\dfrac{24(\cos 165° + j\sin 165°)}{3(\cos 106° + j\sin 106°)} = \dfrac{24}{3}[\cos(165° - 106°) + j\sin(165°)$
   $\qquad\qquad\qquad\qquad\qquad\qquad = 8(\cos 59° + j\sin 59°)$

53. $0.983\underline{/47.2°} + 0.366\underline{/95.1°}$

   $0.983(\cos 47.2° + j\sin 47.2°) + 0.366(\cos 95.1° + j\sin 95.1°)$

   $= 0.6679 + 0.7213j - 0.0325 + 0.3646j$

   $= 0.6354 + 1.0859j$

   Change this to polar form: $r = \sqrt{0.6354^2 + 1.0859^2} = 1.26$

   $\tan\theta = \dfrac{1.0859}{0.6354} = 1.7090;\quad \theta = 59.7°$

   Therefore, $0.983\underline{/47.2°} + 0.366\underline{/95.1°} = 1.26(\cos 59.7° + j\sin 59.7°)$

57. $[2(\cos 16° + j\sin 16°)]^{10} = 2^{10}[\cos 10(16°) + j\sin 10(16°)]$
   $\qquad\qquad\qquad\qquad\qquad\quad = 1024(\cos 160° + j\sin 160°)$

61. $(1 - j)^{10} = [1.41(\cos 315° + j \sin 315°)]^{10}$ (see Exercise 25)
$$= 1.41^{10}[\cos 10(315°) + j \sin 10(315°)]$$
$$= 32(\cos 3150° + j \sin 3150°)$$
$$= 32(\cos 270° + j \sin 270°) = 32[0 + j(-1)] = -32j$$

Rectangular form: $(1 - j)^{10} = [(1 - j)^2]^5$
$$= (1 - 2j + j^2)^5 = (1 - 2j - 1)^5$$
$$= (-2j)^5 = -32j^5$$
$$= -32(j^2)(j^2)(j)$$
$$= -32(-1)(-1)(j)$$
$$= -32j$$

65. $\sqrt[3]{-8} = \sqrt[3]{8(\cos 180° + j \sin 180°)} = [8(\cos 180° + j \sin 180°)]^{1/3}$

First root: $8^{1/3}[\cos \frac{1}{3}(180°) + j \sin \frac{1}{3}(180°)]$
$$= 2(\cos 60° + j \sin 60°) = 2[0.500 + j(0.8660)]$$
$$= 1.00 + 1.73j$$

Second root: $2[\cos \frac{1}{3}(180° + 360°) + j \sin \frac{1}{3}(180° + 360°)]$
$$= 2[\cos \frac{1}{3}(540°) + j \sin \frac{1}{3}(540°)]$$
$$= 2(\cos 180° + j \sin 180°) = 2(-1 + 0j)$$
$$= -2$$

Third root: $2[\cos \frac{1}{3}(180° + 720°) + j \sin \frac{1}{3}(180° + 720°)]$
$$= 2[\cos \frac{1}{3}(900°) + j \sin \frac{1}{3}(900°)]$$
$$= 2(\cos 300° + j \sin 300°) = 2(0.500 - 0.8660j)$$
$$= 1.00 - 1.73j$$

69. $V_{RCL} = V_R + V_L + V_C;\; V_L = +60j,\; V_C = -60j$
$$60 = V_R + 60j - 60j;\; V_R = 60V$$

73. $2\pi fL = \dfrac{1}{2\pi fC};\; 2\pi f(2.65) = \dfrac{1}{2\pi f(18.3 \times 10^{-6})}$

$4\pi^2 f^2 = \dfrac{1}{(18.3 \times 10^{-6})(2.65)} = \dfrac{10^6}{48.5} = 2.06 \times 10^4$

$f^2 = \dfrac{2.06 \times 10^4}{4\pi^2} = 522;\; f = \sqrt{522} = 22.8$ Hz

77. $\dfrac{1}{u + j(\omega n)} = \dfrac{1[u - j(\omega n)]}{[u + j(\omega n)][u - j(\omega n)]} = \dfrac{u - j(\omega n)}{u^2 - (\omega n)^2 j^2} = \dfrac{u - j(\omega n)}{u^2 - (\omega n)^2(-1)}$

$$= \dfrac{u - j(\omega n)}{u^2 + (\omega n)^2} = \dfrac{u - j\omega n}{u^2 + \omega^2 n^2}$$

CHAPTER 12

Exponential and Logarithmic Functions

<u>Exercises 12-1,</u> page 352

1. $y = 9^x$; $x = 0.5$; $y = 9^{0.5} = \sqrt{9} = 3$

5. $3^3 = 27$ has base 3, exponent 3, and number 27

   $\log_3 27 = 3$

9. $4^{-2} = \dfrac{1}{16}$ has base 4, exponent -2, and number $\dfrac{1}{16}$

   $\log_4 \left(\dfrac{1}{16}\right) = -2$

13. $8^{1/3} = 2$ has base 8, exponent $\dfrac{1}{3}$, and number 2

    $\log_8 2 = \dfrac{1}{3}$

17. $\log_3 81 = 4$ has base 3, exponent 4, and number 81

    $81 = 3^4$

21. $\log_{25} 5 = \dfrac{1}{2}$ has base 25, exponent $\dfrac{1}{2}$, and number 5

    $5 = 25^{1/2}$

25. $\log_{10} 0.1 = -1$ has base 10, exponent -1, and number 0.1

    $0.1 = 10^{-1}$

29. $\log_4 16 = x$ has base 4, exponent x, and number 16

    $4^x = 16$, $x = 2$

33. $\log_7 y = 3$ has base 7, exponent 3, and number y

    $7^3 = y$, $y = 343$

37. $\log_b 81 = 2$ has base b, exponent 2, and number 81
     $b^2 = 81$, $b = 9$

41. $\log_{10} 10^{0.2} = x$ has base 10, exponent x, and number $10^{0.2}$
     $10^x = 10^{0.2}$, $x = 0.2$

45. $y_4 = \log x_4 = \log 64$ has base 4, number 64, and exponent y; $4^y = 64$;
     $y = 3$

49. $V = A(1.1)^t$ or $\dfrac{V}{A} = (1.1)^t$ has number of $\dfrac{V}{A}$, exponent t, and base 1.1.
     $t = \log_{1.1}\left(\dfrac{V}{A}\right)$

53. $\log_e\left(N/N_o\right) = -kt$ or $e^{-kt} = N/N_o$; $N_o e^{-kt} = N$

Exercises 12-2, page 355

1. $y = 3^x$

x	-3	-2	-1	0	1	2	3
y	$\frac{1}{27}$	$\frac{1}{9}$	$\frac{1}{3}$	1	3	9	27

5. $y = (1.65)^x$

x	-3	-2	-1	0	1	2	3
y	0.22	0.44	0.61	1	1.65	2.72	4.49

9. $y = 2(2^x) = 2^{x+1}$

x	-3	-2	-1	0	1	2	3
y	$\frac{1}{4}$	$\frac{1}{2}$	1	2	4	8	16

13. $y = \log_2 x$

x	$\frac{1}{8}$	$\frac{1}{4}$	$\frac{1}{2}$	1	2	4	8
y	-3	-2	-1	0	1	2	3

17. $y = \log_{1.65} x$ or $x = 1.65^y$

x	0.22	0.44	0.61	1	1.65	2.72	4.49
y	-3	-2	-1	0	1	2	3

21. $y = 2 \log_3 x$

  $\dfrac{y}{2} = \log_3 x$

  $x = 3^{y/2}$

x	$\frac{1}{9}$	$\frac{1}{3}$	1	3	9
y	-4	-2	0	2	4

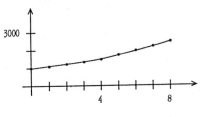

25. $V = P(1 + \sqrt[r]{n})^{nt}$; $P = \$1000$, $r = 12\%$, $n = 2$

  $= 1000(a + \sqrt[0.12]{2})^{2t} = 1000(1 + 0.06)^{2t} = 1000(1.06)^{2t}$, $0 \le t \le 8$

V	t
1000	0
1123.6	1
1262.5	2
1418.5	3

V	t
1593.8	4
1790.8	5
2012.2	6
2260.9	7
2540.4	8

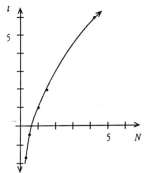

29. $t = N + \log_2 N$

N	t
0.25	-1.75
0.5	-0.5
1	1
1.4	1.9
4	6
8	11

Exercises 12-3, page 360

1. $\log_5 xy = \log_5 x + \log_5 y$ Eq. (12-7)

5. $\log_2 (a^3) = 3 \log_2 a$ Eq. (12-9)

9. $\log_5 \sqrt[4]{y} = \log_5 y^{1/4}$
  $= \dfrac{1}{4} \log_5 y$ Eq. (12-9)

13. $\log_b a + \log_b c = \log_b ac$ Eq. (12-7)

17. $\log_b x^2 - \log_b \sqrt{x} = 2 \log_b x - \dfrac{1}{2} \log_b x = (2 - \dfrac{1}{2}) \log_b x$
  $= \dfrac{3}{2} \log_b x = \log_b x^{3/2}$ Eq. (12-8)

21. $\log_2 \left(\frac{1}{32}\right) = \log_2 \left(\frac{1}{2^5}\right)$

                 $= \log_2 (2^{-5}) = -5$ Eq. (12-10)

25. $\log_7 \sqrt{7} = \log_7 7^{1/2}$

              $= \frac{1}{2}$ Eq. (12-10)

29. $\log_3 18 = \log_3 3^2 (2)$

           $= \log_3 3^2 + \log_3 2 = 2 + \log_3 2$ Eq. (12-7), Eq. (12-10)

33. $\log_3 \sqrt{6} = \log_3 6^{1/2}$

            $= \frac{1}{2}\log_3 6 = \frac{1}{2}\log_3 3(2)$

            $= \frac{1}{2}(\log_3 3 + \log_3 2) = \frac{1}{2}(1 + \log_3 2)$ Eq. (12-9), Eq. (12-7)

37. $\log_{10} 3000 = \log_{10} (3 \times 10^3)$

              $= \log_{10} 3 + \log_{10} 10^3 = \log_{10} 3 + 3$

              $= 3 + \log_{10} 3$ Eq. (12-7)

41. $\log_b y = \log_b 2 + \log_b x$

           $= \log_b 2x; \quad y = 2x$ Eq. (12-7)

45. $\log_{10} y = 2 \log_{10} 7 - 3 \log_{10} x$

           $= \log_{10} 7^2 - \log_{10} x^3 = \log_{10} \frac{49}{x^3}; \quad y = \frac{49}{x^3}$

49. $\log_2 x + \log_2 y = 1; \quad \log_2 (xy) = 1; \quad 2^1 = xy$

                                        $y = \frac{2}{x}$

53. $\log_{10} 4 = \log_{10} 2^2$

            $= 2 \log_{10} 2 = 2(0.301)$

            $= 0.602$

57. $y = 2 \log_2 x$

x	$\frac{1}{8}$	$\frac{1}{4}$	$\frac{1}{2}$	1	2	4	8
y	-6	-4	-2	0	2	4	6

$y = \log_2 x^2$

x	$\frac{1}{8}$	$\frac{1}{4}$	$\frac{1}{2}$	1	2	4	8
y	-6	-4	-2	0	2	4	6

Exercises 12-4, page 364

1. 567;   567, LOG   2.754

5. 9.24×10^6;   6, ⊞ 9.24, LOG , ⊟ 6.966

9. 73.27;   73.27, LOG 1.8649

13. COS 12.5°; 12.5, cos , log, = -0.014

17. 4.437;   4.437, INV , LOG 27,400

21. 3.30112;   3.30112, INV , LOG 2000.4

25. 0.15485;   0.15485, INV , LOG 1.4284

29. (5.98)(14.3); 5.98, log, ⊞, 14.3, log, ⊟, 1.9320, inv, log, = 85.5

33. $(47.3)(22.8)^{250}$;

22.8 log, ✕, 250, ⊟, ⊞, 47.3, log, ⊟, 341.158, ⊟, 341, ⊟,

0.158, inv, log, ⊟ 1.44

$\log (47.3)(22.8)^{250} = 341.158$

$10^{341.158} = 10^{0.158} \times 10^{341} = 1.44 \times 10^{341}$

37. $\log(1.15 \times 10^9) = 9 + \log 1.15 = 9 + 0.0607 = 9.0607$

41. $G = 10 \log (P_o/P_i)$; $P_i = 0.750$, $P_o = 25.0$

$B = 10 \log (25.0/0.750) = 10 [\log 25.0 - \log 0.750]$

$= 10 [1.398 - (-0.125)] = 10 (1.523) = 15.2$ dB

Exercises 12-5, page 368

1. $\ln 26.0 = \dfrac{\log 26.0}{\log e}$

$= \dfrac{1.4150}{0.4343} = 3.258$

5. $\ln 0.5017 = \dfrac{\log 0.5017}{\log e}$

$= \dfrac{-0.2996}{0.4343} = -0.6898$

9. $\log_7 42 = \dfrac{\log 42}{\log 7}$

$= \dfrac{1.6232}{0.8451} = 1.92$

13. $\log_{12} 122 = \dfrac{\log 122}{\log 12}$

$= \dfrac{2.0864}{1.0792} = 1.933$

17. $\ln 51.4 = 2.3026 \log 51.4$

$= 2.3026(1.7110) = 3.940$

21. ln 0.9917 = 2.3026 log 0.9917
                = 2.3026(-0.00362) = -0.00833

25. 45.17, $\boxed{\text{ln } x}$ 3.8104

29. 2.19, $\boxed{\text{INV}}$, $\boxed{\text{ln } x}$ 8.94

33. -0.7429, $\boxed{\text{inv}}$, $\boxed{\text{ln } x}$, = 0.4757

37. ln y - ln x = 1.0986

   ln$(\frac{y}{x})$ = 1.0986

   $\frac{y}{x}$ = 3; $y = 3x$

41. i = 0.1(0.750 A) = 0.075 A

   $t = \dfrac{L(\ln i - \ln I)}{R} = \dfrac{1.25(\ln 0.075 - \ln 0.75)}{7.50}$

   $\ln 0.075 = \dfrac{\log 0.075}{\log e} = \dfrac{8.8751 - 10}{0.4343} = -2.590$

   $\ln 0.75 = \dfrac{\log 0.75}{\log e} = \dfrac{9.8751 - 10}{0.4343} = 0.2876$

   $t = \dfrac{1.25(-2.590 + 0.2876)}{7.50} = \dfrac{1.25(-2.3024)}{7.50} = 0.384$ s

Exercises 12-6, page 372

1. 2^x = 16; $x = \log_2 16 = 4$

5. 3^{-x} = 0.525; $\log(3^{-x})$ = log 0.525; $-x$ log 3 = log 0.525

   $x = \dfrac{-\log 0.525}{\log 3} = \dfrac{(-0.2798)}{0.4771} = 0.587$

9. 6^{x+1} = 10; $\log(6^{x+1})$ = log 10 = 1
   $(x + 1)(\log 6)$ = 1

   $x + 1 = \dfrac{1}{\log 6} = \dfrac{1}{0.7782} = 1.285$

   x = 1.285 - 1 = 0.285

13. 0.8^x = 0.4
    log 0.8^x = log 0.4
    x(log 0.8) = log 0.4

   $x = \dfrac{\log 0.4}{\log 0.8}$

   x = 4.11

17. $3 \log_8 x = -2$; $\log_8 x^3 = -2$; $x^3 = 8^{-2} = [2^3]^{-2} = [2^{-2}]^3$

21. $\log_2 x = + \log_2 7 = \log_2 21$; $\log_2(7x) = \log_2 21$; $7x = 21$; $x = 3$

   $2^{-6} = x^3$; $\dfrac{1}{2^6} = x^3$; $x = \dfrac{1}{2^2}$; $x = \dfrac{1}{4}$

25. $\log 4x + \log x = 2$; $\log(4x \cdot x) = 2$; $\log 4x^2 = 2$; $4x^2 = 100$

   $x^2 = 25$; $x = 5$

29. $3 \ln 2 + \ln(x - 1) = \ln 24$; $\ln 2^3 + \ln(x - 1) = \ln 24$
   $\ln(2^3)(x - 1) = \ln 24$; $2^3(x - 1) = 24$; $8(x - 1) = 24$
   $8x - 8 = 24$; $8x = 32$; $x = 4$

33. $\log_5(x - 3) + \log_5 x = \log_5 4$; $\log_5(x - 3)(x) = \log_5 4$
   $x^2 - 3x - 4 = 0$; $(x - 4)(x + 1) = 0$
   $x = 4$
   $x \neq -1$ since logs are not defined on negatives.

37. $2^x = 2.68 \times 10^8$

   $\log 2^x = 8 + \log 2.68$

   $\log 2^x = 8.428$

   $x \log 2 = 8.423$

      $x = 8.428 \div 0.301 = 28.0$

41. $3.4065 = -\log (H^+)$

   $-3.4065 = \log (H^+)$

   -3.4065, $\boxed{\text{inv}}$, $\boxed{\log}$, $= 3.922 \times 10^{-4}$

45. $\ln n = -0.04t + \ln 20$; $\ln n - \ln 20 = -0.04t$
   $\ln \dfrac{n}{20} = -0.04t$; $e^{-0.04t} = \dfrac{n}{20}$; $n = 20e^{-0.04t}$

49. $2^x + 3^x = 50$
   $2^x + 3^x - 50 = 0$

   If $y = 2^x + 3^x - 50$ is graphed,
   the solutions are where $y = 0$.

x	y
2.0	-37
2.5	-29
3.0	-15
3.5	8
4.0	47

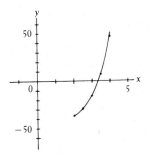

The graph has a zero at
approximately $x = 3.35$.

Exercises 12-7, page 376

1. $y = 2^x$

x	y
0	1
1	2
2	4
4	16
6	64
8	256
10	1024

5. $y = 3^{-x}$

x	y
0	1
1	$\frac{1}{3} = 0.333$
2	$\frac{1}{9} = 0.111$
3	$\frac{1}{27} = 0.037$
4	$\frac{1}{81} = 0.012$
5	$\frac{1}{243} = 0.004$

9. $y = 3x^2$

x	y
0	0
1	3
2	12
4	48
6	108
8	192
10	300

17. $y = x^2 + 2x$

x	y
1	3
3	15
5	35
7	63
10	120
20	440
30	960
40	1680

13. $y = 0.01x^4$

x	y
1	0.01
2	0.16
4	2.56
6	12.96
8	40.96
10	100.00

21. $y^2x = 1; \quad y = \sqrt{\dfrac{1}{x}}$

x	y
0.1	3.16
0.5	1.41
1	1
10	0.316
50	0.141
100	0.1

25. $s = \frac{1}{2} gt^2;\ g = 9.80$

   $s = \frac{1}{2}(9.80)t^2$

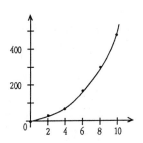

t	0	2	4	6	8	10
s	0	19.6	78.4	176.4	313.6	490

29. $g = 3.99 \times 10^{14}/r^2;\ \log g = \log 3.99 + 14 - 2 \log r = 14.6 - 2 \log r$

r	g
6.37×10^6	9.83
1.0×10^7	3.98
6.0×10^7	1.1×10^{-1}
9.0×10^7	4.9×10^{-2}
3.91×10^8	2.6×10^{-3}

33.

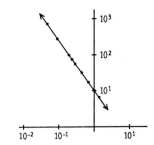

Review Exercises for Chapter 12, page 379

1. $\log_{10} x = 4$
    $x = 10^4 = 10{,}000$

5. $\log_2 64 = x$
    $2^x = 64 = 2^6$
    $x = 6$

9. $\log_x 36 = 2$
    $x^2 = 36$
    $x = 6$

13. $\log_3 2x = \log_3 2 + \log_3 x$

17. $\log_2 28 = \log_2 4 + \log_2 7$
    $= 2 + \log_2 7$

21. $\log_4 \sqrt{48} = \log_4 48^{1/2}$

$\qquad = \frac{1}{2}\log_4 48 = \frac{1}{2}\log_4 (16 \times 3)$

$\qquad = \frac{1}{2}(\log_4 16 + \log_4 3) = \frac{1}{2}(2 + \log_4 3)$

$\qquad = 1 + \frac{1}{2}\log_4 3$

25. $\log_6 y = \log_6 4 - \log_6 x$

$\qquad \log_6 y = \log_6 (\frac{4}{x})$

$\qquad\qquad y = \frac{4}{x}$

29. $\log_5 x + \log_5 y = \log_5 3 + 1$

$\qquad \log_5 x + \log_5 y - \log_5 3 = 1$

$\qquad \log_5 (\frac{xy}{3}) = 1$

$\qquad 5^1 = \frac{xy}{3}$

$\qquad xy = 15; \quad y = \frac{15}{x}$

33. $y = 0.5(5^x)$

x	y
-2	0.02
-1	0.1
0	0.5
1	2.5
2	12.5

37. $y = \log_{3.15} x$; $3.15^y = x$; $y \log 3.15 = \log x$; $y = \dfrac{\log x}{\log 3.15}$

x	0.3	0.6	1	1.8	3.2	5.6	9.9
y	-1	-0.4	0	0.5	0.9	1.5	2

41. $(13.6)(0.693) = x$
$\log x = \log 13.6 + \log 0.693$
$\log x = 1.1335 - 0.1593$
$\log x = 0.9742$
$\qquad x = 9.42$

45. $\ln 8.86 = 2.3026(\log 8.86)$
$\qquad = 2.3026(0.9474)$
$\qquad = 2.18$

49. $e^{2x} = 5$; $\log e^{2x} = \log 5$; $2x \log e = \log 5$
$\qquad x = \dfrac{\log 5}{2 \log e} = \dfrac{0.6990}{0.8685} = 0.805$

53. $\log_4 x + \log_4 6 = \log_4 12$

$\log_4 x + \log_4 6 - \log_4 12 = 0$

$\log_4 \dfrac{6x}{12} = 0; \quad 4^0 = \dfrac{6x}{12}$

$1 = \dfrac{6x}{12}; \quad 6x = 12; \quad x = 2$

57. $y = 6^x$

x	y
0	1
1	6
2	36
3	216
4	1296

61. Since $y = b^x = b^{\log_b y}$; $10^{\log_{10} 4}$ implies $y = 4$

65. $V = 1000\, e^{0.083t}; \quad e^{0.083t} = \sqrt[V]{1000}$

$\ln (e^{0.083t}) = \ln (\sqrt[V]{1000})$

$0.083t = \ln (\sqrt[V]{1000})$

$t = \dfrac{\ln (\sqrt[V]{1000})}{0.083} = 12 \ln (\sqrt[V]{1000})$

69. $N = 1000(1.5)^t$

t	N
0	1000
1	1500
2	2250
3	3375
5	7594
6	11391

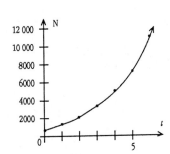

73. $C = B \log_2 (1 + R)$

$c/_B = \log_2 (1 + R)$

$2^{c/_B} = 1 + R$

$R = 2^{c/_B} - 1$

77. $R = 4500(0.750)^{2.50t}, \quad R = 2000$

$2000 = 4500(0.750)^{2.50t}$

$0.750^{2.50t} = \dfrac{2000}{4500} = \dfrac{4}{9}$

$\log (0.750)^{2.50t} = \log (4/9)$

$2.50t \log (0.750) = \log (4/9)$

$2.50t = \dfrac{\log 4 - \log 9}{\log (0.750)} = 2.819$

$t = 1.13 \text{ min.}$

CHAPTER 13

Additional Types of Equations and Systems of Equations

1. $x^2 + y^2 = 16$

   $y = \pm\sqrt{16 - x^2}$ (x restricted to values not less than -4 and not greater than 4)

x	-4	-3	-2	-1	0	1	2	3	4
y	0	$\pm\sqrt{7}$	$\pm\sqrt{12}$	$\pm\sqrt{15}$	±4	$\pm\sqrt{15}$	$\pm\sqrt{12}$	$\pm\sqrt{7}$	0

   $y = 2x$

x	-4	0	4
y	-8	0	8

   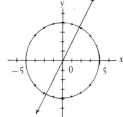

   Points of intersection are $(1.8, 3.6)$ and $(-1.8, -3.6)$. Thus, solutions are $x = 1.8$, $y = 3.6$ and $x = -1.8$, $y = -3.6$.

5. $y = x^2 - 2$

x	-3	-2	$-\sqrt{2}$	-1	0	1	$\sqrt{2}$	2	3
y	7	2	0	-1	-2	-1	0	2	7

   $4y = 12x - 17$

   $y = 3x - \dfrac{17}{4}$

x	-1	0	4
y	$-7\frac{1}{4}$	$-4\frac{1}{4}$	$7\frac{3}{4}$

   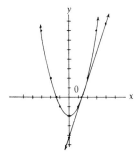

   Point of intersection is $(1.5, 0.2)$. Thus, solution is $x = 1.5$, $y = 0.2$.

9. $y = -x^2 + 4$

x	-4	-3	-2	-1	0	1	2	3	4
y	-12	-5	0	+3	+4	+3	0	-5	-12

$x^2 + y^2 = 9$

x	-3	-2	-1	0	1	2	3
y	0	±2.2	±2.8	±3	±2.8	±2.2	0

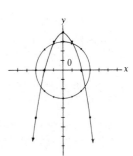

Graphs intersect at four points, (1.1,2.8),
(-1.1,2.8), (2.4,-1.8), (-2.4,-1.8).
Thus, solutions are $x = 1.1$, $y = 2.8$;
$x = -1.1$, $y = 2.8$; $x = 2.4$, $y = -1.8$;
$x = -2.4$, $y = -1.8$.

13. $2x^2 + 3y^2 = 19$

x	-3.1	-3	-2	-1	0	1	2	3.1
y	0	±0.6	±1.9	±2.4	±2.5	±2.4	±1.9	0

$x^2 + y^2 = 9$

x	-3	-2	-1	0	1	2	3
y	0	±2.2	±2.8	±3	±2.8	±2.2	0

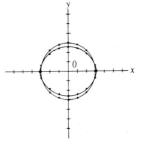

Graphs intersect at four points. Solutions are
$x = -2.8$, $y = 1.0$; $x = -2.8$, $y = -1.0$;
$x = 2.8$, $y = 1.0$; $x = 2.8$, $y = -1.0$.

17. $y = x^2$

x	-3	-2	-1	0	1	2	3
y	9	4	1	0	1	4	9

$y = \sin x$

x	$-\pi$	$-\dfrac{\pi}{2}$	0	$\dfrac{\pi}{2}$	π
y	0	-1	0	1	0

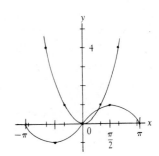

Graphs intersect at two points.
Solutions are $x = 0.0$, $y = 0.0$;
$x = 0.9$, $y = 0.8$.

21. $x^2 - y^2 = 1$

x	-2.0	-2.0	-1.0	1.0	2.0	2.0
y	1.7	-1.7	0.0	0.0	1.7	-1.7

$y = \log_2 x;\quad 2^y = x$

x	0.25	0.5	1.0	2.0	4.0	8.0
y	-2.0	-1.0	0.0	1.0	2.0	3.8

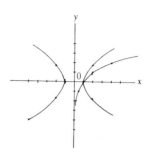

$x = 1.0, \; y = 0.0$

25. $y = 3x;\; x > 0,\; y > 0$
       $x^2 + y^2 = 5.2^2 = 27.04;\; x > 0,\; y > 0$

$y = 3x$

x	0.5	1.0	1.5	2.0	3.0
y	1.5	3.0	4.5	6.0	9.0

$x^2 + y^2 = 27.04$

x	1.0	1.5	2.0	2.5	3.0
y	5.1	5.0	4.8	4.6	4.2

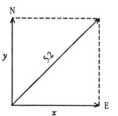 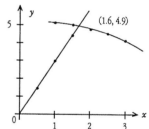

Exercises 13-2, page 391

1. $y = x + 1$ Substituting $x + 1$ for y gives $x + 1 = x^2 + 1$ for the
      $y = x^2 + 1$ second equation $x^2 - x = 0$
                                                  $x(x - 1) = 0$
      Check: $1 = 0^2 + 1$ $x = 0;\; x = 1$
               $2 = 1^2 + 1$ $y = 0 + 1 = 1;\; y = 1 + 1 = 2$
      Thus, solutions are $x = 0,\; y = 1;\; x = 1,\; y = 2$.

5. $x + y = 1;\; y = 1 - x$ Substitute in second equation
      $x^2 - y^2 = 1$ $x^2 - (1 - x)^2 = 1$
                                         $x^2 - (1 - 2x + x^2) = 1$
                                                  $2x - 1 = 1$
                                                     $2x = 2$
                                                      $x = 1$
                                  $y = 1 - x = 1 - 1 = 0$
      The solution is $x = 1,\; y = 0$.

9. $xy = 1$
    $x + y = 2$

    From the first equation, $y = \dfrac{1}{x}$.

    Substitute into second equation $x + \dfrac{1}{x} = 2$; $x^2 + 1 = 2x$; $x^2 - 2x + 1 = 0$

    $(x - 1)(x - 1) = 0$; $x = 1$

    $1 + y = 2$; $y = 1$

13. $y = x^2$ Substitute x^2 for y in second equation
     $y = 3x^2 - 8$ $x^2 = 3x^2 - 8$
                       $2x^2 - 8 = 0$
                         $x^2 - 4 = 0$
                              $x = \pm 2$
                              $x = 2$, $y = 2^2 = 4$; $x = -2$, $y = (-2)^2 = 4$

     Thus, solutions are $x = 2$, $y = 4$; $x = -2$, $y = 4$.

     Check: $3(2)^2 - 8 = 12 - 8 = 4$
             $3(-2)^2 - 8 = 12 - 8 = 4$

17. $x^2 - 1 = y$ Substitute $x^2 - 1$ for y in the second equation
     $x^2 - 2y^2 = 1$ $x^2 - 2(x^2 - 1)^2 = 1$
                     $x^2 - 2x^4 + 4x^2 - 2 - 1 = 0$
                            $2x^4 - 5x^2 + 3 = 0$
                            $(2x^2 - 3)(x^2 - 1) = 0$
                       $(2x^2 - 3)(x + 1)(x - 1) = 0$
                       $2x^2 = 3$, $x^2 = \dfrac{3}{2}$, $x = \pm\dfrac{\sqrt{6}}{2}$; $x = -1$, $x = 1$

     If $x = \dfrac{\sqrt{6}}{2}$, $y = \left(\dfrac{\sqrt{6}}{2}\right)^2 - 1 = \dfrac{1}{2}$

     If $x = -\dfrac{\sqrt{6}}{2}$, $y = \left(-\dfrac{\sqrt{6}}{2}\right)^2 - 1 = \dfrac{1}{2}$

     If $x = -1$, $y = (-1)^2 - 1 = 0$

     If $x = 1$, $y = 1^2 - 1 = 0$

     Solutions are $x = \dfrac{1}{2}\sqrt{6}$, $y = \dfrac{1}{2}$; $x = -\dfrac{1}{2}\sqrt{6}$, $y = \dfrac{1}{2}$; $x = -1$, $y = 0$

     $x = 1$, $y = 0$.

21. $y^2 - 2x^2 = 6$ Multiply first equation by (-3) and add to second
     $3y^2 + 5x^2 = 20$ equation

$$11x^2 = 2$$

$$x^2 = \frac{2}{11}$$

$$x = \pm\sqrt{\frac{2}{11}} = \pm\frac{\sqrt{22}}{11} = \frac{1}{11}\sqrt{22}$$

$$y^2 - 2\left(\pm\frac{\sqrt{22}}{11}\right)^2 = 6$$

$$y^2 - 2\left(\frac{22}{121}\right) = 6$$

$$y^2 = 6 + \frac{44}{121} = 6\frac{4}{11} = \frac{70}{11}$$

$$y = \pm\sqrt{\frac{70}{11}} = \pm\frac{\sqrt{770}}{11} = \pm\frac{1}{11}\sqrt{770}$$

Thus, solutions are as follow:

$x = \frac{1}{11}\sqrt{22}$, $y = \frac{1}{11}\sqrt{770}$; $x = -\frac{1}{11}\sqrt{22}$, $y = \frac{1}{11}\sqrt{770}$; $x = \frac{1}{11}\sqrt{22}$, $y = -\frac{1}{11}\sqrt{770}$;

$x = -\frac{1}{11}\sqrt{22}$, $y = -\frac{1}{11}\sqrt{770}$

Check: $3\left(\frac{\sqrt{770}}{11}\right)^2 + 5\left(\frac{\sqrt{22}}{11}\right)^2 = 3\left(\frac{770}{121}\right) + 5\left(\frac{22}{121}\right)$

$$= 3\left(\frac{70}{11}\right) + 5\left(\frac{2}{11}\right) = \frac{210}{11} + \frac{10}{11} = \frac{220}{11} = 20$$

Similar checks for other solutions

25.
$$v_1^2 + 4v_2^2 = 41; v_2 > 0$$
$$2v_1 + 8v_2 = 12; 2v_1 = 12 - 8v_2; v_1 = 6 - 4v_2$$

Substituting in Eq.1, $(6 - 4v_2)^2 + 4v_2^2 = 41$

$$36 - 48v_2 + 16v_2^2 + 4v_2^2 = 41$$
$$20v_2^2 - 48v_2 - 5 = 0$$
$$(2v_2 - 5)(10v_2 + 1) = 0$$
$$2v_2 = 5 \qquad 10v_2 = -1 \text{ (not valid)}$$
$$v_2 = 2.5; v_1 = 6 - 4(2.5) = -4.0$$

29.

$x + y + 40.0 = 90.0; \ y = 90.0 - 40.0 - x = 50.0 - x$

$\sqrt{x^2 + y^2} = 40.0, \text{ or } x^2 + y^2 = 1600.0$

$$x^2 + (50.0 - x)^2 = 1600.0$$

$$x^2 + 2500.0 - 100.0x + x^2 - 1600.0 = 0$$

$$2x^2 - 100.0x + 900.0 = 0$$

$$x^2 - 50.0x + 450.0 = 0$$

$$x = \frac{-(-50.0) \pm \sqrt{(-50.0)^2 - 4(1)(450.0)}}{2(1)}$$

$$= \frac{50.0 \pm \sqrt{700}}{2}$$

$$x = \frac{50.0 \pm 26.4}{2}; \quad x = 38.2, \ x = 11.8$$

Exercises 13-3, page 395

1. $x^4 - 13x^2 + 36 = 0$; let $y = x^2$ and $y^2 - 13y + 36 = 0$

$$(y - 4)(y - 9) = 0$$

$$y = 4 \qquad y = 9$$

$$x^2 = 4 \qquad x^2 = 9$$

$$x = \pm 2 \qquad x = \pm 3$$

Checking: $(\pm 2)^4 - 13(\pm 2)^2 + 36 = 16 - 52 + 36 = 0$

$(\pm 3)^4 - 13(\pm 3)^2 + 36 = 81 - 117 + 36 = 0$

5. $x^{-2} - 2x^{-1} - 8 = 0$; let $y = x^{-1}$ and $y^2 - 2y - 8 = 0$

$$(y + 2)(y - 4) = 0$$

$$y = -2 \qquad y = 4$$

$$x^{-1} = -2 \qquad x^{-1} = 4$$

$$x = -\frac{1}{2} \qquad x = \frac{1}{4}$$

Checking: $(-\frac{1}{2})^{-2} - 2(-\frac{1}{2})^{-1} - 8 = 4 - 2(-2) - 8 = 4 + 4 - 8 = 0$

$(\frac{1}{4})^{-2} - 2(\frac{1}{4})^{-1} - 8 = 16 - 2(4) - 8 = 16 - 8 - 8 = 0$

9. $2x - 7\sqrt{x} + 5 = 0$; let $y = \sqrt{x}$ and $x = y^2$

$$2y^2 - 7y + 5 = 0$$

$$(2y - 5)(y - 1) = 0$$

$$2y - 5 = 0 \qquad y - 1 = 0$$

$$y = \frac{5}{2} \qquad y = 1$$

$$\sqrt{x} = \frac{5}{2}; \ x = \frac{25}{4} \qquad \sqrt{x} = 1$$

$$x = 1$$

13. $x^{2/3} - 2x^{1/3} - 15 = 0$; let $y = x^{1/3}$ and $y^2 - 2y - 15 = 0$

$(y - 5)(y + 3) = 0$

$y = 5$ $y = -3$

$x^{1/3} = 5$ $x^{1/3} = -3$

$x = 125$ $x = -27$

Checking: $(125)^{2/3} - 2(125)^{1/3} - 15 = 25 - 2(5) - 15 = 25 - 10 - 15 = 0$

$(-27)^{2/3} - 2(-27)^{1/3} - 15 = 9 - 2(-3) - 15 = 9 + 6 - 15 = 0$

17. $(x - 1) - \sqrt{x - 1} - 2 = 0$; let $y = \sqrt{x - 1}$ and $y^2 - y - 2 = 0$

$(y - 2)(y + 1) = 0$

$y = 2$ $y = -1$

$\sqrt{x - 1} = 2$ $\sqrt{x - 1} = -1$

$x - 1 = 4$ $x - 1 = 1$

$x = 5$ $x = 1 + 1 = 2$

Checking: $5 - 1 - \sqrt{5 - 1} - 2 = 4 - \sqrt{4} - 2 = 4 - 2 - 2 = 0$

$2 - 1 - \sqrt{2 - 1} - 2 = 1 - \sqrt{1} - 2 = 1 - 1 - 2 \neq 0$

$x = 2$ is extraneous.   Thus, the only solution is $x = 5$.

21. $x - 3\sqrt{x - 2} = 6$; let $y = \sqrt{x - 2}$; $y^2 = x - 2$; $x = y^2 + 2$
Substituting, $y^2 + 2 - 3y = 6$; $y^2 - 3y - 4 = 0$
$(y - 4)(y + 1) = 0$; $y = 4$ and $y = -1$. Since $y = \sqrt{x - 2}$, we note that
y cannot be negative, so that $y = -1$ cannot lead to a solution. For
$y = 4$, $4 = \sqrt{x - 2}$; $16 = x - 2$; $x = 18$. This satisfies the original
equation. Thus $x = 18$ is the only solution.

25. $R_T^{-1} = R_1^{-1} + R_2^{-1}$; $R_T = 1.00$, $R_2 = \sqrt{R_1}$

$R_T^{-1} = R_1^{-1} + \sqrt{R_1}^{-1} = R_1^{-1} + \left[(R_1)^{1/2}\right]^{-1} = R_1^{-1} + R^{-1/2}$

$1.00 = R_1^{-1} + R_1^{-1/2}$; Let $y = R_1^{-1/2}$; $1.00 = y^2 + y$; $y^2 + y - 1.00 = 0$

$y = \dfrac{-1 \pm \sqrt{1 - 4(1)(-1.00)}}{2} = \dfrac{-1 \pm \sqrt{5}}{2}$

$R_1^{-1/2} = \dfrac{-1 \pm \sqrt{5}}{2}$ (the negative answer is not valid)

$R_1^{1/2} = \dfrac{2}{-1 + \sqrt{5}}$; $R_1 = \left[\dfrac{2}{-1 + \sqrt{5}}\right]^2 = 2.62$; $R_2 = \sqrt{R_1} = 1.62$

Exercises 13-4, page 398

1. $\sqrt{x - 8} = 2$ Squaring both sides gives $x - 8 = 4$, or $x = 12$
    Checking: $\sqrt{12 - 8} = \sqrt{4} = 2$

5. $\sqrt{3x + 2} = 3x$ Squaring both sides gives $3x + 2 = 9x^2$
    $9x^2 - 3x - 2 = 0$ Use the quadratic formula to solve:

$$x = \frac{-(-3) \pm \sqrt{(-3)^2 - 4(9)(-2)}}{2(9)} = \frac{3 \pm \sqrt{81}}{18} = \frac{3 \pm 9}{18}$$

$x = \frac{12}{18} = \frac{2}{3}$ and $x = \frac{-6}{18} = -\frac{1}{3}$

The solution is $x = \frac{2}{3}$ as $x = -\frac{1}{3}$ does not check.

9. $2\sqrt{3 - x} - x = 5;\ 2\sqrt{3 - x} = x + 5;\ 4(3 - x) = (x + 5)^2$

              $12 - 4x = x^2 + 10x + 25$

      $x^2 + 14x + 13 = 0$

  $(x + 13)(x + 1) = 0;\ x = -1$ since $x = -13$ does not check.

13. $\sqrt{x} + 12 = x;\ \sqrt{x} = x - 12;\ x = (x - 12)^2 = x^2 - 24x + 144$

      $x^2 - 25x + 144 = 0$

    $(x - 16)(x - 9) = 0;\ x = 16$ since $x = 9$ doesn't check

17. $\sqrt{x + 4} + 8 = x;\ \sqrt{x + 4} = x - 8;\ x + 4 = (x - 8)^2 = x^2 - 16x + 64$

      $x^2 - 17x + 60 = 0$

  $(x - 5)(x - 12) = 0;\ x = 12$ since $x = 5$ does not check.

21. $3\sqrt{1 - 2t} + 1 = 2t;\ e\sqrt{1 - 2t} = 2t - 1;\ 9(1 - 2t) = (2t - 1)^2$

     $4t^2 - 22t + 10 = 0$

     $2t^2 - 11t + 5 = 0$

  $(2t - 1)(t - 5) = 0;\ t = \frac{1}{2}$ since $t = 5$ does not check.

25. $\sqrt{5x + 1} - 1 = 3\sqrt{x}$ Squaring both sides gives
                    $(5x + 1) - 2\sqrt{5x + 1} + 1 = 9x$
                    $-4x + 2 = 2\sqrt{5x + 1}$
                    Squaring both sides gives
                    $16x^2 - 16x + 4 = 4(5x + 1)$
                    $16x^2 - 36x = 0$
                    $x(16x - 36) = 0$
                    $x = 0;\ x = \frac{36}{16} = \frac{9}{4}$

Checking: $\sqrt{5(0) + 1} - 1 = 1 - 1 = 0 = 3\sqrt{0};\ x = 0$ is a solution.

                $\sqrt{5(\frac{9}{4}) + 1} - 1 = \sqrt{\frac{49}{4}} - 1 = \frac{7}{2} - 1 = \frac{5}{2} \neq 3\sqrt{\frac{9}{4}}$ or $\frac{9}{2};$

          $x = \frac{9}{4}$ is extraneous.

29. $\sqrt[3]{2x - 1} = \sqrt[3]{x + 5}$ Cubing both sides gives $2x - 1 = x + 5$ or $x = 6$
     Checking: $\sqrt[3]{2(6) - 1} = \sqrt[3]{12 - 1} = \sqrt[3]{11} = \sqrt[3]{6 + 5};\ x = 6$ is a solution.

33. $f = \dfrac{1}{2\pi\sqrt{LC}}$; $2\pi f \sqrt{LC} = 1$

$\sqrt{LC} = \dfrac{1}{2\pi f}$; $LC = \left[\dfrac{1}{2\pi f}\right]^2 = \dfrac{1}{4\pi^2 f^2}$

$$L = \dfrac{1}{4\pi^2 f^2 C}$$

37. $x + (x + 1) + (\sqrt{2x + 1}) = 6.0$

$2x + 1 + \sqrt{2x + 1} = 6.0$

$\sqrt{2x + 1} = 6.0 - 2x - 1 = 5.0 - 2x$

$2x + 1 = (5.0 - 2x)^2 = 25.0 - 20.0x + 4x^2$

$4x^2 - 22x + 24 = 0$

$2x^2 - 11x + 12 = 0$

$x = \dfrac{-(-11) \pm \sqrt{(-11)^2 - 4(2)(12)}}{2(2)} = \dfrac{11 \pm \sqrt{25}}{4} = 4,\ 1.5$

$x = 1.5$ ft since $x = 4$ does not check.

$x + 1 = 2.5$ ft; $\sqrt{2x + 1} = \sqrt{4} = 2$ ft

Review Exercises for Chapter 13, page 399

1. $x + 2y = 6$

x	-2	-1	0	1	2
y	4	3.5	3	2.5	2

$y = 4x^2$

x	-1	-0.5	0	0.5	1
y	4	1	0	1	4

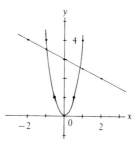

Solutions are $x = -0.9$, $y = 3.5$; $x = 0.8$, $y = 2.6$.

5. $y = x^2 + 1$

x	-2	-1	0	1	2
y	5	2	1	2	5

$2x^2 + y^2 = 4$

x	$-\sqrt{2}$	-1	0	1	$\sqrt{2}$
y	0	$\pm\sqrt{2}$	± 2	$\pm\sqrt{2}$	0

(using 1.4 for $\sqrt{2}$)

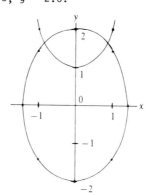

Solutions are $x = -0.8$, $y = 1.6$;
$x = 0.8$, $y = 1.6$.

9. $y = x^2 - 2x$

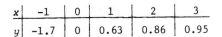

x	-2	-1	0	1	2	3	4
y	8	3	0	-1	0	3	8

$y = 1 - e^{-x}$

x	-1	0	1	2	3
y	-1.7	0	0.63	0.86	0.95

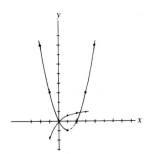

From the graph, the solutions are:
$x = 0$, $y = 0$ and $x = 2.4$, $y = 0.9$.

13. $2y = x^2$
$x^2 + y^2 = 3$
Substitute $2y$ for x^2 in second equation
$2y + y^2 = 3$; $y^2 + 2y - 3 = 0$; $(y + 3)(y - 1) = 0$
$y = -3$, $y = 1$
Substitute to determine x values;

$2(-3) = x^2$ $2(1) = x^2$
 $-6 = x^2$ has no solution $x = \pm\sqrt{2}$
Solutions are $(\sqrt{2},1)$ and $(-\sqrt{2},1)$.

17. $4x^2 - 7y^2 = 21$ Multiplying the second equation by -4 and adding to
$x^2 + 2y^2 = 99$ the first equation gives $-15y^2 = -375$
$\qquad\qquad\qquad\quad y^2 = 25$
$\qquad\qquad\qquad\quad y = \pm 5$
$\qquad\qquad\qquad\quad x^2 = 99 - 2y^2 = 99 - 2(25) = 49$
$\qquad\qquad\qquad\quad x = \pm 7$

Check: $4(49) - 7(25) = 21$
$x = 7$, $y = 5$; $x = -7$, $y = 5$; $x = 7$, $y = -5$; $x = -7$, $y = -5$ are all
solutions.

21. $x^4 - 20x^2 + 64 = 0$; let $y = x^2$ and $y^2 - 20y + 64 = 0$
$\qquad\qquad\qquad\qquad (y - 16)(y - 4) = 0$
$\qquad\qquad\qquad\qquad\quad y = 16 \qquad y = 4$
$\qquad\qquad\qquad\qquad\quad x^2 = 16 \qquad x^2 = 4$
$\qquad\qquad\qquad\qquad\quad x = \pm 4 \qquad x = \pm 2$

Check: $(\pm 4)^4 - 20(\pm 4)^2 + 64 = 256 - 20(16) + 64 = 0$
$\qquad\quad (\pm 2)^4 - 20(\pm 2)^2 + 64 = 16 - 20(4) + 64 = 0$
$x = -4$, $x = -2$, $x = 2$, $x = 4$ are solutions.

25. $x^{-2} + 4x^{-1} - 21 = 0$; let $y = x^{-1}$ and $y^2 + 4y - 21 = 0$

$$(y + 7)(y - 3) = 0$$
$$y = -7 \qquad y = 3$$
$$x^{-1} = -7 \qquad x^{-1} = 3$$
$$x = -\frac{1}{7} \qquad x = \frac{1}{3}$$

Check: $(-\frac{1}{7})^{-2} + 4(-\frac{1}{7})^{-1} - 21 = 49 + 4(-7) - 21 = 49 - 28 - 21 = 0$

$x = -\frac{1}{7}$ is a solution.

$(\frac{1}{3})^{-2} + 4(\frac{1}{3})^{-1} - 21 = 9 + 4(3) - 21 = 9 + 12 - 21 = 0$

$x = \frac{1}{3}$ is a solution.

29. $(\dfrac{1}{x + 1})^2 - \dfrac{1}{x + 1} = 2$

Let $y = \dfrac{1}{x + 1}$; $y^2 - y = 2$; $y^2 - y - 2 = 0$

$(y - 2)(y + 1) = 0$; $y = 2$ and $y = -1$

$2 = \dfrac{1}{x + 1}$; $2x + 2 = 1$; $2x = -1$; $x = -\dfrac{1}{2}$

$-1 = \dfrac{1}{x + 1}$; $-x - 1 = 1$; $-x = 2$; $x = -2$

Both solutions, $x = -\dfrac{1}{2}$ and $x = -2$, check in the equation.

33. $\sqrt{5x - 4} = x$; $(\sqrt{5x - 4})^2 = x^2$; $5x - 4 = x^2$

$x^2 - 5x + 4 = 0$; $(x - 4)(x - 1) = 0$; $x = 4$ and $x = 1$

Both solutions check.

37. $\sqrt{x + 1} + \sqrt{x} = 2$ Squaring both sides gives $x + 1 = 4 - 4\sqrt{x} + x$, or $4\sqrt{x} = 3$

Squaring both sides gives $16x = 9$, or $x = \dfrac{9}{16}$

Check: $\sqrt{\dfrac{9}{16} + 1} + \sqrt{\dfrac{9}{16}} = \sqrt{\dfrac{25}{16}} + \sqrt{\dfrac{9}{16}} = \dfrac{5}{4} + \dfrac{3}{4} = \dfrac{8}{4} = 2$

$x = \dfrac{9}{16}$ is a solution.

41.
$$L = \frac{h}{2\pi} \sqrt{\ell (\ell + 1)} \qquad \text{Let } a = L^2 \left[\frac{4\pi^2}{h^2} \right]$$

$$L \left(\frac{2\pi}{h} \right) = \sqrt{\ell (\ell + 1)} \qquad \text{Then } \ell^2 + \ell = a$$

$$\left[L \left(\frac{2\pi}{h} \right) \right]^2 = \ell (\ell + 1) \qquad \ell^2 + \ell - a = 0$$

$$L^2 \left[\frac{4\pi^2}{h^2} \right] = \ell^2 + \ell \qquad \ell = \frac{-1 \pm \sqrt{1^2 - 4(1)(-a)}}{2}$$

$$\ell = \frac{-1 \pm \sqrt{1 + 4a}}{2}$$

$$\ell = \frac{-1 \pm \sqrt{1 + 4 \left[L^2 \left[\frac{4\pi^2}{h^2} \right] \right]}}{2}$$

$$\ell = -\frac{1}{2} \pm \frac{1}{2} \sqrt{1 + (16L^2 \pi^2)/h^2}$$

$$= \frac{1}{2} \left(-1 + \sqrt{1 + 16\pi^2 L^2/h^2} \right)$$

45.
$$16t_1^2 + 16t_2^2 = 45; \quad t_2 = 2t_1$$
$$16t_1^2 + 16(2t_1)^2 = 45$$
$$16t_1^2 + 64t_1^2 = 45$$
$$80t_1^2 = 45$$
$$t_1^2 = \frac{45}{80} = 0.56$$
$$t_1 = 0.75, \quad t_2 = 2t = 1.5$$

45M.
$$490t_1^2 + 490t_2^2 = 392; \quad t_2 = 2t_1$$
$$490t_1^2 + 490(2t_1)^2 = 392$$
$$490t_1^2 + 1960 \, t_1^2 = 392$$
$$2450 \, t_1^2 = 392$$
$$t_1^2 = 0.16; \quad t_1 = 0.4; \quad t_2 = 2t_1 = 0.8$$

49.
$$Z = \sqrt{R^2 + X^2}; \; Z = 2X^2; \; R = 0.800$$

$$2X^2 = \sqrt{R^2 + X^2}$$

$$4X^4 = R^2 + X^2 = 0.640 + X^2$$

$$4X^4 - X^2 - 0.640 = 0$$

$$X^2 = \frac{-(-1 \pm \sqrt{(-1)^2 - 4(4)(-0.640)}}{2(4)} = \frac{1 \pm \sqrt{1 + 10.24}}{8} = 0.544$$

$$X = \sqrt{0.544} = 0.738; \; Z = 2X^2 = 1.09$$

53.
$$2x + \sqrt{x + 20} = 12.0$$

$$2x - 12.0 = \sqrt{x + 20}$$

$$4x^2 - 48.0x + 144 = x + 20$$

$$4x^2 - 49x + 124 = 0$$

$$x = \frac{-(-49) \pm \sqrt{(-49)^2 - 4(4)(124)}}{2(4)} = \frac{49 \pm \sqrt{417}}{8} = 8.67, \; 3.57$$

$$x = 3.57 \text{ since } x = 8.67 \text{ does not check.}$$

CHAPTER 14

Equations of Higher Degree

Exercises 14-1, page 403

1.
$$
\begin{array}{r}
x^2 + 3x + 2 \\
x - 1{\overline{\smash{\big)}\,x^3 + 2x^2 - x - 2}} \\
\underline{x^3 - x^2} \\
3x^2 - x \\
\underline{3x^2 - 3x} \\
2x - 2 \\
\underline{2x - 2} \\
0
\end{array}
$$

$f(r) = R;\ r = 1$
$f(r) = 1^3 + 2(1^2) - 1 - 2$
$\quad = 1 + 2 - 1 - 2 = 0$
Therefore $R = 0$.

5.
$$
\begin{array}{r}
2x^4 - 4x^3 + 8x^2 - 17x + 42 \\
x + 2{\overline{\smash{\big)}\,2x^5 + 0x^4 + 0x^3 - x^2 + 8x + 44}} \\
\underline{2x^5 + 4x^4} \\
- 4x^4 + 0x^3 \\
\underline{- 4x^4 - 8x^3} \\
8x^3 - x^2 \\
\underline{8x^3 + 16x^2} \\
- 17x^2 + 8x \\
\underline{- 17x^2 - 34x} \\
42x + 44 \\
\underline{42x + 84} \\
-40
\end{array}
$$

$f(r) = R;\ r = -2;\ f(-2)$
$= 2(-2)^5 - (-2)^2 + 8(-2) + 44$
$= -64 - 4 - 16 + 44$
$= -40$

9. $(x^3 + 2x^2 - 3x + 4) \div (x + 1);\ f(r) = R;\ r = -1$
$f(-1) = (-1)^3 + 2(-1)^2 - 3(-1) + 4$
$\qquad = -1 + 2 + 3 + 4 = 8$

13. $(2x^4 - 7x^3 - x^2 + 8) \div (x - 3);\ f(r) = R;\ r = 3$
$f(3) = 2(3^4) - 7(3^3) - (3^2) + 8$
$\qquad = 162 - 189 - 9 + 8 = -28$

17. $x^2 - 2x - 3,\ x - 3;\ r = 3$
$f(3) = 3^2 - 2(3) - 3 = 9 - 6 - 3 = 0$
$x - 3$ is a factor since $f(r) = R = 0$.

161

21. $5x^3 - 3x^2 + 4$, $x - 2$; $r = 2$
     $f(2) = 5(2^3) - 3(2^2) + 4 = 40 - 12 + 4 = 32$
     $x - 2$ is not a factor since $f(r) = R \neq 0$.

25. $f(x) = x^3 - 2x^2 - 9x + 18$
     Is 2 a zero? In order for 2 to be a zero, $x - 2$ must be a factor and
     $f(r) = R = 0$; $f(2) = 2^3 - 2(2)^2 - 9(2) + 18 = 8 - 8 - 18 + 18 = 0$.
     Therefore 2 is a zero of the function.

29.
$$
\begin{array}{r}
2x^2 + 5x + 2 \\
2x - 1\overline{)4x^3 + 8x^2 - x - 2} \\
\underline{4x^3 - 2x^2} \\
10x^2 - x \\
\underline{10x^2 - 5x} \\
4x - 2 \\
\underline{4x - 2} \\
0
\end{array}
$$

We may not conclude that $f(1) = 0$
since $2x - 1$ is not in the form $x - r$.

Exercises 14-2, page 408

1. $(x^3 + 2x^2 - x - 2) \div (x - 1)$
    $= 1x^2 + 3x + 2 = x^2 + 3x + 2$

1	2	-1	-2	$\lfloor 1$
	1	3	2	
1	3	2	0	

5. $(2x^5 - x^2 + 8x + 44) \div (x + 2)$
    $= 2x^4 - 4x^3 + 8x^2 - 17x + 42$
    $R = -40$

2	0	0	-1	8	44	$\lfloor -2$
	-4	8	-16	34	-84	
2	-4	8	-17	42	-40	

9. $(x^3 + 2x^2 - 3x + 4) \div (x + 1)$
    $= x^2 + x - 4$
    $R = 8$

1	2	-3	4	$\lfloor -1$
	-1	-1	4	
1	1	-4	8	

13. $(2x^4 - 7x^3 - x^2 + 8) \div (x - 3)$
     $= 2x^3 - x^2 - 4x - 12$
     $R = -28$

2	-7	-1	0	8	$\lfloor 3$
	6	-3	-12	-36	
2	-1	-4	-12	-28	

17. $(x^6 + 2x^2 - 6) \div (x - 2)$
     $= x^5 + 2x^4 + 4x^3 + 8x^2 + 18x + 36$
     $R = 66$

1	0	0	0	2	0	-6	$\lfloor 2$
	2	4	8	16	36	72	
1	2	4	8	18	36	66	

21. $x^3 + x^2 - x + 2$; $x + 2$
     $x + 2$ is a factor, since $R = 0$.

1	1	-1	2	$\lfloor -2$
	-2	2	-2	
1	-1	1	0	

25. $2x^5 - x^3 + 3x^2 - 4$; $x + 1$
     $x + 1$ is not a factor since
     $R = -2$. R must equal 0 for
     $x + 1$ to be a factor.

2	0	-1	3	0	-4	$\lfloor -1$
	-2	2	-1	-2	2	
2	-2	1	2	-2	-2	

29. $2x^4 - x^3 + 2x^2 - 3x + 1$;

    $2x - 1 = 2(x - \frac{1}{2})$

    $2x - 1$ is a factor since $x - \frac{1}{2}$

    is a factor ($R = 0$), and because
    the quotient is $2x^3 + 2x - 2$,
    which has 2 as a factor.

2	-1	+2	-3	+1	$\lfloor \frac{1}{2}$
	1	0	1	-1	
2	0	2	-2	0	

33. 7 is a zero if $x - 7$ is a
    factor and $x - 7$ is a factor
    because $R = 0$. Therefore, by
    the factor theorem, $f(7) = 0$
    and 7 is a zero.

1	-5	-15	+5	+14	$\lfloor 7$
	7	14	-7	-14	
1	+2	-1	-2	0	

Exercises 14-3, page 413

1. $x^3 + 2x^2 - x - 2 = 0$ $(r_1 = 1)$

1	2	-1	-2	$\lfloor 1$
	1	3	2	
1	3	2	0	

   $(x - 1)(1x^2 + 3x + 2)$
   $= (x - 1)(x + 2)(x + 1)$
   $r_1 = 1, r_2 = -2, r_3 = -1$

5. $2x^3 + 11x^2 + 20x + 12 = 0$ $(r_1 = -\frac{3}{2})$

2	11	20	12	$\lfloor -\frac{3}{2}$
	-3	-12	-12	
2	8	8	0	

   $(x + \frac{3}{2})(2x^2 + 8x + 8)$

   $= 2(x + \frac{3}{2})(x^2 + 4x + 4)$

   $= 2(x + \frac{3}{2})(x + 2)(x + 2)$

   $r_1 = -\frac{3}{2}, r_2 = r_3 = -2$

9. $x^4 + x^3 - 2x^2 + 4x - 24 = 0$
   $(r_1 = 2, r_2 = -3)$

1	1	-2	4	-24	$\lfloor 2$
	2	6	8	24	
1	3	4	12	0	

   or $(x - 2)(1x^3 + 3x^2 + 4x + 12)$

1	3	4	12	$\lfloor -3$
	-3	0	-12	
1	0	4	0	

   or $(x - 2)(x + 3)(1x^2 + 4)$
   $= (x - 2)(x + 3)(x^2 - 4j^2)$
   $r_1 = 2, r_2 = -3, r_3 = 2j, r_4 = -2j$

13. $6x^4 + 5x^3 - 15x^2 + 4 = 0$

$(r_1 = -\frac{1}{2},\ \ r_2 = \frac{2}{3})$

$(x + \frac{1}{2})(6x^3 + 2x^2 - 16x + 8)$

$= 2(x + \frac{1}{2})(3x^3 + x^2 - 8x + 4)$

6	5	-15	0	4	$\lfloor -\frac{1}{2}$
	-3	-1	8	-4	
6	2	-16	8	0	

3	1	-8	4	$\lfloor \frac{2}{3}$
	2	2	-4	
3	3	-6	0	

$2(x + \frac{1}{2})(x - \frac{2}{3})(3x^2 + 3x - 6)$

$= (3)(2)(x + \frac{1}{2})(x - \frac{2}{3})(x^2 + x - 2)$

$= 3(2)(x + \frac{1}{2})(x - \frac{2}{3})(x + 2)(x - 1)$

$r_1 = -\frac{1}{2},\ r_2 = \frac{2}{3},\ r_3 = -2,\ r_4 = 1$

17. $2x^5 + 11x^4 + 16x^3 - 8x^2 - 32x - 16 = 0$
(-2 is a triple root)

2	11	16	-8	-32	-16	$\lfloor -2$
	-4	-14	-4	24	16	
2	7	2	-12	-8	0	

2	7	2	-12	-8	$\lfloor -2$
	-4	-6	8	8	
2	3	-4	-4	0	

2	3	-4	-4	$\lfloor -2$
	-4	2	4	
2	-1	-2	0	

Thus, $2x^5 + 11x^4 + 16x^3 - 8x^2 - 32x - 16$
$= (x + 2)(x + 2)(x + 2)(2x^2 - x - 2)$
The roots from the last factor are found by the quadratic formula:

$x = \dfrac{-(-1) \pm \sqrt{1 + 16}}{4} = \dfrac{1 \pm \sqrt{17}}{4} = \dfrac{1}{4}(1 \pm \sqrt{17})$

Therefore, the roots are $-2,\ -2,\ -2,\ \frac{1}{4}(1 + \sqrt{17}),\ \frac{1}{4}(1 - \sqrt{17})$.

21. $x^5 - 3x^4 - x + 3 = 0$ $(r_1 = 3,\ r_2 = j)$
Since j is a root, $-j$ is also a root.

1	-3	0	0	-1	3	$\lfloor 3$
	3	0	0	0	-3	
1	0	0	0	-1	0	

1	0	0	0	-1	$\lfloor j$
	j	-1	-j	1	
1	j	-1	-j	0	

1	j	-1	-j	$\lfloor -j$
	-j	0	j	
1	0	-1	0	

The quadratic factor $x^2 - 1$ factors into $(x + 1)(x - 1)$. Therefore, the roots of the function are $3,\ j,\ -j,\ -1,\ 1$.

Exercises 14-4, page 419

1. $x^3 + 2x^2 - x - 2 = 0$; there are three roots since the highest degree of
    a term is 3. $f(x) = x^3 + 2x^2 - x - 2 = 0$; there is one positive root
    since there is one sign change. $f(-x) = (-x)^3 + 2(-x)^2 - (-x) - 2$
    $= -x^3 + 2x^2 + x - 2$. There are no more than two negative roots. The
    possible rational roots have numerators which are factors of the last
    term, -2, which are ±1, ±2, and denominators which are factors of the
    coefficient of the first term, 1, which are ±1. Therefore possible
    rational roots are ±1, ±2. Since there is only one positive root, we
    shall look for this first. Trying 1, we have

    $$\begin{array}{rrrr|r}
    1 & 2 & -1 & -2 & \underline{1} \\
      & 1 & 3 & 2 & \\
    \hline
    1 & 3 & 2 & 0 &
    \end{array}$$

    Hence 1 is a root, and the remaining factor is $x^2 + 3x + 2$
    $= (x + 2)(x + 1)$. The remaining roots are -2, -1.

5. $2x^3 - 5x^2 - 28x + 15 = 0$; there are three roots.
    $f(x) = 2x^3 - 5x^2 - 28x + 15$; there are at most two positive roots.
    $f(-x) = -2x^3 - 5x^2 + 28x + 15$; there is one negative root. Possible
    rational roots are $\pm\frac{1}{2}, \pm\frac{3}{2}, \pm\frac{5}{2}, \pm\frac{15}{2}$, ±3, ±5, ±15. We try to find the
    one negative root first. -3 is the first root with 0 remainder.

    $$\begin{array}{rrrr|r}
    2 & -5 & -28 & 15 & \underline{-3} \\
      & -6 & 33 & -15 & \\
    \hline
    2 & -11 & 5 & 0 &
    \end{array}$$

    Thus, -3 is the negative root. The remaining factor, $2x^2 - 11x + 5$
    $= (2x - 1)(x - 5)$. $2x - 1 = 0$ and $x = \frac{1}{2}$ or $x - 5 = 0$ and $x = 5$. The
    roots are -3, $\frac{1}{2}$, 5.

9. $x^4 - 11x^2 - 12x + 4 = 0$; there are four roots.
    $f(x) = x^4 - 11x^2 - 12x + 4$; there are at most two positive roots.
    $f(-x) = x^4 - 11x^2 + 12x + 4$; there are at most two negative roots.
    This does not help in deciding where to start, since the number of
    positive and negative roots is equal. Possible rational roots are
    ±1, ±2, ±4.

    $$\begin{array}{rrrrr|r}
    1 & 0 & -11 & -12 & 4 & \underline{1} \\
      & 1 & 1 & -10 & -22 & \\
    \hline
    1 & 1 & -10 & -22 & -18 & \text{(not a root)}
    \end{array}
    \qquad
    \begin{array}{rrrrr|r}
    1 & 0 & -11 & -12 & 4 & \underline{-1} \\
      &   & -1 & 1 & 10 & 2 \\
    \hline
    1 & -1 & -10 & -2 & 6 & \text{(not a root)}
    \end{array}$$

    $$\begin{array}{rrrrr|r}
    1 & 0 & -11 & -12 & 4 & \underline{2} \\
      & 2 & 4 & -14 & -52 & \\
    \hline
    1 & 2 & -7 & -26 & -48 & \text{(not a root)}
    \end{array}
    \qquad
    \begin{array}{rrrrr|r}
    1 & 0 & -11 & -12 & 4 & \underline{-2} \\
      & -2 & 4 & 14 & -4 & \\
    \hline
    1 & -2 & -7 & 2 & 0 & \text{(-2 is a root)}
    \end{array}$$

We know that 1, -1, and 2 cannot be roots of the remaining factor
$x^3 - 2x^2 - 7x + 2 = 0$ since they were not factors of the original
polynomial. We only need to try -2, ±4:

```
1    -2    -7     2    |-2
      -2     8    -2
1    -4     1     0
```

Therefore -2 is a double root. The remaining factor $x^2 - 4x + 1 = 0$
is solved using the quadratic formula.

$$x = \frac{4 \pm\sqrt{16 - 4}}{2} = \frac{4 \pm\sqrt{12}}{2} = \frac{4 \pm 2\sqrt{3}}{2} = 2 \pm\sqrt{3}$$

The roots are -2, -2, $2 \pm\sqrt{3}$.

13. $2x^4 - 5x^3 - 3x^2 + 4x + 2 = 0$; there are four roots.
$f(x) = 2x^4 - 5x^3 - 3x^2 + 4x + 2$; there are at most two positive roots.
There are at most two roots that are negative. Possible rational roots
are $\pm\frac{1}{2}$, ±1, ±2.

$f(-x)$
$2(-x)^4 - 5(-x)^3 - 3(+x)^2 + 4(-x) + 2$
$2x^4 + 5x - 3x - 4x + 2$

```
2    -5    -3     4     2    |1
      2    -3    -6    -2
2    -3    -6    -2     0  (1 is a root)
```

Checking the remaining factor for rational roots,

```
2    -3    -6    -2    |1              2    -3    -6    -2    |-1
      2    -1    -7                         -2     5     1
2    -1    -7    -9  (not a root)     2    -5    -1    -1  (not a root)
```

```
2    -3    -6    -2    |1/2           2    -3    -6    -2    |-1/2
      1    -1    -7/2                       -1     2     2
2    -2    -7    -11/2  (not a root)  2    -4    -4     0
                                       (-1/2 is a root)
```

The remaining factor $2x^2 - 4x - 4 = 0$ is solved using the quadratic
formula:

$$x = \frac{4 \pm\sqrt{48}}{4} = \frac{4 \pm 4\sqrt{3}}{4} = 1 \pm\sqrt{3}; \text{ the roots are } 1, -\frac{1}{2}, 1 \pm\sqrt{3}.$$

17. $x^5 + x^4 - 9x^3 - 5x^2 + 16x + 12 = 0$; there are five roots.
$f(x) = x^5 + x^4 - 9x^3 - 5x^2 + 16x + 12$; there are at most two positive
roots. $f(-x) = -x^5 + x^4 + 9x^3 - 5x^2 - 16x + 12$; there are at most
three negative roots. We look for positive roots first. Possible
rational roots are ±1, ±2, ±3, ±4, ±6, ±12.

```
1    1    -9    -5    16    12    |1
     1     2    -7   -12     4
1    2    -7   -12     4    16  (not a root)
```

```
1    1   -9    -5    16    12   |2
     2    6    -6   -22   -12
1    3   -3   -11    -6     0
```

2 is a root. The remaining factor is checked for a rational root:

```
1    3   -3   -11    -6   |2
     2   10    14     6
1    5    7     3     0    2 is also a root of the factor.
```

We have found two positive roots, so the remaining roots must be negative.

```
1    5    7    3   |-1
    -1   -4   -3
1    4    3    0
```

-1 is a root. The remaining factor $x^2 + 4x + 3 = (x + 1)(x + 3)$.
$x = -1$ or $x = -3$; the roots are 2, 2, -1, -1, -3.

21. $x^3 - 6x^2 + 10x - 4 = 0$ (0 and 1)
$f(0) = -4$
$f(1) = 1$
The root is between 0 and 1.
$f(0.5) = -0.375$
$f(0.6) = 0.056$
The root is between 0.5 and 0.6.
$f(0.55) = -0.149$
$f(0.56) = -0.106$
$f(0.58) = -0.023$
$f(0.59) = 0.017$
The root is between 0.58 and 0.59. It is closer to 0.59.

25. $V^3 - 6V^2 + 12V - 8 = 0$; There are 3 roots.

$f(V)$ has 3 sign changes and at most 3 positive roots.

$f(-V) = -V^3 - 6V^2 - 12V - 8$ indicates no negative roots.

Possible rational roots are ± 1, ± 2, ± 4, ± 8. Trial,
using synthetic division, yields 2 as a root and $(V - 2)$
as one factor.

```
1    -6    12    -8   |2
      2    -8     8
1    -4     4     0
```

The remaining factor $(V^2 - 4V + 4) = (V - 2)(V - 2)$ or $V = 2$ also.

29. $C = 8x^3 - 36x^2 + 90$; $C = 36$

$8x^3 - 36x^2 + 90 - 36 = 0$; $8x^3 - 36x^2 + 54 = 0$

$f(x)$ has at most 2 positive roots.

$f(-x) = -8x^3 - 36x^2 + 54$ indicates at most 1 negative root.

Possible rational factors are ± 1, ± 2, ± 3, ± 6, ± 9, ± 27,

± 54 and each of these divided by 2, 4, or 8 (factors of 8).

By trial, $x = \frac{3}{2}$ is a root.

$$
\begin{array}{rrrr|l}
8 & -36 & 0 & 54 & \underline{3/2} \\
  & 12 & -36 & -54 & \\
\hline
8 & -24 & -36 & 0 &
\end{array}
$$

$f(x)$ factors into $(x - \frac{3}{2})(8x^2 - 24x - 36)$ or $4(x - \frac{3}{2})(2x^2 - 6x - 9)$.

Solving $2x^2 - 6x - 9 = 0$ gives $x = \dfrac{6 \pm \sqrt{108}}{4}$ or $x = 4.1$, -1.1.

Three solutions of -1.1, 4.1, and $\frac{3}{2}$ or 1.5 give 2 valid answers.

(Negative value is not valid.)

33. $V_1 = x^3$; $V_2 = (x + 1.0)^3$; $V_1 + V_2 = 91.0$

$x^3 + (x + 1)^3 = 91$; $x^3 + x^3 + 3x^2 + 3x + 1 = 91$

$2x^3 + 3x^2 + 3x - 90 = 0$

$f(x) = 2x^3 + 3x^2 + 3x - 90$ has at most 1 positive root.

By trial, $x = 3$ is one root, and $f(x) = (x - 3)(2x^2 + 9x + 30)$.

Negative roots are not valid, so $x = 3.0$ and $x + 1.0 = 4.0$.

Review Exercises for Chapter 14, page 420

1. $(2x^3 - 4x^2 - x + 4) \div (x - 1)$

$$
\begin{array}{rrrr|l}
2 & -4 & -1 & 4 & \underline{1} \\
  & 2 & -2 & -3 & \\
\hline
2 & -2 & -3 & 1 & \text{remainder}
\end{array}
$$

5. $x^4 + x^3 + x^2 - 2x - 3$, $x + 1$
It is a factor since the
remainder is 0.

$$
\begin{array}{rrrrr|l}
1 & 1 & 1 & -2 & -3 & \underline{-1} \\
  & -1 & 0 & -1 & 3 & \\
\hline
1 & 0 & 1 & -3 & 0 &
\end{array}
$$

9. $(x^3 + 3x^2 + 6x + 1) \div x - 1$
The quotient is
$x^2 + 4x + 10$ R 11.

$$
\begin{array}{rrrr|l}
1 & 3 & 6 & 1 & \underline{1} \\
  & 1 & 4 & 10 & \\
\hline
1 & 4 & 10 & 11 &
\end{array}
$$

13. $(x^4 - 2x^3 - 3x^2 - 4x - 8) \div (x + 1)$
$x^3 - 3x^2 - 4$ R -4

$$
\begin{array}{rrrrr|l}
1 & -2 & -3 & -4 & -8 & \underline{-1} \\
  & -1 & +3 & 0 & 4 & \\
\hline
1 & -3 & 0 & -4 & -4 &
\end{array}
$$

17. $x^3 + 8x^2 + 17x - 6$; -3
     -3 is not a zero of the func-
     tion since $R = -12$.

	1	8	17	-6	$\underline{-3}$
		-3	-15	-6	
	1	5	2	-12	

21. $x^3 + 8x^2 + 17x + 6 = 0$ $(r_1 = -3)$
     The remaining factor $x^2 + 5x + 2$
     is solved using the quadratic
     formula:
     $$x = \frac{-5 \pm\sqrt{17}}{2} = \frac{1}{2}(-5 \pm\sqrt{17})$$

	1	8	17	6	$\underline{-3}$
		-3	-15	-6	
	1	5	2	0	

25. $2x^4 + x^3 - 29x^2 - 34x + 24 = 0$
     $(r_1 = -2,\ r_2 = \frac{1}{2})$
     The remaining factor is
     $2x^2 - 2x - 24 = 2(x^2 - x - 12)$
     $= 2(x - 4)(x + 3)$.
     $(x - 4)(x + 3) = 0$; $x - 4 = 0$
     and $x = 4$ or $x + 3 = 0$ and
     $x = -3$.

2	1	-29	-34	24	$\underline{-2}$
	-4	6	46	-24	
2	-3	-23	12	0	

2	-3	-23	12	$\underline{\frac{1}{2}}$
	1	-1	-12	
2	-2	-24	0	

29. $x^5 + 3x^4 - x^3 - 11x^2 - 12x - 4 = 0$ (-1 is a triple root)

1	3	-1	-11	-12	-4	$\underline{-1}$
	-1	-2	3	8	4	
1	2	-3	-8	-4	0	

1	2	-3	-8	-4	$\underline{-1}$
	-1	-1	4	4	
1	1	-4	-4	0	

1	1	-4	-4	$\underline{-1}$
	-1	0	4	
1	0	-4	0	

The remaining factor $x^2 - 4$ can be factored: $(x + 2)(x - 2)$.
Therefore the roots are -1, -1, -1, -2, 2.

33. $x^3 + x^2 - 10x + 8 = 0$; there are three roots. $f(x) = x^3 + x^2 - 10x + 8$;
     there are two possible positive roots. Find one of these first.
     Possible rational roots are ±1, ±2, ±4, ±8.

1	1	-10	8	$\underline{1}$
	1	2	-8	
1	2	-8	0	

1 is a root. The remaining factor $x^2 + 2x - 8 = (x + 4)(x - 2) = 0$;
$x + 4 = 0$ and $x = -4$ or $x - 2 = 0$ and $x = 2$. The roots are -4, 1, 2.

37. $6x^3 - x^2 - 12x - 5 = 0$; there are three roots. $f(x) = 6x^3 - x^2 - 12x - 5$; there is one sign change and one positive root. Look for this first.

Possible rational roots are ± 5, $\pm\frac{5}{2}$, $\pm\frac{5}{3}$, $\pm\frac{5}{6}$, $\pm\frac{1}{2}$, $\pm\frac{1}{3}$, $\pm\frac{1}{6}$, ± 1.

6	-1	-12	-5	$\lfloor 5$
	30	145	665	
6	29	133	660	

6	-1	-12	-5	$\left\lfloor\frac{5}{2}\right.$
	15	35	$\frac{115}{2}$	
6	14	23	$52\frac{1}{2}$	

6	-1	-12	-5	$\left\lfloor\frac{5}{3}\right.$
	10	15	5	
6	9	3	0	

$\frac{5}{3}$ is a root; the remaining factor $6x^2 + 9x + 3 = 3(2x^2 + 3x + 1)$
$= (2x + 1)(x + 1)$. $(2x + 1)(x + 1) = 0$; $2x + 1 = 0$ and $x = -\frac{1}{2}$ or
$x + 1 = 0$ and $x = -1$. The roots are $\frac{5}{3}$, $-\frac{1}{2}$, -1.

41. For what value of k is $x + 2$ a factor of $f(x) = 3x^3 + kx^2 - 8x - 8$?
For $x + 2$ to be a factor, the remainder must be 0, when the function is divided by -2.

3	k	-8	-8	$\lfloor -2$
	-6	$-2k + 12$	$4k - 8$	
3	$k - 6$	$-2k + 4$	$4k - 16$	

$4k - 16 = 0$; $4k = 16$; $k = 4$

45. $3x^3 - x^2 - 8x - 2 = 0$ $f(x) = 3x^3 - x^2 - 8x - 2$

$f(1) \leq f(x) \leq f(2)$

$f(1) = -8$ and $f(2) = 2$ so try values for x close to 1.

$f(1.9) = -0.233$

$f(1.95) = 0.84$ which is too high

$f(1.91) = 0.02$; $x \approx 1.91$

49. $64d^3 - 144d^2 + 108d - 27 = 0$; $f(d) = 64d^3 - 144d^2 + 108d - 27$

$f(d)$ has 3 sign changes so there are at most 3 positive roots.

$f(-d) = -64d^3 - 144d^2 - 108d - 27$ indicates no negative roots.

Possible rational roots are factors of 27 divided by factors of 64.
Trial gives $^3/_4$ as a root.

64	-144	108	-27	$\lfloor ^3/_4$
	48	-72	27	
64	-96	36	0	

Factors are $(d - ^3/_4)(64d^2 - 96d + 36)$
or $4(d - ^3/_4)(16d^2 - 24d + 9)$

Using the quadratic formula to solve for d also gives $d = ^3/_4$.
All solutions are $d = ^3/_4$.

CHAPTER 15

Determinants and Matrices

1.
$$\begin{vmatrix} 3 & 0 & 0 \\ -2 & 1 & 4 \\ 4 & -2 & 5 \end{vmatrix} = (+3)\begin{vmatrix} 1 & 4 \\ -2 & 5 \end{vmatrix} -0\begin{vmatrix} -2 & 4 \\ 4 & 5 \end{vmatrix} +0\begin{vmatrix} -2 & 1 \\ 4 & -2 \end{vmatrix} = +3[5 - (-8)]$$
$$= 3(13) = 39$$

The first row was selected for the expansion since two elements are zero. The first element is assigned a plus sign since the element is in row one, column one and $1 + 1 = 2$.

5.
$$\begin{vmatrix} -6 & -1 & 3 \\ 2 & -2 & -3 \\ 10 & 1 & -2 \end{vmatrix} = +(-6)\begin{vmatrix} -2 & -3 \\ 1 & -2 \end{vmatrix} - (-1)\begin{vmatrix} 2 & -3 \\ 10 & -2 \end{vmatrix} + (+3)\begin{vmatrix} 2 & -2 \\ 10 & 1 \end{vmatrix}$$
$$= -6(4 + 3) + 1(-4 + 30) + 3(2 + 20) = -6(7) + 1(26) + 3(22)$$
$$= -42 + 26 + 66 = 50$$

The first row was selected for the expansion. The first term is positive since $1 + 1 = 2$.

9.
$$\begin{vmatrix} 1 & 0 & 1 & 0 \\ 2 & 4 & -3 & 1 \\ 1 & 1 & 1 & 1 \\ 3 & 5 & 0 & 2 \end{vmatrix} = +1\begin{vmatrix} 4 & -3 & 1 \\ 1 & 1 & 1 \\ 5 & 0 & 2 \end{vmatrix} -0\begin{vmatrix} 2 & -3 & 1 \\ 1 & 1 & 1 \\ 3 & 0 & 2 \end{vmatrix} +(+1)\begin{vmatrix} 2 & 4 & 1 \\ 1 & 1 & 1 \\ 3 & 5 & 2 \end{vmatrix} -0\begin{vmatrix} 2 & 4 & -3 \\ 1 & 1 & 1 \\ 3 & 5 & 0 \end{vmatrix}$$

$$= 1\begin{vmatrix} 4 & -3 & 1 \\ 1 & 1 & 1 \\ 5 & 0 & 2 \end{vmatrix}\begin{matrix} 4 & -3 \\ 1 & 1 \\ 5 & 0 \end{matrix} +1\begin{vmatrix} 2 & 4 & 1 \\ 1 & 1 & 1 \\ 3 & 5 & 2 \end{vmatrix}\begin{matrix} 2 & 4 \\ 1 & 1 \\ 3 & 5 \end{matrix}$$

171

$= 1[8 + (-15) - (+5) - (-6)] + 1[4 + 12 + 5 - (+3) - (+10) - (+8)]$
$= 1(-6) + 1(0) = -6$

The first row was selected for the expansion since it has two zeros.
The sign of the first term is positive.

13.
$$\begin{vmatrix} 1 & 2 & -1 & -2 \\ 3 & 1 & 2 & 1 \\ -1 & 3 & -1 & 2 \\ 2 & 1 & 3 & -3 \end{vmatrix} = +(+1)\begin{vmatrix} 1 & 2 & 1 \\ 3 & -1 & 2 \\ 1 & 3 & -3 \end{vmatrix} -(+2)\begin{vmatrix} 3 & 2 & 1 \\ -1 & -1 & 2 \\ 2 & 3 & -3 \end{vmatrix} +(-1)\begin{vmatrix} 3 & 1 & 1 \\ -1 & 3 & 2 \\ 2 & 1 & -3 \end{vmatrix}$$

$$-(-2)\begin{vmatrix} 3 & 1 & 2 \\ -1 & 3 & -1 \\ 2 & 1 & 3 \end{vmatrix} = 1(29) - 2(-8) - 1(-39) + 2(17) = 29 + 16 + 39 + 34$$

$= 118$

17. $2x + y + z = 6$; $x - 2y + 2z = 10$; $3x - y - z = 4$

$$x = \frac{\begin{vmatrix} 6 & 1 & 1 \\ 10 & -2 & 2 \\ 4 & -1 & -1 \end{vmatrix}}{\begin{vmatrix} 2 & 1 & 1 \\ 1 & -2 & 2 \\ 3 & -1 & -1 \end{vmatrix}} = \frac{+(+6)\begin{vmatrix} -2 & 2 \\ -1 & -1 \end{vmatrix} -(+1)\begin{vmatrix} 10 & 2 \\ 4 & -1 \end{vmatrix} +(+1)\begin{vmatrix} 10 & -2 \\ 4 & -1 \end{vmatrix}}{+(+2)\begin{vmatrix} -2 & 2 \\ -1 & -1 \end{vmatrix} -(+1)\begin{vmatrix} 1 & 2 \\ 3 & -1 \end{vmatrix} +(+1)\begin{vmatrix} 1 & -2 \\ 3 & -1 \end{vmatrix}}$$

$$= \frac{+6(2 + 2) - 1(-10 - 8) + 1(-10 + 8)}{2(2 + 2) - 1(-1 - 6) + 1(-1 + 6)} = \frac{24 + 18 - 2}{8 + 7 + 5}$$

$$= \frac{40}{20} = 2$$

$$y = \frac{\begin{vmatrix} 2 & 6 & 1 \\ 1 & 10 & 2 \\ 3 & 4 & -1 \end{vmatrix}}{20} = \frac{2\begin{vmatrix} 10 & 2 \\ 4 & -1 \end{vmatrix} -6\begin{vmatrix} 1 & 2 \\ 3 & -1 \end{vmatrix} +1\begin{vmatrix} 1 & 10 \\ 3 & 4 \end{vmatrix}}{20} = \frac{2(-18) - 6(-7) + 1(-26)}{20}$$

$$= \frac{-20}{20} = -1$$

Substitute the values for x and y into the first equation, and solve for
z. $2(2) - 1 + z = 6$; $z = 3$.   The solution is $x = 2$, $y = -1$, $z = 3$.

21. (1) $x + 0y + 0z + t = 0$
    (2) $3x + y + z + 0t = -1$
    (3) $0x + 2y - z + 3t = 1$
    (4) $0x + 0y + 2z - 3t = 1$

Each determinant is expanded by minors using the first row. Determinants which are multiplied by zero are omitted since the product is zero. The values for the third-order determinants are found using Eq. (4-13).

$$x = \frac{\begin{vmatrix} 0 & 0 & 0 & 1 \\ -1 & 1 & 1 & 0 \\ 1 & 2 & -1 & 3 \\ 1 & 0 & 2 & -3 \end{vmatrix}}{\begin{vmatrix} 1 & 0 & 0 & 1 \\ 3 & 1 & 1 & 0 \\ 0 & 2 & -1 & 3 \\ 0 & 0 & 2 & -3 \end{vmatrix}} = \frac{(-1)\begin{vmatrix} -1 & 1 & 1 \\ 1 & 2 & -1 \\ 1 & 0 & 2 \end{vmatrix}}{(1)\begin{vmatrix} 1 & 1 & 0 \\ 2 & -1 & 3 \\ 0 & 2 & -3 \end{vmatrix} - (1)\begin{vmatrix} 3 & 1 & 1 \\ 0 & 2 & -1 \\ 0 & 0 & 2 \end{vmatrix}}$$

$$= \frac{-1[(-4) + (-1) - (+2) - (+2)]}{1[(+3) - (+6) - (-6)] - 1(+12)]} = \frac{9}{-9} = -1$$

The value for x is substituted into equation (1); $-1 + t = 0$; $t = 1$. The value for t is substituted into equation (4); $2z - 3(1) = 1$; $2z = 4$; $z = 2$. The values for t and z are substituted into equation 3; $2y - 2 + 3(1) = 1$; $2y = 0$; $y = 0$. The solution is $x = -1$, $y = 0$, $z = 2$, $t = 1$.

25. (1) $I_A + I_B + I_C + I_D = 0$
    (2) $2I_A - I_B + 0I_C + 0I_D = -2$
    (3) $0I_A + 0I_B + 3I_C - 2I_D = 0$
    (4) $0I_A + I_B - 3I_C + 0I_D = 6$

Each determinant is expanded by minors using the first column. Values for the third-order determinants are found using Eq. (4-13).

$$x = \frac{\begin{vmatrix} 0 & 1 & 1 & 1 \\ -2 & -1 & 0 & 0 \\ 0 & 0 & 3 & -2 \\ 6 & 1 & -3 & 0 \end{vmatrix}}{\begin{vmatrix} 1 & 1 & 1 & 1 \\ 2 & -1 & 0 & 0 \\ 0 & 0 & 3 & -2 \\ 0 & 1 & -3 & 0 \end{vmatrix}} = \frac{(2)\begin{vmatrix} 1 & 1 & 1 \\ 0 & 3 & -2 \\ 1 & -3 & 0 \end{vmatrix} - (6)\begin{vmatrix} 1 & 1 & 1 \\ -1 & 0 & 0 \\ 0 & 3 & -2 \end{vmatrix}}{(1)\begin{vmatrix} -1 & 0 & 0 \\ 0 & 3 & -2 \\ 1 & -3 & 0 \end{vmatrix} - (2)\begin{vmatrix} 1 & 1 & 1 \\ 0 & 3 & -2 \\ 1 & -3 & 0 \end{vmatrix}}$$

$$= \frac{2(-2 - 3 - 6) - 6(-3 - 2)}{1(6) - 2(-2 - 3 - 6)} = \frac{2(-11) - 6(-5)}{1(6) - 2(-11)} = \frac{8}{28} = \frac{2}{7}A$$

The value for I_A is substituted into equation (2); $2(\frac{2}{7}) - I_B = -2$; $I_B = \frac{18}{7}A$.

The value for I_B is substituted into equation (4); $\frac{18}{7} - 3I_C = 6$;

$I_C = -\frac{8}{7}A.$

The value for I_C is substituted into equation (3); $3(-\frac{8}{7}) - 2I_D = 0$;

$I_D = -\frac{12}{7}A.$

Exercises 15-2, page 434

1. $\begin{vmatrix} 4 & -5 & 9 \\ 0 & 3 & -8 \\ 0 & 0 & -5 \end{vmatrix} = (4)(3)(-5) = -60;$ property 1

5. $\begin{vmatrix} -2 & 0 & -1 \\ 5 & 0 & 3 \\ 3 & 0 & -4 \end{vmatrix} = 0$

   The determinant is expanded by the second column, which contains all zeros.

9. $\begin{vmatrix} 3 & 1 & 0 \\ -2 & 3 & -1 \\ 4 & 2 & 5 \end{vmatrix} = \begin{vmatrix} 0 & 1 & 0 \\ -11 & 3 & -1 \\ -2 & 2 & 5 \end{vmatrix}$

   $= (-1)\begin{vmatrix} -11 & -1 \\ -2 & 5 \end{vmatrix} = -1(-55 - 2) = 57$

   Each element of the second column is multiplied by -3 and added to corresponding elements of the first column.

13. $\begin{vmatrix} 4 & 3 & 6 & 0 \\ 3 & 0 & 0 & 4 \\ 5 & 0 & 1 & 2 \\ 2 & 1 & 1 & 7 \end{vmatrix} \overset{①}{=} \begin{vmatrix} -2 & 0 & 3 & -21 \\ 3 & 0 & 0 & 4 \\ 5 & 0 & 1 & 2 \\ 2 & 1 & 1 & 7 \end{vmatrix} \overset{②}{=} +(1)\begin{vmatrix} -2 & 3 & -21 \\ 3 & 0 & 4 \\ 5 & 1 & 2 \end{vmatrix} \overset{③}{=} +(1)\begin{vmatrix} -17 & 0 & -27 \\ 3 & 0 & 4 \\ 5 & 1 & 2 \end{vmatrix} \overset{④}{=}$

    $-1\begin{vmatrix} -17 & -27 \\ 3 & 4 \end{vmatrix} = -[(-68) + 81] = -13$

    ① Each element in the fourth row is multiplied by -3 and added to corresponding elements of the first row.
    ② Expand the determinant by the second column.
    ③ Each element in the third row is multiplied by -3 and added to the first row.
    ④ Expand the determinant by the second column.

17. $\begin{vmatrix} 1 & 3 & -3 & 5 \\ 4 & 2 & 1 & 2 \\ 3 & 2 & -2 & 2 \\ 0 & 1 & 2 & -1 \end{vmatrix}$ ① $=$ $\begin{vmatrix} 1 & 3 & -3 & 5 \\ 4 & 2 & 1 & 2 \\ 0 & -7 & 7 & -13 \\ 0 & 1 & 2 & -1 \end{vmatrix}$ ② $=$ $\begin{vmatrix} 1 & 3 & -3 & 5 \\ 0 & -10 & 13 & -18 \\ 0 & -7 & 7 & -13 \\ 0 & 1 & 2 & -1 \end{vmatrix}$

③ $=$ (1) $\begin{vmatrix} -10 & 13 & -18 \\ -7 & 7 & -13 \\ 1 & 2 & -1 \end{vmatrix}$ ④ $=$ $\begin{vmatrix} 0 & 33 & -28 \\ -7 & 7 & -13 \\ 1 & 2 & -1 \end{vmatrix}$ ⑤ $=$ $\begin{vmatrix} 0 & 33 & -28 \\ 0 & 21 & -20 \\ 1 & 2 & -1 \end{vmatrix}$

⑥ $=$ $\begin{vmatrix} 33 & -28 \\ 21 & -20 \end{vmatrix}$ $= 1[-660 - (-588)] = -72$

① Each element in the first row is multiplied by –3 and added to corresponding elements in the third row.
② Each element in the first row is multiplied by –4 and added to corresponding elements in the second row.
③ Expand the determinant by the first column.
④ Each element in the third row is multiplied by 10 and added to corresponding elements in the first row.
⑤ Each element in the third row is multiplied by 7 and added to corresponding elements in the second row.
⑥ Expand the determinant by the first column.

21. $2x - y + z = 5$; $x + 2y + 3z = 10$; $3x + 3y + 2z = 5$;
The denominator is evaluated first:

$\begin{vmatrix} 2 & -1 & 1 \\ 1 & 2 & 3 \\ 3 & 3 & 2 \end{vmatrix}$ ① $=$ $\begin{vmatrix} 0 & -5 & -5 \\ 1 & 2 & 3 \\ 3 & 3 & 2 \end{vmatrix}$ ② $=$ $\begin{vmatrix} 0 & -5 & -5 \\ 1 & 2 & 3 \\ 0 & -3 & -7 \end{vmatrix}$ ③ $=$ $(-1)\begin{vmatrix} -5 & -5 \\ -3 & -7 \end{vmatrix}$

$= -1[35 - 15] = -20$

① Each element in the second row is multiplied by –2 and added to corresponding elements in the first row.
② Each element in the second row is multiplied by –3 and added to corresponding elements in the third row.
③ Expand the determinant by the first column.

$x = \dfrac{\begin{vmatrix} 5 & -1 & 1 \\ 10 & 2 & 3 \\ 5 & 3 & 2 \end{vmatrix}}{-20}$ ① $= \dfrac{\begin{vmatrix} 5 & -1 & 1 \\ -5 & 5 & 0 \\ 5 & 3 & 2 \end{vmatrix}}{-20}$ ② $= \dfrac{\begin{vmatrix} 5 & -1 & 1 \\ -5 & 5 & 0 \\ -5 & 5 & 0 \end{vmatrix}}{-20}$ ③ $= 0$

① Each element in the first row is multiplied by −3 and added to corresponding elements in the second row.

② Each element in the first row is multiplied by −2 and added to corresponding elements in the third row.

③ Property 3

$$y = \frac{\begin{vmatrix} 2 & 5 & 1 \\ 1 & 10 & 3 \\ 3 & 5 & 2 \end{vmatrix}}{-20} \overset{①}{=} \frac{\begin{vmatrix} 0 & -15 & -5 \\ 1 & 10 & 3 \\ 3 & 5 & 2 \end{vmatrix}}{-20} \overset{②}{=} \frac{\begin{vmatrix} 0 & -15 & -5 \\ 1 & 10 & 3 \\ 0 & -25 & -7 \end{vmatrix}}{-20} \overset{③}{=} \frac{-1(105 - 125)}{-20}$$

$$= \frac{20}{-20} = -1$$

① Each element in the second row is multiplied by −2 and added to corresponding elements in the first row.

② Each element in the second row is multiplied by −3 and added to corresponding elements in the third row.

③ Expand the determinant by the first column.

Substitute the values for x and y into the first equation.
$2(0) - (-1) + z = 5;\ z = 4$

25. $2x + y + z = 2;\ 3y - z + 2t = 4;\ y + 2z + t = 0;\ 3x + 2z = 4;$
     The denominator is evaluated first:

$$\begin{vmatrix} 2 & 1 & 1 & 0 \\ 0 & 3 & -1 & 2 \\ 0 & 1 & 2 & 1 \\ 3 & 0 & 2 & 0 \end{vmatrix} \overset{①}{=} \begin{vmatrix} 0 & 1 & 1 & 0 \\ 2 & 3 & -1 & 2 \\ -4 & 1 & 2 & 1 \\ -1 & 0 & 2 & 0 \end{vmatrix} \overset{②}{=} \begin{vmatrix} 0 & 0 & 1 & 0 \\ 2 & 4 & -1 & 2 \\ -4 & -1 & 2 & 1 \\ -1 & -2 & 2 & 0 \end{vmatrix}$$

$$\overset{③}{=} (1) \begin{vmatrix} 2 & 4 & 2 \\ -4 & -1 & 1 \\ -1 & -2 & 0 \end{vmatrix} \overset{④}{=} \begin{vmatrix} 10 & 6 & 0 \\ -4 & -1 & 1 \\ -1 & -2 & 0 \end{vmatrix} \overset{⑤}{=} (-1) \begin{vmatrix} 10 & 6 \\ -1 & -2 \end{vmatrix}$$

$$= -1(-20 + 6) = 14$$

① Each element in column three is multiplied by −2 and added to corresponding elements in the first column.

② Each element in column three is multiplied by −1 and added to corresponding elements in column two.

③ Expand the determinant by the first row.

④ Each element in row two is multiplied by −2 and added to corresponding elements in row one.

⑤ Expand the determinant by column three.

$$x = \frac{\begin{vmatrix} 2 & 1 & 1 & 0 \\ 4 & 3 & -1 & 2 \\ 0 & 1 & 2 & 1 \\ 4 & 0 & 2 & 0 \end{vmatrix}}{14} \; \textcircled{1} = \frac{\begin{vmatrix} 2 & 1 & 1 & 0 \\ -2 & 0 & -4 & 2 \\ 0 & 1 & 2 & 1 \\ 4 & 0 & 2 & 0 \end{vmatrix}}{14} \; \textcircled{2} = (-1)\frac{\begin{vmatrix} 2 & 1 & 1 & 0 \\ -2 & 0 & -4 & 2 \\ -2 & 0 & 1 & 1 \\ 4 & 0 & 2 & 0 \end{vmatrix}}{14}$$

$$\textcircled{3} = (-1)\frac{\begin{vmatrix} -2 & -4 & 2 \\ -2 & 1 & 1 \\ 4 & 2 & 0 \end{vmatrix}}{14} \; \textcircled{4} = (-1)\frac{\begin{vmatrix} 2 & -6 & 0 \\ -2 & 1 & 1 \\ 4 & 2 & 0 \end{vmatrix}}{14} \; \textcircled{5} = (-1)(-1)\frac{\begin{vmatrix} 2 & -6 \\ 4 & 2 \end{vmatrix}}{14}$$

$$= \frac{4 + 24}{14} = 2$$

    ① Each element in row one is multiplied by –3 and added to corresponding elements in row two.
    ② Each element in row one is multiplied by –1 and added to corresponding elements in row three.
    ③ Expand the determinant by column two.
    ④ Each element in row two is multiplied by –2 and added to row one.
    ⑤ Expand the determinant by column three.

The value for x is substituted in the last equation.
$3(2) + 2z = 4$; $6 + 2z = 4$; $z = -1$
The values for x and z are substituted in the first equation.
$2(2) + y - 1 = 2$; $4 + y - 1 = 2$; $y = -1$
The values for y and z are substituted in the second equation.
$3(-1) - (-1) + 2t = 4$; $-3 + 1 + 2t = 4$; $t = 3$

29. (1) $I_A + I_B + I_C + I_D + I_E = 0$

    (2) $-2I_A + 3I_B + 0I_C + 0I_D + 0I_E = 0$

    (3) $0I_A + 3I_B - 3I_C + 0I_D + 0I_E = 6$

    (4) $0I_A + 0I_B - 3I_C + I_D + 0I_E = 0$

    (5) $0I_A + 0I_B + 0I_C - I_D + 2I_E = 0$

Find the denominator first:

$$\begin{vmatrix} 1 & 1 & 1 & 1 & 1 \\ -2 & 3 & 0 & 0 & 0 \\ 0 & 3 & -3 & 0 & 0 \\ 0 & 0 & -3 & 1 & 0 \\ 0 & 0 & 0 & -1 & 2 \end{vmatrix} \textcircled{1} = \begin{vmatrix} 1 & 1 & 1 & 1 & 1 \\ -2 & 3 & 0 & 0 & 0 \\ 0 & 3 & -3 & 0 & 0 \\ 0 & 0 & -3 & 1 & 0 \\ -2 & -2 & -2 & -3 & 0 \end{vmatrix} \textcircled{2} = (1)\begin{vmatrix} -2 & 3 & 0 & 0 \\ 0 & 3 & -3 & 0 \\ 0 & 0 & -3 & 1 \\ -2 & -2 & -2 & -3 \end{vmatrix}$$

$$③ = \begin{vmatrix} -2 & 3 & 0 & 0 \\ 0 & 3 & -3 & 0 \\ 0 & 0 & -3 & 1 \\ -2 & -2 & -11 & 0 \end{vmatrix} \quad ④ = (-1)\begin{vmatrix} -2 & 3 & 0 \\ 0 & 3 & -3 \\ -2 & -2 & -11 \end{vmatrix} \quad ⑤ = (-1)\begin{vmatrix} 0 & 5 & 11 \\ 0 & 3 & -3 \\ -2 & -2 & -11 \end{vmatrix}$$

$$= (-1)(-2)\begin{vmatrix} 5 & 11 \\ 3 & -3 \end{vmatrix} = 2(-15 - 33) = -96$$

① Each element in row one is multiplied by -2 and added to corresponding elements in row five.
② Expand the determinant by column five.
③ Each element in row three is multiplied by 3 and added to corresponding elements in row four.
④ Expand the determinant by column four.
⑤ Each element in row three is multiplied by -1 and added to corresponding elements in row one.
⑥ Expand the determinant using column one.

$$I_A = \frac{\begin{vmatrix} 0 & 1 & 1 & 1 & 1 \\ 0 & 3 & 0 & 0 & 0 \\ 6 & 3 & -3 & 0 & 0 \\ 0 & 0 & -3 & 1 & 0 \\ 0 & 0 & 0 & -1 & 2 \end{vmatrix}}{-96}$$

$$① = (6)\frac{\begin{vmatrix} 1 & 1 & 1 & 1 \\ 3 & 0 & 0 & 0 \\ 0 & -3 & 1 & 0 \\ 0 & 0 & -1 & 2 \end{vmatrix}}{-96} \quad ② = (6)(-3)\frac{\begin{vmatrix} 1 & 1 & 1 \\ -3 & 1 & 0 \\ 0 & -1 & 2 \end{vmatrix}}{-96} \quad ③ = (-18)\frac{\begin{vmatrix} 1 & 1 & 1 \\ -3 & 1 & 0 \\ -2 & -3 & 0 \end{vmatrix}}{-96}$$

$$④ = (-18)(1)\frac{\begin{vmatrix} -3 & 1 \\ -2 & -3 \end{vmatrix}}{-96} = \frac{-18(9 + 2)}{-96}$$

$$= \frac{-198}{-96} = \frac{33}{16}A$$

① Expand the determinant by the first column.
② Expand the determinant by the second row.
③ Each element in row one is multiplied by -2 and added to corresponding elements in row three.
④ Expand the determinant by the third column.

Substitute for I_A in equation (2); $-2\left(\frac{33}{16}\right) + 3I_B = 0$; $I_B = \frac{11}{8}A$

Substitute for I_B in equation (3); $3\left(\frac{11}{8}\right) - 3I_C = 6$; $I_C = -\frac{5}{8}A$

Substitute for I_C in equation (4); $-3\left(-\frac{5}{8}\right) + I_D = 0$; $I_D = -\frac{15}{8}A$

Substitute for I_D in equation (5); $-\left(-\frac{15}{8}\right) + 2I_E = 0$; $I_E = -\frac{15}{16}A$

Exercises 15-3, page 439

1. $a = 1$; $b = -3$; $c = 4$; $d = 7$; see example C.

5. The value of the literal symbols cannot be determined. The first matrix has three rows and the second matrix has two rows.

9. $\begin{pmatrix} 2 + (-1) & 3 + 7 \\ -5 + 5 & 4 + (-2) \end{pmatrix} = \begin{pmatrix} 1 & 10 \\ 0 & 2 \end{pmatrix}$

13. $\begin{pmatrix} -1 + 1 & 4 + 5 & -7 - 6 & 0 + 3 \\ 2 + 4 & -6 - 1 & -1 + 8 & 2 - 2 \end{pmatrix} = \begin{pmatrix} 0 & 9 & -13 & 3 \\ 6 & -7 & 7 & 0 \end{pmatrix}$

17. $2A = \begin{pmatrix} -2 & 8 & -14 & 0 \\ 4 & -12 & -2 & 4 \end{pmatrix}$; $2A + B = \begin{pmatrix} -2 + 1 & 8 + 5 & -14 - 6 & 0 + 3 \\ 4 + 4 & -12 - 1 & -2 + 8 & 4 - 2 \end{pmatrix}$

$= \begin{pmatrix} -1 & 13 & -20 & 3 \\ 8 & -13 & 6 & 2 \end{pmatrix}$

21. $A + B = \begin{pmatrix} -1 + 4 & 2 - 1 & 3 - 3 & 7 + 0 \\ 0 + 5 & -3 + 0 & -1 - 1 & 4 + 1 \\ 9 + 1 & -1 + 11 & 0 + 8 & -2 + 2 \end{pmatrix} = \begin{pmatrix} 3 & 1 & 0 & 7 \\ 5 & -3 & -2 & 5 \\ 10 & 10 & 8 & 0 \end{pmatrix}$

$B + A = \begin{pmatrix} 4 - 1 & -1 + 2 & -3 + 3 & 0 + 7 \\ 5 + 0 & 0 - 3 & -1 - 1 & 1 + 4 \\ 1 + 9 & 11 - 1 & 8 + 0 & 2 - 2 \end{pmatrix} = \begin{pmatrix} 3 & 1 & 0 & 7 \\ 5 & -3 & -2 & 5 \\ 10 & 10 & 8 & 0 \end{pmatrix}$

Therefore, $A + B = B + A$.

25. $V_1 \cos 35.0° - V_2 \cos 51.0° = 71.8 \cos 75.0°$

$V_1 \sin 35.0° + V_2 \sin 51.0° = 71.8 \sin 75.0°$

$0.819V_1 - 0.629V_2 = 0.259 (71.8) = 18.6$

$0.574V_1 + 0.777V_2 = 0.966 (71.8) = 69.4$

$$V_1 = \frac{\begin{vmatrix} 18.6 & -0.629 \\ 69.4 & 0.777 \end{vmatrix}}{\begin{vmatrix} 0.819 & -0.629 \\ 0.574 & 0.777 \end{vmatrix}} = \frac{18.6 (0.777) - (69.4)(-0.629)}{0.819(0.777) - (0.574)(-0.629)} = \frac{58.1}{0.997} = 58.3$$

$$V_2 = \frac{\begin{vmatrix} 0.819 & 18.6 \\ 0.574 & 69.4 \end{vmatrix}}{0.997} = \frac{0.819(69.4) - 0.574(18.6)}{0.997} = \frac{46.2}{0.997} = 46.3$$

Note: Answers may vary slightly with varying numbers of decimals handled.

Exercises 15-4, page 444

1. $(4 - 2)\begin{pmatrix} -1 & 0 \\ 2 & 6 \end{pmatrix} = (4(-1) + (-2)(2) \quad 4(0) + (-2)(6))$

   $= (-8 \quad -12)$

5. $\begin{pmatrix} 2 & -3 & 1 \\ 0 & 7 & -3 \end{pmatrix}\begin{pmatrix} 9 \\ -2 \\ 5 \end{pmatrix} = \begin{pmatrix} 2(9) + -3(-2) + 1(5) \\ 0(9) + 7(-2) + -3(5) \end{pmatrix} = \begin{pmatrix} 29 \\ -29 \end{pmatrix}$

9. $\begin{pmatrix} -1 & 7 \\ 3 & 5 \\ 10 & -1 \\ -5 & 12 \end{pmatrix}\begin{pmatrix} 2 & 1 & 0 \\ 5 & -3 & 1 \end{pmatrix} = \begin{pmatrix} -1(2) + 7(5) & -1(1) + 7(-3) & -1(0) + 7(1) \\ 3(2) + 5(5) & 3(1) + 5(-3) & 3(0) + 5(1) \\ 10(2) - 1(5) & 10(1) - 1(-3) & 10(0) - 1(1) \\ -5(2) + 12(5) & -5(1) + 12(-3) & -5(0) + 12(1) \end{pmatrix}$

   $= \begin{pmatrix} 33 & -22 & 7 \\ 31 & -12 & 5 \\ 15 & 13 & -1 \\ 50 & -41 & 12 \end{pmatrix}$

13. $AB = (1 \ -3 \ 8)\begin{pmatrix} -1 \\ 5 \\ 7 \end{pmatrix} = (1(-1) + -3(5) + 8(7)) = (40)$

$BA = \begin{pmatrix} -1 \\ 5 \\ 7 \end{pmatrix}(1 \ -3 \ 8) = \begin{pmatrix} -1(1) & -1(-3) & -1(8) \\ 5(1) & 5(-3) & 5(8) \\ 7(1) & 7(-3) & 7(8) \end{pmatrix} = \begin{pmatrix} -1 & 3 & -8 \\ 5 & -15 & 40 \\ 7 & -21 & 56 \end{pmatrix}$

17. $AI = \begin{pmatrix} 1 & 8 \\ -2 & 2 \end{pmatrix}\begin{pmatrix} 1 & 0 \\ 0 & 1 \end{pmatrix} = \begin{pmatrix} 1(1) + 8(0) & 1(0) + 8(1) \\ -2(1) + 2(0) & -2(0) + 2(1) \end{pmatrix} = \begin{pmatrix} 1 & 8 \\ -2 & 2 \end{pmatrix}$

$IA = \begin{pmatrix} 1 & 0 \\ 0 & 1 \end{pmatrix}\begin{pmatrix} 1 & 8 \\ -2 & 2 \end{pmatrix} = \begin{pmatrix} 1(1) + 0(-2) & 1(8) + 0(2) \\ 0(1) + 1(-2) & 0(8) + 1(2) \end{pmatrix} = \begin{pmatrix} 1 & 8 \\ -2 & 2 \end{pmatrix}$

Therefore $AI = IA = A$.

21. $AB = \begin{pmatrix} 5 & -2 \\ -2 & 1 \end{pmatrix}\begin{pmatrix} 1 & 2 \\ 2 & 5 \end{pmatrix} = \begin{pmatrix} 5(1) + -2(2) & 5(2) + -2(5) \\ -2(1) + 1(2) & -2(2) + 1(5) \end{pmatrix} = \begin{pmatrix} 1 & 0 \\ 0 & 1 \end{pmatrix}$

Therefore $B = A^{-1}$ since $AB = I$.

25. Represent the coefficients of the equation by a matrix, and multiply this by the solutions matrix, A.

$\begin{pmatrix} 3 & -2 \\ 4 & 1 \end{pmatrix}\begin{pmatrix} 1 \\ 2 \end{pmatrix} = \begin{pmatrix} 3(1) + -2(2) \\ 4(1) + 1(2) \end{pmatrix} = \begin{pmatrix} -1 \\ 6 \end{pmatrix}$

The matrix obtained properly represents the right-side values of the equations and is the proper matrix of solution values.

29. $I = \begin{pmatrix} 1 & 0 \\ 0 & 1 \end{pmatrix}$; $-I = \begin{pmatrix} -1 & 0 \\ 0 & -1 \end{pmatrix}$;

$(-I)^2 = \begin{pmatrix} -1 & 0 \\ 0 & -1 \end{pmatrix}\begin{pmatrix} -1 & 0 \\ 0 & -1 \end{pmatrix} = \begin{pmatrix} 1 + 0 & 0 + 0 \\ 0 + 0 & 0 + 1 \end{pmatrix} = \begin{pmatrix} 1 & 0 \\ 0 & 1 \end{pmatrix} = I$

33. $Sy^2 = \begin{pmatrix} 0 & -j \\ j & 0 \end{pmatrix}\begin{pmatrix} 0 & -j \\ j & 0 \end{pmatrix} = \begin{pmatrix} 0(0) + (-j)(j) & 0(-j) + (-j)(0) \\ j(0) + (0)(j) & (j)(-j) + (0)(0) \end{pmatrix}$

$= \begin{pmatrix} -j^2 & 0 \\ 0 & -j^2 \end{pmatrix} = \begin{pmatrix} -(\sqrt{-1})^2 & 0 \\ 0 & -(\sqrt{-1})^2 \end{pmatrix}$

$= \begin{pmatrix} -(-1) & 0 \\ 0 & -(-1) \end{pmatrix} = \begin{pmatrix} 1 & 0 \\ 0 & 1 \end{pmatrix} = I$

Exercises 15-5, page 449

1. $\begin{pmatrix} 2 & -5 \\ -2 & 4 \end{pmatrix}$

Interchange the elements of the principal diagonal and change the signs of the off-diagonal elements.

$\begin{pmatrix} 4 & 5 \\ 2 & 2 \end{pmatrix}$

Find the determinant of the original matrix.

$\begin{vmatrix} 2 & -5 \\ -2 & 4 \end{vmatrix} = -2$

Divide each element of the second matrix by -2.

$-\dfrac{1}{2}\begin{pmatrix} 4 & 5 \\ 2 & 2 \end{pmatrix} = \begin{pmatrix} -2 & -\frac{5}{2} \\ -1 & -1 \end{pmatrix}$

5. $\begin{pmatrix} 0 & -4 \\ 2 & 6 \end{pmatrix}$

Interchange the elements of the principal diagonal and change the signs of the off-diagonal elements.

$\begin{pmatrix} 6 & 4 \\ -2 & 0 \end{pmatrix}$

Find the determinant of the original matrix.

$\begin{vmatrix} 0 & -4 \\ 2 & 6 \end{vmatrix} = 8$

Divide each element of the second matrix by 8.

$\dfrac{1}{8}\begin{pmatrix} 6 & 4 \\ -2 & 0 \end{pmatrix} = \begin{pmatrix} \frac{3}{4} & \frac{1}{2} \\ -\frac{1}{4} & 0 \end{pmatrix}$

9. ① $\left(\begin{array}{cc|cc} 1 & 2 & 1 & 0 \\ 2 & 3 & 0 & 1 \end{array}\right)$; ② $\left(\begin{array}{cc|cc} 1 & 2 & 1 & 0 \\ 0 & -1 & -2 & 1 \end{array}\right)$; ③ $\left(\begin{array}{cc|cc} 1 & 0 & -3 & 2 \\ 0 & 1 & 2 & -1 \end{array}\right)$

$A^{-1} = \begin{pmatrix} -3 & 2 \\ 2 & -1 \end{pmatrix}$

① Original setup.
② -2 times row one added to row two
③ 2 times row two added to row two, and -1 times row two.

13. $\begin{pmatrix} 2 & 5 & | & 1 & 0 \\ -1 & 2 & | & 0 & 1 \end{pmatrix}$; ② $\begin{pmatrix} 1 & 7 & | & 1 & 1 \\ -1 & 2 & | & 0 & 1 \end{pmatrix}$; ③ $\begin{pmatrix} 1 & 7 & | & 1 & 1 \\ 0 & 9 & | & 1 & 2 \end{pmatrix}$

④ $\begin{pmatrix} 1 & 0 & | & \frac{2}{9} & -\frac{5}{9} \\ 0 & 9 & | & 1 & 2 \end{pmatrix}$; ⑤ $\begin{pmatrix} 1 & 0 & | & \frac{2}{9} & -\frac{5}{9} \\ 0 & 1 & | & \frac{1}{9} & \frac{2}{9} \end{pmatrix}$; $A^{-1} = \begin{pmatrix} \frac{2}{9} & -\frac{5}{9} \\ \frac{1}{9} & \frac{2}{9} \end{pmatrix}$

① Original setup.
② Add row two to row one.
③ Add row one to row two.
④ Row two multiplied by $-\frac{7}{9}$ and added to row one.
⑤ Row two divided by 9.

17. ① $\begin{pmatrix} 1 & -3 & -2 & | & 1 & 0 & 0 \\ -2 & 7 & 3 & | & 0 & 1 & 0 \\ 1 & -1 & -3 & | & 0 & 0 & 1 \end{pmatrix}$; ② $\begin{pmatrix} 1 & -3 & -2 & | & 1 & 0 & 0 \\ -2 & 7 & 3 & | & 0 & 1 & 0 \\ 0 & 2 & -1 & | & -1 & 0 & 1 \end{pmatrix}$

③ $\begin{pmatrix} 1 & -3 & -2 & | & 1 & 0 & 0 \\ 0 & 1 & -1 & | & 2 & 1 & 0 \\ 0 & 2 & -1 & | & -1 & 0 & 1 \end{pmatrix}$; ④ $\begin{pmatrix} 1 & -3 & -2 & | & 1 & 0 & 0 \\ 0 & 1 & -1 & | & 2 & 1 & 0 \\ 0 & 0 & 1 & | & -5 & -2 & 1 \end{pmatrix}$

⑤ $\begin{pmatrix} 1 & 0 & -5 & | & 7 & 3 & 0 \\ 0 & 1 & -1 & | & 2 & 1 & 0 \\ 0 & 0 & 1 & | & -5 & -2 & 1 \end{pmatrix}$; ⑥ $\begin{pmatrix} 1 & 0 & -5 & | & 7 & 3 & 0 \\ 0 & 1 & 0 & | & -3 & -1 & 1 \\ 0 & 0 & 1 & | & -5 & -2 & 1 \end{pmatrix}$

⑦ $\begin{pmatrix} 1 & 0 & 0 & | & -18 & -7 & 5 \\ 0 & 1 & 0 & | & -3 & -1 & 1 \\ 0 & 0 & 1 & | & -5 & -2 & 1 \end{pmatrix}$; $A^{-1} = \begin{pmatrix} -18 & -7 & 5 \\ -3 & -1 & 1 \\ -5 & -2 & 1 \end{pmatrix}$

① Original setup.
② −1 times row one added to row three.
③ 2 times row one added to row two.
④ −2 times row two added to row three.
⑤ 3 times row two added to row one.
⑥ Add row three to row two.
⑦ 5 times row three added to row one.

21. ① $\begin{pmatrix} 1 & 3 & 2 & | & 1 & 0 & 0 \\ -2 & -5 & -1 & | & 0 & 1 & 0 \\ 2 & 4 & 0 & | & 0 & 0 & 1 \end{pmatrix}$; ② $\begin{pmatrix} 1 & 3 & 2 & | & 1 & 0 & 0 \\ 0 & -1 & -1 & | & 0 & 1 & 1 \\ 2 & 4 & 0 & | & 0 & 0 & 1 \end{pmatrix}$; ③ $\begin{pmatrix} 1 & 3 & 2 & | & 1 & 0 & 0 \\ 0 & -1 & -1 & | & 0 & 1 & 1 \\ 0 & -2 & -4 & | & -2 & 0 & 1 \end{pmatrix}$

④ $\begin{pmatrix} 1 & 3 & 2 & | & 1 & 0 & 0 \\ 0 & 1 & 1 & | & 0 & -1 & -1 \\ 0 & -2 & -4 & | & -2 & 0 & 1 \end{pmatrix}$; ⑤ $\begin{pmatrix} 1 & 0 & -1 & | & 1 & 3 & 3 \\ 0 & 1 & 1 & | & 0 & -1 & -1 \\ 0 & -2 & -4 & | & -2 & 0 & 1 \end{pmatrix}$

⑥ $\begin{pmatrix} 1 & 0 & -1 & | & 1 & 3 & 3 \\ 0 & 1 & 1 & | & 0 & -1 & -1 \\ 0 & 0 & -2 & | & -2 & -2 & -1 \end{pmatrix}$; ⑦ $\begin{pmatrix} 1 & 0 & -1 & | & 1 & 3 & 3 \\ 0 & 1 & 1 & | & 0 & -1 & -1 \\ 0 & 0 & 1 & | & 1 & 1 & \frac{1}{2} \end{pmatrix}$

⑧ $\begin{pmatrix} 1 & 0 & -1 & | & 1 & 3 & 3 \\ 0 & 1 & 0 & | & -1 & -2 & -\frac{3}{2} \\ 0 & 0 & 1 & | & 1 & 1 & \frac{1}{2} \end{pmatrix}$; ⑨ $\begin{pmatrix} 1 & 0 & 0 & | & 2 & 4 & \frac{7}{2} \\ 0 & 1 & 0 & | & -1 & -2 & -\frac{3}{2} \\ 0 & 0 & 1 & | & 1 & 1 & \frac{1}{2} \end{pmatrix}$; $A^{-1} = \begin{pmatrix} 2 & 4 & \frac{7}{2} \\ -1 & -2 & -\frac{3}{2} \\ 1 & 1 & \frac{1}{2} \end{pmatrix}$

① Original setup.
② Row three added to row two.
③ -2 times row one added to row three.
④ -1 times row two.
⑤ -3 times row two added to row one.

⑥ 2 times row two added to row three.
⑦ Divide row three by -2.
⑧ -1 times row three added to row two.
⑨ Row three added to row one.

25.

$|A| = \begin{vmatrix} 1 & 3 & 2 \\ -2 & -5 & -1 \\ 2 & 4 & 0 \end{vmatrix} = \begin{vmatrix} -3 & -7 & 0 \\ -2 & -5 & -1 \\ 2 & 4 & 0 \end{vmatrix} = -(-1)\begin{vmatrix} -3 & -7 \\ 2 & 4 \end{vmatrix} = (-12 + 14) = 2$

$A^{-1} = \frac{1}{2} \begin{pmatrix} \begin{vmatrix} -5 & -1 \\ 4 & 0 \end{vmatrix} & -\begin{vmatrix} 3 & 2 \\ 4 & 0 \end{vmatrix} & \begin{vmatrix} 3 & 2 \\ -5 & -1 \end{vmatrix} \\ -\begin{vmatrix} -2 & -1 \\ 2 & 0 \end{vmatrix} & \begin{vmatrix} 1 & 2 \\ 2 & 0 \end{vmatrix} & -\begin{vmatrix} 1 & 2 \\ -2 & -1 \end{vmatrix} \\ \begin{vmatrix} -2 & -5 \\ 2 & 4 \end{vmatrix} & -\begin{vmatrix} 1 & 3 \\ 2 & 4 \end{vmatrix} & \begin{vmatrix} 1 & 3 \\ -2 & -5 \end{vmatrix} \end{pmatrix} = \frac{1}{2}\begin{pmatrix} 4 & 8 & 7 \\ -2 & -4 & -3 \\ 2 & 2 & 1 \end{pmatrix} = \begin{pmatrix} 2 & 4 & \frac{7}{2} \\ -1 & -2 & -\frac{3}{2} \\ 1 & 1 & \frac{1}{2} \end{pmatrix}$

29. $\frac{1}{ad-bc} \begin{pmatrix} a & b \\ c & d \end{pmatrix}\begin{pmatrix} d & -b \\ -c & a \end{pmatrix}$

$= \frac{1}{ad-bc}\begin{pmatrix} ad-bc & -ab+ab \\ cd-cd & -bc+ad \end{pmatrix} = \frac{1}{ad-bc}\begin{pmatrix} ad-bc & 0 \\ 0 & ad-bc \end{pmatrix}$

$= \begin{pmatrix} \frac{ad-bc}{ad-bc} & \frac{0}{ad-bc} \\ \frac{0}{ad-bc} & \frac{ad-bc}{ad-bc} \end{pmatrix} = \begin{pmatrix} 1 & 0 \\ 0 & 1 \end{pmatrix}$

Exercises 15-6, page 453

1. $C = \begin{pmatrix} -14 \\ 11 \end{pmatrix}$

From Exercise 1, Section 15-5, $A^{-1} = \begin{pmatrix} -2 & -\frac{5}{2} \\ -1 & -1 \end{pmatrix}$

$A^{-1}C = \begin{pmatrix} -2 & -\frac{5}{2} \\ -1 & -1 \end{pmatrix}\begin{pmatrix} -14 \\ 11 \end{pmatrix} = \begin{pmatrix} 28 & -\frac{55}{2} \\ 14 & -11 \end{pmatrix} = \begin{pmatrix} \frac{1}{2} \\ 3 \end{pmatrix}$

$x = \frac{1}{2}, \; y = 3$

5. $C = \begin{pmatrix} -8 \\ 19 \\ -3 \end{pmatrix}$

From Exercise 17, Section 15-5, $A^{-1} = \begin{pmatrix} -18 & -7 & 5 \\ -3 & -1 & 1 \\ -5 & -2 & 1 \end{pmatrix}$

$A^{-1}C = \begin{pmatrix} -18 & -7 & 5 \\ -3 & -1 & 1 \\ -5 & -2 & 1 \end{pmatrix}\begin{pmatrix} -8 \\ 19 \\ -3 \end{pmatrix} = \begin{pmatrix} 144 & -133 & -15 \\ 24 & -19 & -3 \\ 40 & -38 & -3 \end{pmatrix} = \begin{pmatrix} -4 \\ 2 \\ -1 \end{pmatrix}$

$x = -4, \; y = 2, \; z = -1$

9. $A = \begin{pmatrix} 2 & 7 \\ 1 & 4 \end{pmatrix}; \; C = \begin{pmatrix} 16 \\ 9 \end{pmatrix}$

Using the method of Example A, Section 15-5 to find A^{-1}:

$\begin{vmatrix} 2 & 7 \\ 1 & 4 \end{vmatrix} = 8 - 7 = 1; \; A^{-1} = \frac{1}{1}\begin{pmatrix} 4 & -7 \\ -1 & 2 \end{pmatrix}; \; A^{-1}C = \begin{pmatrix} 4 & -7 \\ -1 & 2 \end{pmatrix}\begin{pmatrix} 16 \\ 9 \end{pmatrix} = \begin{pmatrix} 64 & -63 \\ -16 & +18 \end{pmatrix} = \begin{pmatrix} 1 \\ 2 \end{pmatrix}$

$x = 1, \; y = 2$

13. $A = \begin{pmatrix} 5 & -2 \\ 3 & 4 \end{pmatrix}; \; C = \begin{pmatrix} -14 \\ -11 \end{pmatrix}$

Using the method of Example A, Section 15-5 to find A^{-1}:

$\begin{vmatrix} 5 & -2 \\ 3 & 4 \end{vmatrix} = 20 + 6 = 26; \; A^{-1} = \frac{1}{26}\begin{pmatrix} 4 & 2 \\ -3 & 5 \end{pmatrix} = \begin{pmatrix} \frac{2}{13} & \frac{1}{13} \\ -\frac{3}{26} & \frac{5}{26} \end{pmatrix}$

$A^{-1}C = \begin{pmatrix} \frac{2}{13} & \frac{1}{13} \\ -\frac{3}{26} & \frac{5}{26} \end{pmatrix}\begin{pmatrix} -14 \\ -11 \end{pmatrix} = \begin{pmatrix} -\frac{28}{13} & -\frac{11}{13} \\ \frac{42}{26} & -\frac{55}{26} \end{pmatrix} = \begin{pmatrix} -3 \\ -\frac{1}{2} \end{pmatrix}$

$x = -3$, $y = -\frac{1}{2}$

17. $A = \begin{pmatrix} 1 & 2 & 2 \\ 4 & 9 & 10 \\ -1 & 3 & 7 \end{pmatrix}$; $C = \begin{pmatrix} -4 \\ -18 \\ -7 \end{pmatrix}$

Using the method of Example D, Section 15-5 to find A^{-1}:

$$\left(\begin{array}{ccc|ccc} 1 & 2 & 2 & 1 & 0 & 0 \\ 4 & 9 & 10 & 0 & 1 & 0 \\ -1 & 3 & 7 & 0 & 0 & 1 \end{array}\right); \left(\begin{array}{ccc|ccc} 1 & 2 & 2 & 1 & 0 & 0 \\ 4 & 9 & 10 & 0 & 1 & 0 \\ 0 & 5 & 9 & 1 & 0 & 1 \end{array}\right); \left(\begin{array}{ccc|ccc} 1 & 2 & 2 & 1 & 0 & 0 \\ 0 & 1 & 2 & -4 & 1 & 0 \\ 0 & 5 & 9 & 1 & 0 & 1 \end{array}\right)$$

$$\left(\begin{array}{ccc|ccc} 1 & 2 & 2 & 1 & 0 & 0 \\ 0 & 1 & 2 & -4 & 1 & 0 \\ 0 & 0 & -1 & 21 & -5 & 1 \end{array}\right); \left(\begin{array}{ccc|ccc} 1 & 0 & -2 & 9 & -2 & 0 \\ 0 & 1 & 2 & -4 & 1 & 0 \\ 0 & 0 & -1 & 21 & -5 & 1 \end{array}\right)$$

$$\left(\begin{array}{ccc|ccc} 1 & 0 & -2 & 9 & -2 & 0 \\ 0 & 1 & 2 & -4 & 1 & 0 \\ 0 & 0 & 1 & -21 & 5 & -1 \end{array}\right); \left(\begin{array}{ccc|ccc} 1 & 0 & -2 & 9 & -2 & 0 \\ 0 & 1 & 0 & 38 & -9 & 2 \\ 0 & 0 & 1 & -21 & 5 & 1 \end{array}\right)$$

$$\left(\begin{array}{ccc|ccc} 1 & 0 & 0 & -33 & 8 & -2 \\ 0 & 1 & 0 & 38 & -9 & 2 \\ 0 & 0 & 1 & -21 & 5 & -1 \end{array}\right); A^{-1}C = \begin{pmatrix} -33 & 8 & -2 \\ 38 & -9 & 2 \\ -21 & 5 & -1 \end{pmatrix}\begin{pmatrix} -4 \\ -18 \\ -7 \end{pmatrix} = \begin{pmatrix} 132 - 144 + 14 \\ -152 + 162 - 14 \\ 84 - 90 + 7 \end{pmatrix}$$

$$\begin{pmatrix} 2 \\ -4 \\ 1 \end{pmatrix}$$

$x = 2$, $y = -4$, $z = 1$

21. $A \sin 47.2° + B \sin 64.4° = 254$

$A \cos 47.2° - B \cos 64.4° = 0$

$0.734A + 0.902B = 254$

$0.679A - 0.432B = 0$ $\dfrac{1}{ad - bc} = \dfrac{1}{-0.929}$

$C = \begin{pmatrix} 154 \\ 0 \end{pmatrix}$; $A^{-1} = \dfrac{1}{-0.929}\begin{pmatrix} -0.432 & -0.902 \\ -0.679 & 0.734 \end{pmatrix} = \begin{pmatrix} 0.465 & 0.989 \\ 0.731 & -0.789 \end{pmatrix}$

$A^{-1}C = \begin{pmatrix} 0.465 & 0.989 \\ 0.731 & -0.789 \end{pmatrix}\begin{pmatrix} 254 \\ 0 \end{pmatrix} = \begin{pmatrix} 0.465(254) + 0 \\ 0.731(254) + 0 \end{pmatrix} = \begin{pmatrix} 118 \\ 186 \end{pmatrix}$

$A = 118N$, $B = 186$

Review Exercises for Chapter 15, page 455

1. $\begin{vmatrix} 1 & 2 & -1 \\ 4 & 1 & -3 \\ -3 & -5 & 2 \end{vmatrix} = (1)\begin{vmatrix} 1 & -3 \\ -5 & 2 \end{vmatrix} - (2)\begin{vmatrix} 4 & -3 \\ -3 & 2 \end{vmatrix} + (-1)\begin{vmatrix} 4 & 1 \\ -3 & -5 \end{vmatrix}$

   $= 1(-13) - 2(-1) - 1(-17) = -13 + 2 + 17 = 6$

   The determinant was expanded by the first row.

5. $\begin{vmatrix} 2 & 6 & 2 & 5 \\ 2 & 0 & 4 & -1 \\ 4 & -3 & 6 & 1 \\ 3 & -1 & 0 & -2 \end{vmatrix} = -2\begin{vmatrix} 6 & 2 & 5 \\ -3 & 6 & 1 \\ -1 & 0 & -2 \end{vmatrix} - 4\begin{vmatrix} 2 & 6 & 5 \\ 4 & -3 & 1 \\ 3 & -1 & -2 \end{vmatrix} - 1\begin{vmatrix} 2 & 6 & 2 \\ 4 & -3 & 6 \\ 3 & -1 & 0 \end{vmatrix}$

   $= -2\left[-1\begin{vmatrix} 2 & 5 \\ 6 & 1 \end{vmatrix} - 2\begin{vmatrix} 6 & 2 \\ -3 & 6 \end{vmatrix}\right] - 4\left[-4\begin{vmatrix} 6 & 5 \\ -1 & -2 \end{vmatrix} + (-3)\begin{vmatrix} 2 & 5 \\ 3 & -2 \end{vmatrix} - 1\begin{vmatrix} 2 & 6 \\ 3 & -1 \end{vmatrix}\right]$

   $-1\left[3\begin{vmatrix} 6 & 2 \\ -3 & 6 \end{vmatrix} - (-1)\begin{vmatrix} 2 & 2 \\ 4 & 6 \end{vmatrix}\right]$

   $= -2[-1(-28) - 2(42)] - 4[-4(-7) - 3(-19) - 1(-20)] - 1[3(42) + 1(4)]$

   $= -2(-56) - 4(105) - 1(130) = -438$

9. $\begin{vmatrix} 1 & 2 & -1 \\ 4 & 1 & -3 \\ -3 & -5 & 2 \end{vmatrix} = \begin{vmatrix} 1 & 2 & 0 \\ 4 & 1 & 1 \\ -3 & -5 & -1 \end{vmatrix} = \begin{vmatrix} 1 & 0 & 0 \\ 4 & -7 & 1 \\ -3 & 1 & -1 \end{vmatrix}$

   $= 1\begin{vmatrix} -7 & 1 \\ 1 & -1 \end{vmatrix} = 1(7 - 1) = 6$

13. $\begin{vmatrix} 2 & 6 & 2 & 5 \\ 2 & 0 & 4 & -1 \\ 4 & -3 & 6 & 1 \\ 3 & -1 & 0 & -2 \end{vmatrix} = \begin{vmatrix} 2 & 6 & -2 & 5 \\ 2 & 0 & 0 & -1 \\ 4 & -3 & -2 & 1 \\ 3 & -1 & -6 & -2 \end{vmatrix} = \begin{vmatrix} 12 & 6 & -2 & 5 \\ 0 & 0 & 0 & -1 \\ 6 & -3 & -2 & 1 \\ -1 & -1 & -6 & -2 \end{vmatrix}$

   $= (-1)\begin{vmatrix} 12 & 6 & -2 \\ 6 & -3 & -2 \\ -1 & -1 & -6 \end{vmatrix} = (-1)\begin{vmatrix} 6 & 6 & -2 \\ 9 & -3 & -2 \\ 0 & -1 & -6 \end{vmatrix} = (-1)\begin{vmatrix} 6 & 6 & -38 \\ 9 & -3 & 16 \\ 0 & -1 & 0 \end{vmatrix}$

   $= -(-1)(-1)\begin{vmatrix} 6 & -38 \\ 9 & 16 \end{vmatrix} = -438$

17. $\begin{vmatrix} 1 & 0 & -3 & -2 \\ 1 & -1 & 2 & 0 \\ -1 & 1 & 1 & 1 \\ 5 & -1 & 2 & -1 \end{vmatrix} = \begin{vmatrix} 1 & 0 & -3 & -2 \\ 0 & 0 & 3 & 1 \\ -1 & 1 & 1 & 1 \\ 5 & -1 & 2 & -1 \end{vmatrix} = \begin{vmatrix} 1 & 0 & -3 & -2 \\ 0 & 0 & 3 & 1 \\ -1 & 1 & 1 & 1 \\ 4 & 0 & 3 & 0 \end{vmatrix}$

$= (-1)\begin{vmatrix} 1 & -3 & -2 \\ 0 & 3 & 1 \\ 4 & 3 & 0 \end{vmatrix} = (-1)\begin{vmatrix} 1 & -3 & -2 \\ 0 & 3 & 1 \\ 0 & 15 & 8 \end{vmatrix} = (-1)\begin{vmatrix} 3 & 1 \\ 15 & 8 \end{vmatrix}$

$= (-1)(24 - 15) = -9$

21. $\begin{pmatrix} 2a \\ a - b \end{pmatrix} = \begin{pmatrix} 8 \\ 5 \end{pmatrix}$; $2a = 8$; $a = 4$; $a - b = 5$; $4 - b = 5$; $b = -1$

25. $A + B = \begin{pmatrix} 2 & -3 \\ 4 & 1 \\ -5 & 0 \\ 2 & -3 \end{pmatrix} + \begin{pmatrix} -1 & 0 \\ 4 & -6 \\ -3 & -2 \\ 1 & -7 \end{pmatrix} = \begin{pmatrix} 1 & -3 \\ 8 & -5 \\ -8 & -2 \\ 3 & -10 \end{pmatrix}$

29. $A - C = \begin{pmatrix} 2 & -3 \\ 4 & 1 \\ -5 & 0 \\ 2 & -3 \end{pmatrix} - \begin{pmatrix} 5 & -6 \\ 2 & 8 \\ 0 & -2 \end{pmatrix}$

Cannot be subtracted since A and C do not have the same number of rows.

33. $\begin{pmatrix} 5 & -1 \\ 3 & 2 \end{pmatrix}\begin{pmatrix} 1 \\ -8 \end{pmatrix} = \begin{pmatrix} 5(1) + (-1)(-8) \\ 3(1) + 2(-8) \end{pmatrix} = \begin{pmatrix} 13 \\ -13 \end{pmatrix}$

37. $\begin{pmatrix} 2 & -5 \\ 2 & -4 \end{pmatrix}$

Interchange the elements of the principal diagonal and change the signs of the off-diagonal elements.

$\begin{pmatrix} -4 & 5 \\ -2 & 2 \end{pmatrix}$

Find the determinant of the original matrix.

$\begin{vmatrix} 2 & -5 \\ 2 & -4 \end{vmatrix} = 2$

Divide each element of the second matrix by 2.

$$\frac{1}{2}\begin{pmatrix} -4 & 5 \\ -2 & 2 \end{pmatrix} = \begin{pmatrix} -2 & \frac{5}{2} \\ -1 & 1 \end{pmatrix}$$

41. ① $\begin{pmatrix} 1 & 1 & -2 & | & 1 & 0 & 0 \\ -1 & -2 & 1 & | & 0 & 1 & 0 \\ 0 & 3 & 4 & | & 0 & 0 & 1 \end{pmatrix}$; ② $\begin{pmatrix} 1 & 1 & -2 & | & 1 & 0 & 0 \\ 0 & -1 & -1 & | & 1 & 1 & 0 \\ 0 & 3 & 4 & | & 0 & 0 & 1 \end{pmatrix}$; ③ $\begin{pmatrix} 1 & 1 & -2 & | & 1 & 0 & 0 \\ 0 & -1 & -1 & | & 1 & 1 & 0 \\ 0 & 0 & 1 & | & 3 & 3 & 1 \end{pmatrix}$

④ $\begin{pmatrix} 1 & 0 & -3 & | & 2 & 1 & 0 \\ 0 & -1 & -1 & | & 1 & 1 & 0 \\ 0 & 0 & 1 & | & 3 & 3 & 1 \end{pmatrix}$; ⑤ $\begin{pmatrix} 1 & 0 & -3 & | & 2 & 1 & 0 \\ 0 & -1 & 0 & | & 4 & 4 & 1 \\ 0 & 0 & 1 & | & 3 & 3 & 1 \end{pmatrix}$; ⑥ $\begin{pmatrix} 1 & 0 & 0 & | & 11 & 10 & 3 \\ 0 & -1 & 0 & | & 4 & 4 & 1 \\ 0 & 0 & 1 & | & 3 & 3 & 1 \end{pmatrix}$

⑦ $\begin{pmatrix} 1 & 0 & 0 & | & 11 & 10 & 3 \\ 0 & 1 & 0 & | & -4 & -4 & -1 \\ 0 & 0 & 1 & | & 3 & 3 & 1 \end{pmatrix}$

$$A^{-1} = \begin{pmatrix} 11 & 10 & 3 \\ -4 & -4 & -1 \\ 3 & 3 & 1 \end{pmatrix}$$

① Original setup.
② Row one added to row two.
③ 3 times row two added to row three.

④ Row two added to row one.
⑤ Row three added to row two.
⑥ 3 times row three added to row one.
⑦ -1 times row two.

45. $A = \begin{pmatrix} 2 & -3 \\ 4 & -1 \end{pmatrix}$; $C = \begin{pmatrix} -9 \\ -13 \end{pmatrix}$; $\begin{vmatrix} 2 & -3 \\ 4 & -1 \end{vmatrix} = 10$; $A^{-1} = \frac{1}{10}\begin{pmatrix} -1 & 3 \\ -4 & 2 \end{pmatrix} = \begin{pmatrix} -\frac{1}{10} & \frac{3}{10} \\ -\frac{4}{10} & \frac{2}{10} \end{pmatrix}$

$$A^{-1}C = \begin{pmatrix} -\frac{1}{10} & \frac{3}{10} \\ -\frac{4}{10} & \frac{2}{10} \end{pmatrix}\begin{pmatrix} -9 \\ -13 \end{pmatrix} = \begin{pmatrix} \frac{9}{10} & -\frac{39}{10} \\ \frac{36}{10} & -\frac{26}{10} \end{pmatrix} = \begin{pmatrix} -3 \\ 1 \end{pmatrix}$$

$x = -3$, $y = 1$

49. $A = \begin{pmatrix} 2 & -3 & 2 \\ 3 & 1 & -3 \\ 1 & 4 & 1 \end{pmatrix}$; $C = \begin{pmatrix} 7 \\ -6 \\ -13 \end{pmatrix}$

Using the method of Example D, Section 15-5 to find A^{-1}:

① $\left(\begin{array}{ccc|ccc} 2 & -3 & 2 & 1 & 0 & 0 \\ 3 & 1 & -3 & 0 & 1 & 0 \\ 1 & 4 & 1 & 0 & 0 & 1 \end{array}\right)$; ② $\left(\begin{array}{ccc|ccc} 2 & -3 & 2 & 1 & 0 & 0 \\ 3 & 1 & -3 & 0 & 1 & 0 \\ 0 & -11 & -6 & 0 & 1 & -3 \end{array}\right)$; ③ $\left(\begin{array}{ccc|ccc} 2 & -3 & 2 & 1 & 0 & 0 \\ 0 & 11 & -12 & -3 & 2 & 0 \\ 0 & -11 & -6 & 1 & 0 & -3 \end{array}\right)$

④ $\left(\begin{array}{ccc|ccc} 1 & -\frac{3}{2} & 1 & \frac{1}{2} & 0 & 0 \\ 0 & 11 & -12 & -3 & 2 & 0 \\ 0 & -11 & -6 & 0 & 1 & -3 \end{array}\right)$; ⑤ $\left(\begin{array}{ccc|ccc} 1 & -\frac{3}{2} & 1 & \frac{1}{2} & 0 & 0 \\ 0 & 11 & -12 & -3 & 2 & 0 \\ 0 & 0 & -18 & -3 & 3 & -3 \end{array}\right)$

⑥ $\left(\begin{array}{ccc|ccc} 1 & -\frac{3}{2} & 1 & \frac{1}{2} & 0 & 0 \\ 0 & 1 & -\frac{12}{11} & -\frac{3}{11} & \frac{2}{11} & 0 \\ 0 & 0 & -18 & -3 & 3 & -3 \end{array}\right)$

⑦ $\left(\begin{array}{ccc|ccc} 1 & 0 & -\frac{7}{11} & \frac{1}{11} & \frac{3}{11} & 0 \\ 0 & 1 & -\frac{12}{11} & -\frac{3}{11} & \frac{2}{11} & 0 \\ 0 & 0 & -18 & -3 & 3 & -3 \end{array}\right)$; ⑧ $\left(\begin{array}{ccc|ccc} 1 & 0 & -\frac{7}{11} & \frac{1}{11} & \frac{3}{11} & 0 \\ 0 & 1 & -\frac{12}{11} & -\frac{3}{11} & \frac{2}{11} & 0 \\ 0 & 0 & 1 & \frac{1}{6} & -\frac{1}{6} & \frac{1}{6} \end{array}\right)$

⑨ $\left(\begin{array}{ccc|ccc} 1 & 0 & -\frac{7}{11} & \frac{1}{11} & \frac{3}{11} & 0 \\ 0 & 1 & 0 & -\frac{1}{11} & 0 & \frac{2}{11} \\ 0 & 0 & 1 & \frac{1}{6} & -\frac{1}{6} & \frac{1}{6} \end{array}\right)$; ⑩ $\left(\begin{array}{ccc|ccc} 1 & 0 & 0 & \frac{13}{66} & \frac{11}{66} & \frac{7}{66} \\ 0 & 1 & 0 & -\frac{1}{11} & 0 & \frac{2}{11} \\ 0 & 0 & 1 & \frac{1}{6} & -\frac{1}{6} & \frac{1}{6} \end{array}\right)$

$$A^{-1}C = \begin{pmatrix} \frac{13}{66} & \frac{11}{66} & \frac{7}{66} \\ -\frac{1}{11} & 0 & \frac{2}{11} \\ \frac{1}{6} & -\frac{1}{6} & \frac{1}{6} \end{pmatrix} \begin{pmatrix} 7 \\ -6 \\ -13 \end{pmatrix} = \begin{pmatrix} \frac{91}{66} - \frac{66}{66} - \frac{91}{66} \\ -\frac{7}{11} + 0 - \frac{26}{11} \\ \frac{7}{6} + 1 - \frac{13}{6} \end{pmatrix} = \begin{pmatrix} -1 \\ -3 \\ 0 \end{pmatrix}$$

① Original.
② Row three multiplied by -3 and added to row two.
③ Row two multiplied by 2. Row one multiplied by -3 and added to row two.
④ Row one divided by 2.
⑤ Row two added to row three.

⑥ Row two divided by 11.

⑦ Row two multiplied by $\frac{3}{2}$ and added to row one.

⑧ Row three divided by −18.

⑨ Row three multiplied by $\frac{12}{11}$ and added to row two.

⑩ Row three multiplied by $\frac{7}{11}$ and added to row one.

53. $3x - 2y + z = 6$
$2x + 0y + 3z = 3$
$4x - y + 5z = 6$

The denominator will be found first.

$$\begin{vmatrix} 3 & -2 & 1 \\ 2 & 0 & 3 \\ 4 & -1 & 5 \end{vmatrix} = - \begin{vmatrix} 3 & -2 & 1 \\ 2 & 0 & 3 \\ -4 & 1 & -5 \end{vmatrix} = (-1) \begin{vmatrix} -5 & 0 & -9 \\ 2 & 0 & 3 \\ -4 & 1 & -5 \end{vmatrix}$$

$$= -(-1) \begin{vmatrix} -5 & -9 \\ 2 & 3 \end{vmatrix} = (1)(-15 + 18) = 3$$

$$x = \frac{\begin{vmatrix} 6 & -2 & 1 \\ 3 & 0 & 3 \\ 6 & -1 & 5 \end{vmatrix}}{3} = (-1)\frac{\begin{vmatrix} 6 & -2 & 1 \\ 3 & 0 & 3 \\ -6 & 1 & -5 \end{vmatrix}}{3} = (-1)\frac{\begin{vmatrix} -6 & 0 & -9 \\ 3 & 0 & 3 \\ -6 & 1 & -5 \end{vmatrix}}{3}$$

$$= (-1)(-1)\frac{\begin{vmatrix} -6 & -9 \\ 3 & 3 \end{vmatrix}}{3} = \frac{-18 + 27}{3} = 3$$

Substitute the value for x into the second equation.
$2(3) + 3z = 3$; $6 + 3z = 3$; $3z = -3$; $z = -1$

Substitute the values of x and z into the first equation.
$3(3) - 2y + (-1) = 6$; $9 - 2y - 1 = 6$; $-2y + 8 = 6$; $-2y = -2$; $y = 1$

57. $\begin{vmatrix} 1+\sqrt{2} & 2-\sqrt{3} & 0 \\ 3+\sqrt{5} & 7+\sqrt{6} & \sqrt{2} \\ 2+\sqrt{3} & 1-\sqrt{2} & 0 \end{vmatrix} = 0 \begin{vmatrix} 3+\sqrt{5} & 7+\sqrt{6} \\ 2+\sqrt{3} & 1-\sqrt{2} \end{vmatrix} - \sqrt{2} \begin{vmatrix} 1+\sqrt{2} & 2-\sqrt{3} \\ 2+\sqrt{3} & 1-\sqrt{2} \end{vmatrix}$

$+ 0 \begin{vmatrix} 1+\sqrt{2} & 2-\sqrt{3} \\ 3+\sqrt{5} & 7+\sqrt{6} \end{vmatrix}$

$= -\sqrt{2}(1+\sqrt{2})(1-\sqrt{2}) - (2+\sqrt{3})(2-\sqrt{3})$

$= -\sqrt{2}[(1-2) - (4-3)] = -\sqrt{2}[-1-1] = -\sqrt{2}(-2)$

$= 2\sqrt{2}$

61. $N = \begin{pmatrix} 0 & -1 \\ 1 & 0 \end{pmatrix}$; $N^{-1} = \dfrac{1}{0 - (-1)} \begin{pmatrix} 0 & 1 \\ -1 & 0 \end{pmatrix} = 1 \begin{pmatrix} 0 & 1 \\ -1 & 0 \end{pmatrix} = -1 \begin{pmatrix} 0 & -1 \\ 1 & 0 \end{pmatrix} = -N$

65. $A = \begin{pmatrix} 1 & -2 \\ 0 & 3 \end{pmatrix}$; $2A = \begin{pmatrix} 2 & -4 \\ 0 & 6 \end{pmatrix}$; $(2A)^{-1} = \dfrac{1}{12 + 0} \begin{pmatrix} 6 & 4 \\ 0 & 2 \end{pmatrix} = \begin{pmatrix} \frac{1}{2} & \frac{1}{3} \\ 0 & \frac{1}{6} \end{pmatrix}$

$\dfrac{A^{-1}}{2} = \dfrac{1}{2} \left(\dfrac{1}{3 - (0)} \right) \begin{pmatrix} 3 & 2 \\ 0 & 1 \end{pmatrix} = \dfrac{1}{6} \begin{pmatrix} 3 & 2 \\ 0 & 1 \end{pmatrix} = \begin{pmatrix} \frac{1}{2} & \frac{1}{3} \\ 0 & \frac{1}{6} \end{pmatrix} = (2A)^{-1}$

69. $0.500F = 0.866T$; $0.500F - 0.866T = 0$

$0.866F + 0.500T = 350$; $C = \begin{pmatrix} 0 \\ 350 \end{pmatrix}$

$A^{-1} = \dfrac{1}{1} \begin{pmatrix} 0.500 & 0.866 \\ -0.866 & 0.500 \end{pmatrix} = \begin{pmatrix} 0.500 & 0.866 \\ -0.866 & 0.500 \end{pmatrix}$

$A^{-1}C = \begin{pmatrix} 0.500 & 0.866 \\ -0.866 & 0.500 \end{pmatrix} \begin{pmatrix} 0 \\ 350 \end{pmatrix} = \begin{pmatrix} 0 + 0.866(350) \\ 0 + 0.500(350) \end{pmatrix} = \begin{pmatrix} 303 \\ 175 \end{pmatrix}$

73. Let A be the number of grams of alloy A and B be the number of grams of alloy B. Let C be the number of grams of alloy C.

Lead: $0.60A + 0.40B + 0.30C = 0.44(100)$; $6A + 4B + 3C = 440$
Zinc: $0.30A + 0.30B + 0.70C = 0.38(100)$; $3A + 3B + 7C = 380$
Copper: $0.10A + 0.30B = 0.18(100)$; $A + 3B = 180$

Using determinants with expansion by minors. Eq. 15-3

$A = \dfrac{\begin{vmatrix} 440 & 4 & 3 \\ 380 & 3 & 7 \\ 180 & 3 & 0 \end{vmatrix}}{\begin{vmatrix} 6 & 4 & 3 \\ 3 & 3 & 7 \\ 1 & 3 & 0 \end{vmatrix}} = \dfrac{440 \begin{vmatrix} 37 \\ 30 \end{vmatrix} - 380 \begin{vmatrix} 4 & 3 \\ 3 & 0 \end{vmatrix} + 180 \begin{vmatrix} 4 & 3 \\ 3 & 7 \end{vmatrix}}{6 \begin{vmatrix} 37 \\ 30 \end{vmatrix} - 3 \begin{vmatrix} 4 & 3 \\ 3 & 0 \end{vmatrix} + 1 \begin{vmatrix} 4 & 3 \\ 3 & 7 \end{vmatrix}} = \dfrac{440(-21) - 380(-9) + 180(19)}{6(-21) - 3(-9) + 1(19)}$

$= \dfrac{-2400}{-80} = 30$ g

Substituting into the third equation, $30 + 3B = 180$; $B = 50$ g; substituting into the first equation $6(30) + 4(50) + 3C = 440$; $C = 20$ g.

77.

		Standard Transmission	Automatic Transmission		Standard Transmission	Automatic Transmission
4 cylinders	$A =$	15 000	10 000	$B =$	18 000	12 000
6 cylinders		20 000	18 000		30 000	22 000
8 cylinders		8 000	30 000		12 000	40 000

$A + B = \begin{pmatrix} 33\ 000 & 22\ 000 \\ 50\ 000 & 40\ 000 \\ 20\ 000 & 70\ 000 \end{pmatrix}$

CHAPTER 16

Inequalities

Exercises 16-1, page 465

1. $4 + 3 < 9 + 3$; $7 < 12$; property (1)

5. $\dfrac{4}{-1} > \dfrac{9}{-1}$; $-4 > -9$; property (3)

9. $x > -2$

13. $1 < x < 7$

17. $x < 1$, or $3 < x \leq 5$

21. x is greater than 0 and less than or equal to 2

25.
-1 0 1 2 3

33.
-4 -3 -2 -1 0 1 2 3 4

41. $0 \leq n \leq 2565$

29.
-1 0 1 2 3 4 5 6

37. $d > 3 \times 10^{12}$

37M. $d > 5 \times 10^{12}$

Exercises 16-2, page 468

1. $x - 3 > -4$

   $x > -4 + 3$

   $x > -1$

   -1 0 1

5. $3x - 5 \leq -11$

   $3x \leq -11 + 5$

   $3x \leq -6$

   $x \leq -2$
   -3 -2 -1 0

9. $4x - 5 \leq 2x$

$4x - 2x \leq 5$

$2x \leq 5$

$x \leq {}^5\!/_2$

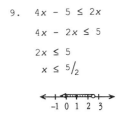

13. $x + 4 \geq 3(x - 3)$

$x + 4 \geq 3x - 9$

$4 + 9 \geq 3x - x$

$13 \geq 2x$

${}^{13}\!/_2 \geq x$

$x \leq {}^{13}\!/_2$

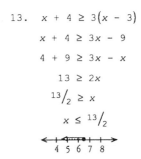

17. $-1 < 2x + 1 < 3$

$-2 < 2x < 2$

$-1 < x < 1$

21. $2x < x - 1 \leq 3x + 5$

$0 < -x - 1 \leq x + 5$

$0 < -x - 1$ and $-x - 1 \leq x + 5$

$x < -1$ and $-6 \leq 2x$

$x < -1$ and $x \geq -3$

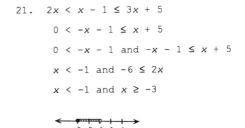

25. $v = 40\ 000 + 4000t; \ v \leq 64\ 000$ *and* $t \geq 0$

$0 \leq 4000t + 40\ 000 \leq 64\ 000$

$0 \leq 4000t + 40\ 000$ and $4000t + 40\ 000 \leq 64\ 000$

$4000t \geq -40\ 000$ $4000t \leq 24\ 000$

$\left\{ \begin{array}{l} t \geq -10 \\ t \geq 0 \end{array} \right\}$ $t \leq 6$

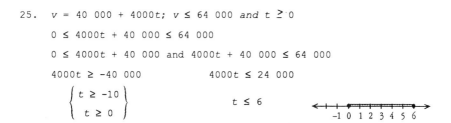

29. $x \leq 800 - 300$ $y = x + 400 - 200$

$x \leq 500$ $0 \leq x \leq 500$

$0 + 400 - 200 \leq x + 400 - 200 \leq 500 + 400 - 200$

$200 \leq y \leq 700$

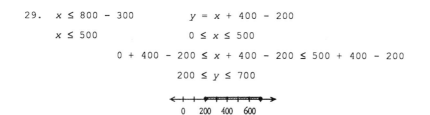

Exercises 16-3, page 474

1. $x^2 - 1 < 0$

   $(x + 1)(x - 1) < 0$

   Critical values are

   $x = -1$ and $x = 1$

	$(x + 1)$	$(x - 1)$	SIGN
$x < -1$	−	−	+
$-1 < x < 1$	+	−	−
$x > 1$	+	+	+

$(x + 1)(x - 1) < 0$ for $-1 < x < 1$

-2 -1 0 1 2

5. $3x^2 + 5x \geq 2$

   $3x^2 + 5x - 2 \geq 0$

   $(3x - 1)(x + 2) \geq 0$

   Critical values are

   $x = \frac{1}{3}$ and $x = -2$

	$(3x - 1)$	$(x + 2)$	SIGN
$x < -2$	−	−	+
$-2 < x < \frac{1}{3}$	−	+	−
$x > \frac{1}{3}$	+	+	+

$(3x - 1)(x + 2) \geq 0$ for $x \leq -2$ or $x \geq \frac{1}{3}$

-3 -2 -1 0 1 2

9. $x^2 + 4x \leq -4$

   $x^2 + 4x + 4 \leq 0$

   $(x + 2)(x + 2) \leq 0$

   Critical value is

   $x = -2$

	$(x + 2)$	$(x + 2)$	SIGN
$x < -2$	−	−	+
$x > -2$	+	+	+

$(x + 2)(x + 2) \leq 0$ has only $x = -2$ for a solution since all other products are positive

-3 -2 -1 0 1 2

13. $x^3 + x^2 - 2x > 0$

    $x(x^2 + x - 2) > 0$

    $x(x - 1)(x + 2) > 0$

    Critical values are

    $x = 0$, $x = 1$, and $x = -2$

	(x)	$(x - 1)$	$(x + 2)$	SIGN
$x < -2$	−	−	−	−
$-2 < x < 0$	−	−	+	+
$0 < x < 1$	+	−	+	−
$x > 1$	+	+	+	+

$x(x - 1)(x + 2) > 0$ for $-2 < x < 0$ or $x > 1$

-3 -2 -1 0 1 2 3

17. $\dfrac{x - 8}{3 - x} < 0$

Critical values are

$x = 8$ and $x = 3$

	$(x - 8) \div (3 - x)$		SIGN
$x < 3$	−	+	−
$3 < x < 8$	−	+	+
$x > 8$	+	−	−

$\dfrac{x - 8}{3 - x} < 0$ for $x < 3$ or $x > 8$

2 3 4 5 6 7 8 9

21. $\dfrac{2}{x^2 - x - 2} < 0$

$\dfrac{2}{(x - 2)(x + 1)} < 0$

Critical values are

$x = 2$ and $x = -1$

	$2 / (x - 2)(x + 1)$			SIGN
$x < 1$	+	−	−	+
$-1 < x < 2$	+	+	−	−
$x > 2$	+	+	+	+

$\dfrac{2}{(x + 2)(x - 1)} < 0$ for $-1 < x < 2$

-2 -1 0 1 2 3 4

25. $\dfrac{6 - x}{3 - x - 4x^2} \ge 0; \; x \ne -1, \; x \ne {}^3/_4$

$\dfrac{(6 - x)}{(1 + x)(3 - 4x)} \ge 0$

Critical values are

$x = 6$, $x = -1$, and $x = {}^3/_4$

	$(6 - x) / (1 + x)(3 - 4x)$			SIGN
$x < -1$	+	−	+	−
$-1 < x < {}^3/_4$	+	+	+	+
${}^3/_4 < x < 6$	+	+	−	−
$x > 6$	−	+	−	+

$\dfrac{6 - x}{(1 + x)(3 - 4x)} \ge 0$ for $-1 < x < {}^3/_4$ or $x \ge 6$

-2 -1 0 1 2 3 4 5 6 7 8

29. $\sqrt{(x - 1)(x + 2)}$ is real if

$(x - 1)(x + 2) \ge 0$

Critical values are

$x = 1$ and $x = -2$

	$(x - 1)(x + 2)$		SIGN
$x < -2$	−	−	+
$-2 < x < 1$	−	+	−
$x > 1$	+	+	+

$(x - 1)(x + 2) \ge 0$ for $x \le -2$ or $x \ge 1$

-3 -2 -1 0 1 2

33. $x^3 > 2; \; x^3 - 2 > 0$

Graph $y = x^3 - 2$

$x^3 - 2 > 0$ for $x > 1.3$

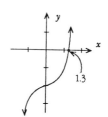

37. $2^x > 3$; graph $y = 2^x$

x	0	0.5	1	1.5	2	2.5
2^x	1	1.4	2	2.8	4	5.7

$x = 1.6$

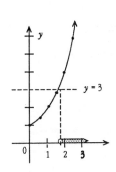

41. $p = 6i - 4i^2$; $6i - 4i^2 > 2$ and $i > 0$

$4i^2 - 6i + 2 < 0$

$2i^2 - 3i + 1 < 0$

$(2i - 1)(i - 1) < 0$

Critical values are

$i = 0.5$ and $i = 1$

	$(2i - 1)$	$(i - 1)$	SIGN
$i < 0.5$	−	−	+
$0.5 < i < 1$	+	−	−
$i > 1$	+	+	+

$(2i - 1)(i - 1) < 0$ for $0.5 < i < 1$

45. $\ell = w + 2.0$; $w(w + 2.0) < 35$; $w \geq 3.0$

$w^2 + 2.0w - 35 < 0$

$(w + 7)(w - 5) < 0$

Critical values are

$w = -7$ and $w = 5$

	$(w + 7)$	$(w - 5)$	SIGN
$w < -7$	−	−	+
$-7 < w < 5$	+	−	−
$w > 5$	+	+	+

$(w + 7)(w - 5) < 0$ for $-7 < w < 5$; however, w must be positive and $w \geq 3.0$, so $3.0 \leq w \leq 5.0$

Exercises 16-4,   page 477

1. $|x - 4| < 1$

$x - 4 < 1$ or $x - 4 > -1$

$x < 5$ or $x > 3$

$3 < x < 5$

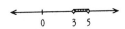

5. $|6x - 5| \leq 4$

$6x - 5 \leq 4$ or $6x - 5 \geq -4$

$6x \leq 9$ $6x \geq 1$

$x \leq {}^2/_3$ $x \geq {}^1/_6$

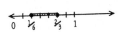

9. $\left|\dfrac{x + 1}{5}\right| < 3$

$\dfrac{x + 1}{5} < 3$ or $\dfrac{x + 1}{5} > -3$

$x + 1 < 15$ $x + 1 > -15$

$x < 14$ $x > -16$

-16   0   14

13. $2\,|x - 4| > 8$

$|x - 4| > 4$

$x - 4 > 4$ or $x - 4 < -4$

$x > 8$ $x < 0$

0   8

17. $\left|\dfrac{x}{2} + 1\right| < 8$

$\dfrac{x}{2} + 1 < 8$ or $\dfrac{x}{2} + 1 > -8$

$\dfrac{x}{2} < 7$ $\dfrac{x}{2} > -9$

$x < 14$ $x > -18$

-18   0   14

21. $\left|x^2 + x - 4\right| > 2$

$x^2 + x - 4 > 2$ or $x^2 + x - 4 < -2$

$x^2 + x - 6 > 0$ $x^2 + x - 2 < 0$

(A) $(x + 3)(x - 2)$

(B) $(x - 1)(x + 2)$

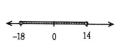

(A)   Critical values are $x = -3$, $x = 2$.

	$(x + 3)$	$(x - 2)$	SIGN
$x < -3$	−	−	+
$-3 < x < 2$	+	−	−
$x > 2$	+	+	+

$(x + 3)(x - 2) > 0$ for $x < -3$ or
for $x > 2$

(B)   Critical values are $x = 1$, $x = -2$

	$(x - 1)$	$(x + 2)$	SIGN
$x < -2$	−	−	+
$-2 < x < 1$	−	+	−
$x > 1$	+	+	+

$(x - 1)(x + 2) < 0$ for $-2 < x < 1$

The solution is (A) and (B)
combined: $x < -3$, $-2 < x < 1$,
and $x > 2$

25. $0.2537 \pm 0.0003 = d$ implies $0.2537 - 0.0003 \le d \le 0.2537 + 0.0003$

$-0.0003 \le d - 0.2537 \le 0.0003$

$|d - 0.2537| \le 0.0003$

Exercises 16-5, page 482

1. $y > x - 1$; graph $y = x - 1$. Use a dashed line
    to indicate that points on it do not satisfy the
    inequality. Shade the region above the line.

5. $2x + y < 5$; $y < -2x + 5$; graph $y = -2x + 5$.
    Use a dashed line to indicate that points on it
    do not satisfy the inequality. Shade the region
    below the line.

9. $y < x^2$; graph $y = x^2$. Use a dashed line to
    indicate that points on it do not satisfy the
    inequality. Shade the region below the curve.

13. $x^2 + 2x + y < 0$; $y < -x^2 - 2x$. Graph
    $y = -x^2 - 2x$. Use a dashed line to
    indicate that points on it do not
    satisfy the inequality. Shade the region
    below the curve.

17. $y > x^4 - 8$, Graph $y = x^4 - 8$,
    using a dashed line. Shade the
    region above the line.

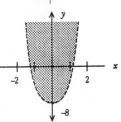

21. $y < \ln x$. Graph $y = \ln x$. Use a dashed
    line to show that the points on it do not
    satisfy the inequality. Shade the region
    below the line.

25. $y > x$ and $y > 1 - x$. Graph $y = x$ using a
    dashed line. Shade the region above the
    line. Graph $y = 1 - x$ using a dashed line.
    Shade the region above the line. The
    region where the shadings overlap satisfies
    both inequalities.

29. $y > \frac{1}{2}x^2$ and $y \le 4x - x^2$
    $$y - 4 \le -(x^2 - 4x + 4)$$
    $$y - 4 \le -(x - 2)^2$$
    Graph $y = \frac{1}{2}x^2$ with a dashed line, and shade
    the region above the line. Graph $y = 4x - x^2$
    with a solid line, and shade below it. The
    region in the intersection of the two shaded
    regions satisfies both inequalities.

The profit is calculated using the coordinates of the vertices. $p = 8(80) + 10(0) = 640$; $p = 8(0) + 10(80) = 800$; $p = 8(40) + 10(60) = 920$. The maximum profit occurs if 40 business models and 60 scientific models are produced.

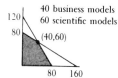

40 business models
60 scientific models
(40,60)

33. $A \leq 300$; $200 \leq B \leq 400$

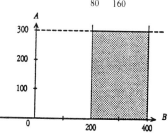

37.   Let x = the number of business models produced. Let y = the number of scientific models produced. Let p = the profit. $p = 8x + 10y$. The maximum time worked = 8 hours or 480 minutes. For the first operation, $3x + 6y \leq 480$. For the second operation, $6x + 4y \leq 480$. $3x + 6y \leq 480$; $y \leq -\frac{1}{2}x + 80$. Graph $y = -\frac{1}{2}x + 80$ and shade the region below the graph. $6x + 4y \leq 480$; $y \leq -\frac{3}{2}x + 120$; graph $y = -\frac{3}{2}x + 120$ and shade the region below the graph. The overlapping shaded regions satisfy both inequalities. Vertices of the region are (80,0), (0,80) and the point p. p is the intersection of $y = -\frac{1}{2}x + 80$ and $y = -\frac{3}{2}x + 120$. Solve these equations simultaneously by substitution. $-\frac{1}{2}x + 80 = -\frac{3}{2}x + 120$; $x = 40$, $y = 60$.

Review Exercises for Chapter 16, page 484

1. $2x - 12 > 0$; $2(x - 6) > 0$. The critical value is 6. If $x > 6$, $2x - 12 > 0$. Thus the values which satisfy the inequality are $x > 6$.

5. $3(x - 7) \geq 5x + 8$
      $3x - 21 \geq 5x + 8$
      $-2x \geq 29$
      $x \leq -\frac{29}{2}$ or $-14\frac{1}{2}$

9. $2x < x + 1 < 4x + 7$
      $2x < x + 1$ and $x + 1 < 4x + 7$
      $x < 1$ and $-3x < 6$
      $x < 1$ and $x > -2$
      $-2 < x < 1$

13. $x^2 + 2x > 63$; $x^2 + 2x - 63 > 0$; $(x + 9)(x - 7) > 0$. The critical values are -9, 7. Determine the sign for the intervals $x < -9$, $-9 < x < 7$, and $x > 7$.

interval	$(x + 9)(x - 7)$		sign of $(x + 9)(x - 7)$
$x < -9$	$-$	$-$	$+$
$-9 < x < 7$	$+$	$-$	$-$
$x > 7$	$+$	$+$	$+$

The inequality is positive for $x < -9$ or $x > 7$.

17. $\dfrac{x - 8}{2x + 1} \leq 0$. Critical values are $-\dfrac{1}{2}$, 8. Determine the sign for the intervals $x < -\dfrac{1}{2}$, $-\dfrac{1}{2} < x < 8$, and $x > 8$.

interval	$\dfrac{x - 8}{2x + 1}$	sign of $\dfrac{x - 8}{2x + 1}$
$x < -\dfrac{1}{2}$	$\dfrac{-}{-}$	$+$
$-\dfrac{1}{2} < x < 8$	$\dfrac{-}{+}$	$-$
$x > 8$	$\dfrac{+}{+}$	$+$

The inequality is zero for $x = 8$. It is undefined for $x = -\dfrac{1}{2}$. Therefore, the solution is $-\dfrac{1}{2} < x \leq 8$.

21. $x^4 + x^2 \leq 0$. The left side of the inequality is positive or zero for all values of x. There are no values, except $x = 0$.

25. $|x - 2| > 3$; $x - 2 < -3$ or $x - 2 > 3$; $x < -1$ or $x > 5$

29. $|3 - 5x| > 7$
    $3 - 5x > 7$ or $3 - 5x < -7$
    $-5x > 4$ or $-5x < -10$
    $x < {}^{-4}/5$ or $x > 2$

$\longleftarrow$ ⚬━━⚬┼━┼━⚬⚬⟶
-2 -1 0 1 2

33. $x^3 + x + 1 < 0$; Graph $x^3 + x + 1 = y$.

x	-1	0	1	...
y	-1	1	3	

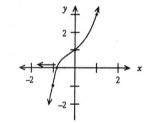

37. $\sqrt{3 - x}$ will be a real number if $3 - x \geq 0$. The critical value is $+3$. If $x \leq 3$, $3 - x \geq 0$. Thus the values which satisfy the inequalities are $x \leq 3$.

41. $y > 4 - x$. Graph $y = 4 - x$. Use a dashed line to indicate that points on it do not satisfy the inequality. Shade the region above the line.

45. $y > x^2 + 1$. Graph $y = x^2 + 1$. Use a
     dashed line to indicate that points
     on it do not satisfy the inequality.
     Shade the region above the line.

49. If $x > 0$, $\dfrac{1}{x} > 0$. If $y < 0$, $\dfrac{1}{y} < 0$. Therefore $\dfrac{1}{x} > \dfrac{1}{y}$.

53. $L = 150 + 0.00025T$
     $150 + 0.00025T \geq 151$
     $\qquad 0.00025T \geq 1$
     $\qquad\qquad\quad T \geq 4000°C$

57. $p = Ri^2 = 12.0i^2$; $2.50 < 12.0i^2 < 8.00$

$$\frac{2.50}{12.0} < i^2 < \frac{8.00}{12.0}$$

$$\sqrt{\frac{2.50}{12.0}} < i < \sqrt{\frac{8.00}{12.0}}$$

$$0.46 < i < 0.82$$

61. Let R = minutes research time
     $\quad\;\; D$ = minutes development time

     $R \leq 1200$
     $D \leq 1000$

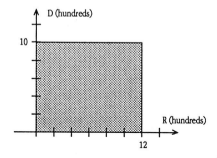

CHAPTER 17

Variation

1. $\dfrac{18V}{3V} = 6$ 5. $\dfrac{20 \text{ qt}}{25 \text{ gal}} = \dfrac{20 \text{ qt}}{100 \text{ qt}} = \dfrac{1}{5}$ 5M. $\dfrac{20 \text{ h}}{25 \text{ min}} = \dfrac{1200 \text{ min}}{25 \text{ min}} = 48$

9. $\dfrac{1250 \text{ mi/h}}{740 \text{ mi/h}} = 1.69$ 9M. $\dfrac{2000 \text{ km/h}}{1200 \text{ km/h}} = 1.67$

13. $\dfrac{2540 - 2450}{2450} = \dfrac{90}{2450} = 0.037 = 3.7\%$

17. $\dfrac{6.00}{R_2} = \dfrac{62.5}{15.0}$

   $62.5R_2 = 90; \quad R_2 = 1.44 \ \Omega$

21. $\dfrac{1.00}{6.45} = \dfrac{x}{36.3}; \quad x = \dfrac{1.00(36.3)}{6.45} = 5.63$

21M. $\dfrac{1 \text{ kg}}{1000 \text{ g}} = \dfrac{20.0 \text{ kg}}{x}; \quad x = 1000(20.0) = 20\ 000$

25M. $\dfrac{360}{2\pi} = \dfrac{x}{5.00}; \quad x = \dfrac{360(5.00)}{2\pi} = 286°$

25M. $\dfrac{360}{2\pi} = \dfrac{x}{5.00}; \quad x = \dfrac{360(5.00)}{2\pi} = 286°$

29. $\dfrac{60\ 000}{2.00} = \dfrac{x}{0.75}; \quad x = \dfrac{60\ 000(0.75)}{2.00} = 22\ 500$

33. Let x be the number of mg in the dose for the 60-kg patient. The total mass of both patients is 132 kg. Let y be the dosage for the other patient.

   $\dfrac{60}{132} = \dfrac{x}{220}; \quad x = 100 \text{ mg}$ $x + y = 220; \ y = 120 \text{ mg}$

Exercises 17-2, page 495

1. $y = kz$

5. $f = k\sqrt{x}$

9. $r = k/y;\ r = 2,\ y = 8$
    $2 = k/8$
    $k = 2(8) = 16$
    $r = \dfrac{16}{y}$ is the equation.

13. $s = \dfrac{k}{\sqrt{t}};\ s = \dfrac{1}{2},\ t = \sqrt{49}$
     $\dfrac{1}{2} = \dfrac{k}{\sqrt{49}}$
     $2k = \sqrt{49} = 7$
     $k = \dfrac{7}{2};$ therefore $s = \dfrac{7}{2\sqrt{t}}$

17. $y = kx;\ 20 = k(8);\ k = 2.5;\ y = 2.5x$
     $y = 2.5(10) = 25$

21. $y = \dfrac{kx}{z};\ 60 = \dfrac{k(4)}{10};\ k = 150;\ y = \dfrac{150x}{z}$
     $y = \dfrac{150(6)}{5} = 180$

25. $\dfrac{75}{160} = \dfrac{x}{130};\ x = \dfrac{75(130)}{160} = 61$

29. $\dfrac{I}{E} = \dfrac{3600}{5400} = 0.67$

33. $\dfrac{7500}{t + 1} = \dfrac{7500}{6.0} = 1250$

37. $F = kAv^2;\ 19.2 = k(3.72)(31.4)^2;\ k = 5.23 \times 10^{-3}$
     $F = 5.23 \times 10^{-3}Av^2$

37M. $F = kAv^2;\ 76.5 = k(0.372)(9.42)^2;\ k = 2.32$
      $F = 2.32\ Av^2$

41. $R = \dfrac{k\ell}{A};\ 0.200 = \dfrac{k(200)}{0.0500}$
     $k = 5.00 \times 10^{-5}\ \dfrac{\Omega \cdot \text{in.}^2}{\text{ft}}$
     $R = \dfrac{5.00 \times 10^{-5}\ \ell}{A}$

45. $G = \dfrac{kd^2}{\lambda^2};\ 5.5 \times 10^4 = \dfrac{k(2.9)^2}{(0.030)^2};\ k = \dfrac{5.5 \times 10^4(0.030)^2}{(2.9)^2} = 5.9$
     $G = \dfrac{5.9d^2}{\lambda^2}$

Review Exercises for Chapter 17, page 497

1. $\dfrac{4\ mg}{20\ kg} = \dfrac{4000\ kg}{20\ kg} = 200$ 5. $\dfrac{28\ kN}{5000\ N} = \dfrac{28\ kN}{5kN} = 5.6$

9. $\dfrac{1.00\ in.}{16.0\ mi} = \dfrac{x}{52.0\ mi}$ 9M. $\dfrac{1.00\ cm}{16.0\ km} = \dfrac{x}{52.0\ km}$

$x = 3.25\ in.$ $x = 3.25\ cm$

13. $\dfrac{3600}{30\ s} = \dfrac{x}{300\ s}$ since 5 min = 300 s

$30x = (360)(300)$
$x = 36000$ characters in 5 min

17. $\dfrac{25.0\ ft}{2.00\ in.} = \dfrac{x}{5.75\ in.}$; $x = 71.9\ ft$ 17M. $\dfrac{25.0\ m}{20.0\ mm} = \dfrac{x}{57.5\ mm}$; $x = 71.9\ m$

21. $y = kx^2$; $27 = k(3^2)$; $k = 3$; $y = 3x^2$

25. $\dfrac{F_1}{F_2} = \dfrac{L_2}{L_1}$; $F_1 = 4.50$, $F_2 = 6.75$, $L_1 = 17.5$ 25M. $L_2 = 11.7$ cm

$\dfrac{4.50}{6.75} = \dfrac{L_2}{17.5}$

$(6.75)L_2 = (4.50)(17.5)$
$L_2 = 11.7$ in.

29. $p = kA$; $30.0 = k(8.00)$; $k = 3.75\ \dfrac{hp}{in.^2}$
$p = 3.75A$; $p = 3.75(6.00) = 22.5$ hp

29M. $p = kA$; $22.5 = k(50.0)$; $k = 0.450\ \dfrac{kW}{cm^2}$
$p = 0.450(40.0) = 18.0$ kW

33. $\dfrac{(\log 8000)^2}{(\log 2000)^2} = 8000$ ⬚log ⬚x^2 ⬚+ 2000 ⬚log ⬚x^2 ⬚= 1.4

37. $f = \dfrac{k}{\lambda}$; $v = kf\lambda = 90.9 \times 10^6(3.29) = 299 \times 10^6 = 2.99 \times 10^8$

41. $1200 = \dfrac{k(30)}{8.0}$; $k = \dfrac{9600}{30} = 320$; $x = \dfrac{320(90)}{6.0} = 4800$

41M. $1.2 = \dfrac{k(30)}{20}$; $k = \dfrac{24}{30} = 0.8$; $x = \dfrac{0.8(90)}{15} = 4.8$

45. $V = \dfrac{kr^4}{d}$; $V^1 = \dfrac{k(1.25r)^4}{0.98d} = \dfrac{2.44kr^4}{0.98d} = 2.49\left(\dfrac{kr^4}{d}\right) = 2.49V$

An increase of $V^1 - V = 2.49V - V = 1.49V$ or 149% increase.

CHAPTER 18

Progressions and the Binomial Theorem

Exercises 18-1, page 505

1. 4, 6, 8, 10, 12

5. $a_8 = 1 + (8 - 1)(3) = 22$

9. $a_{12} = -7 + (11)4 = 37$

13. $S_{20} = \dfrac{20}{2}(4 + 40) = 440$

17. $45 = 5 + (n - 1) 8$
$45 = 8n - 3$
$n = 6$
$S_6 = \dfrac{6}{2}(5 + 45) = 150$

21. $a_{30} = a_1 + (29)(3) = a_1 + 87$
$1875 = \dfrac{30}{2}(a_1 + a_1 + 87)$
$125 = 2a, + 87$
$a_1 = 19; \; a_{30} = 106$

25. $a_n = -5k + (n - 1)(\tfrac{1}{2}k)$
$S_n = \dfrac{n}{2}\left(-5k + (-5k + (n - 1)(\tfrac{1}{2}k))\right)$
$\dfrac{23}{2}k = \dfrac{n}{2}\left(-5k - 5k + \tfrac{1}{2}kn - \tfrac{1}{2}kn\right)$
$23k = n\left(-\dfrac{21}{2}k + \tfrac{1}{2}kn\right)$
$46k = n(-21k + kn)$
$46k = -21kn + kn^2$
$kn^2 - 21kn - 46k = 0$
$n^2 - 21n - 46 = 0$
$(n + 2)(n - 23) = 0$
$n = -2$ (not valid)
$n = 23; \; a_n = -5k + (22)(\tfrac{1}{2}k) = -5k + 11k = 6k$

29. $d = \dfrac{72 - 56}{10 - 6} = 4; \; a_6 = 56 = a_1 + (5)(4); \; a_1 = 56 - 20 = 36$
$S_{10} = 5(36 + 72) = 540$

33. $a_1 = 1, \; a_n = 100, \; n = 100; \; S_{100} = \dfrac{100}{2}(1 + 100) = 50(101) = 5050$

37. $a_1 = 16.5°$, $d = -1.25°$, $n = 12$; $a_{12} = 16.5° + (11)(-1.25°) = 2.75°$

41. $a_1 = 1800$, $d = -150$, $a_n = 0$;

$0 = 1800 + (n - 1)(-150) = 1800 - 150n + 150$

$150n = 1800 + 150 = 1950$;

$n = 13$ (12 more years)

$S_{13} = \dfrac{13}{2}(1800 + 0) = \$11,700$, the sum of all depreciations which is

the cost of the car

45. $S_n = \dfrac{n}{2}(a_1 + a_n) = \dfrac{n}{2}\left[a_1 + \left(a_1 + (n - 1)\right)d\right] = \dfrac{n}{2}\left[2a_1 + (n - 1)d\right]$

Exercises 18-2, page 509

1. $45\left(\dfrac{1}{3}\right)^{1-1}$, $45\left(\dfrac{1}{3}\right)^{2-1}$, $45\left(\dfrac{1}{3}\right)^{3-1}$, $45\left(\dfrac{1}{3}\right)^{4-1}$, $45\left(\dfrac{1}{3}\right)^{5-1}$ is 45, 15, 5, $\dfrac{5}{3}$, $\dfrac{5}{9}$.

5. $r = 1 \div \dfrac{1}{2} = 2$, $a = \dfrac{1}{2}$, $n = 6$

$a_n = \dfrac{1}{2}(2)^{6-1} = \dfrac{1}{2}(32) = 16$

9. $a_n = -27\left(-\dfrac{1}{3}\right)^{6-1} = -27\left(-\dfrac{1}{243}\right) = \dfrac{1}{9}$

13. $S_n = \dfrac{\frac{1}{8}\left(1 - 4^5\right)}{1 - 4} = \dfrac{-1023}{8(-3)} = \dfrac{341}{8}$

17. From Eq. (18-3), $a_n = \left(\dfrac{1}{16}\right)(4)^{6-1} = \left(\dfrac{1}{16}\right)(4)^5$

$= (4^{-2})(4^5) = 4^3 = 64$

From Eq. (18-4), $S_n = \dfrac{\frac{1}{16}(1 - 4^6)}{1 - 4} = \dfrac{\frac{1}{16}(1 - 4096)}{-3}$

$= \dfrac{\frac{-4095}{16}}{-3} = \dfrac{4095}{48} = \dfrac{1365}{16}$

21. From Eq. (18-3), $27 = a_1 r^{4-1}$; $a_1 r^3 = 27$; $a_1 = \dfrac{27}{r^3}$

From Eq. (18-4), $40 = a_1\dfrac{(1 - 4^4)}{1 - r} = a_1\dfrac{(1 + r^2)(1 + r)(1 - r)}{1 - r}$

$= a_1(1 + r^2)(1 + r)$

Substitute the value of a from the first equation in the second equation:

$40 = \dfrac{27}{r^3}(1 + r^2)(1 + r)$; $40r^3 = 27 + 27r + 27r^2 + 27r^3$

$13r^3 - 27r^2 - 27r - 27 = 0$

Possible rational roots are ±1, ±3, ±9, ±27, $\pm\dfrac{1}{13}$, $\pm\dfrac{3}{13}$, $\pm\dfrac{9}{13}$, $\pm\dfrac{27}{13}$.

Using synthetic division, 3 gives a remainder of zero. Therefore, $r = 3$. From Eq. (18-3), $27 = a(3^{4-1})$; $27a = 27$; $a = 1$

25. ___ ___ ___ 8 ___ ___ 16 . Find r by letting $a = 8$, $a_n = 16$, $n = 4$.

From Eq. (18-3), $16 = 8r^{4-1}$; $8r^3 = 16$; $r = \sqrt[3]{2}$
From Eq. (18-3), find a by letting $a_n = 8$, $\sqrt[3]{2}$, $n = 4$; $8 = a_1^3 \sqrt{2}^{\,4-3}$
$2a = 8$; $a = 4$
From Eq. (18-3), find the 10th term by letting $n = 10$, $a_1 = 4$, $r = \sqrt[3]{2}$
$a_n = 4 \sqrt[3]{2}^{\,10-1}$; $a_n = 32$

29. $8.2 \mu s$ will be 8.2 more terms, so $n = 9.2$.

12.5% decrease means $r = (1 - 0.125) = 0.875$.

$a_1 = 3.27$

$a_n = 3.27(0.875)^{8.2} = 0.875$, $\boxed{x^y}$, 8.2, $\boxed{=}$, $\boxed{\times}$ 3.27, $\boxed{=}$ 1.09

33. $a_1 = 80.8$, $n = 12$, $r = 0.85$

$$S_{12} = \frac{80.8(1 - 0.85^{12})}{1 - 0.85} = \frac{80.8(1 - 0.142)}{0.15} = \frac{80.8(0.858)}{0.15} = 462$$

37. $a_1 = 0.01$ (dollars); $r = 2$; $n = 27 (a_1 + 26$ more$)$

$$S_n = \frac{0.01(1 - 2^{27})}{1 - 2} = \frac{0.01(1 - 1.3 \times 10^8)}{-1} = \frac{0.01(1.3 \times 10^8)}{-1}$$

$S_n = 0.01(1.3 \times 10^8) = 1.3 \times 10^6$ or 1.3 million dollars.

Answer will vary according to the number of decimal places the equipment is capable of handling. Theoretically, numbers in this problem should not be rounded at all.

Exercises 18-3, page 513

1. $a_1 = 4$, $r = \dfrac{1}{2}$

By Eq. (18-5), $s = \dfrac{4}{1 - \dfrac{1}{2}} = 8$

5. $a_1 = 20$, $r = -\dfrac{1}{20}$

By Eq. (18-5), $s = \dfrac{20}{1 + 1/20} = \dfrac{20}{21/20} = \dfrac{400}{21}$

9. $a_1 = 1$, $r = \dfrac{1}{10,000}$ 9M. $s = \dfrac{10\ 000}{9999}$

By Eq. (18-5), $s = \dfrac{1}{1 - 1/10,000} = \dfrac{1}{9999/10,000} = \dfrac{10,000}{9999}$

13. $a_1 = 0.3$, $r = 0.1$

By Eq. (18-5), $s = \dfrac{0.3}{1 - 0.1} = \dfrac{0.3}{0.9} = \dfrac{1}{3}$

17. $a_1 = 0.18$, $r = 0.01$

$$s = \frac{0.18}{1 - 0.01} = \frac{2}{11}$$

21. $0.36666... = 0.3 + 0.06666... = \frac{3}{10} + 0.06666...$

For the GP $0.06666...$, $a = 0.06$, $r = 0.01$

$$s = \frac{0.06}{1 - 0.1} = \frac{0.06}{0.9} = \frac{1}{15}$$

Therefore $0.36666... = \frac{3}{10} + \frac{1}{15} = \frac{11}{30}$

25. $r = 0.92$, $a_1 = 28.0$

$$s = \frac{28.0}{1 - 0.92} = \frac{28.0}{0.08} = 350$$

Exercises 18-4, page 517

1. $(t + 1)^3 = t^3 + 3t^2(1) + \dfrac{3(2)}{2}t(1)^2 + \dfrac{3(2)(1)(1)}{3(2)}$

$$= t^3 + 3t^2 + 3t + 1$$

5. $(2x + 3)^5$

$(2x)^5 + 5(2x)^4(3) + \dfrac{5(4)}{2}(2x)^3(3)^2 + \dfrac{5(4)(3)}{2(3)}(2x)^2(3)^3$

$\qquad + \dfrac{5(4)(3)(2)}{2(3)(4)}(2x)(3)^4 + (3)^5$

$= 32x^5 + 240x^4 + 720x^3 + 1080x^2 + 810x + 243$

9.   From Pascal's triangle, the coefficients for $n = 4$ are 1, 4, 6, 4, 1.
$(5x - 3)^4 = [5x + (-3)]^4$
$\qquad\qquad = 1(5x)^4 + 4(5x)^3(-3) + 6(5x)^2(-3)^2 + 4(5x)(-3)^3 + (-3)^4$
$\qquad\qquad = 625x^4 - 1500x^3 + 1350x^2 - 540x + 81$

13. $(x + 2)^{10} = x^{10} + 10x^9(2) + \dfrac{10 \times 9}{2}x^8(2^2) + \dfrac{10 \times 9 \times 8}{3 \times 2}x^7(2)^3$

$$= x^{10} + 20x^9 + 180x^8 + 960x^7 + \cdots$$

17. $(x^2 - \frac{y}{2})^{12}$

$= (x^2)^{12} + 12(x^2)^{12-1}(-\frac{y}{2}) + \frac{12(12-1)}{2!}(x^2)^{12-2}(-\frac{y}{2})^2$

$\quad + \frac{12(12-1)(12-2)}{3!}(x^2)^{12-3}(-\frac{y}{2})^3 + \cdots$

$= x^{24} + 12x^{22}(-\frac{y}{2}) + 66x^{20}(-\frac{y}{2})^2 + 220x^{18}(-\frac{y}{2})^3 + \cdots$

$= x^{24} - 6x^{22}y + \frac{33}{2}x^{20}y^2 - \frac{55}{2}x^{18}y^3 + \cdots$

21. $(1 + x)^8 = 1 + 8x + \frac{8(8-1)}{2!}x^2 + \frac{8(8-1)(8-2)}{3!}x^3 + \cdots$

$\quad = 1 + 8x + 28x^2 + 56x^3 + \cdots$

25. $\sqrt{1 + x} = (1 + x)^{1/2}$

$\quad = 1 + \frac{1}{2}x + \frac{\frac{1}{2}(\frac{1}{2} - 1)}{2!}x^2 + \frac{\frac{1}{2}(\frac{1}{2} - 1)(\frac{1}{2} - 2)}{3!}x^3 + \cdots$

$\quad = 1 + \frac{1}{2}x - \frac{1}{8}x^2 + \frac{1}{16}x^3 + \cdots$

29. (a) $17! + 4! = 17$, $\boxed{x!}$, $\boxed{+}$, $\boxed{4}$, $\boxed{x!}$, $\boxed{=}$, 3.557×10^{14}

33. The term involving b^5 will be the sixth term.
   $r = 5, n = 8$
   The sixth term is $\dfrac{8(7)(6)(5)(4)}{5(4)(3)(2)}a^3b^5 = 56a^3b^5$.

37. $V = A(1 - r)^5$; expand $(1 - r)^5$

$\quad 1^5 + 5(1)^4(-r) + \frac{5(4)}{2}(1)^3(-r)^2 + \frac{5(4)(3)}{2(3)}(1)^2(-r)^3$

$\quad\quad + \frac{5(4)(3)(2)}{2(3)(4)}(1)(-r)^4 + (-r)^5$

$\quad 1 - 5r + 10r^2 - 10r^3 + 5r^4 - r^5$

$\quad A(1 - r)^5 = A(1 - 5r + 10r^2 - 10r^3 + 5r^4 - r^5)$

Review Exercises for Chapter 18, page 519

1. $d = 6$; $a_n = 1 + (17 - 1)5 = 1 + 80 = 81$ Eq. (18-1)

5. $d = \frac{7}{2} - 8 = -\frac{9}{2}$; $a_n = 8 + (16 - 1)(-\frac{9}{2}) = 8 - \frac{135}{2} = -\frac{119}{2}$ Eq. (18-1)

9. $s_n = \dfrac{15}{2}(-4 + 17) = \dfrac{15}{2}(13) = \dfrac{195}{2}$ Eq. (18-2)

13. From Eq. (18-1), $a_n = 17 + (9 - 1)(-2) = 17 - 16 = 1$
    From Eq. (18-2), $s_n = \dfrac{9}{2}(1 + 17) = 81$

17. $s_n = \dfrac{n}{2}(a + a_n)$; $a_1 = 8$, $a_n = -2.5$, $s_n = 22$

    $22 = \dfrac{n}{2}(8 - 2.5)$ $\qquad\qquad a_n = a_1 + (n - 1)d$

    $44 = n(8 - 2.5) = 5.5n$ $\qquad -2.5 = 8 + (8 - 1)d = 8 + 7d$

    $n = \dfrac{44}{5.5} = 8$ $\qquad\qquad\qquad d = \dfrac{-10.5}{7} = -1.5$

21. From Eq. (18-2), $s = \dfrac{12}{2}(-1 + 32) = 6(31) = 186$

25. $r = {}^6\!/_9 = {}^2\!/_3$; $s = \dfrac{a_1}{1 - r} = \dfrac{9}{1 - {}^2\!/_3} = \dfrac{9}{{}^1\!/_3} = 27$

29. $0.030303... = 0.03 + 0.0003 + 0.000003...$
    $a_1 = 0.03$; $r = 0.01$

    $s = \dfrac{0.03}{1 - 0.01} = \dfrac{0.03}{0.99} = \dfrac{3}{99} = \dfrac{1}{3}$

33. $(x - 2)^4 = [x + (-2)]^4$

    $\qquad = x^4 + 4x^3(-2) + \dfrac{4(3)x^2(-2)^2}{2} + \dfrac{4(3)(2)}{3(2)}x(-2)^3 + (-2)^4$

    $\qquad = x^4 - 8x^3 + 24x^2 - 32x + 16$

37. $(a + 2b^2)^{10} = a^{10} + 10a^{10-1}(2b^2) + \dfrac{10(10 - 1)}{2!}a^{10-2}(2b^2)^2$

    $\qquad + \dfrac{10(10 - 1)(10 - 2)}{3!}a^{10-3}(2b^2)^3 + \cdots$

    $\qquad = a^{10} + 10a^9(2b^2) + 45a^8(4b^4) + 120a^7(8b^6) + \cdots$
    $\qquad = a^{10} + 20a^9b^2 + 180a^8b^4 + 960a^7b^6 + \cdots$

41. $(1 + x)^{12} = 1 + 12x + \dfrac{12(12 - 1)}{2!}x^2 + \dfrac{12(12 - 1)(12 - 2)}{3!}x^3 + \cdots$

    $\qquad = 1 + 12x + 66x^2 + 220x^3 + \cdots$

45. $[1 + (-a^2)]^{1/2} = 1 + \dfrac{1}{2}(-a^2) + \dfrac{(\frac{1}{2})(-\frac{1}{2})(-a^2)^2}{2} + \dfrac{\frac{1}{2}(-\frac{1}{2})(-\frac{3}{2})(-a^2)}{3(2)}$

    $\qquad = 1 - \dfrac{1}{2}a^2 - \dfrac{1}{8}a^4 - \dfrac{1}{16}a^6 + \cdots$ Eq. (18-7)

49. $a = 2$, $d = 2$, $n = 1000$

By Eq. (18-1), $a_n = 2 + (1000 - 1)2 = 2 + 999(2) = 2000$

By Eq. (18-2), $s = \dfrac{1000}{2}(2 + 2000) = 1,001,000$

53. n = lens thickness in mm; $a_n = a_1 (r^{n-1})$

$r = (1 - 0.12) = 0.88$ $0.20 = 1.00(0.88^{n-1}) = 0.88^{n-1}$

$a_1 = 100\% = 1.00$ $\log 0.20 = (-1)\log 0.88$

$a_n = 20\% = 0.20$ $n - 1 = \dfrac{\log 0.20}{\log 0.88}$

$n - 1 = 12.6$; $n = 13.6$

57. 6% annually = 1.5% quarterly; $r = 1.015$

3 yrs = 12 quarters; $n = 12$

$a_n = 2000(1.015)^{11} = 2000(1.178) \approx \$2356$

Answers will vary according to the number of decimal places the equipment can handle.

61. The value of each investment is $a_n = 1000(1.035)^n$;

(n = twice the number of invested years.) The total value is the sum of each investment value. Investment values (20 terms) are as follows:

$1000(1.035)^{40}$, $1000(1.035)^{38}$, $1000(1.035)^{36}$, $\ldots$ $1000(1.035)^2$

These terms are changing by a factor of 1.035^{-2}

$$s = \frac{\left[1000(1.035)^{40}\right]\left[1 - \left(1.035^{-2}\right)^{20}\right]}{1 - (1.035)^{-2}} = \frac{\left[1000(1.035)^{40}\right]\left[1 - 1.035^{-40}\right]}{\left[1 - 1.035^{-2}\right]}$$

$$\approx \frac{\left[3959.26\right]\left[0.7474275\right]}{0.0664893} = \$44,507.31$$

Value increases as additional decimal places are used.

65. Let $x = \dfrac{a - 1}{2}\, m^2$ and $y = \dfrac{a}{a - 1}$.

$(1 + x)^y = 1 + yx + \dfrac{y(y - 1)}{2} x^2 \ldots$ (3 terms)

$$= 1 + \left(\frac{a}{a - 1}\right)\left(\frac{a - 1}{2} m^2\right) + \frac{\left(\dfrac{a}{a - 1}\right)\left(\dfrac{a}{a - 1} - 1\right)}{2}\left(\frac{a - 1}{2} m^2\right)^2$$

$$= 1 + \frac{a}{2} m^2 + \frac{\left(\dfrac{a}{a - 1}\right)\left(\dfrac{1}{a - 1}\right)}{2}\left(\frac{(a - 1)^2}{2^2} m^4\right)$$

$$= 1 + \frac{a}{2} m^2 + \frac{a}{2(a - 1)^2}\left(\frac{(a - 1)^2}{2^2} m^4\right)$$

$$= 1 + \frac{a}{2} m^2 + \frac{a}{2^3} m^4 = 1 + \frac{1}{2} am^2 + \frac{1}{8} am^4$$

CHAPTER 19

Additional Topics in Trigonometry

Exercises 19-1, page 528

1. Verify $\tan \theta = \dfrac{1}{\cot \theta}$ for $\theta = 56°$.

   $\tan 56° = 1.483$; $\cot 56° = 0.6745$

   $\dfrac{1}{0.6745} = 1.483 = \tan 56°$

5. $\dfrac{\cot \theta}{\cos \theta} = \cot \theta \times \dfrac{1}{\cos \theta}$

   $= \dfrac{\cos \theta}{\sin \theta} \times \dfrac{1}{\cos \theta} = \dfrac{1}{\sin \theta} = \csc \theta$

9. $\sin y \cot y = \dfrac{\sin y}{1} \times \dfrac{\cos y}{\sin y} = \cos y$

13. $\csc^2 x (1 - \cos^2 x) = \dfrac{1}{\sin^2 x} \times \dfrac{(\sin^2 x)}{1}$

    $= \dfrac{\sin^2 x}{\sin^2 x} = 1$

17. $\sin x (\csc x - \sin x) = \sin x (\dfrac{1}{\sin x} - \sin x)$

    $= 1 - \sin^2 x = \cos^2 x$

21. $\sin x \tan x + \cos x = \dfrac{\sin x}{1} \times \dfrac{\sin x}{\cos x} + \dfrac{\cos x}{1}$

    $= \dfrac{\sin^2 x}{\cos x} + \dfrac{\cos x}{1} = \dfrac{\sin^2 x + \cos^2 x}{\cos x}$

    $= \dfrac{1}{\cos x} = \sec x$

213

25. $\sec \theta \tan \theta \csc \theta = \dfrac{1}{\cos \theta} \times \dfrac{\sin \theta}{\cos \theta} \times \dfrac{1}{\sin \theta}$

$$= \dfrac{1}{\cos^2 \theta} \times \dfrac{\sin \theta}{\sin \theta} = \sec^2 \theta = \tan^2 \theta + 1$$

29. $\tan x + \cot x = \dfrac{\sin x}{\cos x} + \dfrac{\cos x}{\sin x}$

$$= \dfrac{\sin^2 x + \cos^2 x}{\cos x \sin x} = \dfrac{1}{\cos x \sin x}$$

$$= \sec x \csc x$$

33. $\dfrac{\sin x}{1 - \cos x} = \dfrac{\sin x(1 + \cos x)}{(1 - \cos x)(1 + \cos x)}$

$$= \dfrac{\sin x(1 + \cos x)}{1 - \cos^2 x} = \dfrac{\sin x(1 + \cos x)}{\sin^2 x}$$

$$= \dfrac{1 + \cos x}{\sin x} = \dfrac{1}{\sin x} + \dfrac{\cos x}{\sin x} = \csc x + \cot x$$

37. $\tan^2 x \cos^2 x + \cot^2 x \sin^2 x = \dfrac{\sin^2 x}{\cos^2 x} \times \dfrac{\cos^2 x}{1} + \dfrac{\cos^2 x}{\sin^2 x} \times \dfrac{\sin^2 x}{1}$

$$= \sin^2 x + \cos^2 x = 1$$

41. $\dfrac{1 - 2\cos^2 x}{\sin x \cos x} = \dfrac{(\sin^2 x + \cos^2 x) - 2\cos^2 x}{\sin x \cos x}$

$$= \dfrac{\sin^2 x - \cos^2 x}{\sin x \cos x} = \dfrac{\sin^2 x}{\sin x \cos x} - \dfrac{\cos^2 x}{\sin x \cos x}$$

$$= \dfrac{\sin x}{\cos x} - \dfrac{\cos x}{\sin x} = \tan x - \cot x$$

45. $\sec x + \tan x + \cot x = \dfrac{1}{\cos x} + \dfrac{\sin x}{\cos x} + \dfrac{\cos x}{\sin x}$

$$= \dfrac{\sin x + \sin^2 x + \cos^2 x}{\cos x \sin x} = \dfrac{1 + \sin x}{\cos x \sin x}$$

49. $(\tan x + \cot x) \sin x \cos x = (\tan x) \sin x \cos x + (\cot x) \sin x \cos x$

$$= (\dfrac{\sin x}{\cos x}) \sin x \cos x + (\dfrac{\cos x}{\sin x}) \sin x \cos x$$

$$= \sin^2 x + \cos^2 x = 1$$

53. $\sec x \, (\sec x - \cos x) + \dfrac{\cos x - \sin x}{\cos x} + \tan x = \sec^2 x$

$\sec x \left(\sec x - \dfrac{1}{\sec x}\right) + \dfrac{\cos x}{\cos x} - \dfrac{\sin x}{\cos x} + \tan x =$

$\sec^2 x - 1 + 1 - \dfrac{\sin x}{\cos x} + \dfrac{\sin x}{\cos x} = \sec^2 x$

57. $\cos\theta = \cos A \cos B \cos C + \sin A \sin B; \quad \theta = 90°, \ \cos\theta = 0$

$0 = \cos A \cos B \cos C + \sin A \sin B$

$\dfrac{\cos A \cos B}{\cos A \cos B}\cos C = \dfrac{-\sin A \sin B}{\cos A \cos B}$

(1) $\cos C = -\tan A \tan B$

61. $x = \cos\theta; \ \sqrt{1 - x^2} = \sqrt{1 - \cos^2\theta} = \sqrt{\sin^2\theta} = \sin\theta$

Exercises 19-2, page 533

1. Given: $105° = 60° + 45°$
    $\sin(\alpha + \beta) = \sin\alpha \cos\beta + \cos\alpha \sin\beta$
    $\sin(105°) = \sin(60° + 45°)$

    $\sin 60° \cos 45° + \cos 60° \sin 45° = \dfrac{\sqrt{3}}{2} \times \dfrac{\sqrt{2}}{2} + \dfrac{1}{2} \times \dfrac{\sqrt{2}}{2}$

    $= \dfrac{\sqrt{6}}{4} + \dfrac{\sqrt{2}}{4} = \dfrac{\sqrt{6} + \sqrt{2}}{4} = 0.9659$

5. $\sin\alpha = \dfrac{4}{5}$ (Q I); $\cos^2\alpha = 1 - \sin^2\alpha$; $\cos^2\alpha = 1 - \dfrac{16}{25}$; $\cos^2\alpha = \dfrac{9}{25}$;

    $\cos\alpha = \dfrac{3}{5}$

    $\cos\beta = -\dfrac{12}{13}$ (Q II); $\sin^2\beta = 1 - \cos^2\beta$

    $\sin^2\beta = 1 - \dfrac{144}{169}$; $\sin^2\beta = \dfrac{25}{169}$

    $\sin\beta = \dfrac{5}{13}$ ($\sin\beta$ is + in Q II.)

    $\sin(\alpha + \beta) = \sin\alpha \cos\beta + \cos\alpha \sin\beta$

    $\qquad = \dfrac{4}{5}(-\dfrac{12}{13}) + \dfrac{3}{5}(\dfrac{5}{13}) = -\dfrac{48}{65} + \dfrac{15}{65} = -\dfrac{33}{65}$

9. $\sin\alpha \cos\beta + \sin\beta \cos\alpha = \sin(\alpha + \beta)$
    $\sin x \cos 2x + \sin 2x \cos x = \sin(x + 2x) = \sin 3x$

13. $\cos 1 \cos(1 - x) - \sin 1 \sin(1 - x)$; Let $\alpha = 1$ and $\beta = (1 - x)$
    $\cos\alpha \cos\beta - \sin\alpha \sin\beta = \cos(\alpha + \beta) = \cos(1 + 1 - x)$
    $\qquad\qquad\qquad\qquad\qquad\qquad = \cos(2 - x)$

17. sin 122° cos 32° − cos 122° sin 32° is of the form:
     sin α cos β − cos α sin β where α = 122° and β = 32°.
     sin α cos β − cos α sin β = sin (α − β) so
     sin 122° cos 32° − cos 122° sin 32° = sin(122° − 32°)
     $$= \sin 90° = 1$$

21. sin (180° + x) = sin 180° cos x + sin x cos 180°
     $$= (0) \cos x + \sin x(-1)$$
     $$= -\sin x$$

25. sin(270° − x) = sin 270° cos x − sin x cos 270°
     $$= (-1) \cos x - (\sin x)(0) = -\cos x - 0 = -\cos x$$

29. $\cos (30° + x) = \cos 30° \cos x - \sin 30° \sin x = \dfrac{\sqrt{3}}{2} \cos x - \dfrac{1}{2} \sin x$
     $$= \frac{\sqrt{3} \cos x}{2} - \frac{\sin x}{2} = \frac{\sqrt{3} \cos x - \sin x}{2}$$
     $$= \frac{1}{2}(\sqrt{3} \cos x - \sin x)$$

33. $\sin(x + y) \sin(x - y)$
     $$= (\sin x \cos y + \cos x \sin y)(\sin x \cos y - \cos x \sin y)$$
     $$= \sin^2 x \cos^2 y - \cos^2 x \sin^2 y = \sin^2 x(1 - \sin^2 y) - (1 - \sin^2 x)(\sin^2 y)$$
     $$= \sin^2 x - \sin^2 x \sin^2 y - \sin^2 y + \sin^2 x \sin^2 y$$
     $$= \sin^2 x - \sin^2 y$$

37. $\tan(\alpha + \beta) = \dfrac{\sin(\alpha + \beta)}{\cos(\alpha + \beta)} = \dfrac{\sin \alpha \cos \beta + \cos \alpha \sin \beta}{\cos \alpha \cos \beta - \sin \alpha \sin \beta}$

     (divide numerator and denominator by cos α cos β)

     $$= \frac{\dfrac{\sin \alpha \cos \beta}{\cos \alpha \cos \beta} + \dfrac{\cos \alpha \sin \beta}{\cos \alpha \cos \beta}}{\dfrac{\cos \alpha \cos \beta}{\cos \alpha \cos \beta} - \dfrac{\sin \alpha \sin \beta}{\cos \alpha \cos \beta}} = \frac{\dfrac{\sin \alpha}{\cos \alpha} + \dfrac{\sin \beta}{\cos \beta}}{1 - \dfrac{\sin \alpha}{\cos \alpha} \times \dfrac{\sin \beta}{\cos \beta}}$$

     $$= \frac{\tan \alpha + \tan \beta}{1 - \tan \alpha \tan \beta}$$

41. $\alpha + \beta = x$; $\alpha - \beta = y$; $\alpha = \dfrac{1}{2}(x + y)$; $\beta = \dfrac{1}{2}(x - y)$ Eq. (19-13)

     sin x + sin y = sin (α + β) + sin (α − β)
     $$= 2[\frac{1}{2} \sin (\alpha + \beta) + \frac{1}{2} \sin(\alpha - \beta)]$$
     $$= 2 \sin \alpha \cos \beta = 2 \sin \frac{1}{2}(x + y) \cos \frac{1}{2}(x - y)$$

45. $i_0 \sin(\omega t + \alpha) = i_0[\sin \omega t \cos \alpha + \sin \alpha \cos \omega t]$
     $$= i_0 \cos \alpha \sin \omega t + i_0 \sin \alpha \cos \omega t$$
     $$= i_1 \sin \omega t + i_2 \cos \omega t$$

Exercises 19-3, page 538

1. $60° = 2(30°)$; $\sin 2\alpha = 2 \sin \alpha \cos \alpha$
     $\sin 2(30°) = 2 \sin 30° \cos 30°$
     $$= 2\left(\frac{1}{2}\right)\left(\frac{\sqrt{3}}{2}\right) = \frac{\sqrt{3}}{2} = \frac{1}{2}\sqrt{3}$$

5. $\sin 258° = -0.9781476$
     $\sin 258° = \sin 2(129°) = 2 \sin 129° \cos 129° = -0.9781476$
     using calculator sequence 129, $\boxed{\sin}$, $\boxed{\times}$, 129, $\boxed{\cos}$, $\boxed{\times}$, 2 $\boxed{=}$.

9.   Given: $\cos x = \frac{4}{5}$ (Q I)
     $\sin^2 x = 1 - \cos^2 x = 1 - \left(\frac{4}{5}\right)^2 = 1 - \frac{16}{25} = \frac{9}{25}$
     $\sin x = \frac{3}{5}$
     $\sin 2x = 2 \sin x \cos x = 2\left(\frac{3}{5}\right)\left(\frac{4}{5}\right) = \frac{24}{25}$

13. $\sin 2(4x) = 2 \sin 4x \cos 4x$
     $2 \sin 2(4x) = 2(2 \sin 4x \cos 4x)$
     $2 \sin 8x = 4 \sin 4x \cos 4x$

17. $2 \cos^2 \frac{1}{2} x - 1$; Let $\theta = \frac{1}{2}x$
     $2 \cos^2 \theta - 1 = \cos 2\theta = \cos 2\left(\frac{1}{2}x\right) = \cos x$

21. $\cos^2 \alpha - \sin^2 \alpha = \cos^2 \alpha - (1 - \cos^2 \alpha)$
     $ = \cos^2 \alpha - 1 + \cos^2 \alpha$
     $ = 2 \cos^2 \alpha - 1$

25. $\cos^4 x - \sin^4 x = \cos 2x$
     $(\cos^2 x + \sin^2 x)(\cos^2 x - \sin^2 x) =$
     $1(\cos^2 x - \sin^2 x) = \cos 2x$

29. $\tan \theta = \dfrac{\sin 2\theta}{1 + \cos 2\theta} = \dfrac{2 \sin \theta \cos \theta}{1 + (2 \cos^2 \theta - 1)}$
     $ = \dfrac{2 \sin \theta \cos \theta}{2 \cos^2 \theta} = \dfrac{\sin \theta}{\cos \theta} = \tan \theta$

33. $2 \csc 2x \tan x = \dfrac{2}{1} \times \dfrac{1}{\sin 2x} \times \dfrac{\sin x}{\cos x}$
     $ = \dfrac{2 \sin x}{2 \sin x \cos x(\cos x)} = \dfrac{1}{\cos^2 x} = \sec^2 x$

37. $\sin 3x = \sin(2x + x) = \sin 2x \cos x + \cos 2x \sin x$
$= (2 \sin x \cos x)(\cos x) + (\cos^2 x - \sin^2 x)(\sin x)$
$= 2 \sin x \cos^2 x + \sin x \cos^2 x - \sin^3 x$
$= 3 \sin x \cos^2 x - \sin^3 x$

41. $R = vt \cos \alpha; \quad t = (2v \sin \alpha/g)$
$= v\left(\dfrac{2v \sin \alpha}{g}\right) \cos \alpha$
$= \dfrac{v^2(2 \sin \alpha \cos \alpha)}{g} = \dfrac{v^2 \sin 2\alpha}{g}$

Exercises 19-4, page 542

1. $\cos \dfrac{\alpha}{2} = \sqrt{\dfrac{1 + \cos \alpha}{2}}; \quad \cos \dfrac{30°}{2} = \sqrt{\dfrac{1 + \cos 30°}{2}}$
$= \sqrt{\dfrac{1 + 0.8660}{2}} = \sqrt{0.9330} = 0.9659$

$\cos 15° = 0.9659$

5. $\sqrt{\dfrac{1 - \cos 236°}{2}} = \sin \dfrac{1}{2}(236°)$
$= \sin 118° = 0.8829476$

The calculator sequence 1, $\boxed{-}$, 236, $\boxed{\text{COS}}$, $\boxed{=}$, $\boxed{\div}$, 2 $\boxed{=}$ $\boxed{\sqrt{x}}$
also gives 0.8829476.

9. $\sin \dfrac{\alpha}{2} = \sqrt{\dfrac{1 - \cos \alpha}{2}}$
$\sqrt{\dfrac{1 - \cos 6\alpha}{2}} = \sin \dfrac{6\alpha}{2} = \sin 3\alpha$

13. $\sin \dfrac{\alpha}{2} = \sqrt{\dfrac{1 - \cos \alpha}{2}} = \sqrt{\dfrac{1 - \dfrac{12}{13}}{2}} = \sqrt{\dfrac{1}{13} \times \dfrac{1}{2}} = \sqrt{\dfrac{1}{26}}$
$= \sqrt{\dfrac{1}{26} \times \dfrac{26}{26}} = \dfrac{1}{26}\sqrt{26}$

17. $\csc \dfrac{\alpha}{2} = \dfrac{1}{\sin \dfrac{\alpha}{2}} = \dfrac{1}{\pm\sqrt{\dfrac{1 - \cos \alpha}{2}}}$
$= \dfrac{1}{\pm\sqrt{1 - \cos \alpha} \times \dfrac{1}{\sqrt{2}}} = \dfrac{\sqrt{2}}{\pm\sqrt{1 - \cos \alpha}}$
$= \pm\sqrt{\dfrac{2}{1 - \cos \alpha}}$

21. $\dfrac{1 - \cos \alpha}{2 \sin \frac{\alpha}{2}} = \dfrac{1 - \cos \alpha}{2 \sqrt{\dfrac{1 - \cos \alpha}{2}}} \times \dfrac{\sqrt{\dfrac{1 - \cos \alpha}{2}}}{\sqrt{\dfrac{1 - \cos \alpha}{2}}}$

$= \dfrac{(1 - \cos \alpha) \sqrt{\dfrac{1 - \cos \alpha}{2}}}{2 (\dfrac{1 - \cos \alpha}{2})} = \sqrt{\dfrac{1 - \cos \alpha}{2}}$

$= \sin \dfrac{\alpha}{2}$

25. $\cos \dfrac{\theta}{2} = \dfrac{\sin \theta}{2 \sin \frac{\theta}{2}} = \dfrac{\sin \theta}{2 \sqrt{\dfrac{1 - \cos \theta}{2}}} \times \dfrac{\sqrt{\dfrac{1 + \cos \theta}{2}}}{\sqrt{\dfrac{1 + \cos \theta}{2}}}$

$= \dfrac{\sin \theta \sqrt{\dfrac{1 + \cos \theta}{2}}}{2 \sqrt{\dfrac{1 - \cos^2 \theta}{4}}} = \dfrac{\sin \theta \sqrt{\dfrac{1 + \cos \theta}{2}}}{2 \sqrt{\dfrac{\sin^2 \theta}{4}}}$

$= \dfrac{\sin \theta \cos \dfrac{\theta}{2}}{\sin \theta} = \cos \dfrac{\theta}{2}$

29. $\sin^2 \omega t = \sin^2 \left(\dfrac{1}{2}\right)(2\omega t) = \left(\sqrt{\dfrac{1 - \cos(2\omega t)}{2}}\right)^2 = \dfrac{1 - \cos 2\omega t}{2}$

Exercises 19-5, page 546

1. $\sin x - 1 = 0$, $0 \le x < 2\pi$; $\sin x = 1$; $x = \dfrac{\pi}{2}$

5. $2(2 + \cos x) = 3 + \cos x$
    $4 + 2 \cos x = 3 + \cos x$
    $2 \cos x - \cos x = 3 - 4$
    $\qquad \cos x = -1$
    $\qquad\qquad x = \pi$

9. $4 \cos^2 x - 1 = 0$; $0 \le x < 2\pi$
    $4 \cos^2 x = 1$; $\cos^2 x = \dfrac{1}{4}$; $\cos x = \pm\dfrac{1}{2}$
    $x = \dfrac{\pi}{3}, \dfrac{2\pi}{3}, \dfrac{4\pi}{3}, \dfrac{5\pi}{3}$

13. $2 \sin^2 x - \sin x = 0$; $0 \leq x < 2\pi$
$\sin x (2 \sin x - 1) = 0$; $\sin x = 0$; $x = 0, \pi$
$2 \sin x - 1 = 0$; $2 \sin x = 1$
$\sin x = \dfrac{1}{2}$; $x = \dfrac{\pi}{6}, \dfrac{5\pi}{6}$
Thus, $x = 0, \dfrac{\pi}{6}, \dfrac{5\pi}{6}, \pi$

17. $\sin 2x \sin x + \cos x = 0$, $0 \leq x < 2\pi$
$(2 \sin x \cos x)(\sin x) + \cos x = 0$ Eq. (19-20)
$2 \sin^2 x \cos x + \cos x = 0$; $\cos x (2 \sin^2 x + 1) = 0$
$\cos x = 0$; $x = \dfrac{\pi}{2}, \dfrac{3\pi}{2}$; $2 \sin^2 x + 1 = 0$; $2 \sin^2 x = -1$
$\sin^2 x = -\dfrac{1}{2}$, which has no real solution; thus, $x = \dfrac{\pi}{2}, \dfrac{3\pi}{2}$

21. $2 \cos^2 x - 2 \cos 2x - 1 = 0$; $2 \cos^2 x - 2(2 \cos^2 x - 1) - 1 = 0$
$2 \cos^2 x - 4 \cos^2 x + 2 - 1 = 0$; $-2 \cos^2 x + 1 = 0$
$\cos^2 x = \dfrac{-1}{-2} = \dfrac{1}{2}$; $\cos x = \dfrac{1}{\pm\sqrt{2}}$; $x = \dfrac{\pi}{4}, \dfrac{3\pi}{4}, \dfrac{5\pi}{4}, \dfrac{7\pi}{4}$

25. $4 \tan x - \sec^2 x = 0$; $4 \tan x - (1 + \tan^2 x) = 0$
$4 \tan x - 1 - \tan^2 x = 0$; $\tan^2 x - 4 \tan x + 1 = 0$
$\tan^2 x - 4 \tan x = -1$; $\tan^2 x - 4 \tan x + 4 = -1 + 4$ (completing the square)
$(\tan x - 2)^2 = 3$; $\tan x - 2 = \pm\sqrt{3}$
$\tan x = 2 \pm\sqrt{3} = 3.732, 0.268$
$x = 1.309$ (Q I), $3.142 + 1.309$ (Q III) $= 4.451$ radians
0.262 (Q I), $3.142 + 0.262$ (Q III) $= 3.404$ radians
$x = 0.262, 1.309, 3.404, 4.451$

29. $\sin 2x + \cos 2x = 0$; $0 \leq x < 2\pi$; $\sin 2x = -\cos 2x$
$\dfrac{\sin 2x}{\cos 2x} = -1$; $\tan 2x = -1$ (Q II) or (Q IV)
$2x = \dfrac{3\pi}{4}, \dfrac{7\pi}{4}, \dfrac{11\pi}{4}, \dfrac{15\pi}{4}$
$x = \dfrac{3\pi}{8}, \dfrac{7\pi}{8}, \dfrac{11\pi}{8}, \dfrac{15\pi}{8}$

33.
$$\dfrac{p^2 \tan \theta}{0.0063 + p \tan \theta} = 1.6; p = 4.8$$

$$\dfrac{4.8^2 \tan \theta}{0.0063 + 4.8 \tan \theta} = 1.6$$

$$4.8^2 \tan \theta = 1.6(0.0063 + 4.8 \tan \theta) = 0.01008 + 7.68 \tan \theta$$

$$4.8^2 \tan \theta - 7.68 \tan \theta = 0.01008$$

$$15.36 \tan \theta = 0.01008$$

$$\tan \theta = \dfrac{0.01008}{15.36} = 6.5625 \times 10^{-4}$$

$$\theta = 0.0376°$$

37. $3 \sin x - 1 = 0$
$3 \sin x = 1$

$\sin x = \dfrac{1}{3}$;   1, $\boxed{\div}$, 3, $\boxed{\text{ARCSIN}}$ on radian setting

$x = 0.3398,\ 2.802$ radians since sine is positive in quadrants
I and II.

41.   This can be solved by
the method of Section
2-5, or we can find
the points of inter-
section of the graphs
of $y = x$ and $y = \sin 2x$.
Apparent intersections
are approximately
$x = 0.95$, $x = -0.95$,
and $x = 0.00$.
Check:
$\sin 2(0.95) = 0.946$
$\sin 2(-0.95) = -0.946$
$\sin 2(0.00) = 0.00$

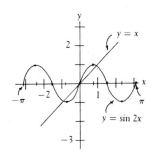

Exercises 19-6, page 552

1. y is an angle whose tangent is x.

5. y is twice the angle whose sine is x.

9.   The smallest positive angle whose cosine is $\dfrac{1}{2}$ is $60°$ or $\dfrac{\pi}{3}$.

13.   The smallest positive angle whose cotangent is $-\sqrt{3}$ is $150°$ or $\dfrac{5\pi}{6}$.

17. $\text{Arctan} \left(\dfrac{\sqrt{3}}{3}\right) = \dfrac{\pi}{6}$ since $\tan \dfrac{\pi}{6} = \dfrac{\sqrt{3}}{3}$ and $-\dfrac{\pi}{2} < \dfrac{\pi}{6} < \dfrac{\pi}{2}$

21. $\text{Arccsc} \sqrt{2} = \dfrac{\pi}{4}$ since $\csc \dfrac{\pi}{4} = \sqrt{2}$ and $-\dfrac{\pi}{2} \le \dfrac{\pi}{4} \le \dfrac{\pi}{2}$

   or $\text{Arccsc} \sqrt{2} = \text{Arcsin} \dfrac{1}{\sqrt{2}} = \text{Arcsin} \dfrac{\sqrt{2}}{2} = \dfrac{\pi}{4}$ since $\sin \dfrac{\pi}{4} = \dfrac{\sqrt{2}}{2}$ and

   $-\dfrac{\pi}{2} \le \dfrac{\pi}{4} \le \dfrac{\pi}{2}$

25. $\text{Arcsec} (-\sqrt{2}) = \dfrac{3\pi}{4}$ since $\sec \dfrac{3\pi}{4} = -\sqrt{2}$

29. $\cos[\text{Arctan}(-1)] = \cos -0.7853 = 0.7071 = \dfrac{1}{2}\sqrt{2}$

33. Arctan(-3.7321) = -1.3090 using the sequence 3.7321, $\boxed{+/-}$, $\boxed{\text{ARCTAN}}$ on radian setting

37. Arccos 0.1291 = 1.4413 using the sequence 0.1291, $\boxed{\text{ARCCOS}}$ on radian setting

41. tan[Arccos(-0.6281)] = tan 2.250 = -1.2389 using the sequence 0.6281, $\boxed{+/-}$, $\boxed{\text{ARCCOS}}$, $\boxed{\text{TAN}}$ on radian setting

45. $y = \sin 3x$; $3x = $ Arcsin y; $x = \frac{1}{3}$ Arcsin y

49. $y = 1 + \sec 3x$; $y - 1 = \sec 3x$

$3x = $ Arcsec$(y - 1)$; $x = \frac{1}{3}$ Arcsec$(y - 1)$

53. tan(Arcsin x) $= \tan \theta = \dfrac{x}{\sqrt{1 - x^2}}$

In this triangle, θ is set up such that its sine is x and is therefore Arcsin x. From the Pythagorean relation, using x for the opposite and 1 for the hypotenuse requires $\sqrt{1 - x^2}$ for the adjacent side.

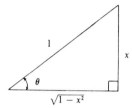

57. sec(Arccsc $3x$) $= \sec \theta = \dfrac{3x}{\sqrt{9x^2 - 1}}$

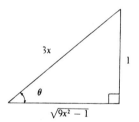

61.
$$y = A \cos 2(\omega t + \emptyset)$$

$$\frac{y}{A} = \cos 2(\omega t + \emptyset)$$

$$\text{Arccos } \frac{y}{A} = 2(\omega t + \emptyset) = 2\omega t + 2\emptyset$$

$$\text{Arccos } \frac{y}{A} - 2\emptyset = 2\omega t$$

$$\frac{\text{Arccos } \dfrac{y}{A} - 2\emptyset}{2\omega} = t$$

$$t = \frac{1}{2\omega} \text{ Arccos } \frac{y}{A} - \frac{\emptyset}{\omega}$$

65. Let $\alpha = \text{Arcsin } \dfrac{3}{5}$ and $\beta = \text{Arcsin } \dfrac{5}{13}$; $\sin \alpha = \dfrac{3}{5}$

$\cos \alpha = \sqrt{1 - \dfrac{9}{25}} = \sqrt{\dfrac{16}{25}} = \dfrac{4}{5}$; $\sin \beta = \dfrac{5}{13}$

$\cos \beta = \sqrt{1 - \dfrac{25}{169}} = \sqrt{\dfrac{144}{169}} = \dfrac{12}{13}$

$\text{Arcsin } \dfrac{3}{5} + \text{Arcsin } \dfrac{5}{13} = \alpha + \beta$

$\sin(\alpha + \beta) = \sin \alpha \cos \beta + \cos \alpha \sin \beta$ Eq. (19-9)

$\dfrac{3}{5}(\dfrac{12}{13}) + \dfrac{4}{5}(\dfrac{5}{13}) = \dfrac{36}{65} + \dfrac{20}{65} = \dfrac{56}{65}$

69. Since $\sin A = \dfrac{a}{c}$, $A = \text{Arcsin}\left(\dfrac{a}{c}\right)$

Review Exercises for Chapter 19, page 555

1. $\sin 120° = \sin(90° + 30°)$
    $= \sin 90° \cos 30° + \cos 90° \sin 30°$
    $= 1(\dfrac{\sqrt{3}}{2}) + 0(\dfrac{1}{2}) = \dfrac{\sqrt{3}}{2} = \dfrac{1}{2}\sqrt{3}$

5. $\cos 180° = \cos 2(90°)$
    $= \cos^2 90° - \sin^2 90°$
    $= 0 - (1)^2 = -1$

9. $\sin 14° \cos 38° + \cos 14° \sin 38° = \sin(14° + 38°)$ by Eq. (19-19)
    $= \sin 52°$
    $= 0.7880108$

    The calculator sequence 14, $\boxed{\text{SIN}}$, $\boxed{\times}$, 38 , $\boxed{\text{COS}}$ $\boxed{=}$ $\boxed{+}$, 14,
    $\boxed{\text{COS}}$, $\boxed{\times}$, 38, $\boxed{\text{SIN}}$ $\boxed{=}$ also gives 0.7880108.

13. $\sqrt{\dfrac{1 + \cos 12°}{2}} = \cos \dfrac{12°}{2}$ by Eq. (19-25)
    $= \cos 6°$
    $= 0.9945219$

    The calculator sequence 12, $\boxed{\text{COS}}$, $\boxed{+}$, 1 $\boxed{=}$ $\boxed{\div}$, 2 $\boxed{=}$ $\boxed{\sqrt{x}}$ also gives
    0.9945219.

17. $\sin 2x \cos 3x + \cos 2x \sin 3x$
    $= \sin \alpha \cos \beta + \cos \alpha \sin \beta$
    $= \sin(\alpha + \beta)$ where $\alpha = 2x$, $\beta = 3x$
    $\sin(\alpha + \beta) = \sin(2x + 3x) = \sin 5x$

21. $2 - 4 \sin^2 6x = 2(1 - 2 \sin^2 6x)$
    $= 2(1 - 2 \sin^2 \alpha)$ where $\alpha = 6x$
    $2(\cos 2\alpha) = 2 \cos 12x$
    $2(1 - 2 \sin^2 \alpha) = 2(\cos 2\alpha) = 2 \cos 12x$

25. $\text{Arcsin}(-1) = -\dfrac{\pi}{2}$ since $\sin(-\dfrac{\pi}{2}) = -1$ and $-\dfrac{\pi}{2} \le -\dfrac{\pi}{2} \le \dfrac{\pi}{2}$

29. $\tan[\text{Arcsin}(-\dfrac{1}{2})] = \tan(-\dfrac{\pi}{6}) = -\dfrac{\sqrt{3}}{3} = -\dfrac{1}{3}\sqrt{3}$

33. $\dfrac{\sec y}{\csc y} = \sec y \times \dfrac{1}{\csc y}$

$= \dfrac{1}{\cos y} \times \dfrac{\sin y}{1} = \dfrac{\sin y}{\cos y} = \tan y$

37. $\dfrac{1}{\sin \theta} - \dfrac{\sin \theta}{1} = \dfrac{1 - \sin^2 \theta}{\sin \theta}$

$= \dfrac{\cos^2 \theta}{\sin \theta} = \dfrac{\cos \theta}{\sin \theta} \times \dfrac{\cos \theta}{1}$

$= \cot \theta \, \cos \theta$

41. $\dfrac{\sec^4 x - 1}{\tan^2 x} = \dfrac{(\sec^2 x - 1)(\sec^2 x + 1)}{\tan^2 x}$

$\dfrac{(\sec^2 x + 1)(\tan^2 x)}{\tan^2 x} = \sec^2 x + 1$

$1 + \tan^2 x + 1 = 2 + \tan^2 x$

45. $\dfrac{1 - \sin^2 \theta}{1 - \cos^2 \theta} = \dfrac{\cos^2 \theta}{\sin^2 \theta}$ since $\sin^2 \theta + \cos^2 \theta = 1$

$= (\dfrac{\cos \theta}{\sin \theta})^2 = (\cot \theta)^2 = \cot^2 \theta$

49. $\sin \dfrac{\theta}{2} \cos \dfrac{\theta}{2} = \dfrac{1}{2}(2 \sin \dfrac{\theta}{2} \cos \dfrac{\theta}{2}) = \dfrac{1}{2}[\sin 2(\dfrac{\theta}{2})]$

$\dfrac{1}{2} \sin \theta = \dfrac{\sin \theta}{2}$

53. Let $\alpha = (x - y)$ and $\beta = y$
$\cos(x - y) \cos y - \sin(x - y) \sin y$
$= \cos \alpha \cos \beta - \sin \alpha \sin \beta$
$= \cos(\alpha + \beta) = \cos(x - y + y) = \cos x$

57. $\dfrac{\sin x}{\csc x - \cot x} = \dfrac{\sin x}{1} \times \dfrac{1}{\csc x - \cot x}$

$= \dfrac{\sin x}{1} \times \dfrac{1}{\dfrac{1}{\sin x} - \dfrac{\cos x}{\sin x}}$

$= \dfrac{\sin x}{1} \times \dfrac{1}{\dfrac{1 - \cos x}{\sin x}} = \dfrac{\sin x}{1} \times \dfrac{\sin x}{1 - \cos x}$

$= \dfrac{\sin^2 x}{1 - \cos x} = \dfrac{1 - \cos^2 x}{1 - \cos x}$

$= \dfrac{(1 - \cos x)(1 + \cos x)}{(1 - \cos x)} = 1 + \cos x$

61. $y = 2 \cos 2x; \dfrac{y}{2} = \cos 2x$

$\text{Arccos} \dfrac{y}{2} = 2x; \dfrac{1}{2} \text{Arccos} \dfrac{1}{2}y = x$

65. $3(\tan x - 2) = 1 + \tan x$
$3 \tan x - 6 = 1 + \tan x$
$2 \tan x = 7$
$\tan x = \dfrac{7}{2}$

$x = \arctan \dfrac{7}{2} = 1.2925$

Since $\tan x$ is positive in quadrant III also, $\pi + 1.2925 = 4.4341$ is also a value for x that is within the specified range of values for x.

69. $\cos^2 2x - 1 = 0; \cos^2 2x = 1$
$\cos 2x = \pm\sqrt{1} = \pm 1; 2x = 0, \pi, 2\pi, 3\pi$

$x = 0, \dfrac{\pi}{2}, \pi, \dfrac{3\pi}{2}$

73. $\sin^2 x - \cos^2 x + 1 = 0$
$-(\cos^2 x - \sin^2 x) + 1 = 0$
$-\cos 2x = -1$ Eq. (19-21)
$\cos 2x = 1; 2x = 0, 2\pi; x = 0, \pi$

77. $\tan(\text{Arccot } x) = \tan \theta = \dfrac{1}{x}$

$\theta = \text{Arccot } x$

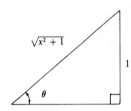

81. $x = 2 \cos \theta; \sqrt{4 - x^2} = \sqrt{4 - 4 \cos^2 \theta} = \sqrt{4(1 - \cos^2 \theta)} = 2\sqrt{1 - \cos^2 \theta}$

$= 2\sqrt{\sin^2 \theta} = 2 \sin \theta$

85. $R = \sqrt{Rx^2 + Ry^2} = \sqrt{(A \cos\theta - B \sin\theta)^2 + (A \sin\theta + B \cos\theta)^2}$

$= \sqrt{(A^2 \cos^2\theta - 2AB \cos\theta \sin\theta + B^2 \sin^2\theta) + (A^2 \sin^2\theta + 2AB \cos\theta \sin\theta + B^2 \cos^2\theta)}$

$= \sqrt{A^2 \cos^2\theta + A^2 \sin^2\theta + B^2 \cos^2\theta + B^2 \sin^2\theta}$

$= \sqrt{A^2 (\cos^2\theta + \sin^2\theta) + B^2 (\cos^2\theta + \sin^2\theta)}$

$= \sqrt{(\cos^2\theta + \sin^2\theta) (A^2 + B^2)} = \sqrt{1(A^2 + B^2)}$

89. $\omega t = \text{Arc sin } \dfrac{\theta - \alpha}{R}$

$\sin(\omega t) = \dfrac{\theta - \alpha}{R}$

$R \sin(\omega t) = \theta - \alpha$

$\theta = R \sin(\omega t) + \alpha$

93. $1.20 \cos \theta + 0.135 \cos 2\theta = 0; \quad 0 < \theta < 180$

$1.20 \cos \theta + 0.135 \ (2 \cos^2 \theta - 1) = 0$

$1.20 \cos \theta + 0.270 \cos^2 \theta - 0.135 = 0$

$0.270 \cos^2 \theta + 1.20 \cos \theta - 0.135 = 0$

$\cos \theta = \dfrac{-1.20 \pm \sqrt{(1.20)^2 - 4(0.270)(-0.135)}}{2(0.270)}$

$\cos \theta = \dfrac{-1.20 \pm \sqrt{1.44 + 0.146}}{0.54} = 0.1099$

$\theta = 83.7°$

CHAPTER 20

Plane Analytic Geometry

Exercises 20-1, page 563

1. Given: $(x_1, y_1) = (3, 8)$; $(x_2, y_2) = (-1, -2)$

   Using Eq. (20-1) $d = \sqrt{(x_2 - x_1)^2 + (y_2 - y_1)^2}$

   $\qquad = \sqrt{(-1 - 3)^2 + (-2 - 8)^2} = \sqrt{(-4)^2 + (-10)^2}$

   $\qquad = \sqrt{16 + 100} = \sqrt{116}$

   $\qquad = \sqrt{4 \times 29} = 2\sqrt{29}$

5. Given: $(x_1, y_1) = (-12, 20)$; $(x_2, y_2) = (32, -13)$

   Using Eq. (20-1) $d = \sqrt{(x_2 - x_1)^2 + (y_2 - y_1)^2}$

   $\qquad = \sqrt{(32 + 12)^2 + (-13 - 20)^2}$

   $\qquad = \sqrt{(44)^2 + (-33)^2} = \sqrt{1936 + 1089}$

   $\qquad = \sqrt{3025} = 55$

9. Given: $(x_1, y_1) = (1.22, -3.45)$; $(x_2, y_2) = (-1.07, -5.16)$

   Using Eq. (20-1) $d = \sqrt{(x_2 - x_1)^2 + (y_2 - y_1)^2}$

   $\qquad = \sqrt{(-1.07 - 1.22)^2 + (-5.16 - (-3.45))^2}$

   $\qquad = \sqrt{(-2.29)^2 + (-5.16 + 3.45)^2} = \sqrt{(-2.29)^2 + (-1.71)^2}$

   $\qquad = \sqrt{8.1682} = 2.86$

13. Given: $(x_1, y_1) = (4, -5)$; $(x_2, y_2) = (4, -8)$

    $m = \dfrac{y_2 - y_1}{x_2 - x_1} = \dfrac{-8 - (-5)}{4 - 4}$ $\qquad$ Eq. (20-2)

    Since $x_2 - x_1 = 4 - 4 = 0$, the slope is undefined.

17. Given: $(x_1,y_1) = (-4,-3)$; $(x_2,y_2) = (3,-3)$

$$m = \frac{y_2 - y_1}{x_2 - x_1} = \frac{-3 - (-3)}{3 - (-4)}$$

$$= \frac{-3 + 3}{3 + 4} = \frac{0}{7} = 0 \qquad \text{Eq. (20-2)}$$

21. Given: $\alpha = 30°$; $m = \tan \alpha$, $0° < \alpha < 180°$

$$\tan 30° = \frac{\sqrt{3}}{3} \text{ or } \frac{1}{3}\sqrt{3} \qquad \text{Eq. (20-3)}$$

25. Given: $m = 0.3640$; $m = \tan \alpha$; $0.3640 = \tan \alpha$; $\alpha = 20.0°$

29. Given: (x,y) is $(6,-1)$, (x_1,y_1) is $(4,3)$, (x_2,y_2) is $(-5,2)$,

(x_3,y_3) is $(-7,6)$

$$m_1 = \frac{y - y_1}{x - x_1} = \frac{-1 - 3}{6 - 4} = \frac{-4}{2} = -2$$

$$m_2 = \frac{y_2 - y_3}{x_2 - x_3} = \frac{2 - 6}{-5 - (-7)} = \frac{-4}{-5 + 7} = \frac{-4}{2} = -2 \qquad \text{Eq. (20-2)}$$

$m_1 = m_2$ for all $||$ lines   Eq. (20-4)

33. Given: distance between $(-1,3)$ and $(11,k)$ is 13.

$$d = \sqrt{(x_1 - x_2)^2 + (y_1 - y_2)^2}$$

$$13 = \sqrt{(-1 - 11)^2 + (3 - k)^2} = \sqrt{(-12)^2 + (3 - k)^2} = \sqrt{144 + (3 - k)^2}$$

$169 = 144 + (3 - k)^2$; $(3 - k)^2 = 25$; $3 - k = \pm 5$

$-k = -3 \pm 5$; $k = -2, 8$ Eq. (20-1)

37. $d_1 = \sqrt{(9 - 7)^2 + [4 - (-2)]^2} = \sqrt{2^2 + 6^2} = \sqrt{40} = 2\sqrt{10}$

$d_2 = \sqrt{(9 - 3)^2 + (4 - 2)^2} = \sqrt{6^2 + 2^2} = \sqrt{40} = 2\sqrt{10}$

$d_1 = d_2$ so the triangle is isosceles.

41. $d_1 = \sqrt{(3 - 5)^2 + (-1 - 3)^2} = \sqrt{(-2)^2 + (-4)^2} = \sqrt{4 + 16} = \sqrt{20}$

$$m_1 = \frac{y - y_1}{x - x_1} = \frac{5 - 3}{3 - (-1)} = \frac{5 - 3}{3 + 1} = \frac{2}{4} = \frac{1}{2}$$

$d_2 = \sqrt{(5 - 1)^2 + (3 - 5)^2} = \sqrt{(4)^2 + (-2)^2}$

$$= \sqrt{16 + 4} = \sqrt{20}$$

$$m_2 = \frac{y - y_1}{x - x_1} = \frac{5 - 1}{3 - 5} = \frac{4}{-2} = -2$$

$m_1 = \frac{-1}{m^2}$, $m_1 \perp m_2$

$A = \frac{1}{2}d_1 d_2 = \frac{1}{2}\sqrt{20}\sqrt{20} = \frac{1}{2}(20) = 10$

45. $\left(\dfrac{-4 + 6}{2}, \dfrac{9 + 1}{2}\right) = \left(\dfrac{2}{2}, \dfrac{10}{2}\right) = (1,5)$

Exercises 20-2, page 569

1. Given: $m = 4$; (x_1, y_1) is $(-3, 8)$.
$y - y_1 = m(x - x_1)$ Eq. (20-6)
$y - 8 = 4[x - (-3)] = 4(x + 3) = 4x + 12$
$y = 4x + 20$ or $4x - y + 20 = 0$ Eq. (20-9) or Eq. (20-10)

5. Given: (x_1, y_1) is $(1, 3)$, $\alpha = 45°$
$m = \tan \alpha = \tan 45° = 1$ Eq. (20-3)
$y - y_1 = m(x - x_1)$ Eq. (20-6)
$y - 3 = 1(x - 1) = x - 1$; $y = x + 2$ or $x - y + 2 = 0$ Eq. (20-9) or
Eq. (20-10)

9. Given: $\parallel$ to y-axis and 3 units left of y-axis

$x = -3$ Eq. (20-7)

13. Given: $\perp$ to line with slope 3, (x_1, y_1) is $(1, -2)$.
$m = -\dfrac{1}{3}$ Eq. (20-5)
$y - y_1 = m(x - x_1)$ Eq. (20-6)
$y - (-2) = -\dfrac{1}{3}(x - 1)$; $y + 2 = -\dfrac{1}{3}x + \dfrac{1}{3}$
$y = -\dfrac{1}{3}x - \dfrac{5}{3}$ or $-\dfrac{1}{3}x - y - \dfrac{5}{3} = 0$ Eq. (20-9) or Eq. (20-10)
$x + 3y + 5 = 0$ (Simplify by multiplying through by -3.)

17. Given: $\parallel$ to a line through $(7, -1)$ and $(4, 3)$;
         y-intercept is -2.
$m = \dfrac{y_2 - y_1}{x_2 - x_1} = \dfrac{3 - (-1)}{4 - 7} = \dfrac{4}{-3} = -\dfrac{4}{3}$; y-intercept is $b = -2$.
$y = mx + b$ Eq. (20-9)
$y = -\dfrac{4}{3}x - 2$
$4x + 3y + 6 = 0$

21. Given: $4x - y = 8$; $(0,-8)$ and $(2,0)$ are intercepts.

$4(0) - y = 8$; $y = -8$

$4x - y = 8$; $x = 2$

25. Given: $3x - 2y - 1 = 0$

$3x - 2y - 1 = 0$; $-2y = -3x + 1$

$y = \frac{-3}{-2}x + \frac{1}{-2}$; $y = \frac{3}{2}x - \frac{1}{2}$

slope $= \frac{3}{2} = m$ From Eq. (20-9) y-intercept $= -\frac{1}{2} = b$

29. Given: $4x - ky = 6 \parallel 6x + 3y + 2 = 0$

$6x + 3y + 2 = 0$; $3y = -6x - 2$

$y = \frac{-6}{3}x - \frac{2}{3}$; $y = -2x - \frac{2}{3}$; slope is -2

$4x - ky = 6$; $-ky = -4x + 6$

$y = \frac{-4}{-k}x + \frac{6}{-k}$; $y = \frac{4}{k}x - \frac{6}{k}$; slope is $\frac{4}{k}$

Since the lines are parallel, the slopes are equal.

$\frac{4}{k} = -2$; $4 = -2k$; $k = -2$

33. $3x - 2y + 5 = 0$; $-2y = -3x - 5$; $y = \frac{-3}{-2}x + \frac{-5}{-2}$; $y = \frac{3}{2}x + \frac{5}{2}$; slope $= \frac{3}{2} = m_1$

$4y = 6x - 1$; $y = \frac{6}{4}x - \frac{1}{4}$

$y = \frac{3}{2}x - \frac{1}{4}$; slope $= \frac{3}{2} = m_2$

$m_1 = m_2$ for $\parallel$ lines   Eq. (20-4)

37. $2y - 3x - 4 = 0$; $2y = 3x + 4$

$y = \frac{3}{2}x + \frac{4}{2}$; $y = \frac{3}{2}x + 2$

y-intercept is 2 and (x_2,y_2) is $(0,2)$, $b = 2$
x-intercept is 4; gives (x_1,y_1) as $(4,0)$

$m = \frac{2 - 0}{0 - 4} = \frac{2}{-4} = \frac{-1}{2}$ Eq. (20-2)

$y = mx + b = -\frac{1}{2}x + 2$ or $-\frac{1}{2}x - y + 2 = 0$

or $x + 2y - 4 = 0$ (Multiply by -2.) Eq. (20-9)

41. $v = v_0 + at$; $35.4 = 12.2 + a(4.50)$

$4.50a = 35.4 - 12.2$

$a = 5.16$

$v = 12.2 + 5.16t$

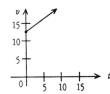

41M. $v = v_0 + at$; $9.87 = 3.35 + a(4.50)$

$4.50a = 9.87 - 3.35$

$a = 1.45$

$v = 3.35 + 1.45t$

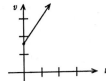

45. $50x + 60y = 12\ 200$; $5x + 6y = 1220$

49. $m = \tan(180° - 0.0032°)$ $b = 24\mu m = 24 \times 10^{-6}m = 2.4 \times 10^{-5}$

$m = -5.6 \times 10^{-5}$

$y = mx + b = -5.6 \times 10^{-5}x + 2.4 \times 10^{-5} = (-5.6x + 2.4)10^{-5}$

53. $n = 1200\sqrt{t} + 0$

$m = 1200$

$b = 0$

57. Slope is found by measuring between
points. The vertical displacement
and the horizontal displacement between
the extreme points is in a 1 to 2 ratio;
$m = \frac{1}{2}$. Since the graph is linear, the
log equation is of the form
$\log y = m \log x + \log a$, where a is
the intercept $(1, a)$.

Solution and graph are shown on the
following page.

$y = ax^n$
$y = 3x^4$; $a = 3$, $n = 4$

x	y
1.0	3.0
1.1	4.4
1.2	6.2
1.3	8.6

$\log y = \log a + n \log x$
$\log y = \log 3 + 4 \log x$

Verify:

1) slope is $\dfrac{\log y - \log a}{\log x} = 4$.

   Vertical and horizontal measures
in millimeters between points
are shown. Each slope is 4.

2) The intercept is $a = 3$.
The line crosses the vertical
axis at $x = 1.0$, $y = 3.0$.

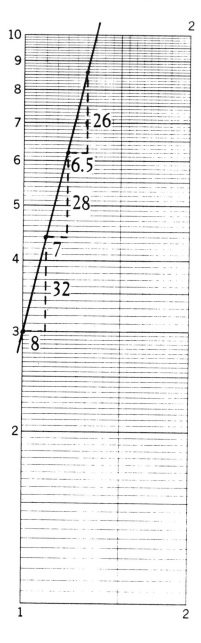

LOGARITHMIC 1 X 1 CYCLES

Exercises 20-3, page 574

1. Given: $(x - 2)^2 + (y - 1)^2 = 25$; center is (2,1); radius is 5.
   Eq. (20-11)

5. Given: center at (0,0); radius 3
   $(x - h)^2 + (y - k)^2 = r^2$; $(x - 0)^2 + (y - 0)^2 = 3^2$; $x^2 + y^2 = 9$
   Eq. (20-11)

9. Given: center at (-2,5); radius $\sqrt{5}$
   $(x - h)^2 + (y - k)^2 = r^2$ Eq. (20-11)
   $[x - (-2)]^2 + (y - 5)^2 = (\sqrt{5})^2$; $(x + 2)^2 + (y - 5)^2 = \sqrt{25}$
   $x^2 + 4x + 4 + y^2 - 10y + 25 = 5$
   $x^2 + y^2 + 4x - 10y + 4 + 25 - 5 = 0$
   $x^2 + y^2 + 4x - 10y + 24 = 0$

13. Given: center at (2,1), passes through (4,-1)

   $r^2 = (2 - 4)^2 + (1 + 1)^2 = (-2)^2 + (2)^2 = 8$

   $(x - h)^2 + (y - k)^2 = r^2$
   $(x - 2)^2 + (y - 1)^2 = 8$
   $x^2 - 4x + 4 + y^2 - 2y + 1 = 8$
   $x^2 - 4x + y^2 - 2y - 3 = 0$

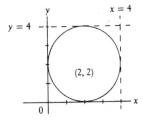

17. The center is (2,2) and radius is 2.
    $(x - h)^2 + (y - k)^2 = r^2$
    $(x - 2)^2 + (y - 2)^2 = 2^2$
    $x^2 - 4x + 4 + y^2 - 4y + 4 = 4$
    $x^2 + y^2 - 4x - 4y + 4 = 0$

21. $x^2 + (y - 3)^2 = 4$ is the same
    as $(x - 0)^2 + (y - 3)^2 = 2^2$, so
    $(h,k) = (0,3)$ and $r = 2$.

25. $x^2 + y^2 = r^2$ Eq. (20-12)
    $x^2 + y^2 - 25 = 0$
    $x^2 + y^2 = 25$
    $x^2 + y^2 = 5^2$

    center (0,0); radius 5

29. $x^2 + y^2 + Dx + Ey + F = 0$ Eq. (20-14)
    $x^2 + y^2 + 8x - 10y - 8 = 0$
    $x^2 + 8x + y^2 - 10y = 8$
    $(x^2 + 8x + 16) + (y^2 - 10y + 25)$
    $= 8 + 16 + 25$
    $(x + 4)^2 + (y - 5)^2 = 49$ Eq. (20-11)
    $[x - (-4)]^2 + (y - 5)^2 = 7^2$

    center (-4,5); radius 7

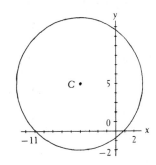

33. $(-x)^2 + y^2 = 100$ $x^2 + (-y)^2 = 100$
     $x^2 + y^2 = 100$ $x^2 + y^2 = 100$
     symmetrical to y-axis (Ex. A)   symmetrical to x-axis (Ex. A)

     $$(-x)^2 + (-y)^2 = 100$$
     $$x^2 + y^2 = 100$$
     symmetrical to origin

37. Find all points for which $y = 0$.
     $x^2 - 6x + (0)^2 - 7 = 0$; $x^2 - 6x - 7 = 0$; $(x + 1)(x - 7) = 0$
     $x = -1$ or $x = 7$; $(-1,0)$ and $(7,0)$

41.   60 Hz = 60 cycles/s = 37.7 m/s $(h,k) = (0,0)$

     60 cycles = 37.7 m $x^2 + y^2 = (0.10)^2$

      1 cycle = 0.628 m $x^2 + y^2 = 0.0100$

            $r = 0.628$ m $\div 2\pi$

            $r = 0.10$

Exercises 20-4, page 579

1. Given: $y^2 = 4x$
    $y^2 = 4px$ Eq. (20-15)
    $y^2 = 4x = 4(1)x$; $p = 1$
    focus at $(1,0)$; directrix $x = -1$

5. Given: $x^2 = 8y$
    $x^2 = 4py$ Eq. (20-16)
    $x^2 = 8y = 4(2)y$; $p = 2$
    focus at $(0,2)$; directrix $y = -2$

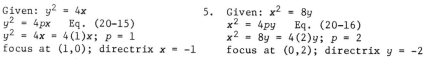

9. Given: $y^2 = 2x$
    $y^2 = 4px$ Eq. (20-15)
    $y^2 = 2x = 4(\frac{1}{2})x$; $p = \frac{1}{2}$
    focus at $(\frac{1}{2},0)$; directrix $x = -\frac{1}{2}$

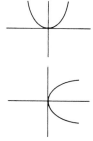

13. Given: focus $(3,0)$
            directrix $x = -3$; $p = 3$
    $y^2 = 4px$ Eq. (20-15)
    $y^2 = 4(3)x$; $y^2 = 12x$

17. Given: vertex $(0,0)$
            directrix $y = -1$; $p = 1$
    $x^2 = 4py$ Eq. (20-16);
    $x^2 = 4(1)y$; $x^2 = 4y$

21. Given: focus $(6,1)$; directrix $x = 0$; vertex $(3,1)$

$d_1 = d_2$

$d_1 = x$

$d_2 = \sqrt{(x - 6)^2 + (y - 1)^2}$

$x = \sqrt{(x - 6)^2 + (y - 1)^2}$

$x^2 = (x - 6)^2 + (y - 1)^2$

$x^2 = x^2 - 12x + 36 + y^2 - 2y + 1$

$0 = -12x + 36 + y^2 - 2y + 1$

$0 = y^2 - 2y - 12x + 37$

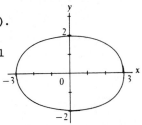

25. Focus is $(p,0)$; latus rectum is from (p,y) to $(p,-y)$

where $y = \pm\sqrt{4px}$. If $x = p$, then $y = \pm\sqrt{4p\,(p)} = \pm\sqrt{4p^2} = \pm 2p$.

The distance from $(p,2p)$ to $(p,-2p)$ is $4p$.

29.

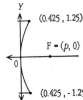

(0.425, 1.25)

F = $(p, 0)$

(0.425, -1.25)

$y^2 = 4px$

$(1.25)^2 = 4p\,(0.425)$

$p = \dfrac{(1.25)^2}{4(0.425)} = 0.919$

33. The graph is parabolic since it can be transformed into the form $f^2 = 4pA$.

$f = 0.065\sqrt{A}$

$= 0.065\sqrt{200}$

$= 0.92$

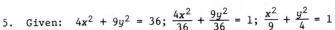

Exercises 20-5, page 585

1. Given: $\dfrac{x^2}{4} + \dfrac{y^2}{1} = 1$, $a^2 = 4$, $b^2 = 1$

$\dfrac{x^2}{a^2} + \dfrac{y^2}{b^2} = 1$ Eq. (20-17)

$c^2 = a^2 - b^2$ From Eq. (20-18)

$c^2 = 4 - 1 = 3$, $c = \sqrt{3}$

$V(a,0)$ and $(-a,0)$ are $(2,0)$ and $(-2,0)$.

$F(c,0)$ and $(-c,0)$ are $(\sqrt{3},0)$ and $(-\sqrt{3},0)$.

y-intercepts $(0,b)$ and $(0,-b)$ are $(0,1)$ and $(0,-1)$.

5. Given: $4x^2 + 9y^2 = 36$; $\dfrac{4x^2}{36} + \dfrac{9y^2}{36} = 1$; $\dfrac{x^2}{9} + \dfrac{y^2}{4} = 1$

$a^2 = 9$, $b^2 = 4$, $c^2 = 9 - 4 = 5$

$F(c,0)$ and $(-c,0)$ are $(-\sqrt{5},0)$ and $(\sqrt{5},0)$.

$V(a,0)$ and $(-a,0)$ are $(3,0)$ and $(-3,0)$.

y-intercepts are $(0,2)$ and $(0,-2)$.

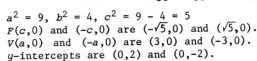

9. Given: $8x^2 + y^2 = 16$; $\dfrac{8x^2}{16} + \dfrac{y^2}{16} = 1$; $\dfrac{x^2}{2} + \dfrac{y^2}{16} = 1$; $\dfrac{y^2}{16} + \dfrac{x^2}{2} = 1$

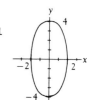

$a^2 = 16$, $b^2 = 2$, $c^2 = 16 - 2 = 14$
$V(0,4)$ and $(0,-4)$; $F(0,\sqrt{14})$ and $(0,-\sqrt{14})$;
x-intercepts are $(2,0)$ and $(-2,0)$.

13. Given: $V(15,0)$; $F(9,0)$
$a = 15$, $a^2 = 225$; $c = 9$, $c^2 = 81$; $a^2 - c^2 = b^2$ From Eq. (20-18)

$b^2 = 144$; $\dfrac{x^2}{a^2} + \dfrac{y^2}{b^2} = 1$; $\dfrac{x^2}{225} + \dfrac{y^2}{144} = 1$

$144x^2 + 225y^2 = 32{,}400$

17. Given: vertex $(8,0)$; (x,y) is $(2,3)$; $a^2 = 8^2 = 64$

$\dfrac{x^2}{a^2} + \dfrac{y^2}{b^2} = 1$; $\dfrac{x^2}{64} + \dfrac{y^2}{b^2} = 1$; $\dfrac{(2)^2}{64} + \dfrac{(3)^2}{b^2} = 1$

$\dfrac{4}{64} + \dfrac{9}{b^2} = 1$; $\dfrac{9}{b^2} = 1 - \dfrac{1}{16}$; $\dfrac{9}{b^2} = \dfrac{16}{16} - \dfrac{1}{16}$

$\dfrac{9}{b^2} = \dfrac{15}{16}$; $15b^2 = 144$; $b^2 = \dfrac{144}{15}$; $b^2 = \dfrac{48}{5}$

$\dfrac{x^2}{64} + \dfrac{5y^2}{48} = 1$; $48x^2 + 320y^2 = 3{,}072$; $3x^2 + 20y^2 = 192$

21. Given: foci $(-2,1)$ and $(4,1)$, major axis 10

$\sqrt{[x - (-2)]^2 + (y - 1)^2} + \sqrt{(x - 4)^2 + (y - 1)^2} = 10$

$\sqrt{(x + 2)^2 + (y - 1)^2} = 10 - \sqrt{(x - 4)^2 + (y - 1)^2}$

$(x + 2)^2 + (y - 1)^2 = 100 - 20\sqrt{(x - 4)^2 + (y - 1)^2} + (x - 4)^2 + (y - 1)^2$

$x^2 + 4x + 4 + y^2 - 2y + 1$

$= 100 - 20\sqrt{(x - 4)^2 + (y - 1)^2} + x^2 - 8x + 16 + y^2 - 2y + 1$

$x^2 + 4x + 4 + y^2 - 2y + 1 - 100 - x^2 + 8x - 16 - y^2 + 2y - 1$

$= -20\sqrt{(x - 4)^2 + (y - 1)^2}$

$12x - 112 = -20\sqrt{(x - 4)^2 + (y - 1)^2}$

$3x - 28 = -5\sqrt{(x - 4)^2 + (y - 1)^2}$

$(3x - 28)^2 = 25[(x - 4)^2 + (y - 1)^2]$

$9x^2 - 168x + 784 = 25(x^2 - 8x + 16 + y^2 - 2y + 1)$

$9x^2 - 168x + 784 = 25x^2 - 200x + 400 + 25y^2 - 50y + 25$

$9x^2 - 25x^2 - 25y^2 - 168x + 200x + 50y + 784 - 400 - 25 = 0$

$-16x^2 - 25y^2 + 32x + 50y + 359 = 0$

$16x^2 + 25y^2 - 32x - 50y - 359 = 0$

25. Given: $2x^2 + 3y^2 - 8x - 4 = 0$
$2x^2 + 3(-y)^2 - 8x - 4 = 2x^2 + 3y^2 - 8x - 4$ (See page 504.)

29.

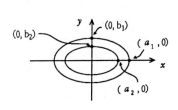

$x^2 + 4y^2 = 100$

$\dfrac{x^2}{10^2} + \dfrac{y^2}{5^2} = 1$

$a_1 = 10, \; b_1 = 5$

$t_b = a_2 - a_1 = 5.8$

$t_c = b_2 - b_1 = 5.0$

$2x^2 + 5y^2 = 500$

$\dfrac{x^2}{\sqrt{250}^2} + \dfrac{y^2}{10^2} = 1$

$a_2 = \sqrt{250} = 15.8, \; b_2 = 10$

33. If the two vertices of each base are fixed at $(-3,0)$ and $(3,0)$, and the sum of the two leg lengths is also fixed, the third vertex lies on an ellipse. The base is 6 cm, so $d_1 + d_2 = 14$ cm $- 6$ cm $= 8$ cm.

$(-3,0)$ and $(3,0)$ are foci $(-c,0)$ and $(c,0)$
$d_1 + d_2 = 2a = 8; \; a = 4$
$a^2 - c^2 = b^2$
$4^2 - 3^2 = b^2$
$b^2 = 7, \; a^2 = 16$, and
the equation is
$\dfrac{x^2}{16} + \dfrac{y^2}{7} = 1$, or
$7x^2 + 16y^2 = 112$.

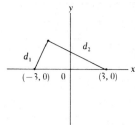

Exercises 20-6, page 592

1. Given $\dfrac{x^2}{25} - \dfrac{y^2}{144} = 1$

$a^2 = 25 \quad b^2 = 144 \quad V(-5,0), \; (5,0)$
$a = \pm 5 \quad b = \pm 12 \quad F(-13,0), \; (13,0)$
$c^2 = a^2 + b^2$ Eq. (20-21)
$c^2 = 169, \; c = \pm 13$
$V(5,0)(-5,0); \; F(\sqrt{13},0)(-\sqrt{13},0)$

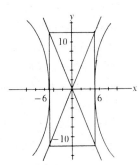

5. Given: $4x^2 - y^2 = 4$

$\dfrac{4x^2}{4} - \dfrac{y^2}{4} = 1$; $\dfrac{x^2}{1} - \dfrac{y^2}{4} = 1$

$a^2 = 1$; $b^2 = 4$; $c^2 = 5$

$V (\sqrt{1},0)$, $(-\sqrt{1},0)$; $F (\sqrt{5},0)$, $(-\sqrt{5},0)$

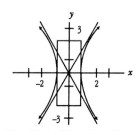

9. Given: $4x^2 - y^2 + 4 = 0$

$4x^2 - y^2 + 4 = 0$; $4x^2 - y^2 = -4$

$\dfrac{4x^2}{-4} - \dfrac{y^2}{-4} = \dfrac{-4}{-4}$

$-x^2 + \dfrac{y^2}{4} = 1$; $\dfrac{y^2}{4} - \dfrac{x^2}{1} = 1$

$a^2 = 4$; $b^2 = 1$; $c^2 = 5$

$V(0,2)$, $(0,-2)$; $F(0,\sqrt{5})$, $(0,-\sqrt{5})$

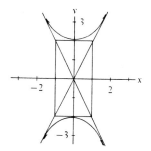

13. Given: $V(3,0)$; $F(5,0)$

$a = 3$; $c = 5$; $a^2 = 9$; $c^2 = 25$

$b^2 = c^2 - a^2 = 25 - 9 = 16$

$\dfrac{x^2}{a^2} - \dfrac{y^2}{b^2} = 1$; $\dfrac{x^2}{9} - \dfrac{y^2}{16} = 1$

$16x^2 - 9y^2 = 144$

17. Given: (x,y) is $(2,3)$; $F(2,0)$, $(-2,0)$; $c = \pm 2$, $c^2 = 4$

$d_1 = \sqrt{(2 - [-2])^2 + (3 - 0)^2} = \sqrt{4^2 + 3^2} = \sqrt{16 + 9} = \sqrt{25} = 5$

$d_2 = \sqrt{(2 - 2)^2 + (3 - 0)^2} = \sqrt{0 + 9} = \sqrt{9} = 3$

$d_1 - d_2 = 2a$; $5 - 3 = 2a$; $2 = 2a$; $1 = a$; $a^2 = 1$

$c^2 = 4$; $b^2 = c^2 - a^2 = 3$

$\dfrac{x^2}{a^2} - \dfrac{y^2}{b^2} = 1$; $\dfrac{x^2}{1} - \dfrac{y^2}{3} = 1$

$3x^2 - y^2 = 3$

21. Given: $xy = 2$; $y = \dfrac{2}{x}$

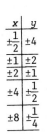

x	y
$\pm\dfrac{1}{2}$	± 4
± 1	± 2
± 2	± 1
± 4	$\pm\dfrac{1}{2}$
± 8	$\pm\dfrac{1}{4}$

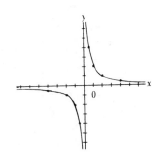

25. Given: $F(1,2)$, $(11,2)$; transverse axis of 8

$d_1 - d_2 = 8$; $d_1 = \sqrt{(x-1)^2 + (y-2)^2}$; $d_2 = \sqrt{(x-11)^2 + (y-2)^2}$

$\sqrt{(x-1)^2 + (y-2)^2} - \sqrt{(x-11)^2 + (y-2)^2} = 8$

$\sqrt{(x-1)^2 + (y-2)^2} = 8 + \sqrt{(x-11)^2 + (y-2)^2}$

$(x-1)^2 + (y-2)^2 = 64 + 16\sqrt{(x-11)^2 + (y-2)^2} + (x-11)^2 + (y-2)^2$

$x^2 - 2x + 1 + y^2 - 4y + 4$

$= 64 + 16\sqrt{(x-11)^2 + (y-2)^2} + x^2 - 22x + 121 + y^2 - 4y + 4$

$x^2 - x^2 + y^2 - y^2 - 2x + 22x - 4y + 4y + 1 + 4 - 64 - 121 - 4$

$= 16\sqrt{(x-11)^2 + (y-2)^2}$

$20x - 184 = 16\sqrt{(x-11)^2 + (y-2)^2}$; $5x - 46 = 4\sqrt{(x-11)^2 + (y-2)^2}$

$(5x - 46)^2 = 16[(x-11)^2 + (y-2)^2]$

$25x^2 - 460x + 2116 = 16(x^2 - 22x + 121 + y^2 - 4y + 4)$

$25x^2 - 460x + 2116 = 16x^2 - 352x + 1936 + 16y^2 - 64y + 64$

$25x^2 - 16x^2 - 16y^2 - 460x + 352x + 64y + 2116 - 1936 - 64 = 0$

$9x^2 - 16y^2 - 108x + 64y + 116 = 0$

29. $V(0,1)$, $F(0,\sqrt{3})$; $c^2 = a^2 + b^2$ where $c = \sqrt{3}$ and $a = 1$

$$b^2 = \sqrt{3}\,^2 - 1^2 = 2$$

$$\frac{y^2}{1^2} - \frac{x^2}{\sqrt{2}\,^2} = 1$$

The transverse axis of the first equation is length $2a = 2\sqrt{1}$ along the y-axis. Its conjugate axis is length $2b = 2\sqrt{2}$ along the x-axis.

The transverse axis of the conjugate hyperbola is length $2\sqrt{2}$ along the x-axis, and its conjugate axis is length $2\sqrt{1}$ along the y-axis.

The equation, then, is

$$\frac{x^2}{\sqrt{2}\,^2} - \frac{y^2}{\sqrt{1}\,^2} = 1$$

$$\frac{x^2}{2} - \frac{y^2}{1} = 1 \text{ or } x^2 - 2y^2 = 2$$

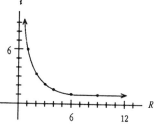

33. $i R = V$ See Eq. (20-25).

   $i R = 6.00$

   $i = \dfrac{6.00}{R}$

R	0	1	2	3	4	6	9	12
i	∞	6	3	2	1.5	1	0.7	0.5

Exercises 20-7, page 596

1.   Given: $(y - 2)^2 = 4(x + 1)$
   $y - 2 = y'$; $x + 1 = x'$; $y'^2 = 4x'$; parabola, Eq. (20-15)
   $y' = y - k$ $x' = x - h$
   $y - 2 = y - k$, $2 = k$ $x + 1 = x - h$, $h = -1$
   origin $0'$ at $(h,k) = (-1,2)$
   $y'^2 = 4(1)x'$; $p = 1$;
   focus $(-1 + p, 2)$ or $(0,2)$
   directrix $x' = -p$; $x + 1 = -1$, $x = -2$
   vertex $(-1,2)$

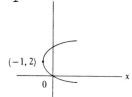

$(-1,2)$

5.   Given: $\dfrac{(x + 1)^2}{1} + \dfrac{y^2}{9} = 1$

   $x' = x + 1$; $x - h = x + 1$; $h = -1$
   $y' = y - 0$; $y - k = y - 0$; $k = 0$

   $\dfrac{x'^2}{1} + \dfrac{y'^2}{9} = 1$; $\dfrac{y'^2}{9} + \dfrac{x'^2}{1} = 1$

   ellipse, center (h,k) at $(-1,0)$
   $a^2 = 9$, $a = 3$; $b^2 = 1$, $b = 1$;
   $c^2 = 8$, $c = 2\sqrt{2}$

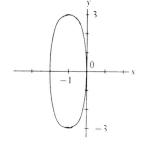

9.   Given: parabola; vertex $(-1,3)$; $p = 4$; parallel to x-axis
            center (h,k) at $(-1,3)$

   $y'^2 = 4px'$; $(y - k)^2 = 4p(x - h)$
   $(y - 3)^2 = 4(4)[x - (-1)]$
   $(y - 3)^2 = 16(x + 1)$; $y^2 - 6y + 9 = 16x + 16$
   $y^2 - 6y - 16x + 9 - 16 = 0$; $y^2 - 6y - 16x - 7 = 0$

13.   Given: ellipse; center $(-2,2)$; foci $(-5,2)$, $(1,2)$
            vertices $(-7,2)$, $(3,2)$

   (h,k) at $(-2,2)$; $c = 3$, $c^2 = 9$; $a = 5$, $a^2 = 25$; $b^2 = a^2 - c^2 = 16$

   $\dfrac{x'^2}{a^2} + \dfrac{y'^2}{b^2} = 1$; $\dfrac{[x - (-2)]^2}{25} + \dfrac{(y - 2)^2}{16} = 1$

   $\dfrac{(x + 2)^2}{25} + \dfrac{(y - 2)^2}{16} = 1$; $16(x + 2)^2 + 25(y - 2)^2 = 400$

   $16(x^2 + 4x + 4) + 25(y^2 - 4y + 4) = 400$

   $16x^2 + 64x + 64 + 25y^2 - 100y + 100 - 400 = 0$

   $16x^2 + 25y^2 + 64x - 100y - 236 = 0$

17. Given: center $(h,k) = (-1,2)$; $F_1 = (-1,4)$ and $V_1 = (-1,1)$

$c^2 = a^2 + b^2$ where $c = \sqrt{(-1 + 1)^2 + (4 - 2)^2} = 2$ is the distance between F_1 and (h,k).

Substituting known values,

$\dfrac{(y - 2)^2}{a^2} - \dfrac{(x + 1)^2}{b^2} = 1$ (h,k) substituted

$\dfrac{(1 - 2)^2}{a^2} - \dfrac{(-1 + 1)^2}{b^2} = 1$ V_1 substituted

$\dfrac{1}{a^2} - \dfrac{0}{b^2} = 1$; $a^2 = 1$

Since $c^2 = a^2 + b^2$

$2^2 = 1^2 + b^2$

$b^2 = 3$; so,

$\dfrac{(y - 2)^2}{1} - \dfrac{(x + 1)^2}{3} = 1$; $3(y - 2)^2 - (x + 1)^2 = 3$ or

$x^2 - 3y^2 + 2x + 12y - 8 = 0$

21. Given: $x^2 + 2x - 4y - 3 = 0$; $x^2 + 2x - 3 = 4y$

$x^2 + 2x = 4y + 3$; $x^2 + 2x + 1 = 4y + 3 + 1$

$(x + 1)^2 = 4y + 4$; $(x + 1)^2 = 4(y + 1)$

parabola, $p = 1$; $x - h = x + 1$, $h = -1$; $y - k = y + 1$, $k = -1$

vertex is $(-1,-1)$

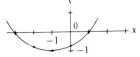

25. Given: $9x^2 - y^2 + 8y - 7 = 0$; $9x^2 - (y^2 - 8y + 7 + 9) = -9$

$9x^2 - (y - 4)^2 = -9$; $\dfrac{9x^2}{-9} - \dfrac{(y - 4)^2}{-9} = 1$

$-x^2 + \dfrac{(y - 4)^2}{9} = 1$; $\dfrac{(y - 4)^2}{9} - \dfrac{x^2}{1} = 1$

hyperbola, (h,k) is $(0,4)$; $a = 3$; $b = 1$

29. Given: hyperbola; asymptotes: $x - y = -1$ or $x + 1 = y$, and $x + y = -3$ or $y = -x - 3$

vertices $(3,-1)$ and $(-7,-1)$

The center is at the point of intersection of the asymptotes. The equations for the asymptotes are solved simultaneously by adding; $2y = -2$, $y = -1$; $-1 = x + 1$, $x = -2$. Therefore the coordinates of the center are $(-2,-1)$. Since the slopes are 1 and -1, $a = b$, where a is the distance from the center $(-2,-1)$ to the vertex $(3,-1)$; $a = 5$, $b = 5$.

$\dfrac{(x - h)^2}{a^2} - \dfrac{(y - k)^2}{b^2} = 1$;

$\dfrac{[x - (-2)]^2}{25} - \dfrac{[y - (-1)]^2}{25} = 1$; $\dfrac{(x + 2)^2}{25} - \dfrac{(y + 1)^2}{25} = 1$

$x^2 + 4x + 4 - (y^2 + 2y + 1) = 25$

$x^2 + 4x + 4 - y^2 - 2y - 1 = 25$

$x^2 - y^2 + 4x - 2y - 22 = 0$

33.

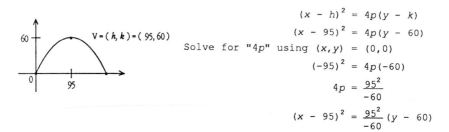

$$(x - h)^2 = 4p(y - k)$$
$$(x - 95)^2 = 4p(y - 60)$$
Solve for "$4p$" using $(x, y) = (0, 0)$
$$(-95)^2 = 4p(-60)$$
$$4p = \frac{95^2}{-60}$$
$$(x - 95)^2 = \frac{95^2}{-60}(y - 60)$$

33M.

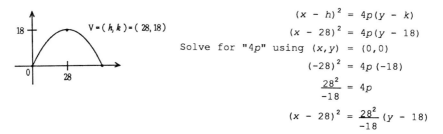

$$(x - h)^2 = 4p(y - k)$$
$$(x - 28)^2 = 4p(y - 18)$$
Solve for "$4p$" using $(x, y) = (0, 0)$
$$(-28)^2 = 4p(-18)$$
$$\frac{28^2}{-18} = 4p$$
$$(x - 28)^2 = \frac{28^2}{-18}(y - 18)$$

Exercises 20-8, page 600

1. $x^2 + 2y^2 - 2 = 0$
    $A \neq C$, they have the same sign, and $B = 0$; ellipse

5. Given: $2x^2 + 2y^2 - 3y - 1 = 0$
    $A = C$; $B = 0$; equation for circle

9. Given: $x^2 = y^2 - 1$
    $x^2 - y^2 + 1 = 0$; A and C have different signs; $B = 0$; equation for
    hyperbola

13. Given: $y(3 - 2y) = 2(x^2 - y^2)$
    $$3y - 2y^2 = 2x^2 - 2y^2$$
    $$3y = 2x^2 \text{ or } x^2 = \frac{3}{2}y, \text{ a parabola}$$

17. Given: $2x(x - y) = y(3 - y - 2x)$
    $2x^2 - 2xy = 3y - y^2 - 2xy$
    $2x^2 - 2xy + 2xy + y^2 - 3y = 0$
    $2x^2 + y^2 - 3y = 0$
    $A \neq C$, same sign; equation for ellipse; $B = 0$

21. Given: $x^2 = 8(y - x - 2)$
$x^2 = 8y - 8x - 16$
$x^2 + 8x - 8y + 16 = 0$
$A \neq 0$; $C = 0$, $B = 0$, parabola

$x^2 + 8x - 8y + 16 = 0$
$x^2 + 8x + 16 = 8y$
$(x + 4)^2 = 4(2)y$; $p = 2$
vertex $(-4,0)$
focus $(-4,2)$

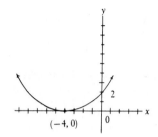

25. Given: $y^2 + 42 = 2x(10 - x)$

$y^2 + 42 = 20x - 2x^2$
$y^2 + 2x^2 - 20x + 42 = 0$; ellipse
$\dfrac{y^2}{2} + x^2 - 10x = -21$

$\dfrac{y^2}{2} + x^2 - 10x + 25 = -21 + 25$

$\dfrac{y^2}{2} + (x - 5)^2 = 4$

$\dfrac{y^2}{8} + \dfrac{(x - 5)^2}{4} = 1$ (h,k) at $(5,0)$

$a = \sqrt{8} = 2\sqrt{2}$; $b = 2$

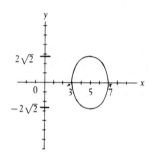

29. (a) If $k = 1$, $x^2 + ky^2 = a^2$
$\qquad\qquad\qquad x^2 + (1)y^2 = a^2$
$\qquad\qquad\qquad x^2 + y^2 = a^2$ (circle)

(b) If $k < 0$, $x^2 + ky^2 = a^2$
$\qquad\qquad\qquad x^2 - |k|y^2 = a^2$
$\qquad\qquad\qquad \dfrac{x^2}{a^2} - \dfrac{y^2}{a^2/|k|} = 1$ (hyperbola)

(c) If $k > 0$ $(k \neq 1)$, $x^2 + ky^2 = a^2$
$\qquad\qquad\qquad\qquad \dfrac{x^2}{a^2} + \dfrac{y^2}{a^2/k} = 1$ (ellipse)

33.

$x^2 + y^2 = (x + 3)^2$
$x^2 + y^2 = x^2 + 6x + 9$

$y^2 = 6x + 9 = 6(x + \tfrac{3}{2})$; a parabola

Exercises 20-9, page 604

1. Given: $(3, \frac{\pi}{6})$; $r = 3$, $\theta = \frac{\pi}{6}$

5. Given: $(-2, \frac{7\pi}{6})$ (Negative r is reversed in direction from positive r.)

9. Given: $(0.5, -\frac{8\pi}{3})$

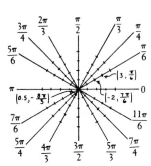

13. Given: $(\sqrt{3}, 1)$ is (x, y), quadrant I

$\tan \theta = \frac{y}{x}$ Eq. (20-30)

$\theta = \arctan \frac{y}{x} = \arctan \frac{1}{\sqrt{3}} = \arctan \frac{\sqrt{3}}{3}$; $\theta = 30° = \frac{\pi}{6}$

Eq. (20-30) $r = \sqrt{x^2 + y^2} = \sqrt{(\sqrt{3})^2 + 1^2}$

$$= \sqrt{3 + 1} = \sqrt{4} = 2$$

(r, θ) is $(2, \frac{\pi}{6})$

17. Given: (r, θ) is $(8, \frac{4\pi}{3})$, quadrant III

$x = r \cos \theta = 8 \cos \frac{4\pi}{3} = 8(-\frac{1}{2}) = -4$

$y = r \sin \theta = 8(-\frac{\sqrt{3}}{2}) = -4\sqrt{3}$ Eq. (20-29)

(x, y) is $(-4, -4\sqrt{3})$

21. Given: $x = 3$

$r \cos \theta = x = 3$ Eq. (20-29)

$r = \frac{3}{\cos \theta} = 3 \frac{1}{\cos \theta} = 3 \sec \theta$

25. Given: $y^2 = 4x$

$(r \sin \theta)^2 = 4r \cos \theta$; $r^2 \sin^2 \theta = 4r \cos \theta$

$r \sin^2 \theta = 4 \cos \theta$; $r = \frac{4 \cos \theta}{\sin^2 \theta}$

$r = \frac{4 \cos \theta}{\sin \theta} \times \frac{1}{\sin \theta}$ or $r = 4 \cot \theta \csc \theta$

29. Given: $r = \sin \theta$; $r^2 = r \sin \theta$; $r^2 = x^2 + y^2$ Eq. (20-30)

$x^2 + y^2 = r^2 = r \sin \theta = y$; Eq. (20-29)

$x^2 + y^2 - y = 0$

33. Given: $r = 2(1 + \cos \theta)$; $x = r \cos \theta$; $\dfrac{x}{r} = \cos \theta$

$r^2 = x^2 + y^2$; $r = \sqrt{x^2 + y^2}$

$r = 2(1 + \cos \theta) = 2(1 + \dfrac{x}{r}) = 2 + \dfrac{2x}{r}$; $r^2 = 2r + 2x$ (Multiply through by r.)

$x^2 + y^2 = 2\sqrt{x^2 + y^2} + 2x$; $x^2 + y^2 - 2x = 2\sqrt{x^2 + y^2}$

$(x^2 + y^2 - 2x)^2 = 4(x^2 + y^2)$

$x^4 + y^4 - 4x^3 + 2x^2y^2 - 4xy^2 + 4x^2 = 4x^2 + 4y^2$

$x^4 + y^4 - 4x^3 + 2x^2y^2 - 4xy^2 + 4x^2 - 4x^2 - 4y^2 = 0$

$x^4 + y^4 - 4x^3 + 2x^2y^2 - 4xy^2 - 4y^2 = 0$

37. $B_x = \dfrac{-ky}{x^2 + y^2} = \dfrac{-ky}{r^2} = -\dfrac{kr \sin \theta}{r^2} = -\dfrac{k \sin \theta}{r}$

$B_y = \dfrac{kx}{x^2 + y^2} = \dfrac{kx}{r^2} = \dfrac{kr \cos \theta}{r^2} = \dfrac{k \cos \theta}{r}$

Exercises 20-10, page 607

1. Given: $r = 4$ for all θ.

Graph is a circle with radius 4.

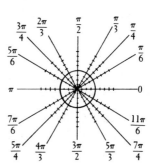

5. Given: $r = 4 \sec \theta = \dfrac{4}{\cos \theta}$

θ	0	$\frac{\pi}{6}$	$\frac{\pi}{4}$	$\frac{\pi}{3}$	$\frac{\pi}{2}$	$\frac{2\pi}{3}$	$\frac{3\pi}{4}$	$\frac{5\pi}{6}$	π	$\frac{5\pi}{4}$	$\frac{3\pi}{2}$	$\frac{7\pi}{4}$	2π
r	4	4.6	5.7	8	*	-8	-5.7	4.6	-4	-5.7	*	5.7	4

* denotes undefined

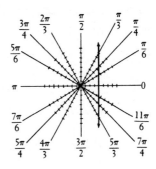

9. Given $r = 1 - \cos \theta$

θ	0	$\frac{\pi}{4}$	$\frac{\pi}{2}$	$\frac{3\pi}{4}$	π	$\frac{5\pi}{4}$	$\frac{3\pi}{2}$	$\frac{7\pi}{4}$
r	0	0.3	1	1.7	2	1.7	1	0.3

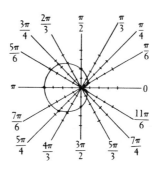

13. Given: $r = 4 \sin 2\theta$

θ	0	$\frac{\pi}{8}$	$\frac{\pi}{4}$	$\frac{3\pi}{8}$	$\frac{\pi}{2}$	$\frac{5\pi}{8}$	$\frac{3\pi}{4}$	$\frac{7\pi}{8}$	π	$\frac{9\pi}{8}$	$\frac{5\pi}{4}$	$\frac{11\pi}{8}$	$\frac{3\pi}{2}$	$\frac{13\pi}{8}$	$\frac{7\pi}{4}$	2π
r	0	2.8	4	2.8	0	-2.8	-4	-2.8	0	2.8	4	2.8	0	-2.8	-4	-2.8

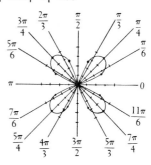

17. Given: $r = 2^{\theta}$

θ	0	$\frac{\pi}{4}$	$\frac{\pi}{2}$	$\frac{3\pi}{4}$	π	$\frac{5\pi}{4}$	$\frac{3\pi}{2}$	$\frac{7\pi}{4}$	2π
r	1	1.7	3.0	5.1	8.8	15.2	26.2	45.2	77.9

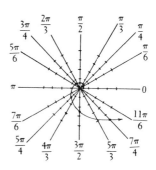

21. Given: $r = \dfrac{1}{2 - \cos \theta}$

θ	0	$\frac{\pi}{4}$	$\frac{\pi}{2}$	$\frac{3\pi}{4}$	π	$\frac{5\pi}{4}$	$\frac{3\pi}{2}$	$\frac{7\pi}{4}$	2π
r	1	0.77	0.50	0.37	0.33	0.37	0.50	0.77	1

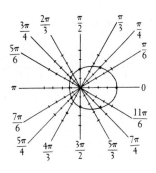

25. Given: $r = 4 \cos \frac{1}{2} \theta$

θ	r
0	4.0
$\frac{\pi}{6}$	3.9
$\frac{\pi}{4}$	3.7
$\frac{\pi}{3}$	3.5
$\frac{\pi}{2}$	2.8
$\frac{2\pi}{3}$	2.0
$\frac{3\pi}{4}$	1.5
$\frac{5\pi}{6}$	1.0
π	0
$\frac{7\pi}{6}$	-1.0
$\frac{5\pi}{4}$	-1.5
$\frac{4\pi}{3}$	-2.0

θ	r
$\frac{3\pi}{2}$	-2.8
$\frac{5\pi}{3}$	-3.5
$\frac{7\pi}{4}$	-3.7
$\frac{11\pi}{6}$	-3.9
2π	-4.0
$\frac{13\pi}{6}$	-3.9
$\frac{9\pi}{4}$	-3.7
$\frac{7\pi}{3}$	-3.5
$\frac{5\pi}{2}$	-2.8
$\frac{8\pi}{3}$	-2.0
$\frac{11\pi}{4}$	-1.5
$\frac{17\pi}{6}$	-1.0

θ	r
3π	0
$\frac{19\pi}{6}$	1.0
$\frac{13\pi}{4}$	1.5
$\frac{10\pi}{3}$	2.0
$\frac{7\pi}{2}$	2.8
$\frac{11\pi}{3}$	3.5
$\frac{15\pi}{4}$	3.7
$\frac{23\pi}{6}$	3.9
4π	4.0

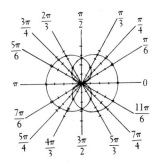

29. $r = 4.0 - \sin \theta$

θ	0	$\pi/4$	$\pi/2$	$3\pi/4$	π	$5\pi/4$	$3\pi/2$	$7\pi/4$
r	4.0	3.3	3.0	3.3	4.0	4.7	5.0	4.7

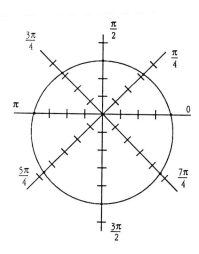

Review Exercises for Chapter 20, page 609

1. Given: straight line; (x_1, y_1) is $(1,-7)$; $m = 4$
    $y - y_1 = m(x - x_1)$; $y - (-7) = 4(x - 1)$
    $y + 7 = 4x - 4$; $y = 4x - 4 - 7$
    $y = 4x - 11$ or $4x - y - 11 = 0$

5. Given: circle; (h,k) is $(1,-2)$; (x,y) is $(4,-3)$
    $r = \sqrt{(1 - 4)^2 + [-2 - (-3)]^2} = \sqrt{(-3)^2 + (-2 + 3)^2}$
    $\quad = \sqrt{9 + 1} = \sqrt{10}$
    $(x - h)^2 + (y - k)^2 = r^2$; $(x - 1)^2 + (y + 2)^2 = \sqrt{10}^2$
    $x^2 - 2x + 1 + y^2 + 4y + 4 = 10$
    $x^2 + y^2 - 2x + 4y + 1 + 4 - 10 = 0$
    $x^2 + y^2 - 2x + 4y + 1 + 4 - 10 = 0$; $x^2 + y^2 - 2x + 4y - 5 = 0$

9. Given: ellipse; vertex (10,0); focus (8,0); (h,k) is (0,0)

$$\frac{(x-h)^2}{a^2} + \frac{(y-k)^2}{b^2} = 1 \qquad\qquad a = 10 \qquad\qquad c = 8$$
$$\frac{x^2}{100} + \frac{y^2}{36} = 1 \qquad\qquad \begin{aligned} b^2 &= a^2 - c^2 \\ b^2 &= 100 - 64 \\ b^2 &= 36 \end{aligned}$$

$$9x^2 + 25y^2 = 900$$

13. Given: $x^2 + y^2 + 6x - 7 = 0$

$(x^2 + 6x) + (y^2) = 7; \ (x^2 + 6x + 9) + y^2 = 7 + 9$

$(x + 3)^2 + (y + 0)^2 = 16$

$[x - (-3)]^2 + (y - 0)^2 = 4^2$

center $(h,k) = (-3,0)$; radius $r = 4$

17. Given: $16x^2 + y^2 = 16$

$$\frac{16x^2}{16} + \frac{y^2}{16} = 1; \ \frac{x^2}{1^2} + \frac{y^2}{4^2} = 1; \ \frac{y^2}{4^2} + \frac{x^2}{1^2} = 1$$

$a = 4, \ b = 1, \ c = \sqrt{16 - 1} = \sqrt{15}$

vertices $(0,a)$, $(0,-a)$ or $(0,4)$, $(0,-4)$
foci $(0,c)$, $(0,-c)$ or $(0,\sqrt{15})$, $(0,-\sqrt{15})$

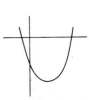

21. Given: $x^2 - 8x - 4y - 16 = 0$

$x^2 - 8x = 4y + 16; \ x^2 - 8x + 16 = 4y + 16 + 16$

$(x - 4)^2 = 4y + 32; \ (x - 4)^2 = 4(y + 8)$

$(x - 4)^2 = 4(1)(y + 8); \ p = 1$

vertex (h,k) is $(4,-8)$; focus is $(4,-7)$

25. Given: $r = 4(1 + \sin \theta)$

θ	0	$\frac{\pi}{4}$	$\frac{\pi}{2}$	$\frac{3\pi}{4}$	π	$\frac{5\pi}{4}$	$\frac{3\pi}{2}$	$\frac{7\pi}{4}$
r	4	6.8	8	6.8	4	1.2	0	1.2

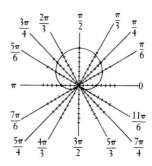

29. Given: $r = \cot \theta$

θ	0	$\dfrac{\pi}{12}$	$\dfrac{\pi}{6}$	$\dfrac{\pi}{4}$	$\dfrac{\pi}{3}$	$\dfrac{5\pi}{12}$
r	und.	3.73	1.73	1	0.58	0.27

$\dfrac{\pi}{2}$	$\dfrac{2\pi}{3}$	$\dfrac{3\pi}{4}$	$\dfrac{5\pi}{6}$	π	$\dfrac{7\pi}{6}$
0.00	-0.58	-1.0	-1.73	und.	1.73

$\dfrac{5\pi}{4}$	$\dfrac{4\pi}{3}$	$\dfrac{3\pi}{2}$	$\dfrac{5\pi}{3}$	$\dfrac{7\pi}{4}$	$\dfrac{11\pi}{6}$	2π
1.00	0.58	und.	-0.58	-1.0	-1.73	und.

und. = undefined

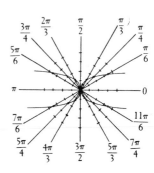

33. Given: $y = 2x$

$\dfrac{y}{x} = 2$; $\tan \theta = 2$; $\theta = \arctan 2 = 1.11$ Eq. (20-30)

37. Given: $r = 2 \sin 2\theta$; $r = \sqrt{x^2 + y^2}$

$\sin \theta = \dfrac{y}{r} = \dfrac{y}{\sqrt{x^2 + y^2}}$; $\cos \theta = \dfrac{x}{r} = \dfrac{x}{\sqrt{x^2 + y^2}}$

$r = 2(2 \sin \theta \cos \theta)$ Eq. (19-20)
$r = 4 \sin \theta \cos \theta$

$\sqrt{x^2 + y^2} = 4 \dfrac{y}{\sqrt{x^2 + y^2}} \times \dfrac{x}{\sqrt{x^2 + y^2}}$; $\sqrt{x^2 + y^2} = \dfrac{4xy}{x^2 + y^2}$

$x^2 + y^2 = \dfrac{(4xy)^2}{(x^2 + y^2)^2} = \dfrac{16\,x^2 y^2}{(x^2 + y^2)^2}$

$(x^2 + y^2)^3 = 16x^2 y^2$

41. $x^2 + y^2 = 9$ circle; center (0,0); radius 3
$x^2 + y^2 = 3^2$

$4x^2 + y^2 = 16$ ellipse; centered at (0,0)
$\dfrac{x^2}{4} + \dfrac{y^2}{16} = 1$ $(\pm 2,0)$ and $(0,\pm 4)$ vertices

$\dfrac{x^2}{2^2} + \dfrac{y^2}{4^2} = 1$

four real solutions

45. Given: $25x^2 + 4y^2 = 100$ and $4x^2 + 9y^2 = 36$

$9 \begin{bmatrix} 25x^2 + 4y^2 = 100 \\ 4x^2 + 9y^2 = 36 \end{bmatrix}$ $\quad\quad$ $4 \begin{bmatrix} 25x^2 + 4y^2 = 100 \\ 4x^2 + 9y^2 = 36 \end{bmatrix}$
-4 $\quad\quad\quad\quad\quad\quad\quad\quad\quad$ -25

$\quad\quad\quad$ $225x^2 + 36y^2 = 900$ $\quad\quad\quad\quad$ $100x^2 + 16y^2 = 400$
$\quad\quad$ $\underline{- 16x^2 - 36y^2 = -144}$ $\quad\quad\quad$ $\underline{-100x^2 - 225y^2 = -900}$
$\quad\quad\quad$ $209x^2 \quad\quad\quad = 756$ $\quad\quad\quad\quad\quad$ $- 209y^2 = -500$

$\quad\quad\quad\quad\quad\quad\quad$ $x^2 = \dfrac{756}{209}$ $\quad\quad\quad\quad\quad\quad\quad\quad$ $y^2 = \dfrac{-500}{-209}$

$\quad\quad\quad\quad\quad\quad\quad$ $x^2 = 3.62$ $\quad\quad\quad\quad\quad\quad\quad\quad$ $y^2 = 2.39$

$\quad\quad\quad\quad\quad\quad\quad$ $x = \pm1.90$ $\quad\quad\quad\quad\quad\quad\quad\quad$ $y = \pm1.55$

$\quad$ $(1.90,1.55)$, $(-1.90,1.55)$, $(1.90,-1.55)$, and $(-1.90,-1.55)$

49. Given: focus $(3,1)$; directrix $y = -3$; vertex $(3,-1)$ is (h,k); $p = 2$
By definition, $d_1 = d_2$ where d_1 is from (x,y) to $(x,-3)$, a point on the
directrix, and d_2 is from (x,y) to $(3,1)$, the focus.

$d_1 = \sqrt{(x - x)^2 + [y - (-3)]^2}$; $d_2 = \sqrt{(x - 3)^2 + (y - 1)^2}$

$\sqrt{0^2 + (y + 3)^2} = \sqrt{(x - 3)^2 + (y - 1)^2}$; $0 + (y + 3)^2 = (x - 3)^2 + (y - 1)^2$

$y^2 + 6y + 9 = x^2 - 6x + 9 + y^2 - 2y + 1$

$0 = x^2 + y^2 - y^2 - 6x - 2y - 6y + 9 + 1 - 9$; $0 = x^2 - 6x - 8y + 1$

By translation, $(x - h)^2 = 4(p)(y - k)$

$(x - 3)^2 = 4(2)[y - (-1)]$; $(x - 3)^2 = 8(y + 1)$

$x^2 - 6x + 9 = 8y + 8$; $x^2 - 6x - 8y + 9 - 8 = 0$

$x^2 - 6x - 8y + 1 = 0$

53. $2500x + 1500y = 37\ 500$

$\quad\quad\quad 1500y = -2500x + 37\ 500$

$\quad\quad\quad\quad\quad y = -\dfrac{5}{3} x + 25$

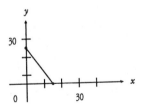

57. If circumference is 4.50 m, then
radius is $\dfrac{c}{2\pi} = \dfrac{4.50}{2\pi} = 0.716$. The
equation of the circle is
$x^2 + y^2 = (0.716)^2$
$x^2 + y^2 = 0.513$

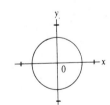

61.

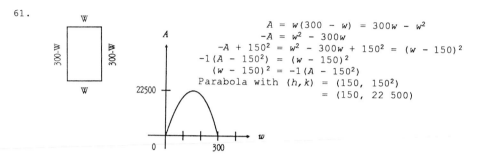

$$A = w(300 - w) = 300w - w^2$$
$$-A = w^2 - 300w$$
$$-A + 150^2 = w^2 - 300w + 150^2 = (w - 150)^2$$
$$-1(A - 150^2) = (w - 150)^2$$
$$(w - 150)^2 = -1(A - 150^2)$$
Parabola with $(h, k) = (150, 150^2)$
$$= (150, 22\ 500)$$

65.   If the curve is a quarter ellipse, the ellipse will have a semimajor
axis of 10 and a semiminor axis of 1, with $(h, k) = (10, 0)$. Therefore,

$$a = 10 \text{ and } b = 1, \text{ so } \frac{(d - 10)^2}{10^2} + \frac{f^2}{1^2} = 1$$

$$\text{or } \frac{(d - 10)^2}{100} + \frac{f^2}{1} = 1$$

69.

$$\frac{x^2}{a^2} - \frac{y^2}{b^2} = 1$$

$y = 40$ when $x = 40$

$y = 100$ when $x = 50$

$$\begin{cases} \dfrac{40^2}{a^2} - \dfrac{40^2}{b^2} = 1 \text{ or } 40^2 b^2 - 40^2 a^2 = a^2 b^2 \\[3mm] \dfrac{50^2}{a^2} - \dfrac{100^2}{b^2} = 1 \text{ or } 50^2 b^2 - 100^2 a^2 = a^2 b^2 \end{cases}$$

M by 100^2 $\quad \begin{cases} 40^2 b^2 - 40^2 a^2 = a^2 b^2 \\ 50^2 b^2 - 100^2 a^2 = a^2 b^2 \end{cases}$
M by -40^2

$$\begin{cases} 100^2 40^2 b^2 - 100^2 40^2 a^2 = 100^2 a^2 b^2 \\ -40^2 50^2 b^2 + 40^2 100^2 a^2 = -40^2 a^2 b^2 \end{cases}$$

Add $\begin{cases} 16 \times 10^6 b^2 - 1.6 \times 10^7 a^2 = 10 \times 10^3 \ a^2 b^2 \\ 4 \times 10^6 b^2 + 1.6 \times 10^7 a^2 = -1.6 \times 10^3 \ a^2 b^2 \end{cases}$

$$12 \times 10^6 b^2 = 8.4 \times 10^3 \ a^2 b^2$$

$$12 \times 10^6 = 8.4 \times 10^3 \ a^2$$

$$a^2 = \frac{12 \times 10^6}{8.4 \times 10^3} = 1.42 \times 10^3$$

$$a = 37.8$$

73. $r^2 = R^2 \cos 2\left(\theta + \frac{\pi}{2}\right)$

$r = R\sqrt{\cos 2\left(\theta + \frac{\pi}{2}\right)} = R\sqrt{\cos(2\theta + \pi)}$

Since the square root is real only when $\cos 2\left(\theta + \frac{\pi}{2}\right)$

is not negative, the range is $\frac{\pi}{4} \leq \frac{3\pi}{4}$ or $\frac{5\pi}{4} \leq r \leq \frac{7\pi}{4}$.

(See Chapter 9.)

θ	$\frac{\pi}{4}$	$\frac{\pi}{3}$	$\frac{\pi}{2}$	$\frac{2\pi}{3}$	$\frac{3\pi}{4}$	$\frac{5\pi}{4}$	$\frac{4\pi}{3}$	$\frac{3\pi}{2}$	$\frac{5\pi}{3}$	$\frac{7\pi}{4}$
r	0	0.5R	1R	0.5R	01	0	0.5R	1R	0.5R	0

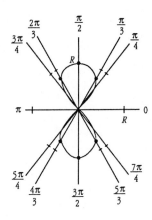

73M.   From Exercise 29, Section 20-5, $6.91x^2 + 7.06y^2 = 4.88 \times 10^8$

$x = r \cos \theta; \ y = r \sin \theta;$ Eq. (20-29)

$6.91(r \cos \theta)^2 + 7.06(r \sin \theta)^2 = 4.88 \times 10^8$

$r^2(6.91 \cos^2 \theta + 7.06 \sin^2 \theta) = 4.88 \times 10^8$

$r^2[6.91(1 - \sin^2 \theta) + 7.06 \sin^2 \theta] = 4.88 \times 10^8$

$r^2(6.91 - 6.91 \sin^2 \theta + 7.06 \sin^2 \theta) = 4.88 \times 10^8$

$r^2(6.91 + 0.15 \sin^2 \theta) = 4.88 \times 10^8$

$r^2 = \dfrac{4.88 \times 10^8}{6.91 + 0.15 \sin^2 \theta}$

CHAPTER 21

Introduction to Statistics and Empirical Curve Fitting

Exercises 21-1, page 622

1. | Number | 2 | 3 | 4 | 5 | 6 | 7 |
    |---|---|---|---|---|---|---|
    | Frequency | 1 | 3 | 4 | 2 | 3 | 2 |

5. | Interval | 2 – 3 | 4 – 5 | 6 – 7 |
    |---|---|---|---|
    | Frequency | 4 | 6 | 5 |

9.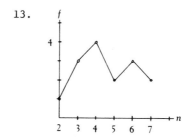

13.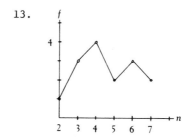

17. | Number | 18 | 19 | 20 | 21 | 22 | 23 | 25 |
    |---|---|---|---|---|---|---|---|
    | Frequency | 1 | 3 | 2 | 4 | 3 | 1 | 1 |

21. | Time (s) | 2.21 | 2.22 | 2.23 | 2.24 | 2.25 | 2.26 | 2.27 | 2.28 | 2.29 |
    |---|---|---|---|---|---|---|---|---|---|
    | Frequency | 2 | 7 | 18 | 41 | 56 | 32 | 8 | 3 | 3 |

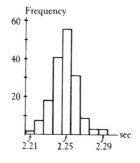

25.

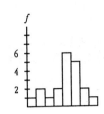

29.

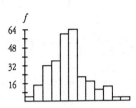

Exercises 21-2, page 627

1. Arrange the numbers in numerical order:
    2, 3, 3, 3, 4, 4, 4, 4, 5, 5, 6, 6, 6, 7, 7
    There are 15 numbers. The middle number is eighth. Since the eighth number is 4, the median is 4.

5. The arithmetic mean is:

$$\overline{X} = \frac{2 + 3 + 3 + 3 + 4 + 4 + 4 + 4 + 5 + 5 + 6 + 6 + 6 + 7 + 7}{15} = \frac{69}{15} = 4.6$$

9. The mode is the number which occurs most frequently, which is 4 since it occurs four times.

13. Arranged in numerical order, find the value of the 8th (middle) number.
     18, 19, 19, 19, 20, 20, 21, 21, 21, 21, 22, 22, 22, 23, 25
     The median value is 21.

17. $\overline{T} = (2(2.21) + 7(2.22) + 18(2.23) + 41(2.24) + 56(2.25)$

     $+ 32(2.26) + 8(2.27) + 3(2.28) + 3(2.29))/170$

     $= 382.13/170 = 2.248$

21. $\overline{d} = (3.83 + 3.90 + 3.96 + 4.09 + 4.15 + 4.18 + 4.21 + 4.23 + 4.25 + 4.26$
     $+ 4.27 + 4.29 + 4.33 + 4.34 + 2(4.36) + 4.37 + 4.41 + 4.44 + 4.51)/20 = 4.237$

25. 25, 28, 28, 29, 29, 30, 30, 30, 30, 30, 31, 31, 31, 32, 32, 32, 33, 33,
     34, 34, 34, 35, 36.
     The median value is the 12th number, which is 31.

29. Arrange the salaries in order:

     175, 200, 225, 225, 250, 250, 275, 275, 300, 300, 300, 350, 350, 400

     There are 14 numbers. The middle number is between the seventh and the eighth, which are both 275. Therefore, the median is $275. The mode is $300, which occurs three times.

33. The lowest is 2 and the highest is 7. Therefore, the midrange is
$\frac{2 + 7}{2} = \frac{9}{2} = 4.5.$

Exercises 21-3, page 628

1.

x	$x - \bar{x}$	$(x - \bar{x})^2$
2	-2.60	6.76
3	-1.60	2.56
3	-1.60	2.56
3	-1.60	2.56
4	-0.60	0.36
4	-0.60	0.36
4	-0.60	0.36
4	-0.60	0.36
5	0.40	0.16
5	0.40	0.16
6	1.40	1.96
6	1.40	1.96
6	1.40	1.96
7	2.40	5.76
7	2.40	5.76
69		33.60

$\bar{x} = \frac{69}{15} = 4.60$

$\overline{(x - \bar{x})^2} = \frac{33.60}{15} = 2.24$

$s = \sqrt{2.24} = 1.50 \qquad \text{Eq. (21-6)}$

5.

x	x^2
2	4
3	9
3	9
3	9
4	16
4	16
4	16
4	16
5	25
5	25
6	36
6	36
6	36
7	49
7	49
69	351

$\bar{x} = \frac{69}{15} = 4.60$

$\bar{x}^2 = 4.60^2 = 21.16$

$\overline{x^2} = \frac{351}{15} = 23.40$

$s = \sqrt{23.40 - 21.16}$

$\quad = \sqrt{2.24} = 1.50 \qquad \text{Eq. (21-9)}$

9.

x	x^2
18	324
19	361
19	361
19	361
20	400
20	400
21	441
21	441
21	441
21	441
22	484
22	484
22	484
23	529
25	625
313	6577

$$\bar{x} = \frac{\Sigma x}{15} = \frac{313}{15} = 20.867$$

$$\bar{x}^2 = 435.419$$

$$\overline{x^2} = 6577/15 = 438.467$$

$$s = \sqrt{\overline{x^2} - \bar{x}^2} = \sqrt{438.467 - 435.419} = \sqrt{3.0477} = 1.7$$

13.

x	f	x^2	fx^2
2.21	2	4.8841	9.7682
2.22	7	4.9284	34.4988
2.23	18	4.9729	89.5122
2.24	41	5.0176	205.7216
2.25	56	5.0625	283.5
2.26	32	5.1076	163.4432
2.27	8	5.1529	41.2232
2.28	3	5.1984	15.5952
2.29	3	5.2441	15.7323
382.13	170		858.9947

$$\bar{x} = 382.13/170 = 2.2478$$

$$\bar{x}^2 = (382.13/170)^2 = 5.0527$$

$$\overline{x^2} = 858.9947/170$$

$$= 5.05291$$

$$s = \sqrt{\overline{x^2} - \bar{x}^2} = \sqrt{0.00020}$$

$$= 0.014$$

17. From Exercise 1, $\bar{x} = 4.60$ and $s = 1.50$

$\bar{x} - s = 4.60 - 1.50 = 3.10$

$\bar{x} + s = 4.60 + 1.50 = 6.10$

The interval from 3.10 to 6.10 contains the following values: 4, 4, 4, 4, 5, 5, 6, 6, 6, or 9 of the 15 values. The percent in this interval is

$$\frac{9}{15} = 0.60 = 60\%.$$

21. From Exercise 9 Section 21-3, $\bar{x} = 20.867$ and $s = 1.7$

$\bar{x} - s = 20.867 - 1.7 = 19.167$
$\bar{x} + s = 20.867 + 1.7 = 22.567$

The interval 19.167 to 22.567 contains $2(20) + 4(21) + 3(22) = 190$ for

$\dfrac{190}{313} = 60\%$

Exercises 21-4, page 632

1.
x	y	xy	x^2
4	1	4	16
6	4	24	36
8	5	40	64
10	8	80	100
12	9	108	144
40	27	256	360

$\bar{x} = \dfrac{40}{5} = 8 \quad \bar{x}^2 = 64 \quad \bar{y} = \dfrac{27}{5} \quad .4 \quad \bar{x}\bar{y} = 8(5.4)$
$= 43.2$

$\overline{xy} = \dfrac{256}{5} = 51.2 \qquad \overline{x^2} = \dfrac{360}{5} = 72$

$s_x^2 = \overline{x^2} - \bar{x}^2 = 72 - 64 = 8$

$m = \dfrac{\overline{xy} - \bar{x}\bar{y}}{s_x^2} = \dfrac{51.2 - 43.2}{8} = 1.0$

$b = \bar{y} - m\bar{x} = 5.4 - 1.0(8)$
$= 5.4 - 8.0 = -2.6$

$y = mx + b; \quad y = 1.0x - 2.6$

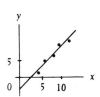

5.
i	V	iV	i^2
15.00	3.00	45.00	225.00
10.80	4.10	44.28	116.64
9.30	5.60	52.08	86.49
3.55	8.00	28.40	12.60
4.60	10.50	48.30	21.16
43.25	31.20	218.06	461.89

$\bar{I} = \dfrac{43.25}{5} = 8.65 \qquad \bar{I}^2 = 74.82$

$\bar{V} = \dfrac{31.2}{5} = 6.24 \qquad \bar{I}\bar{V} = 8.65(6.24) = 53.98$

$\overline{iV} = \dfrac{218.06}{5} = 43.61 \qquad \overline{i^2} = \dfrac{461.89}{5} = 92.38$

$s_i^2 = \overline{i^2} - \bar{i}^2$
$= 92.38 - 74.82$
$= 17.56$

$m = \dfrac{\overline{iV} - \bar{i}\bar{V}}{s_i^2}$

$= \dfrac{43.61 - 53.98}{17.56} = -0.590$

$b = \bar{V} - m\bar{i}$

$= 6.24 + (0.590)(8.65)$

$= 6.24 + 5.10 = 11.34$

$V = mi + b = -0.590i + 11.3$

9. $p = f(x)$

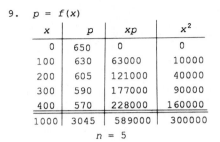
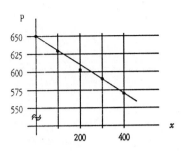

x	p	xp	x^2
0	650	0	0
100	630	63000	10000
200	605	121000	40000
300	590	177000	90000
400	570	228000	160000
1000	3045	589000	300000

$$n = 5$$

$$m = \frac{n\Sigma xp - (\Sigma x)(\Sigma p)}{n\Sigma x^2 - (\Sigma x)^2} = \frac{5(589000) - (1000)(3045)}{5(300000) - (1000)^2}$$

$$= \frac{2945000 - 3045000}{1500000 - 1000000} = \frac{-100000}{500000} = -0.200$$

$$b = \frac{(\Sigma x^2)(\Sigma p) - (\Sigma xp)(\Sigma x)}{n\Sigma x^2 - (\Sigma x)^2}$$

$$= \frac{(300000)(3045) - (589000)(1000)}{5(300000) - (1000)^2}$$

$$= \frac{913500000 - 589000000}{1500000 - 1000000} = \frac{324500000}{500000} = 649 \qquad p = -0.200x + 649$$

9M. $p = f(x)$

x	p	xp	x^2
0	4370	0	0
50	4240	21200	2500
100	4070	407000	10000
150	3970	595500	22500
200	3840	768000	40000
500	20490	1791700	75000

$$n = 5$$

$$m = \frac{n\Sigma xp - (\Sigma x)(\Sigma p)}{n\Sigma x^2 - (\Sigma x)^2} = \frac{5(1791700) - 500(20490)}{5(75000) - (500)^2}$$

$$= \frac{8958500 - 10245000}{375000 - 250000}$$

$$= \frac{-1286500}{125000} = \frac{-2573}{250} = -10.292$$

$$b = \frac{(\Sigma x^2)(\Sigma p) - (\Sigma xp)(\Sigma x)}{n\Sigma x^2 - (\Sigma x)^2} = \frac{75000(20490) - 1791700(500)}{5(75000) - (500)^2}$$

$$= \frac{1536750000 - 895850000}{375000 - 250000}$$

$$= \frac{640900000}{125000} = \frac{25636}{5} = 5127.2$$

$$p = -10.3x + 5127$$

13.

x	4	6	8	10	12
y	1	4	5	8	9

x	x^2
4	16
6	36
8	64
10	100
12	144
40	360

y	$y2$
1	1
4	16
5	25
8	64
9	81
27	187

$$\overline{x} = \frac{40}{5} = 8 \qquad\qquad \overline{y} = \frac{27}{5} = 5.4$$

$$\overline{x}^2 = 64 \qquad\qquad \overline{y}^2 = 5.4^2 = 29.16$$

$$\overline{x^2} = \frac{360}{5} = 72 \qquad\qquad \overline{y^2} = \frac{187}{5} = 37.4$$

$$s_x = \sqrt{72 - 64} = \sqrt{8} \qquad\qquad s_y = \sqrt{37.4 - 29.16} = \sqrt{8.24}$$

$$r = m\left(s_x/s_y\right) = 1.0\left(\sqrt{8}/\sqrt{8.24}\right) = 0.985$$

Exercises 21-5, page 637

1.

x	y	x^2	yx^2	$(x^2)^2$
2	12	4	48	16
4	38	16	608	256
6	72	36	2,592	1,296
8	135	64	8,640	4,096
10	200	100	20,000	10,000
	457	220	31,888	15,664

$$\overline{x^2} = \frac{220}{5} = 44.0 \qquad \overline{x^2}^2 = 1936$$

$$\overline{y} = \frac{457}{5} = 91.40 \qquad \overline{x^2}\,\overline{y} = 44.0(91.40) = 4022$$

$$\overline{yx^2} = \frac{31,888}{5} \qquad \overline{(x^2)^2} = \frac{15,644}{5} = 3133$$

$$= 6378$$

$$s_{x^2}^2 = \overline{(x^2)^2} - \overline{x^2}^2$$

$$= 3133 - 1936 = 1197$$

$$m = \frac{\overline{yx^2} - \overline{x^2}\,\overline{y}}{s_{x^2}^2} = \frac{6378 - 4022}{1197} = 1.97$$

$$b = \overline{y} - m\overline{x^2} = 91.40 - 1.97(44.0) = 4.7$$

$$y = 1.97x^2 + 4.7$$

(Answers may vary due to rounding.)

Answers will
differ somewhat
depending on the
number of digits
carried from step
to step.

5.

t	y	t^2	yt^2	$(t^2)^2$
1.0	6.0	1.0	6.0	1.0
2.0	23	4.0	92	16.0
3.0	55	9.0	495	81.0
4.0	98	16.0	1568	256
5.0	148	25.0	3700	625
	330	55.0	5861	979

$\overline{t^2} = \dfrac{55}{5} = 11.0$ $\overline{(t^2)}^2 = 121.0$

$\overline{y} = \dfrac{330}{5} = 66.00$ $\overline{t^2}\,\overline{y} = 11.0(66.00) = 726$

$\overline{yt^2} = \dfrac{5861}{5} = 1172$ $\overline{(t^2)^2} = \dfrac{979}{5} = 195.8$

$s_{t^2}^2 = \overline{(t^2)^2} - \overline{t^2}^2 = 196 - 121.0 = 75$

$m = \dfrac{\overline{yt^2} - \overline{t^2}\,\overline{y}}{s_{t^2}^2} = \dfrac{1172 - 726}{75} = \dfrac{446}{75} = 5.95$

$b = \overline{y} - m\overline{t^2} = 66.00 - 5.95(11.0) = 0.55$

$y = mt^2 + b; \quad y = 5.95t^2 + 0.55$

(Answers may vary due to rounding.)

9.

	x	y	xy	x^2
s	$1/s$	p	$(1/s)\,(p)$	$(1/s)^2$
240	0.00416666667	5.60	0.0233333333	1.736111×10^{-5}
305	0.00327868852	4.40	0.01442622951	1.07498×10^{-5}
420	0.00238095238	3.20	0.00761904762	5.668935×10^{-6}
480	0.0020833333	2.80	0.00583333333	4.340279×10^{-6}
560	0.00178571429	2.40	0.00428571429	3.188776×10^{-6}
2005	0.01369536	18.40	0.05549766	4.13089×10^{-5}

$m = \dfrac{n\Sigma xy - (\Sigma x)(\Sigma y)}{n\Sigma x^2 - (\Sigma x)^2} = \dfrac{5(0.05549766) - (0.01369536)(18.40)}{5(4.13089 \times 10^{-5}) - (0.01369536)^2}$

$= 1343$

$b = \dfrac{(\Sigma x^2)(\Sigma y) - (\Sigma xy)(\Sigma x)}{n\Sigma x^2 - (\Sigma x)^2}$

$= \dfrac{(4.13089 \times 10^{-5})(18.40) - (0.05549766)(0.01369536)}{5(4.13089 \times 10^{-5}) - (0.01369536)^2}$

$= 1.226612 \times 10^{-3}$ or 0

$p = 1343(\tfrac{1}{s}) + 0 = \dfrac{1343}{s}$

Review Exercises for Chapter 21, page 639

1. 2.3, 2.5, 2.6, 3.0, 3.5, 3.6, 3.8, 4.1, 4.1, 4.2, 4.8

↑
median

5.

Interval	101–103	104–106	107–109	110–112	113–115
Frequency	5	4	3	3	5

9.

x	f	fx	fx^2
102	5	510	52,020
105	4	420	44,100
108	3	324	34,992
111	3	333	36,963
114	5	570	64,980
	20	2157	233,055

$$\bar{x} = \frac{\Sigma fx}{\Sigma f} = \frac{2157}{20} = 107.85$$

$$\overline{x}^2 = 11,631.623$$

$$\overline{x^2} = \frac{233,055}{20} = 11,652.75$$

$$s = \sqrt{11,652.75 - 11,631.623} = 4.6$$

13. $\bar{x} = (0.24 + 0.28 + 0.29 + 0.26 + 0.27 + 0.26 + 0.25 + 0.27 + 0.28$

$+ 0.26 + 0.26 + 0.25) \div 12 = 0.264 \text{ Pa} \cdot \text{s}$

17.

x	f
0.24	1
0.25	2
0.26	4
0.27	2
0.28	2
0.29	1

21. The mode is the most prevalent value, which is 700.

25. $\Sigma f = 200$; the median is the mean of the 100th and 101st entries.

Counts	0	1	2	3	4	5	6	7	8	9	10
Intervals	3	10	25	45	29	39	26	11	7	2	3

83

112

The 100th and 101st are both 4.
Therefore, the median is 4.

29.

T	R	TR	T^2
0.0	25.0	0	0
20.0	26.8	536	400
40.0	28.9	1156	1600
60.0	31.2	1872	3600
80.0	32.8	2624	6400
100	34.7	3470	10000
300	179.4	9658	22000

$\overline{T} = \dfrac{300}{6} = 50.0$ $(\overline{T})^2 = 2500$

$\overline{R} = \dfrac{179.4}{6} = 29.90$ $\overline{T}\,\overline{R} = 50.0(29.9) = 1495$

$\overline{TR} = \dfrac{9658}{6} = 1610$ $\overline{T^2} = \dfrac{22\,000}{6} = 3667$

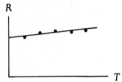

$s_T^{\,2} = \overline{T^2} - (\overline{T})^2 = 3667 - 2500 = 1167$

$m = \dfrac{\overline{TR} - \overline{T}\,\overline{R}}{s_T^{\,2}} = \dfrac{1610 - 1495}{1167} = 0.0985$

$b = \overline{R} - m\overline{T} = 29.90 - 0.0977(50.0) = 25.02$

$R = mT + b;\quad R = 0.0985T + 25.0$

(Answers may vary due to rounding.)

33.

	x	y	xy	x^2
t	t^2	s	$(t^2)(s)$	$(t^2)^2$
0.0	0.0	3000	0	0
3.0	9.0	2960	26640	81
6.0	36.0	2820	101520	1296
9.0	81.0	2600	210600	6561
12.0	144.0	2290	329760	20736
15.0	225.0	1900	427500	50625
18.0	324.0	1410	456840	104976
63.0	819.0	16980	1552860	184275

$m = \dfrac{n\Sigma xy - (\Sigma x)(\Sigma y)}{n\Sigma x^2 - (\Sigma x)^2} = \dfrac{7(1552860) - (819)(16980)}{7(184275) - (819)^2} = -4.90$

$b = \dfrac{(\Sigma x^2)(\Sigma y) - (\Sigma xy)(\Sigma x)}{n\Sigma x^2 - (\Sigma x)^2} = \dfrac{(184275)(16980) - 1552860(819)}{7(184275) - (819)^2} = 3000$

$s = -4.90t^2 + 3000$

CHAPTER 22

The Derivative

1. $f(x) = 3x - 2$ is continuous for all real x since it is defined for all x, and any small change in x will produce only a small change in $f(x)$.

5. $f(x) = \dfrac{1}{\sqrt{x}}$ is not defined for $x = 0$ since $\sqrt{0} = 0$ and division by zero is undefined, and is not defined for $x < 0$ since square roots of negative numbers are not real. The function is continuous for all $x > 0$ since it is defined, and any small change in x will produce only a small change in $f(x)$.

9. The graph is not continuous at $x = 1$ since $f(0.9)$ is a value between -1 and -2, and $f(1.1)$ is a value greater than $+2$. This amount of change in y for a 0.2 change in x is not consistent with equivalent changes in x at other values in the function domain. A small change in x does not produce a small change in y at $x = 1$. The function is continuous for $x < 1$, and continuous for $x > 1$.

13.

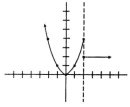

Not continuous. As $x \to 2$
from below, $f(x) \to 4$. As
$x \to 2$ from above, $f(x) \to 2$.

17. $f(x) = 3x - 2$

x	4.000	3.500	3.100	3.010	3.001	2.000	2.500	2.900	2.990	2.999
$f(x)$	10.000	8.500	7.300	7.030	7.003	4.000	5.500	6.700	6.970	6.997

Therefore, $\lim\limits_{x \to 3} (3x - 2) = 7$.

21. $f(x) = \dfrac{2 - \sqrt{x + 2}}{x - 2}$

x	2.100	2.010	2.001	1.900	1.990	1.999
$f(x)$	-0.2484567	-0.2498440	-0.2499850	-0.2515823	-0.2501564	-0.2500755

We see that $f(x) \to -0.25$ as $x \to 2$ from above 2 and from below 2.

25. $\lim_{x \to 3} (x + 4) = 3 + 4 = 7$ [$f(x)$ is continuous]

29. $\lim_{x \to 0} \dfrac{x^2 + x}{x} = \lim_{x \to 0} \dfrac{x(x + 1)}{x}$

$= \lim_{x \to 0} (x + 1) = 0 + 1 = 1$

33. $\lim_{x \to 1} \dfrac{x^3 - x}{x - 1} = \lim_{x \to 1} \dfrac{x(x^2 - 1)}{x - 1}$

$= \lim_{x \to 1} \dfrac{x(x + 1)(x - 1)}{x - 1} = \lim_{x \to 1} x(x + 1) = 1(2) = 2$

37. $\lim_{x \to -1} \sqrt{x}\,(x + 1)$ does not exist since x cannot approach -1 without the function going through imaginary values. $f(0)$, which is real, is the only defined value for the function near $x = -1$.

41. $\lim_{x \to \infty} \dfrac{3x^2 + 5}{x^2 - 2} = \lim_{x \to \infty} \dfrac{(3x^2 + 5) \div x^2}{(x^2 - 2) \div x^2} = \lim_{x \to \infty} \dfrac{3 + \dfrac{5}{x^2}}{1 - \dfrac{2}{x^2}} = \dfrac{3 + 0}{1 - 0} = 3$

45. $\lim_{x \to 0} \dfrac{x^2 - 3x}{x}$

x	-0.1	-0.01	-0.001	0.001	0.01	0.1
$f(x)$	-3.10	-3.01	-3.001	-2.999	-2.99	-2.9

We see that $f(x) \to -3$ as $x \to 0$. We now find the limit by changing the algebraic form.

$\lim_{x \to 0} \dfrac{x^2 - 3x}{x} = \lim_{x \to 0} \dfrac{x(x - 3)}{x}$

$= \lim_{x \to 0} x - 3 = 0 - 3 = -3$

49.

d	0.480000	0.280000	0.029800	0.002998	0.00029998
t	0.200000	0.100000	0.010000	0.001000	0.00010000
$v = d/t$	2.40000	2.80000	2.98000	2.99800	2.999800

$\lim_{t \to 0} v = 3$

53.

x	0.1	0.01	0.001	0.0001	0.00001
$(1+x)^{1/x}$	2.6	2.70	2.716	2.718	2.71827

$\lim_{x \to 0} (1 + x)^{1/x} = 2.71827$ or e

Exercises 22-2, page 655

1. $y = x^2$; $P = (2,4)$

	Q_1	Q_2	Q_3	Q_4	P
x_2	1.5	1.9	1.99	1.999	2
y_2	2.25	3.61	3.9601	3.996001	4
$y_2 - 4$	-1.75	-0.39	-0.0399	-0.003999	
$x_2 - 2$	-0.5	-0.1	-0.01	-0.001	
$m = \dfrac{y_2 - 4}{x_2 - 2}$	3.5	3.9	3.99	3.999	

$m_{tan} = 4$

5. $y = x^2$; $P = (2,4)$

$4 + \Delta y = (2 + \Delta x)^2$; $4 + \Delta y = 4 + 4\Delta x + (\Delta x)^2$; $\Delta y = 4 - 4 + 4\Delta x + (\Delta x)^2$

$\Delta y = 4\Delta x + (\Delta x)^2$; $m_{PQ} = \dfrac{\Delta y}{\Delta x} = \dfrac{4\Delta x + (\Delta x)^2}{\Delta x} = \dfrac{\Delta x(4 + \Delta x)}{\Delta x} = 4 + \Delta x$

As Δx becomes smaller and approaches zero, $m_{tan} = 4 + 0 = 4$.

9. $y = x^2$; $x = 2$, $x = -1$

$y_1 + \Delta y = (x_1 + \Delta x)^2$; $y_1 + \Delta y = x_1{}^2 + 2x_1\Delta x + (\Delta x)^2$

$y_1 + \Delta y - y_1 = x_1{}^2 + 2x_1\Delta x + (\Delta x)^2 - x_1{}^2$

$\Delta y = 2x_1\Delta x + (\Delta x)^2$

$m_{tan} = \dfrac{\Delta y}{\Delta x} = \dfrac{2x_1\Delta x + (\Delta x)^2}{\Delta x} = \dfrac{(\Delta x)(2x_1 + \Delta x)}{\Delta x} = 2x_1 + \Delta x$;

As $\Delta x \to 0$, $m_{tan} = 2x_1$.

If $x_1 = 2$, $m_{tan} = 2(2) = 4$.

If $x_1 = -1$, $m_{tan} = 2(-1) = -2$.

13. $y = x^2 + 4x + 5$; $x = -3$; $x = 2$; $y_1 = x_1{}^2 + 4x_1 + 5$

$y_1 + \Delta y = (x_1 + \Delta x)^2 + 4(x_1 + \Delta x) + 5$

$y_1 + \Delta y = x_1{}^2 + 2x_1\Delta x + (\Delta x)^2 + 4x_1 + 4\Delta x + 5$

$y_1 + \Delta y - y_1 = \Delta y = 2x_1\Delta x + (\Delta x)^2 + 4\Delta x$

$\dfrac{\Delta y}{\Delta x} = \dfrac{\Delta x(2x_1 + \Delta x + 4)}{\Delta x} = 2x_1 + 4 + \Delta x$

As $\Delta x \to 0$, $m_{tan} = 2x_1 + 4$.

$m_{tan} = 2(-3) + 4 = -2$ at $x = -3$

$m_{tan} = 2(2) + 4 = 8$ at $x = 2$

17. $y = x^3 - 2x$; $x = -1$; $x = 0$; $x = 1$; $y_1 = x_1^3 - 2x^1$

$y_1 + \Delta y = (x_1 + \Delta x)^3 - 2(x_1 + \Delta x)$

$y_1 + \Delta y = x_1^3 + 3x_1^2\Delta x + 3x_1(\Delta x)^2 + (\Delta x)^3 - 2x_1 - 2\Delta x$

$y_1 + \Delta y - y_1 = \Delta y$

$\Delta y = 3x_1^2\Delta x + 3x_1(\Delta x)^2 + (\Delta x)^3 - 2\Delta x$

$\dfrac{\Delta y}{\Delta x} = \dfrac{\Delta x[3x_1^2 + 3x_1\Delta x + (\Delta x)^2 - 2]}{\Delta x} = 3x_1^2 + 3x_1\Delta x + (\Delta x)^2 - 2$

As $\Delta x \to 0$, $m_{tan} = 3x_1^2 - 2$.

$m_{tan} = 3(-1)^2 - 2 = 1$ at $x = -1$

$m_{tan} = 3(0)^2 - 2 = -2$ at $x = 0$

$m_{tan} = 3(1)^2 - 2 = 1$ at $x = 1$

21. $y = x^2 + 2$, $P(2,6)$, $Q(2.1, 6.41)$. From P to Q, x changes from 0.1 unit and Q by 0.41 unit. The average change in y for 1 unit change in x is $0.41/0.1 = 4.1$ units.

$y + \Delta y = (x + \Delta x)^2 + 2$; $y + \Delta y = x^2 + 2x\Delta x + \Delta x^2 + 2$

$y + \Delta y - y = (x^2 + 2x\Delta x + \Delta x^2 + 2) - (x^2 + 2)$

$\Delta y = 2x\Delta x + \Delta x^2$

$\dfrac{\Delta y}{\Delta x} = 2x + \Delta x$; $m_{tan} = \lim\limits_{\Delta x \to 0} 2x + \Delta x = 2x$

If $x = 2$, $m_{tan} = 2(2) = 4$.

The average rate of change from P to Q is 4.1. The instantaneous rate of change at P is 4.

Exercises 22-3, page 660

1. $y = 3x - 1$; $y + \Delta y = 3(x + \Delta x) - 1$; $y + \Delta y = 3x + 3\Delta x - 1$

$y + \Delta y - y = 3x + 3\Delta x - 1 - (3x - 1)$

$\Delta y = 3\Delta x$; $\dfrac{\Delta y}{\Delta x} = \dfrac{3\Delta x}{\Delta x} = 3$; $\lim\limits_{\Delta x \to 0} 3 = 3$; $\dfrac{dy}{dx} = 3$

5. $y = x^2 - 1$; $y + \Delta y = (x + \Delta x)^2 - 1$

$y + \Delta y = x^2 + 2x\Delta x + (\Delta x)^2 - 1$

$y + \Delta y - y = x^2 + 2x\Delta x + (\Delta x)^2 - 1 - (x^2 - 1)$

$\Delta y = 2x\Delta x + (\Delta x)^2$; $\dfrac{\Delta y}{\Delta x} = \dfrac{\Delta x(2x + \Delta x)}{\Delta x} = 2x + \Delta x$

$\lim\limits_{\Delta x \to 0} (2x + \Delta x) = 2x + 0 = 2x$; $\dfrac{dy}{dx} = 2x$

9. $y = x^2 - 7x$; $y + \Delta y = (x + \Delta x)^2 - 7(x + \Delta x)$

$y + \Delta y = x^2 + 2x\Delta x + (\Delta x)^2 - 7x - 7\Delta x$

$y + \Delta y - y = x^2 + 2x\Delta x + (\Delta x)^2 - 7x - 7\Delta x - (x^2 - 7x)$

$\Delta y = 2x\Delta x + (\Delta x)^2 - 7\Delta x$

$\dfrac{\Delta y}{\Delta x} = \dfrac{\Delta x(2x + \Delta x - 7)}{\Delta x} = 2x + \Delta x - 7$

$\lim\limits_{\Delta x \to 0} (2x + \Delta x - 7) = 2x - 7$; $\dfrac{dy}{dx} = 2x - 7$

13. $y = x^3 + 4x - 6$

$y + \Delta y = (x + \Delta x)^3 + 4(x + \Delta x) - 6$

$y + \Delta y = x^3 + 3x^2\Delta x + 3x(\Delta x)^2 + (\Delta x)^3 + 4x + 4\Delta x - 6$

$y + \Delta y - y = x^3 + 3x^2\Delta x + 3x(\Delta x)^2 + (\Delta x)^3 + 4x + 4\Delta x - 6 - (x^3 + 4x - 6)$

$\Delta y = 3x^2\Delta x + 3x(\Delta x)^2 + (\Delta x)^3 + 4\Delta x$

$\dfrac{\Delta y}{\Delta x} = \dfrac{\Delta x[3x^2 + 3x\Delta x + (\Delta x)^2 + 4]}{\Delta x} = 3x^2 + 3x\Delta x + (\Delta x)^2 + 4$

$\lim\limits_{\Delta x \to 0} [3x^2 + 3x\Delta x + (\Delta x)^2 + 4] = 3x^2 + 4$; $\dfrac{dx}{dy} = 3x^2 + 4$

17. $y = x + \dfrac{1}{x}$; $y + \Delta y = x + \Delta x + \dfrac{1}{x + \Delta x}$

$y + \Delta y - y = x + \Delta x + \dfrac{1}{x + \Delta x} - x - \dfrac{1}{x}$

$\Delta y = \Delta x + \dfrac{1}{x + \Delta x} - \dfrac{1}{x}$

$\Delta y = \dfrac{\Delta x(x^2) + x(\Delta x)^2 + x - x - \Delta x}{x^2 + x(\Delta x)}$; $\dfrac{\Delta y}{\Delta x} = \dfrac{x^2 + x(\Delta x) - 1}{x^2 + x(\Delta x)}$

$\lim\limits_{\Delta x \to 0} \dfrac{[x^2 + x(\Delta x) - 1]}{x^2 + x(\Delta x)} = \dfrac{x^2 - 1}{x^2} = 1 - \dfrac{1}{x^2}$; $\dfrac{dy}{dx} = 1 - \dfrac{1}{x^2}$

21. $y = x^4 + x^3 + x^2 + x$; $y + \Delta y = (x + \Delta x)^4 + (x + \Delta x)^3 + (x + \Delta x)^2 + (x + \Delta x)$

$= [x^4 + 4x^3\Delta x + 6x^2(\Delta x)^2 + 4x(\Delta x)^3 + (\Delta x)^4]$
$+ [x^3 + 3x^2\Delta x + 3x(\Delta x)^2 + (\Delta x)^3] + [x^2 + 2x\Delta x + (\Delta x)^2] + (x + \Delta x)$

$y + \Delta y - y = 4x^3\Delta x + 6x^2(\Delta x)^2 + 4x(\Delta x)^3 + (\Delta x)^4 + 3x^2\Delta x + 3x(\Delta x)^2$
$+ (\Delta x)^3 + 2x\Delta x + (\Delta x)^2 + \Delta x$

$\Delta y = (\Delta x)[4x^3 + 6x^2\Delta x + 4x(\Delta x)^2 + (\Delta x)^3 + 3x^2 + 3x(\Delta x) + (\Delta x)^2 + 2x + \Delta x + 1]$

$\dfrac{\Delta y}{\Delta x} = (\Delta x)[4x^3 + 6x^2\Delta x + 4x(\Delta x)^2 + (\Delta x)^3 + 3x^2 + 3x(\Delta x) + (\Delta x)^2 + 2x + \Delta x + 1]/\Delta x$

$\dfrac{\Delta y}{\Delta x} = 4x^3 + 6x^2\Delta x + 4x(\Delta x)^2 + (\Delta x)^3 + 3x^2 + 3x(\Delta x) + (\Delta x)^2 + 2x + \Delta x + 1$

$\lim\limits_{\Delta x \to 0} \dfrac{\Delta y}{\Delta x} = 4x^3 + 3x^2 + 2x + 1$; $\dfrac{dy}{dx} = 4x^3 + 3x^2 + 2x + 1$

25. $y = 3x^2 - 2x$; $y + \Delta y = 3(x + \Delta x)^2 - 2(x + \Delta x)$

$y + \Delta y = 3x^2 + 6x\Delta x + 3\Delta x^2 - 2x - 2\Delta x$

$y + \Delta y - y = 3x^2 + 6x\Delta x + 3\Delta x^2 - 2x - 2\Delta x - 3x^2 + 2x$

$\Delta y = 6x\Delta x + 3\Delta x^2 - 2\Delta x$

$\dfrac{\Delta y}{\Delta x} = 6x + 3\Delta x - 2$

$\dfrac{dy}{dx} = \lim\limits_{\Delta x \to 0} \dfrac{\Delta y}{\Delta x} = 6x - 2$

At point $(-1,5)$, $\dfrac{dy}{dx} = 6(-1) - 2 = -8$

29. $y = \sqrt{x + 1}$; $y^2 = x + 1$; $(y + \Delta y)^2 = x + \Delta x + 1$

$y^2 + 2y\Delta y + (\Delta y)^2 = x + \Delta x + 1$

$y^2 + 2y\Delta y + (\Delta y)^2 - y^2 = x + \Delta x + 1 - (x + 1)$; $2y\Delta y + (\Delta y)^2 = \Delta x$

$\dfrac{2y\Delta y}{\Delta x} + \dfrac{\Delta y \Delta y}{\Delta x} = 1$; $\dfrac{\Delta y}{\Delta x}(2y + \Delta y) = 1$; $\dfrac{\Delta y}{\Delta x} = \dfrac{1}{2y + \Delta y}$

$\lim\limits_{\Delta x \to 0} \dfrac{1}{2y + \Delta y} = \dfrac{1}{2y} = \dfrac{1}{2\sqrt{x + 1}}$; $\dfrac{dy}{dx}\ \dfrac{1}{2\sqrt{x + 1}}$

Exercises 22-4, page 664

1. $y = x^2 - 1$ at $(2,3)$

$y + \Delta y = (x + \Delta x)^2 - 1$

$y + \Delta y = x^2 + 2x\Delta x + (\Delta x)^2 - 1$

$y + \Delta y - y = x^2 + 2x\Delta x + (\Delta x)^2 - 1 - (x^2 - 1)$

$\Delta y = 2x\Delta x + (\Delta x)^2$

$\dfrac{\Delta y}{\Delta x} = 2x + \Delta x$

$\lim\limits_{\Delta x \to 0} \dfrac{\Delta y}{\Delta x} = 2x = 4$ at $(2,3)$

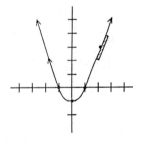

5.

	Q_1	Q_2	Q_3	Q_4	Q_5	P
t_2	2.0	2.5	2.9	2.99	2.999	3
s_2	18	20	21.6	21.96	21.996	22
$t_2 - 3$	-1	-0.5	-0.1	-0.01	-0.001	
$s_2 - 22$	-4	-2	-0.4	-0.04	-0.004	
$v = \dfrac{s_2 - 22}{t_2 - 3}$	4.00	4.00	4.00	4.00	4.00	

$s = 4t + 10$

$\lim\limits_{t \to 3} \dfrac{\Delta s}{\Delta t} = 4$

9. $s = 4t + 10;\ t = 3$

    $s + \Delta s = 4(t + \Delta t) + 10;\ s + \Delta s = 4t + 4\Delta t + 10$

    $s + \Delta s - s = 4t + 4\Delta t + 10 - (4t + 10)$

    $\Delta s = 4\Delta t;\ \dfrac{\Delta s}{\Delta t} = \dfrac{4\Delta t}{\Delta t} = 4$

    $\lim\limits_{\Delta t \to 0} \dfrac{\Delta s}{\Delta t} = 4$ at $t = 3$ s

13. $v = \lim\limits_{\Delta t \to 0} \dfrac{\Delta s}{\Delta t}$ Eq. (22-7); $s + \Delta s = 3(t + \Delta t) - \dfrac{2}{t + \Delta t}$

    $s + \Delta s - s = 3t + 3\Delta t - \dfrac{2}{t + \Delta t} - (3t - \dfrac{2}{t})$

    $\Delta s = 3\Delta t - \dfrac{2}{t + \Delta t} + \dfrac{2}{t};\ \Delta s = 3\Delta t + \dfrac{-2t + 2(t + \Delta t)}{(t + \Delta t)(t)}$

    $\Delta s = 3\Delta t + \dfrac{2\Delta t}{(t + \Delta t)(t)};\ \dfrac{\Delta s}{\Delta t} = \dfrac{3\Delta t}{\Delta t} + \dfrac{2\Delta t}{(t + \Delta t)(t)} \times \dfrac{1}{\Delta t}$

    $\dfrac{\Delta s}{\Delta t} = 3 + \dfrac{2}{(t + \Delta t)(t)};\ \lim\limits_{\Delta t \to 0} 3 + \dfrac{2}{(t + \Delta t)(t)} = 3 + \dfrac{2}{(t)(t)}$

    $= 3 + \dfrac{2}{t^2};\ v = 3 + \dfrac{2}{t^2}$

17. $a = \lim\limits_{\Delta t \to 0} \dfrac{\Delta v}{\Delta t};\ v + \Delta v = 6(t + \Delta t)^2 - 4(t + \Delta t) + 2$

    $v + \Delta v = 6[t^2 + 2t\Delta t + (\Delta t)^2] - 4t - 4\Delta t + 2$

    $v + \Delta v = 6t^2 + 12t\Delta t + 6(\Delta t)^2 - 4t - 4\Delta t + 2$

    $v + \Delta v - v = 6t^2 + 12t\Delta t + 6(\Delta t)^2 - 4t - 4\Delta t + 2 - (6t^2 - 4t + 2)$

    $\Delta v = 12t\Delta t + 6(\Delta t)^2 - 4\Delta t$

    $\dfrac{\Delta v}{\Delta t} = \dfrac{\Delta t(12t + 6\Delta t - 4)}{\Delta t} = \dfrac{\Delta v}{\Delta t} = 12t + 6\Delta t - 4$

    $\lim\limits_{\Delta t \to 0} 12t + 6\Delta t - 4 = 12t - 4;\ a = 12t - 4$

21. $q = 30 - 2t;\ q + \Delta q = 30 - 2(t + \Delta t);\ q + \Delta q = 30 - 2t - 2\Delta t$

    $q + \Delta q - q = (30 - 2t - 2\Delta t) - (30 - 2t)$

    $\qquad\qquad\quad = 30 - 2t - 2\Delta t - 30 + 2t$

    $\Delta q = -2\Delta t;\ \dfrac{\Delta q}{\Delta t} = -2$

    Therefore $i = \dfrac{\Delta q}{\Delta t} = -2.$

25. $P = 500 + 250m^2$

$P + \Delta P = 500 + 250(m + \Delta m)^2$

$P + \Delta P = 500 + 250(m^2 + 2m\Delta m + (\Delta m)^2)$

$P + \Delta P = 500 + 250m^2 + 500m\Delta m + (\Delta m)^2$

$P + \Delta P - P = 500 + 250m^2 + 500m\Delta m + (\Delta m)^2 - (500 + 250m^2)$

$\Delta P = 500m\Delta m + (\Delta m)^2$

$\dfrac{\Delta P}{\Delta m} = 500m + \Delta m$

$\lim\limits_{\Delta m \to 0} \dfrac{\Delta P}{\Delta m} = 500m = 500(0.92) = 460$

29.   The volume of a cone $= \frac{1}{3}\pi r^2 h$. For this cone, the radius and height are
equal to 4 cm. Due to the similarity of the figures, as the level of the
oil decreases, the radius and height will still be equal; $r = h = d$.
Therefore, $v = \frac{1}{3}\pi d^2(d) = \frac{1}{3}\pi d^3$.

$v + \Delta v = \frac{1}{3}\pi(d + \Delta d)^3 = \frac{1}{3}\pi(d^3 + 3d\Delta d^2 + 3d^2\Delta d + \Delta d^3)$

$v + \Delta v - v = \frac{1}{3}\pi d^3 + \pi d\Delta d^2 + \pi d^2\Delta d + \frac{1}{3}\pi\Delta d^3 - \frac{1}{3}\pi d^3$

$\Delta v = \pi d\Delta d^2 + \pi d^2\Delta d + \frac{1}{3}\pi\Delta d^3$

$\dfrac{\Delta V}{\Delta d} = \pi d\Delta d + \pi d^2 + \frac{1}{3}\pi\Delta d^2$; $\lim\limits_{\Delta d \to 0} \dfrac{\Delta V}{\Delta d} = \pi d^2$

Exercises 22-5, page 669

1. $y = x^5$; $\dfrac{dy}{dx} = 5x^{5-1} = 5x^4$ Eq. (22-9)

5. $y = x^4 - 6$; $\dfrac{dy}{dx} = 4x^{4-1} - 0 = 4x^3$ Eq. (22-9), (22-8)

9. $y = 5x^3 - x - 1$; $\dfrac{dy}{dx} = 5(3x^2) - 1 - 0 = 15x^2 - 1$ Eq. (22-11)

13. $y = -6x^7 + 5x^3 + 2^3$; $\dfrac{dy}{dx} = -6(7x^6) + 5(3x^2) + 0 = -42x^6 + 15x^2$ Eq. (22-11)

17. $y = 6x^2 - 8x + 1$; $\dfrac{dy}{dx} = \dfrac{d(6x^2)}{dx} - \dfrac{d(8x)}{dx} + \dfrac{d(1)}{dx} = 12x - 8 + 0$
Since the derivative is a function of only x, we now evaluate it for
$x = 2$.

$\dfrac{dy}{dx}\Big|_{x=2} = 12(2) - 8 = 24 - 8 = 16$

21. $y = 2x^6 - 6x^2$; $x = 2$

$\dfrac{dy}{dx} = 2(6x^5) - 6(2x) = 12x^5 - 12x$

At $x = 2$, $m_{tan} = 12(2)^5 - 12(2) = 360$.

25. $s = 6t^5 - 5t + 2$; $v = \dfrac{ds}{dt} = 30t^4 - 5$

29. $s = 2t^3 - 4t^2$; $t = 4$

$v = \dfrac{ds}{dt} = 2(3t^2) - 4(2t) = 6t^2 - 8t$

$v\Big|_{t = 4} = 6(4^2) - 8(4) = 64$

33. $y = 3x^2 - 6x$; $m_{tan} = \dfrac{dy}{dx} = 6x - 6$

Tangent is parallel where slope is zero. Therefore, $6x - 6 = 0$; $x = 1$.

37. $P = 16i^2 + 60i$

$\dfrac{dP}{di} = 16(2i) + 60 = 32i + 60$

$\dfrac{dP}{di}\Big|_{i = 0.75} = 32(0.75) + 60 = 84$

41. $F = x^4 - 12x^3 + 46x^2 - 60x + 25$

$\dfrac{dF}{dx} = 4x^3 - 12(3x^2) + 46(2x) - 60 = 4x^3 - 36x^2 + 92x - 60$

$\dfrac{dF}{dx}\Big|_{x = 4} = 4(4^3) - 36(4^2) + 92(4) - 60 = 256 - 576 + 368 - 60$
$= -12$ N/cm

Exercises 22-6, page 674

1. $y = x^2(3x + 2)$; $u = x^2$, $\dfrac{du}{dx} = 2x$, $v = 3x + 2$, $\dfrac{dv}{dx} = 3$

By Eq. (22-12), $\dfrac{dy}{dx} = x^2(3) + (3x + 2)(2x)$
$= 3x^2 + 6x^2 + 4x = 9x^2 + 4x$

Check: $y = 3x^3 + 2x^2$; $\dfrac{dy}{dx} = 9x^2 + 4x$

5. $y = (x + 2)(2x - 5)$; $u = x + 2$, $\dfrac{du}{dx} = 1$, $v = 2x - 5$, $\dfrac{dv}{dx} = 2$

By Eq. (29-12), $\dfrac{dy}{dx} = (x + 2)(2) + (2x - 5)(1)$
$= 2x + 4 + 2x - 5 = 4x - 1$

Check: $y = 2x^2 - x - 10$; $\dfrac{dy}{dx} = 4x - 1$

9. $y = (2x - 7)(5 - 2x)$; $u = (2x - 7)$; $v = (5 - 2x)$

Using Eq. (22-12), $\dfrac{dy}{dx} = (2x - 7)(-2) + (5 - 2x)(2)$
$= -4x + 14 + 10 - 4x = -8x + 24$

Multiplying out the function, $y = (2x - 7)(5 - 2x)$
$= 10x - 4x^2 - 35 + 14x$
$= -4x^2 + 24x - 35$

$\dfrac{dy}{dx} = -8x + 24$

13. $y = \dfrac{x}{2x + 3}$; $u = x$, $\dfrac{du}{dx} = 1$, $v = 2x + 3$, $\dfrac{dv}{dx} = 2$

By Eq. (29-13), $\dfrac{dy}{dx} = \dfrac{(2x + 3)(1) - x(2)}{(2x + 3)^2}$

$= \dfrac{2x + 3 - 2x}{(2x + 3)^2} = \dfrac{3}{(2x + 3)^2}$

17. $y = \dfrac{x^2}{3 - 2x}$; $u = x^2$, $\dfrac{du}{dx} = 2x$, $v = 3 - 2x$, $\dfrac{dv}{dx} = -2$

By Eq. (29-13), $\dfrac{dy}{dx} = \dfrac{(3 - 2x)(2x) - (x^2)(-2)}{(3 - 2x)^2}$

$$= \dfrac{6x - 4x^2 + 2x^2}{(3 - 2x)^2} = \dfrac{6x - 2x^2}{(3 - 2x)^2}$$

21. $y = \dfrac{x + 8}{x^2 + x + 2}$; $u = x + 8$; $\dfrac{du}{dx} = 1$; $v = x^2 + x + 2$; $\dfrac{dv}{dx} = 2x + 1$

By Eq. (29-13), $\dfrac{dy}{dx} = \dfrac{(x^2 + x + 2)1 - (x + 8)(2x + 1)}{(x^2 + x + 2)^2}$

$$= \dfrac{x^2 + x + 2 - 2x^2 - 17x - 8}{(x^2 + x + 2)^2} = \dfrac{-x^2 - 16x - 6}{(x^2 + x + 2)^2}$$

25. $y = (3x - 1)(4 - 7x)$, $x = 3$

$u = 3x - 1$; $\dfrac{du}{dx} = 3$; $v = (4 - 7x)$; $\dfrac{dv}{dx} = -7$

By Eq. (22-12), $\dfrac{dy}{dx} = (3x - 1)(-7) + (4 - 7x)(3)$
$$= -21x + 7 + 12 - 21x = -42x + 19$$

$\dfrac{dy}{dx}\Big|_{x = 3} = -42(3) + 19 = -126 + 19 = -107$

29. $y = \dfrac{x^2(1 - 2x)}{3x - 7}$; $u = x^2(1 - 2x)$, $\dfrac{du}{dx} = x^2(-2) + (1-2x)(2x) = -6x^2 + 2x$

$v = 3x - 7$, $\dfrac{dv}{dx} = 3$

By Eq. (29-13), $\dfrac{dy}{dx} = \dfrac{(3x - 7)(-6x^2 + 2x) - x^2(1 - 2x)(3)}{(3x - 7)^2}$

$$= \dfrac{-18x^3 + 6x^2 + 42x^2 - 14x - 3x^2 + 6x^3}{(3x - 7)^2}$$

$$= \dfrac{-12x^3 + 45x^2 - 14x}{(3x - 7)^2}$$

33. $y = \dfrac{x}{x^2 + 1}$; $u = x$; $\dfrac{du}{dx} = 1$, $v = x^2 + 1$, $\dfrac{dv}{dx} = 2x$

$\dfrac{dy}{dx} = \dfrac{(x^2 + 1)(1) - (x)(2x)}{(x^2 + 1)^2} = \dfrac{x^2 + 1 - 2x^2}{(x^2 + 1)^2} = \dfrac{-x^2 + 1}{(x^2 + 1)^2}$

Therefore, $m_{tan} = 0$ when $\dfrac{-x^2 + 1}{(x^2 + 1)^2} = 0$; $\dfrac{-x^2 + 1}{(x^2 + 1)^2} = 0$

$-x^2 + 1 = 0$; $x^2 = 1$; $x = 1, -1$

37. $V = \dfrac{6R + 25}{R + 3}$

$\dfrac{dV}{dR} = \dfrac{6(R + 3) - (6R + 25)}{(R + 3)^2} = \dfrac{6R + 18 - 6R - 25}{(R + 3)^2} = \dfrac{-7}{(R + 3)^2}$

$\dfrac{dV}{dR}\Big|_{R = 7} = \dfrac{-7}{(7 + 3)^2} = \dfrac{-7}{100} = -0.07$

41. $r_f = \dfrac{2(R^2 + Rr + r^2)}{3(R + r)}$

$\dfrac{dR}{dr_f} = \dfrac{2(2R + r)(3)(R + r) - 2(R^2 + Rr + r^2)(3)}{9(R + r)^2}$

$= \dfrac{6(2R + r)(R + r) - 6(R^2 + Rr + r^2)}{9(R + r)^2}$

$= \dfrac{12R^2 + 18Rr + 6r^2 - 6R^2 - 6Rr - 6r^2}{9(R + r)^2}$

$= \dfrac{6R^2 + 12Rr}{9(R + r)^2} = \dfrac{6R(R + 2r)}{9(R + r)^2} = \dfrac{2R(R + 2r)}{3(R + r)^2}$

Exercises 22-7, page 680

1. $y = \sqrt{x} = x^{1/2}$; $\dfrac{dy}{dx} = \dfrac{1}{2}x^{1/2-1} = \dfrac{1}{2}x^{-1/2}$

$= \dfrac{1}{2}\left(\dfrac{1}{x^{1/2}}\right) = \dfrac{1}{2x^{1/2}}$ Eq. (22-16)

5. $y = \dfrac{3}{3\sqrt{x}} = \dfrac{3}{x^{1/3}} = 3x^{-1/3}$; $\dfrac{dy}{dx} = 3(-\dfrac{1}{3}x^{-1/3-1})$

$= -1x^{-4/3} = -1\left(\dfrac{1}{x^{4/3}}\right) = -\dfrac{1}{x^{4/3}}$ Eq. (22-15)

9. $y = (x^2 + 1)^5$; $u = x^2 + 1$; $\dfrac{du}{dx} = 2x$

By Eq. (22-15), $\dfrac{dy}{dx} = 5(x^2 + 1)^4(2x) = 10x(x^2 + 1)^4$

13. $y = (2x^3 - 3)^{1/3}$; $u = 2x^3 - 3$; $\dfrac{du}{dx} = 6x^2$

By Eq. (22-15), $\dfrac{dy}{dx} = \dfrac{1}{3}(2x^3 - 3)^{-2/3}(6x^2) = \dfrac{2x^2}{(2x^3 - 3)^{2/3}}$

17. $y = 4(2x^4 - 5)^{3/4}$; $u = 2x^4 - 5$; $\dfrac{du}{dx} = 8x^3$

$\dfrac{dy}{dx} = 4[\dfrac{3}{4}(2x^4 - 5)^{-1/4}(8x^3)] = \dfrac{24x^3}{(2x^4 - 5)^{1/4}}$

21. $y = x\sqrt{8x + 5} = x(8x + 5)^{1/2}$

$\dfrac{dy}{dx} = x\left(\dfrac{1}{2}\right)(8x + 5)^{-1/2}(8) + (8x + 5)^{1/2}(1)$

$= 4x(8x + 5)^{-1/2} + (8x + 5)^{1/2}$

$= \dfrac{4x}{(8x + 5)^{1/2}} + \dfrac{(8x + 5)}{(8x + 5)^{1/2}} = \dfrac{12x + 5}{(8x + 5)^{1/2}}$

25. $y = \sqrt{3x + 4}$; $x = 7$

$y = (3x + 4)^{1/2}$; $u = 3x + 4$; $n = \dfrac{1}{2}$; $\dfrac{du}{dx} = 3$

$\dfrac{dy}{dx} = \dfrac{1}{2}(3x + 4)^{-1/2}(3) = \dfrac{3}{2}(3x + 4)^{-1/2} = \dfrac{3}{2\sqrt{3x + 4}}$

$\dfrac{dy}{dx}\Big|_{x = 7} = \dfrac{3}{2\sqrt{3(7) + 4}} = \dfrac{3}{2\sqrt{25}} = \dfrac{3}{2(5)} = \dfrac{3}{10}$

29. $y = \dfrac{1}{x^3}$

(a) $u = 1$; $\dfrac{du}{dx} = 0$; $v = x^3$; $\dfrac{dv}{dx} = 3x^2$

$\dfrac{dy}{dx} = \dfrac{x^3(0) - 1(3x^2)}{(x^3)^2} = \dfrac{-3x^2}{x^6} = \dfrac{-3}{x^4}$

(b) $y = x^{-3}$; $\dfrac{dy}{dx} = -3x^{-3-1} = -3x^{-4}$

33. $y^2 = 4x$; $y = \sqrt{4x} = 2\sqrt{x} = 2x^{1/2}$

$m_{\tan} = \dfrac{dy}{dx} = \dfrac{d(2x^{1/2})}{dx} = 2\left(\dfrac{1}{2}\right)x^{-1/2} = \dfrac{1}{\sqrt{x}}$

$m_{\tan}\Big|_{x=1} = \dfrac{1}{\sqrt{1}} = \dfrac{1}{1} = 1$

37. $P = \dfrac{k}{V^{3/2}}$; $P = 300$ kPa when $V = 100$ cm^3

$300 = \dfrac{k}{100^{3/2}}$; $k = 300(100)^{3/2} = 300(10^3) = 300(1000) = 300\ 000$

$P = \dfrac{300\ 000}{V^{3/2}} = 300\ 000\,v^{-3/2}$; $\dfrac{dP}{dV} = 300\ 000\left(-\dfrac{3}{2}\right)v^{-5/2}$

$= -450\ 000 v^{-5/2} = \dfrac{-450\ 000}{v^{5/2}}$

$\dfrac{dP}{dV}\Big|_{V=100} = \dfrac{-450\ 000}{100^{5/2}} = \dfrac{-450\ 000}{100\ 000} = -4.50$ kPa/cm^3

41. $\lambda_r = \dfrac{2a\lambda}{\sqrt{4a^2 - \lambda^2}} = \dfrac{2a\lambda}{(4a^2 - \lambda^2)^{1/2}}$

$\dfrac{d\lambda_r}{d\lambda} = \dfrac{(4a^2 - \lambda^2)^{1/2}(2a) - 2a\lambda\left(\dfrac{1}{2}\right)(4a^2 - \lambda^2)^{-1/2}(-2\lambda)}{(4a^2 - \lambda^2)}$

$= \dfrac{2a(4a^2 - \lambda^2)^{1/2} + 2a\lambda^2(4a^2 - \lambda^2)^{-1/2}}{(4a^2 - \lambda^2)}$

$= \dfrac{(4a^2 - \lambda^2)^{-1/2}[(2a)(4a^2 - \lambda^2) + 2a\lambda^2]}{(4a^2 - \lambda^2)} = \dfrac{-8a^3}{(4a^2 - \lambda^2)^{3/2}}$

Exercises 22-8, page 684

1. $3x + 2y = 5$; $\dfrac{d(3x)}{dx} + \dfrac{d(2y)}{dx} = \dfrac{d(5)}{dx}$

$3 + \dfrac{2dy}{dx} = 0$; $\dfrac{2dy}{dx} = -3$; $\dfrac{dy}{dx} = -\dfrac{3}{2}$

5. $x^2 - y^2 - 9 = 0$; $\dfrac{d(x^2)}{dx} - \dfrac{d(y^2)}{dx} - \dfrac{d(9)}{dx} = \dfrac{d(0)}{dx}$

$2x - \dfrac{2y\,dy}{dx} - 0 = 0$; $\dfrac{-2y\,dy}{dx} = -2x$; $\dfrac{dy}{dx} = \dfrac{x}{y}$

9. $y^2 + y = x^2 - 4$; $\dfrac{d(y^2)}{dx} + \dfrac{d(y)}{dx} = \dfrac{d(x^2)}{dx} - \dfrac{d(4)}{dx}$

$\dfrac{2y\,dy}{dx} + \dfrac{dy}{dx} = 2x - 0$

$\dfrac{dy}{dx}(2y + 1) = 2x$; $\dfrac{dy}{dx} = \dfrac{2x}{2y + 1}$

13. $xy^3 + 3y + x^2 = 9$; $\dfrac{d(xy^3)}{dx} + \dfrac{d(3y)}{dx} + \dfrac{d(x^2)}{dx} = \dfrac{d(9)}{dx}$

$\dfrac{x\,dy^3}{dx} + \dfrac{y^3\,dx}{dx} + \dfrac{3\,dy}{dx} + 2x = 0$

$x(3y^2)\dfrac{dy}{dx} + y^3(1) + \dfrac{3\,dy}{dx} + 2x = 0$

$3xy^2\dfrac{dy}{dx} + y^3 + \dfrac{3\,dy}{dx} + 2x = 0$

$3xy^2\dfrac{dy}{dx} + \dfrac{3\,dy}{dx} = -y^3 - 2x$

$\dfrac{dy}{dx}(3xy^2 + 3) = -y^3 - 2x$; $\dfrac{dy}{dx} = \dfrac{-2x - y^3}{3xy^2 + 3}$

17. $(2y - x)^4 + x^2 = y + 3$

$4(2y - x)^3(2\dfrac{dy}{dx} - 1) + 2x = \dfrac{dy}{dx} + 0$

$4(2y - x)^3(2\dfrac{dy}{dx}) - 4(2y - x)^3 + 2x = \dfrac{dy}{dx}$

$8\dfrac{dy}{dx}(2y - x)^3 - \dfrac{dy}{dx} = 4(2y - x)^3 - 2x$

$\dfrac{dy}{dx}[8(2y - x)^3 - 1] = 4(2y - x)^3 - 2x$

$\dfrac{dy}{dx} = \dfrac{4(2y - x)^3 - 2x}{8(2y - x)^3 - 1}$

21. $3x^3y^2 - 2y^3 = -4$

$3x^3(2y)\dfrac{dy}{dx} + y^2(9x^2) - 6y^2\dfrac{dy}{dx} = 0$

$6x^3y\dfrac{dy}{dx} + 9x^2y^2 - 6y^2\dfrac{dy}{dx} = 0$

$6x^3y\dfrac{dy}{dx} - 6y^2\dfrac{dy}{dx} = -9x^2y^2$

$\dfrac{dy}{dx}(6x^3y - 6y^2) = -9x^2y^2$

$\dfrac{dy}{dx} = \dfrac{-9x^2y^2}{6x^3y - 6y^2} = \dfrac{-9x^2y^2}{3(2x^3y - 2y^2)}$

$\dfrac{dy}{dx} = \dfrac{-3x^2y^2}{2x^3y - 2y^2}$

$\dfrac{dy}{dx} = \dfrac{-3(1^2)(2^2)}{2(1^3)(2) - 2(2^2)} = \dfrac{-12}{-4} = 3$

25. $xy + y^2 + 2 = 0;$ $\dfrac{d(xy)}{dx} + \dfrac{d(y^2)}{dx} + \dfrac{d(2)}{dx} = \dfrac{d(0)}{dx}$

$\dfrac{xdy}{dx} + \dfrac{ydx}{dx} + \dfrac{2ydy}{dx} + 0 = 0$

$\dfrac{xdy}{dx} + y(1) + \dfrac{2ydy}{dx} = 0;$ $\dfrac{xdy}{dx} + \dfrac{2ydy}{dx} = -y$

$\dfrac{dy}{dx}(x + 2y) = -y;$ $\dfrac{dy}{dx} = \dfrac{-y}{x + 2y}$

$\dfrac{dy}{dx}\bigg|(-3,1) = \dfrac{-1}{-3 + 2(1)} = \dfrac{-1}{-1} = 1$

29. $r^2 = 2rR + 2R - 2r$

$2r = 2R + \dfrac{dR}{dr}(2r) + 2\dfrac{dR}{dr} - 2$

$2r - 2R + 2 = 2r\dfrac{dR}{dr} + 2\dfrac{dR}{dr}$

$2r - 2R + 2 = \dfrac{dR}{dr}(2r + 2)$

$\dfrac{2(r - R + 1)}{2(r + 1)} = \dfrac{r - R + 1}{r + 1} = \dfrac{dR}{dr}$

Exercises 22-9, page 688

1. $y = x^3 + x^2$

$y' = 3x^2 + 2x$

$y'' = 6x + 2$

$y''' = 6$

$y^{(4)} = 0$

5. $y = (1 - 2x)^4$

$y' = 4(1 - 2x)^3(-2) = -8(1 - 2x)^3$

$y'' = -24(1 - 2x)^2(-2) = 48(1 - 2x)^2$

$y'' = 96(1 - 2x)(-2) = -192(1 - 2x)$

$y^{(4)} = -2(-192) = 384$

$y^{(5)} = 0$

9. $y = 2x^7 - x^6 - 3x$

$y' = 14x^6 - 6x^5 - 3$

$y'' = 84x^5 - 30x^4$

13. $f(x) = \sqrt[4]{8x - 3} = (8x - 3)^{1/4}$

$f'(x) = \dfrac{1}{4}(8x - 3)^{-3/4}(8) = 2(8x - 3)^{-3/4}$

$f''(x) = -\dfrac{3}{2}(8x - 3)^{-7/2}(8) = -12(8x - 3)^{-7/2}$

$= \dfrac{-12}{(8x - 3)^{7/2}}$

17. $y = 2(2 - 5x)^4$

$y' = 8(2 - 5x)^3(-5) = -40(2 - 5x)^3$

$y'' = -120(2 - 5x)^2(-5) = 600(2 - 5x)^2$

21. $f(x) = \dfrac{2x}{1-x}$

$f'(x) = \dfrac{(1-x)(2) - (2x)(-1)}{(1-x)^2} = \dfrac{2}{(1-x)^2}$

$f''(x) = \dfrac{(1-x)^2(0) - 2(2)(1-x)(-1)}{(1-x)^4} = \dfrac{4(1-x)}{(1-x)^4} = \dfrac{4}{(1-x)^3}$

25. $x^2 - y^2 = 9$

$2x - 2yy' = 0$

$2x = 2\,yy'$

$\dfrac{2x}{2y} = \dfrac{x}{y} = y'$

$y'' = \dfrac{y(1) - xy'}{y^2} = \dfrac{y - x\left(\frac{x}{y}\right)}{y^2} = \dfrac{\frac{y^2 - x^2}{y}}{y^2} = \dfrac{y^2 - x^2}{y^3} = \dfrac{-9}{y^3}$

29. $f(x) = \sqrt{x^2 + 9} = (x^2 + 9)^{1/2}$

$f'(x) = \dfrac{1}{2}(x^2 + 9)^{-1/2}(2x) = x(x^2 + 9)^{-1/2}$

$f''(x) = x\left[-\dfrac{1}{2}(x^2 + 9)^{-3/2}(2x)\right] + (x^2 + 9)^{-1/2}$

$= -x^2(x^2 + 9)^{-3/2} + (x^2 + 9)^{-1/2}$

$= \dfrac{-x^2}{\sqrt{(x^2 + 9)^3}} + \dfrac{1}{\sqrt{x^2 + 9}}$

$f''(4) = \dfrac{-16}{(\sqrt{25})^3} + \dfrac{1}{\sqrt{25}} = \dfrac{-16}{5^3} + \dfrac{1}{5} = \dfrac{-16}{125} + \dfrac{25}{125} = \dfrac{9}{125}$

33. $s = 2250t - 16.1t^2$

$s' = v = 2250 - 32.2t$

$s'' = a = -32.2$

33M. $s = 670t - 4.9t^2$

$s' = v = 670 - 9.8t$

$s'' = a = -9.8$

37. $y = \dfrac{4}{x} + 2\sqrt[3]{x}$, $x = 8$

$y = 4x^{-1} + 2x^{1/3}$; $\dfrac{dy}{dx} = 4(-1)x^{-2} + 2\left(\frac{1}{3}x^{-2/3}\right) = \dfrac{-4}{x^2} + \dfrac{2}{3x^{2/3}}$

$\dfrac{dy}{dx}\bigg|_{x=8} = \dfrac{-4}{8^2} + \dfrac{2}{3(8)^{2/3}} = \dfrac{-4}{64} + \dfrac{2}{3(4)}$

$= \dfrac{-1}{16} + \dfrac{1}{6} = \dfrac{-3}{48} + \dfrac{8}{48} = \dfrac{5}{48}$

41. $y = 3x^4 - \dfrac{1}{x} = 3x^4 - x^{-1}$

$y' = 12x^3 + x^{-2}$

$y'' = 36x^2 - 2x^{-3}$

45. $f = \dfrac{f_1 f_2}{f_1 + f_2 - d}$; f_2 and d are constant

$f' = \dfrac{(f_1 + f_2 - d)(f_2) - (f_1 f_2)}{(f_1 + f_2 - d)^2} = \dfrac{f_1 f_2 + f_2^2 - f_2 d - f_1 f_2}{(f_1 + f_2 - d)^2}$

$f'' = \dfrac{f_2^2 - f_2 d}{(f_1 + f_2 - d)^2}$; f_2 is the limiting factor

49. $R = 1 - kt + \dfrac{k^2 t^2}{2} - \dfrac{k^3 t^3}{6} = 1 - kt + \dfrac{1}{2}(k^2 t^2) - \dfrac{1}{6}(k^3 t^3)$

$R' = -k + k^2 t - \dfrac{1}{2} k^3 t^2$

53. $E = \dfrac{L dI}{dt}$; $I = t(0.01t + 1)^3$; $\dfrac{dI}{dt} = (0.01t + 1)^2 (0.04t + 1)$; $L = 0.4H$

$E = 0.4 \dfrac{dI}{dt}$; substituting the value for $\dfrac{dI}{dt}$, $E = 0.04(0.01t + 1)^2 (0.04t + 1)$

57. $f = \dfrac{1}{2\pi \sqrt{c(L + 2)}} = \dfrac{1}{2\pi}(CL + 2C)^{-\frac{1}{2}}$

$\dfrac{df}{dL} = -\dfrac{1}{2}\left(\dfrac{1}{2\pi}\right)(CL + 2C)^{-\frac{3}{2}} (C) = -\dfrac{C}{4\pi}(CL + 2C)^{-\frac{3}{2}}$

$= -\dfrac{C}{4\pi \sqrt{c^3(L + 2)^3}} = -\dfrac{C}{4\pi c \sqrt{C(L + 2)^3}} = -\dfrac{1}{4\pi \sqrt{C(L + 2)^3}} = -\dfrac{1}{4\pi \sqrt{c}\,(L + 2)^{\frac{1}{2}}}$

61. $y = 0.0001(x^5 - 25x^2) = 0.0001 x^5 - 0.0025 x^2$

$\dfrac{dy}{dx} = 0.0005 x^4 - 0.0050x$

$\dfrac{d^2 y}{dx^2} = 0.0020 x^3 - 0.0050$

$\dfrac{d^2 y}{dx^2} \bigg|_{x = 3.00} = 0.0020(27) - 0.0050 = 0.0540 - 0.0050$

$= 0.049$

65. $A = xy = x(4 - x^2) = 4x - x^3$

$\dfrac{dA}{dx} = 4 - 3x^2$

Review Exercises for Chapter 22, page 689

1. $\lim\limits_{x \to 4} (8 - 3x) = 8 - 3(4) = -4$

5. $\lim\limits_{x \to 2} \dfrac{4x - 8}{x^2 - 4} = \lim\limits_{x \to 2} \dfrac{4(x - 2)}{(x - 2)(x + 2)} = \lim\limits_{x \to 2} \dfrac{4}{x + 2} = \dfrac{4}{2 + 2} = 1$

9. $\lim\limits_{x \to \infty} \dfrac{2 + \dfrac{1}{x + 4}}{3 - \dfrac{1}{x^2}} = \dfrac{2 + 0}{3 - 0} = \dfrac{2}{3}$

13. $y = 7 + 5x$

   $y + \Delta y = 7 + 5(x + \Delta x) = 7 + 5x + 5\Delta x$

   $\Delta y = 7 + 5x + 5\Delta x - (7 + 5x) = 5\Delta x$

   $\dfrac{\Delta y}{\Delta x} = \dfrac{5\Delta x}{\Delta x} = 5$

   $\lim\limits_{\Delta x \to 0} 5 = 5$

17. $y = \dfrac{2}{x^2}$

   $y + \Delta y = \dfrac{2}{(x + \Delta x)^2}$

   $\Delta y = \dfrac{2}{(x + \Delta x)^2} - \dfrac{2}{x^2} = \dfrac{2x^2 - 2(x + \Delta x)^2}{x^2(x + \Delta x)^2}$

   $= \dfrac{2x^2 - 2x^2 - 4x\Delta x - 2(\Delta x)^2}{x^2(x + \Delta x)^2}$

   $= \dfrac{\Delta x(-4x - 2\Delta x)}{x^2(x + \Delta x)^2}$

   $\dfrac{\Delta y}{\Delta x} = \dfrac{-4x - 2\Delta x}{x^2(x + \Delta x)^2}$

   $\lim\limits_{\Delta x \to 0} \dfrac{-4x - 2\Delta x}{x^2(x + \Delta x)^2} = \dfrac{-4x - 0}{x^2(x + 0)^2} = \dfrac{-4x}{x^2(x^2)} = \dfrac{-4}{x^3}$

21. $y = 2x^7 - 3x^2 + 5$

   $\dfrac{dy}{dx} = 2(7x^6) - 3(2x) + 0$

   $= 14x^6 - 6x$

25. $y = \dfrac{x}{1 - x}$

$\dfrac{dy}{dx} = \dfrac{(1 - x)(1) - x(-1)}{(1 - x)^2} = \dfrac{1 - x + x}{(1 - x)^2} = \dfrac{1}{(1 - x)^2}$

29. $y = \dfrac{3}{(5 - 2x^2)^{3/4}};$ $u = 3,\ \dfrac{du}{dx} = 0;$ $v = (5 - 2x^2)^{3/4}$

$\dfrac{dy}{dx} = \dfrac{(5 - 2x^2)^{3/4}(0) - 3\left[-3x(5 - 2x^2)^{-1/4}\right]}{(5 - 2x^2)^{3/2}}$ $\dfrac{dv}{dx} = \dfrac{3}{4}(5 - 2x^2)^{-1/4}(-4x)$

$= \dfrac{9x(5 - 2x^2)^{-1/4}}{(5 - 2x^2)^{3/2}}$ $= -3x(5 - 2x^2)^{-1/4}$

$= \dfrac{9x}{(5 - 2x^2)^{3/2}(5 - 2x^2)^{1/4}} = \dfrac{9x}{(5 - 2x^2)^{7/4}}$

33. $y = \dfrac{\sqrt{4x + 3}}{2x};$ $u = \sqrt{4x + 3} = (4x + 3)^{1/2};$ $\dfrac{du}{dx} = \dfrac{1}{2}(4x + 3)^{-1/2}(4)$

$= 2(4x + 3)^{-1/2}$

$v = 2x,\ \dfrac{dv}{dx} = 2$

$\dfrac{dy}{dx} = \dfrac{2x(2)(4x + 3)^{-1/2} - (4x + 3)^{1/2}(2)}{(2x)^2}$

$= \dfrac{(4x + 3)^{-1/2}[4x - (4x + 3)(2)]}{4x^2} = \dfrac{-4x - 6}{(4x + 3)^{1/2}(4x^2)} = \dfrac{2(-2x - 3)}{2(2x^2)(4x + 3)^{1/2}}$

$= \dfrac{-2x - 3}{2x^2(4x + 3)^{1/2}}$

CHAPTER 23

Applications of the Derivative

1. $y = x^2 + 2$ at $(2,6)$

$\frac{dy}{dx} = 2x; \left.\frac{dy}{dx}\right|_{(2,6)} = 4$

$m = 4, (x_1, y_1) = (2,6)$

$y - y_1 = m(x - x_1)$ Eq. (20-6)

$y - 6 = 4(x - 2) = 4x - 8$

$y = 4x - 8 + 6 = 4x - 2; \quad 4x - y - 2 = 0$

5. $\qquad\qquad y = 6x - 2x^2$ at $(2,4)$

$\qquad\qquad \frac{dy}{dx} = 6 - 4x; \quad m = 2$ at $(2,4)$

$\qquad\qquad -\frac{1}{m} = \frac{1}{2}$, the normal slope

$\qquad y - y_1 = m(x - x_1)$

$\qquad\quad y - 4 = \frac{1}{2}(x - 2)$

$\qquad\quad 2y - 8 = x - 2$

$\quad x - 2y + 6 = 0$ or $y = \frac{1}{2}x + \frac{13}{2}$, the normal at $(2,4)$

9. $y = \dfrac{1}{\sqrt{x^2 + 1}}$ where $x = \sqrt{3}, \ y = \frac{1}{2}$

$y = (x^2 + 1)^{-1/2}$

$\frac{dy}{dx} = -\frac{1}{2}(x^2 + 1)^{-3/2}(2x) = -\dfrac{x}{(x^2 + 1)^{3/2}}$

At $(\sqrt{3}, \frac{1}{2})$, $m = \dfrac{-\sqrt{3}}{(3 + 1)^{3/2}} = \dfrac{-\sqrt{3}}{8}$ for the slope of the tangent.

$-\dfrac{1}{m} = \dfrac{8}{\sqrt{3}}$ for the slope of the normal line

$y - \dfrac{1}{2} = -\dfrac{\sqrt{3}}{8}(x - \sqrt{3})$

$8y - 4 = -\sqrt{3}x + 3$

$\sqrt{3}x + 8y - 7 = 0$, tangent line

$y - \dfrac{1}{2} = \dfrac{8}{\sqrt{3}}(x - \sqrt{3})$

$2\sqrt{3}y - \sqrt{3} = 16x - 16\sqrt{3}$

$16x - 2\sqrt{3}y - 15\sqrt{3} = 0$, the normal line

13. $y = x^2 - 2x$, tangent line with slope of 2

slope $= \dfrac{dy}{dx} = 2x - 2$

$2 = 2x - 2$; $x = 2$; $y = 2^2 - 2(2) = 0$

Therefore the point at which the slope is 2 is $(2,0)$. Using the point slope formula for the equation of a line, $y - 0 = 2(x - 2)$; $y = 2x - 4$.

17. $y = 4x^2 - 8x$; $\dfrac{dy}{dx} = 8x - 8$

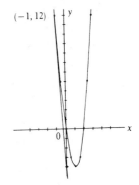

$\dfrac{dy}{dx}\bigg|_{(-1,12)} = 8(-1) - 8 = -16$

$m = -16$, slope of tangent line

$y - 12 = -16(x + 1) = -16x - 16$

$y + 16x + 4 = 0$, equation of tangent line

$0 + 16x + 4 = 0$

$16x = -4$

$x = -\dfrac{1}{4}$, x-intercept of tangent line

21. $y = \sqrt{2x^2 + 8} = (2x^2 + 8)^{1/2}$

$m = \tan 135° = -1$

$\dfrac{dy}{dx} = \dfrac{1}{2}\left(2x^2 + 8\right)^{-1/2}(4x) = \dfrac{2x}{\sqrt{2x^2 + 8}} = 1$

$2x = \sqrt{2x^2 + 8}$

$4x^2 = 2x^2 + 8$

$2x^2 = 8$

$x^2 = 4$

$x = \pm 2$, $y = \sqrt{2(4) + 8} = \sqrt{16} = 4$

$y - y_1 = m(x - x_1)$

$y - 4 = -1(x - 2)$

$y + x - 6 = 0$

Exercises 23-2, page 700

1. $x^2 - 2x - 5 = 0$ (between 3 and 4)

   $f(x) = x^2 - 2x - 5;\ f'(x) = 2x - 2$

   $f(3) = 3^2 - 2(3) - 5 = -2;\ f(4) = 4^2 - 2(4) - 5 = 3$

   The root is possibly closer to 3 than 4. Thus, let $x_1 = 3.3$:

   $f(x_1) = 3.3^2 - 2(3.3) - 5 = -0.71;\ f'(x_1) = 2(3.3) - 2 = 4.6$

   $x_2 = x_1 - \dfrac{f(x_1)}{f'(x_1)} = 3.3 - \dfrac{(-0.71)}{4.6} = 3.4543478$

   $f(x_2) = 3.4543478^2 - 2(3.4543478) - 5 = 0.0237627$

   $f'(x_2) = 2(3.4543478) - 2 = 4.9086956$

   $x_3 = 3.4543478 - \dfrac{0.0237627}{4.9086956} = 3.4495069$

   Using the quadratic formula, $x = \dfrac{-(-2) \pm \sqrt{(-2)^2 - 4(1)(-5)}}{2} = \dfrac{2 \pm \sqrt{24}}{2}$.

   The positive root is the one between 3 and 4. $x = \dfrac{2 + \sqrt{24}}{2} = 3.4494897$.

   The results agree to four (rounded off) decimal places.

1M. $x = \dfrac{2 + \sqrt{24}}{2} = 3.449\ 489\ 7$

5. $x^3 - 6x^2 + 10x - 4 = 0$ (between 0 and 1)

   $f(x) = x^3 - 6x^2 + 10x - 4;\ f'(x) = 3x^2 - 12x + 10$

   $f(0) = -4;\ f(1) = 1$

   The root is possibly closer to 1. Let $x_1 = 0.7$:

n	x_n	$f(x_n)$	$f'(x_n)$	$x_n - \dfrac{f(x_n)}{f'(x_n)}$
1	0.7	0.403	3.07	0.5687296
2	0.5687296	-0.0694666	4.1456049	0.5854863
3	0.5854863	-0.0012009	4.002547	0.5857863
4	0.5857863	-0.0000005	4.0000012	0.5857864

   $x_4 = x_3 = 0.5857864$ to seven decimal places

5M. $x_4 = x_3 = 0.585\ 786\ 4$

9. $x^4 - x^3 - 3x^2 - x - 4 = 0$ (between 2 and 3)

  $f(x) = x^4 - x^3 - 3x^2 - x - 4$; $f'(x) = 4x^3 - 3x^2 - 6x - 1$

  $f(2) = -10$; $f(3) = 20$; the root is possibly closer to 2. Let $x_1 = 2.3$:

n	x_n	$f(x_n)$	$f'(x_n)$	$x_n - \dfrac{f(x_n)}{f'(x_n)}$
1	2.3	-6.3529	17.998	2.6529781
2	2.6529781	3.0972725	36.657001	2.5684848
3	2.5684848	0.2175007	31.576097	2.5615967
4	2.5615967	0.0013683	-31.179599	2.5615528

  $x_3 = x_4 = 2.5615528$ to seven decimal places.

13. $2x^2 = \sqrt{2x + 1}$ or

$2x^2 - \sqrt{2x + 1} = 0$
$4x^4 - 2x - 1 = 0$ (Square both sides.)
$f(x) = 4x^4 - 2x - 1$

From the sketch, the intersections
lie between 0 and 1, and between 0
and -1. Approximate the positive root
at 0.8.

n	x_n	$f(x_n)$	$f'(x_n)$	$x_n - \dfrac{f(x_n)}{f'(x_n)}$
1	0.8	-0.9616	6.192	0.9552972
2	0.9552972	0.4207071	11.948757	0.9200879
3	0.9200879	0.0264914	10.462579	0.9175559
4	0.9175559	0.0001301	10.3601	0.9175433

The positive root is approximately 0.9175433.

17. $f(x) = x^3 - 2x^2 - 5x + 4$. From the sketch, one root lies between -1 and -2, and the other between 0 and 1.

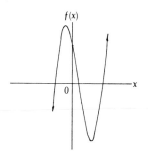

$f'(x) = 3x^2 - 4x - 5$

$f(-1) = 6$; and $f(-2) = -2$

One root is possibly closer to -2.
Let $x_1 = -1.7$:

$f(0) = 4$; $f(1) = -2$. The second root is possibly closer to 1. Let $x_1 = 0.7$.

$f(3) = -2$; $f(4) = 16$. The third root is closer to 3.

Let $x_1 = 3.1$.

n	x_n	$f(x_n)$	$f'(x_n)$	$x_n - \dfrac{f(x_n)}{f'(x_n)}$
1	-1.7	1.807	10.47	-1.8725883
2	-1.8725883	-0.2166261	13.010114	-1.8559377
3	-1.8559377	-0.0021072	12.757265	-1.8557725
4	-1.8557725	-0.000000057	11.562575	-1.8557725
1	0.7	-0.137	-6.33	0.678357
2	0.678357	0.0000369	-6.332923	0.6783628
1	3.1	-0.929	11.43	3.1812773
2	3.1812773	0.0487615	12.636467	3.1774185
3	3.1774185	0.000111	12.578291	3.1774097

The roots are -1.8557725, 0.6783628, and 3.1774097.

17M. $-1.855\ 77\ 25$, $0.678\ 36\ 28$, and $3.177\ 40\ 97$

21. $v = \frac{1}{6}\pi h\,(h^2 + 3r^2) = \frac{1}{6}\pi h^3 + \frac{1}{2}\pi r^2 h$

    $180\ 000 = \frac{1}{6}\pi h\,[h^2 + 3(60)^2] = \frac{1}{6}\pi h^3 + 1800\pi h$

      $f(h) = \frac{1}{6}\pi h^3 + 1800\pi h - 180\ 000$

     $f'(h) = \frac{1}{2}\pi h^2 + \frac{1}{2}\pi r^2 = \frac{1}{2}\pi h^2 + 1800\pi$

n	h_n	F $f(h_n)$	FF $f'(h_n)$	FG $h_n - \dfrac{f(h_n)}{f'(h_n)}$
1	29	-3238.953	6975.901	29.46431
2	29.46431	9.890625	7018.541	29.4629
3	29.4629	0	7018.411	29.4629

$h = 29.4629$ (calculated using a computer and $\pi = 3.14159$)

<u>Exercises 23-3,</u> page 705

1. $x = 3t$, $y = 1 - t$, $t = 4$

$v_x = \dfrac{dx}{dt} = 3$, $v_y = \dfrac{dy}{dt} = -1$

$v_x\Big|_{t\,=\,4} = 3$, $v_y\Big|_{t\,=\,4} = -1$

$v\Big|_{t\,=\,4} = \sqrt{3^2 + (-1)^2} = \sqrt{9 + 1} = \sqrt{10} = 3.16$

$\tan \theta = \dfrac{-1}{3} = -0.3333$, $\theta = -18.4°$
$\phantom{\tan \theta = \dfrac{-1}{3} = -0.3333, \theta } = 341.6°$

5. $x = 3t$, $y = 1 - t$, $t = 4$

$v_x = \dfrac{dx}{dt} = 3$, $a_x = \dfrac{d^2x}{dt^2} = \dfrac{d(3)}{dt} = 0$

$a_x\Big|_{t\,=\,4} = 0$

$v_y = \dfrac{dy}{dt} = -1$, $a_x = \dfrac{d^2y}{dt^2} = \dfrac{d}{dt}(-1) = 0$

$a_y\Big|_{t\,=\,4} = 0$

The particle is not accelerating since $a = \sqrt{0^2 + 0^2} = 0$.

9. $y = 4.0 - 0.20\,x^2$; $v_x = \dfrac{dx}{dt} = 5.0$

$\dfrac{dy}{dx} = -0.40x$; $\dfrac{dy}{dx}\Big|_{x\,=\,4.0} = -1.60$

$v_y = \dfrac{dy}{dt}\dfrac{dy}{dx}\cdot\dfrac{dx}{dt} = -1.60(5.0) = -8.0$

$v = \sqrt{(5.0)^2 + (-8.0)^2} = \sqrt{25 + 64} = \sqrt{89} = 9.4$

$\tan \theta = \dfrac{v_y}{v_x} = \dfrac{-8.0}{5.0} = -1.60$; $\theta = -58° = 302°$

13. $x = 120t$; $\dfrac{dx}{dt} = 120 = v_x$

$y = 160t - 16t^2$; $\dfrac{dy}{dt} = 160 - 32t\Big|_{t\,=\,6.0} = -32 = v_y$

$v = \sqrt{(120)^2 + (-32)^2} = 124$

$\tan \theta = \dfrac{-32}{120} = -0.2667 = 345°$

$\dfrac{d^2x}{dt^2} = 0 = a_x$; $\dfrac{d^2y}{dt^2} = -32 = a_y$

$a = \sqrt{0^2 + (-32)^2} = 32$

$\tan \theta$ is undefined; $\theta = 270°$

13M.

$$x = 40t; \ \frac{dx}{dt} = 40 = v_x$$

$$y = 50t - 4.9t^2; \ \frac{dy}{dt} = 50 - 9.8t \ \Big|_{t = 6.0} \ = -8.8$$

$$v = \sqrt{(40)^2 + (-8.8)^2} = 41$$

$$\tan \theta = \frac{-8.8}{40} = -0.2200; \ \theta = 348°$$

$$\frac{d^2x}{dt^2} = 0 = a_x; \ \frac{d^2y}{dt^2} = -9.8 = a_y$$

$$a = \sqrt{0^2 + (-9.8)^2} = 9.8$$

$\tan \theta$ is undefined; $\theta = 270°$

17. $\quad x = 10(\sqrt{1 + t^4} - 1) = 10(1 + t^4)^{1/2} - 10; \ y = 40t^{3/2}$

$$\frac{dx}{dt} = 5(1 + t^4)^{-1/2} (4t^3) = 20t^3(1 + t^4)^{-1/2} = \frac{20t^3}{\sqrt{1 + t^4}}$$

$$\frac{dy}{dt} = 60t^{1/2} = 60\sqrt{t}$$

$$a_x = \frac{(1 + t^4)^{1/2} (60t^2) - (20t^3)(1/2)(1 + t^4)^{-1/2} (4t^3)}{1 + t^4}$$

$$= \frac{(1 + t^4)^{-1/2} \left[60t^2 (1 + t^4) - 40t^6\right]}{(1 + t^4)}$$

$$= \frac{60t^2 (1 + t^4) - 40t^6}{(1 + t^4)^{3/2}}$$

$$a_y = 30 \ t^{-1/2}$$

$$a_x \Big|_{t = 10.0} = \frac{6000(10001) - 40000000}{(10001)^{3/2}} = \frac{20006000}{100150} = 20.0$$

$$a_y \Big|_{t = 10.0} = 30(10)^{-1/2} = \frac{30}{\sqrt{10}} = 9.5$$

$$a = \sqrt{(20)^2 + (9.5)^2} = 22.1$$

$$\tan \theta = \frac{9.5}{20.0} = 0.475; \ \theta = 25.4°$$

$$a_x \big|_{t = 100} = \frac{60(10^4)(10^8) - 40(10^{12})}{(10^8)^{3/2}} = \frac{6 \times 10^{13} - 4 \times 10^{13}}{10^{12}}$$

$$= \frac{2 \times 10^{13}}{10^{12}} = 20.0$$

$$a_y \big|_{t = 100} = 30(10^2)^{-1/2} = 30(10^{-1}) = 3.0$$

$$a = \sqrt{(20.0)^2 + (3.0)^2} = 20.2$$

$$\tan \theta = \frac{3.0}{20.0} = 0.1500, \quad \theta = 8.5°$$

21. $\quad d = 3.50, \quad r = 1.75; \quad x^2 + y^2 = 1.75^2; \quad \dfrac{dy}{dx} = \dfrac{-x}{y}$

3600 r/min $= 7200\pi$ rad/min $= \omega$
$\qquad V = \omega r = 7200\pi \ (1.75) - 12600 \ \pi$ in./min

$\qquad x^2 + y^2 = 1.75^2$

$\qquad\qquad y^2 = 1.75^2 - x^2 = 1.75^2 - 1.20^2 = 3.062 - 1.44 = 1.622$
$\qquad\qquad y = 1.27$

$\qquad\qquad \dfrac{dy}{dx} = \dfrac{-x}{y} = -\dfrac{1.20}{1.27} = -0.945 = \dfrac{V_y}{V_x}; \ V_y = -0.945 V_x$

$\qquad\qquad V = 12600\pi = \sqrt{V_x^2 + V_y^2} = \sqrt{(-0.945 V_x)^2 + V_x^2} = \sqrt{1.893 V_x^2} = 1.376 V_x$

$\qquad\qquad V_x = 9158\pi = 28800$
$\qquad\qquad V_y = -0.945 V_x = 8654\pi = -27200$

21M. $\quad d = 8.89$ cm, $\ r = 4.45$ cm; $\ x^2 + y^2 = 4.45^2$

$\qquad \dfrac{dy}{dx} = \dfrac{-x}{y}$

3600 r/min $= 7200\pi$ rad/min $= \omega$
$\qquad V = \omega r = 7200\pi \ (4.45) = 32000\pi$ cm/min

$\qquad x^2 + y^2 = 4.45^2$

$\qquad\qquad y^2 = 4.45^2 - x^2 = 4.45^2 - 3.05^2 = 19.80 - 9.30 = 10.50$
$\qquad\qquad y = 3.24$

$\qquad\qquad \dfrac{dy}{dx} = -\dfrac{x}{y} = \dfrac{-3.05}{3.24} = -0.941 - \dfrac{V_y}{V_x}; \ V_y = -0.941 V_x$

$\qquad\qquad V = 32000\pi = \sqrt{V_x^2 + V_y^2} = \sqrt{(-0.941 V_x)^2 + V_x^2} = \sqrt{1.885 V_x^2}$

$\qquad\qquad\quad = 1.373 V_x$

$\qquad\qquad V_x = \dfrac{32000\pi}{1.373} = 23300\pi = 73200$

$\qquad\qquad V_y = -0.941 V_y = -21900\pi = -68900$

Exercises 23-4, page 709

1. $R = 4.000 + 0.003T^2$, $\dfrac{dT}{dt} = 0.100°$ C/s

   $\dfrac{dR}{dt} = 0 + 0.006T\dfrac{dT}{dt}$

   $\dfrac{dR}{dt}\Big|T = 150°$ C $= 0 + 0.006(150)(0.1) = 0.0900$ Ω/s

5. $r = \sqrt{0.4\lambda}$; $\dfrac{d\lambda}{dt} = 0.10 \times 10^{-7}$

   $r = (0.4\lambda)^{\frac{1}{2}}$

   $\dfrac{dr}{dt} = \dfrac{1}{2}(0.4\lambda)^{-\frac{1}{2}}(0.4)\dfrac{d\lambda}{dt} = 0.2(0.4\lambda)^{-\frac{1}{2}}\dfrac{d\lambda}{dt} = \dfrac{2 \times 10^{-9}}{\sqrt{24 \times 10^{-4}}} = 4.1 \times 10^{-6}$

   $\dfrac{dr}{dt}\Big|\lambda = 6.0 \times 10^{-7} = 0.2\left[0.4(6.0 \times 10^{-7})\right]^{-\frac{1}{2}}(0.10 \times 10^{-7})$

9. $A = \pi r^2$; $\dfrac{dr}{dt} = 0.020$ mm/mo

   $\dfrac{dA}{dt} = 2\pi r \dfrac{dr}{dt}$

   $\dfrac{dA}{dt}\Big|r = 3.0 = 2\pi(3.0)(0.020) = 0.12\pi = 0.38$ mm²/mo

13. $p = \dfrac{k}{V}$; $\dfrac{dV}{dt} = 20$ cm³/min; $V = 800$ cm³

    $200 = \dfrac{k}{600}$; $k = 1.20 \times 10^5$ kPa × cm³

    $p = \dfrac{120,000}{V} = 120,000\ V^{-1}$

    $\dfrac{dp}{dt} = -120,000\ V^{-2}\dfrac{dV}{dt}$

    $\dfrac{dp}{dt}\Big|V = 800 = -120,000(800)^{-2}(20) = -3.75$ kPa/min

17. $V = \pi r^2 h$; at $h = 1.80$, $r = 0.575$ found by $h:r$ proportion

    $\dfrac{dV}{dt} = \pi r^2 \dfrac{dh}{dt}$; $\dfrac{dv}{dt} = 0.50$

    $0.50 = \pi(0.575)^2\dfrac{dh}{dt}$

    $\dfrac{dh}{dt} = \dfrac{0.50}{\pi(0.575)^2} = 0.48$

21. Let x be the distance traveled by the jet going due east, and y be the distance traveled by the jet going north of east. Since the second jet remains due north of the first jet, we have a right triangle and can use the Pythagorean theorem.

$$x^2 + z^2 = y^2$$

Taking the derivative of this expression,

$$2x\frac{dx}{dt} + 2z\frac{dz}{dt} = 2y\frac{dy}{dt}$$

$$x\Big|_{t = \frac{1}{2}} = 1600(\tfrac{1}{2}) = 800; \quad y\Big|_{t = \frac{1}{2}} = 1800(\tfrac{1}{2}) = 900$$

$$z = \sqrt{y^2 - x^2} = \sqrt{900^2 - 800^2} = 412$$

$$\frac{dx}{dt} = 1600; \quad \frac{dy}{dt} = 1800; \text{ Substituting, } 2(800)(1600) + 2(412)\frac{dz}{dt} = 2(900)(1800)$$

$$\frac{dz}{dt} = 825 \text{ mi/h}$$

21M. $\frac{dz}{dt} = 825$ km/h

Exercises 23-5, page 716

1. $y = x^2 + 2x; \ y' = 2x + 2; \ 2x + 2 > 0$

2$x > -2; \ x > -1; \ f(x)$ increases.

2$x + 2 < 0; \ 2x < -2; \ x < -1; \ f(x)$ decreases.

5. $y = x^2 + 2x; \ y' = 2x + 2; \ y' = 0$ at $x = 1$

$y'' = 2 > 0$ at $x = -1$ and $(-1,-1)$ is a relative minimum.

9. $y = x^2 - 2x; \ y' = 2x - 2; \ y'' = 2$

Thus, $y'' > 0$ for all x. The graph is concave up for all x and has no points of inflection.

13.

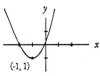

$y = x^2 + 2x$

$(-1, 1)$

17. $y = 12x - 2x^2; \ y' = 12 - 4x; \ y' = 0$ at $x = 3$; for $x = 3$, $y = 12(3) - 2(3)^2 = 18$, and $(3,18)$ is a critical point. $12 - 4x < 0$ for $x > 3$ and the graph decreases; $12 - 4x > 0$ for $x < 3$ and the graph increases; $y'' = -4$; thus $y'' < 0$ for all x. There are no inflections; the graph is concave down for all x, and $(3,18)$ is a maximum point.

21. $y = x^3 + 3x^2 + 3x + 2;$ $y' = 3x^2 + 6x + 3 = 3(x^2 + 2x + 1)$
$$= 3(x + 1)(x + 1)$$

$3(x + 1)(x + 1) = 0$ for $x = -1$

$(-1,1)$ is a critical point

$3(x + 1)(x + 1) > 0$ for $x < -1$ and the slope is positive

$3(x + 1)(x + 1) > 0$ for $x > -1$ and the slope is positive

$y'' = 6x + 6;$ $6x + 6 = 0$ for $x = 1,$ and $(-1,1)$ is an inflection point

$6x + 6 < 0$ for $x < -1$ and the graph is concave down

$6x + 6 > 0$ for $x > -1$ and the graph is concave up

Since there is no change in slope from positive to negative or vice versa, there are no maximum or minimum points.

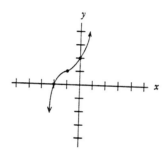

25. $y = 4x^3 - 3x^4;$ $y' = 12x^2 - 12x^3 = 12x^2(1 - x) = 0$

$12x^2(1 - x) = 0$ for $x = 0$ and $x = 1$

$(0,0)$ and $(1,1)$ are critical points.

$12x^2 - 12x^3 > 0$ for $x < 0$ and the slope is positive

$12x - 12x^3 > 0$ for $0 < x < 1$ and the slope is positive

$12x - 12x^3 < 0$ for $x > 1$ and the slope is negative

$y'' = 24x - 36x^2;$ $24x - 36x^2 = 12x(2 - 3x) = 0$ for $x = 0,$ $x = \frac{2}{3}$

$(0,0)$ and $(\frac{2}{3}, \frac{16}{27})$ are possible inflection points.

$24x - 36x^2 < 0$ for $x < 0$ and the graph is concave down

$24x - 36x^2 > 0$ for $0 < x < \frac{2}{3}$ and the graph is concave up

$24x - 36x^2 < 0$ for $x > \frac{2}{3}$ and the graph is concave down

$(1,1)$ is a relative maximum point since $y' = 0$ at $(1,1)$ and the slope is positive for $x < 1$ and negative for $x > 1.$

$(0,0)$ and $(\frac{2}{3}, \frac{16}{27})$ are inflection points since there is a concavity change.

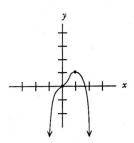

29. $R = 75 - 18i^2 + 8i^3 - i^4$; $R' = -36i + 24i^2 - 4i^3 = -4i\left(9 - 6i + i^2\right)$

$-4i\left(9 - 6i - i^2\right) = -4i(i - 3)(i - 3) = 0$ for $i = 0$ and $i = 3$

$(0,75)$ and $(3,48)$ are critical points.

$-4i(i - 3)^2 < 0$ for $0 < x < 3$ and the slope is negative

$-4i(i - 3)^2 < 0$ for $x > 3$ and the slope is negative

$y'' = -36 + 48i - 12i^2 = -12\left(3 - 4i + i^2\right) = -12(i - 3)(i - 1)$

$-12(i - 3)(i - 1) = 0$ for $i = 3$ and $i = 1$

$(1,64)$ and $(3,48)$ are possible inflection points

$-12(i - 3)(i - 1) < 0$ for $0 < x < 1$ and the graph is concave down

$-12(i - 3)(i - 1) > 0$ for $1 < x < 3$ and the graph is concave up

$-12(i - 3)(i - 1) < 0$ for $x > 3$ and the graph is concave down

$(1,64)$ and $(3,48)$ are inflection points.

There are no relative maximum or minimum points since there is no change of slope, other than $(0,75)$ where the graph is not continuous.

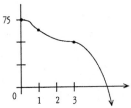

33.

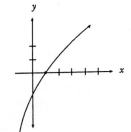

$f(1) = 0 \rightarrow (1,0)$ is an x-intercept

$f'(x) > 0$ for all $x \rightarrow$ curve rises left to right

$f''(x) < 0$ for all $x \rightarrow$ concave down

Exercises 23-6, page 720

1. $y = \dfrac{4}{x^2}$

    Intercepts:

    (1) y is undefined for $x = 0$, so the graph is not continuous at $x = 0$ and cannot cross the y-axis.

    (2) Since 4 is positive and x^2 is positive for all non-zero x, $\dfrac{4}{x^2}$ is always positive, and the graph does not cross into quadrants III or IV.

    (3) Since $\dfrac{4}{x^2} > 0$ for any x, $y \neq 0$ and the graph does not intersect the x-axis at any point.

    Symmetry:

    (4) Symmetrical about y-axis since replacing x with $-x$ produces no change. Not symmetrical about x-axis.

    Behavior as x becomes large:

    (5) $x \to -\infty$, $\dfrac{4}{x^2}$ approaches 0, and the negative x-axis is an asymptote.

    (6) As $x \to +\infty$, $\dfrac{4}{x^2}$ approaches 0, and the positive x-axis is an asymptote.

    Derivatives:

    (7) $y' = -8x^{-3} = \dfrac{-8}{x^3}$. For negative x, $\dfrac{-8}{x^3}$ is positive and the graph rises. For positive x, $\dfrac{-8}{x^3}$ is negative and the graph falls.

    (8) $y'' = 24x^{-4} = \dfrac{24}{x^4}$. y'' is positive for all x. There are no inflections and the graph is everywhere concave up.

    Inc. $x < 0$, dec. $x > 0$
    Conc. up $x < 0$, $x > 0$
    Asym. $x = 0$, $y = 0$

5. $y = x^2 + \dfrac{2}{x} = \dfrac{x^3 + 2}{x}$

    (1) $\dfrac{2}{x}$ is undefined for $x = 0$, so the graph is not continuous at the y-axis; i.e., no y-intercept exists.

    (2) $\dfrac{x^3 + 2}{x} = 0$ at $x = \sqrt[3]{-2} = -\sqrt[3]{2}$. There is an x-intercept at $(-\sqrt[3]{2}, 0)$.

    (3) As $x \to \infty$, $x^2 \to \infty$ and $\dfrac{2}{x} \to 0$, so $x^2 + \dfrac{2}{x} \to \infty$.

    (4) As $x \to 0$ through positive x, $x^2 \to 0$ and $\dfrac{2}{x} \to \infty$, so $x^2 + \dfrac{2}{x} \to \infty$.

    (5) As $x \to -\infty$, $x^2 \to \infty$ and $\dfrac{2}{x} \to 0$ so $x^2 + \dfrac{2}{x} \to \infty$.

    (6) As $x \to 0$ through negative numbers, $x^2 \to 0$ and $\dfrac{2}{x} \to -\infty$, so $x^2 + \dfrac{2}{x} \to -\infty$.

(7) $y' = 2x - 2x^{-2} = 0$ at $x = 1$ and the slope is zero at $(1,3)$.

(8) $y'' = 2 + 4x^{-3} = 0$ at $x = -\sqrt[3]{2}$ and $(-\sqrt[3]{2},0)$ is an inflection point.

(9) $y'' > 0$ at $x = 1$, so the graph is concave up and $(1,3)$ is a relative minimum.

(10) Since $(-\sqrt[3]{2},0)$ is an inflection, $f''(-1) < 0$ and the graph is concave down. $f''(-2) > 0$ and the graph is concave up.

(11) Not symmetrical about the x- or y-axis.

9. $y = \dfrac{x^2}{x + 1}$

Intercepts:

(1) Function undefined at $x = -1$; not continuous at $x = -1$.

(2) At $x = 0$, $y = 0$. The origin is the only intercept.

Behavior as x becomes large:

(3) As $x \to +\infty$, $y \to x$, so $y = x$ is an asymptote. As $x \to -\infty$, $y \to -\infty$.

Vertical asymptotes:

(4) As $x \to -1$ from the left, $x + 1 \to 0$ through negative values and $\dfrac{x^2}{x + 1} \to -\infty$ since $x^2 > 0$ for all x. As $x \to -1$ from the right, $x + 1 \to 0$ through positive values and $\dfrac{x^2}{x + 1} \to +\infty$. $x = -1$ is an asymptote.

Symmetry:

(5) The graph is not symmetrical about the y-axis or the x-axis.

Derivatives:

(6) $y' = \dfrac{x^2 + 2x}{(x + 1)^2}$; $y' = 0$ at $x = -2$, $x = 0$. $(-2,-4)$ and $(0,0)$ are critical points. Checking the derivative at $x = -3$, the slope is positive, and at $x = -1.5$ the slope is negative. $(-2,-4)$ is a relative maximum point. Checking the derivative at $x = -0.5$, the slope is negative, and at $x = 1$ the slope is positive, so $(0,0)$ is a relative minimum point.

Int. $(0,0)$, max. $(-2,-4)$, min. $(0,0)$, asym. $x = -1$

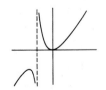

13. $y = \dfrac{4}{x} - \dfrac{4}{x^2}$

Intercepts:

(1) There are no y intercepts since $x = 0$ is undefined.

(2) $y = 0$ when $x = 1$ so $(1,0)$ is an x-intercept

Asymptotes:

(3) $x = 0$ is an asymptote; the denominator is 0.

Symmetry:

(4) Not symmetrical about y-axis since

$\dfrac{4}{x} - \dfrac{4}{x^2}$ is different from $\dfrac{4}{(-x)} - \dfrac{4}{(-x)^2}$

(5) Not symmetrical about x-axis since

$y = \dfrac{4}{x} - \dfrac{4}{x^2}$ is different from $-y = \dfrac{4}{x} - \dfrac{4}{x^2}$

(6) Not symmetrical about the origin since

$y = \dfrac{4}{x} - \dfrac{4}{x^2}$ is different from $-y = \dfrac{4}{-x} - \dfrac{4}{(-x)^2}$

Derivatives:

(7) $y' = -4x^{-2} + 8x^{-3} = 0$ at $x = 2$; $(2,1)$ is a relative maximum.

(8) $y'' = 8x^{-3} - 24x^{-4} = 0$ at $x = 3$ so $\left(3, \frac{8}{9}\right)$ is a possible inflection.

$y'' < 0$ (concave down) for $x < 3$ and 0 (concave up) for $x > 3$

so $\left(3, \frac{8}{9}\right)$ is an inflection.

Behavior as x becomes large:

(9) As $x \to \infty$ or $-\infty$, $\dfrac{4}{x}$ and $-\dfrac{4}{x^2}$ each approach 0

As $x \to 0$, $\dfrac{4}{x} - \dfrac{4}{x^2} = \dfrac{4x - 4}{x^2}$ approaches $-\infty$ through positive

or negative values of x

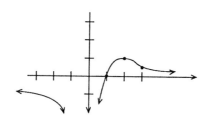

17. $y = \dfrac{9x}{9 - x^2}$

Intercept:

(1) Intercept at $x = 0$, $y = 0$ only

(2) Asymptotes at $x = -3$, $x = 3$

Derivatives:

(3) $y' = \dfrac{(9 - x^2)(9) - (9x)(-2x)}{(9 - x^2)^2} = \dfrac{81 + 9x^2}{(9-x^2)^2}$

$81 + 9x^2 = 0$

$9x^2 = -81$ No real value max. or min.

$x^2 = \sqrt{-9}$ (imaginary) ± 3 are critical values.

(4) $y'' = \dfrac{(9 - x^2)^2(18x) - (81 + 9x^2)(2)(9 - x^2)(-2x)}{(9-x^2)^4} = \dfrac{-18x^5 - 324x^3 + 4374x}{(9 - x^2)^4}$

$-18x^5 - 324x^3 + 4374x = 0$

$-18x(x^4 + 18x^2 - 243) = 0$

$-18x = 0$ $x^4 + 18x^2 - 243 = 0$

$x = 0$ $(x^2 + 27)(x^2 - 9) = 0$

                   (imaginary) $x^2 = 9$

                        $x = \pm 3$ (these are asymptotes)

$y \big|_{x=0} = 0$; a possible inflection is $(0,0)$

at $(-1, -\frac{9}{8})$ $y'' = -4032$; concave down

at $(1, \frac{9}{8})$ $y'' = 4032$; concave up, and $(0,0)$ is an inflection point

Symmetry:

(5) There is no symmetry.

As x becomes large:

(6) $\lim\limits_{x \to -\infty} \dfrac{9x}{9 - x^2} = 0$ and $y = 0$ is an asymptote

21. $R = \dfrac{200}{\sqrt{t^2 + 40000}}$ Intercepts:

(1) not continuous at $t = 0$ (not defined at $t < 0$).
(2) no t-intercept; R-intercept at $(0,1)$

Symmetry:

(3) No symmetry about either axis (R is undefined for $t < 0$)

Derivatives:

(4) $R = 200(t^2 + 40000)^{-1/2}$

$R' = -100(t^2 + 40000)^{-3/2} (2t) = -200t(t^2 + 40000)^{-3/2}$

$\dfrac{-200t}{(t^2 + 40000)^{3/2}} = 0$

$-200t = 0$

$t = 0$ is a max since $R\Big|_{t=0} = 1$

and $R\Big|_{t=1} < 1$ (R is undefined for $t<0$)

(5) $R'' = \dfrac{(t^2 + 40000)^{3/2}(-200) - (-200t)(3/2)(t^2 + 40000)^{1/2}(2t)}{[(t^2 + 40000)^{3/2}]^2} = 0$

$= -200(t^2 + 40000)^{3/2} + 600t^2(t^2 + 40000)^{1/2} = 0$

$t^2 + 40000)^{1/2}[-200(t^2 + 40000) + 600t^2] = 0$
$(t^2 + 40000)^{1/2} = 0$ $-200(t^2 + 40000) + 600t^2 = 0$

$t^2 + 40000 = 0$ $-200[(t^2 + 40000) - 3t^2] = 0$

$t^2 = -40000$ $t^2 + 40000 - 3t^2 = 0$

imaginary $-2t^2 = -40000$

$t^2 = 20000$

$t = 141$ possible inflection

$R''\Big|_{t=140} \le 0 \; ; \; R''\Big|_{t=142} > 0; \; R\Big|_{t=141} = 0.82$ is an inflection

As x becomes large:

(6) As $x \to \infty$, $\sqrt{t^2 + 40000}$ becomes infinitely large

and $\dfrac{200}{\sqrt{t^2 + 40000}}$ is a positive value that becomes infinitely small but never zero.

Exercises 23-7, page 726

1. $s = 112t - 16t^2$. Find maximum s.
    $s' = 112 - 32t = 0$; $-32t = -112$; $t = 3.5$
    $s'' = -32 < 0$ for all t, so the graph is concave down and $t = 3.5$ is a
    maximum.
    $s = 112(3.5) - 16(3.5)^2 = 196$ ft

1M. $s = 34.3t - 4.9t^2$. Find maximum s.
    $s' = 34.3 - 9.8t$; $34.3 - 9.8t = 0$; $-9.8t = -34.3$; $t = 3.5$;
    $s'' = -9.8 < 0$ for all t, so the graph is concave down and $t = 3.5$ is a
    maximum.
    $s = 34.3(3.5) - 4.9(3.5)^2 = 60.0$ m

5. $S = 360A - 0.1A^3$ Find maximum S

    $S' = 360 - 0.3A^2$; $A^2 = 1200$; $A = 34.6$

    $S'' = -0.6A < 0$ for all valid (positive) A so the graph is concave down
        and $A = 34.6$ is a max. Maximum savings $S = 360(34.6) - 0.1(34.6)^2$
        $= 8310$

9. Let S = sum of numbers x and $\dfrac{64}{x}$

    $S = x + \dfrac{64}{x} = x + 64x^{-1}$

    $S' = 1 - 64x^{-2}$; $1 = 64x^{-2}$; $x^{-2} = \dfrac{1}{64}$

    $\qquad\qquad\qquad\qquad\quad x^2 = 64$

    $\qquad\qquad\qquad\qquad\quad x = 8$

    $S'' = 128x^{-3} = \dfrac{128}{x^3} > 0$ for all pos. x so the graph is concave up and

    $S = 8 + {64}/{8} = 16$ is a minimum

13. $A = \frac{1}{2}bh$; let $b = 2x$ so that half of b is x.
    By the Pythag. Th. $h = \sqrt{144-x^2}$
    $A = x(\sqrt{144 - x^2}) = x(144 - x^2)^{1/2}$

    $A' = (144 - x^2)^{1/2} + x(\frac{1}{2})(144 - x^2)^{-1/2}(-2x)$

    $\quad = (144 - x^2)^{1/2} - x^2(144 - x^2)^{-1/2}$

    $\quad = (144 - x^2)^{-1/2}[(144 - x^2) - x^2] = 0$

    $(144 - x^2)^{-1/2} = 0 \qquad\qquad 144 - 2x^2 = 0$

    $\dfrac{1}{(144 - x^2)} = 0 \qquad\qquad 2x^2 = 144$

    $\qquad\qquad\qquad\qquad\qquad x^2 = 72$

    $\qquad\qquad\qquad\qquad\qquad x = 8.5$; $b = 2x = 17.0$

    $\qquad$ imaginary

17. $A = \frac{1}{2}(a + b)h;\ a = 6.00$

$A = \frac{1}{2}(6 + 2x)(\sqrt{9-x^2}) = (3 + x)(9 - x^2)^{1/2}$

$A' = (3 + x)(\frac{1}{2})(9 - x^2)^{-1/2}(-2x) + (1)(9 - x^2)^{1/2}$

$\quad = (-3x - x^2)(9 - x^2)^{-1/2} + (9 - x^2)^{1/2}$

$\quad = \dfrac{-3x - x^2}{(9 - x^2)^{1/2}} + \dfrac{9 - x^2}{(9 - x^2)^{1/2}} = \dfrac{9 - x^2 - 3x - x^2}{(9 - x^2)^{1/2}} = \dfrac{-2x^2 - 3x + 9}{(9 - x^2)^{1/2}}$

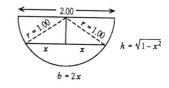

$-2x^2 - 3x + 9 = 0$

$2x^2 + 3x - 9 = 0$

$(2x - 3)(x + 3) = 0$

$\qquad 2x - 3 = 0 \qquad\qquad x + 3 = 0$

$\qquad\qquad x = 1.50 \qquad\qquad x = -3\ \text{(not valid)}$

$\qquad\qquad b = 2x = 3.00$

17M. $A = \frac{1}{2}(a + b)h;\ a = 2.00$

$\quad = \frac{1}{2}(2 + 2x)(\sqrt{1 - x^2}) = (1 + x)(1 - x^2)^{1/2}$

$A' = (1 + x)(\frac{1}{2})(1 - x^2)^{-1/2}(-2x) + (1 - x^2)^{1/2}$

$\quad = (-x - x^2)(1 - x^2)^{-1/2} + (1 - x^2)^{1/2}$

$\quad = \dfrac{-x - x^2}{(1 - x^2)^{1/2}} + \dfrac{1 - x^2}{(1 - x^2)^{1/2}} = \dfrac{-2x^2 - x + 1}{(1 - x^2)^{1/2}} = 0$

$-2x^2 - x + 1 = 0$

$2x^2 + x - 1 = 0$

$(2x - 1)(x + 1) = 0$

$\qquad\qquad 2x = 1 \qquad\qquad\qquad x = -1\ \text{(not valid)}$

$\qquad\qquad\ x = 0.500$

$\qquad\qquad\ b = 2x = 1.00$

21. $y = k(2x^4 - 5Lx^3 + 3L^2x^2) = 2kx^4 - 5kLx^3 + 3kL^2x^2$

$y' = 8kx^3 - 15kLx^2 + 6kL^2x = 0$

$kx(8x^2 - 15Lx + 6L^2) = 0$

$kx = 0 \qquad 8x^2 - 15Lx + 6L^2 = 0$

$x = 0 \qquad x = \dfrac{-(-15) \pm \sqrt{(-15)^2 - 4(8)(6)}}{2(8)} = \dfrac{15 \pm \sqrt{33}}{16} = \dfrac{15 \pm 5.75}{16}$

$x = 0.58L,\ 1.30L$ (not valid-this distance is greater than L,
    the length of the beam.)

25. $2x + \pi d = 400;\ \pi d = 400 - 2x;\ d = \dfrac{400 - 2x}{\pi}$

$$A = x(d) = x\left(\dfrac{400 - 2x}{\pi}\right) = \dfrac{400x - 2x^2}{\pi}$$

$$A' = \dfrac{400 - 4x}{\pi} = 0;\ 400 - 4x = 0;\ x = 100$$

29. Let C = total cost;

$$C = 50\,000(10 - x) + 80\,000\left(\sqrt{x^2 + 2.5^2}\right)$$

$$= 500\,000 - 50\,000x + 80\,000(x^2 + 6.25)^{1/2}$$

$$C' = -50\,000 + 40\,000(x^2 + 6.25)^{-1/2}(2x)$$

$$= -50\,000 + 80\,000x(x^2 + 6.25)^{-1/2}$$

$$= -50\,000 + \dfrac{80\,000x}{\sqrt{x^2 + 6.25}}$$

$$= \dfrac{-50\,000\sqrt{x^2 + 6.25} + 80\,000x}{\sqrt{x^2\ 6.25}} = 0$$

$$-50\,000\sqrt{x^2 + 6.25} + 80\,000x = 0$$

$$\sqrt{x^2 + 6.25} = \dfrac{-80\,000x}{-50\,000} = \dfrac{8x}{5}$$

$$x^2 + 6.25 = \dfrac{64}{25}x^2$$

$$6.25 = \dfrac{64}{25}x^2 - x^2 = \dfrac{39}{25}x^2$$

$$x^2 = 6.25\left(\dfrac{25}{39}\right) = 4.00$$

$$x = 2,00$$

$$10 - x = 8.00$$

Review Exercises for Chapter 23, page 728

1. $y = 3x - x^2$ at $(-1,-4);\ y' = 3 - 2x$
$y'\big|_{x = -1} = 3 - 2(-1) = 5$
$m = 5$ for tangent line
$y - y_1 = 5(x - x_1)$
$y - (-4) = 5[x - (-1)];\ y + 4 = 5x + 5$
$5x - y + 1 = 0$

5. $y = \sqrt{x^2 + 3}$; $m = \dfrac{1}{2}$

   $y = (x^2 + 3)^{1/2}$; $\dfrac{dy}{dx} = \dfrac{1}{2}(x^2 + 3)^{-1/2}(2x) = \dfrac{x}{\sqrt{x^2 + 3}}$

   $m_{tan} = \dfrac{dy}{dx} = \dfrac{x}{\sqrt{x^2 + 3}} = \dfrac{1}{2}$

   $2x = \sqrt{x^2 + 3}$
   Squaring both sides, $4x^2 = x^2 + 3$; $3x^2 = 3$; $x^2 = 1$; $x = 1$ and $x = -1$.
   Therefore, the abscissa of the point at which $m = \frac{1}{2}$ is 1 or -1. If
   $x = 1$, $y = \sqrt{1^2 + 3} = \sqrt{4} = 2$ or -2. If $x = -1$, $y = \sqrt{(-1)^2 + 3} = 2$ or -2.
   The possible points where the slope of the tangent line is $\frac{1}{2}$ are (1,2),
   (1,-2), (-1,2), (-1,-2). A sketch of the curve shows that the only
   relative maximum or minimum point is at (0,1.7). Therefore the point
   is (1,2).
   $y - 2 = \dfrac{1}{2}(x - 1)$; $y = \dfrac{1}{2}x + \dfrac{3}{2}$ is the equation of the tangent line.

9. $y = 0.5x^2 + x$; $V_y = \dfrac{dy}{dt} = \dfrac{x \, dx}{dt} + \dfrac{dx}{dt}$; $V_x = 0.5\sqrt{x}$
   Substituting, $V_y = x(0.5\sqrt{x}) + 0.5\sqrt{x}$
   Find V_y at (2,4):
   $V_y \big|_{x = 2} = 2(0.5\sqrt{2}) + 0.5\sqrt{2} = \sqrt{2} + 0.5\sqrt{2} = 1.5\sqrt{2} = 2.12$

13. $x^3 - 3x^2 - x + 2 = 0$ (between 0 and 1)
   $f(x) = x^3 - 3x^2 - x + 2$; $f'(x) = 3x^2 - 6x - 1$
   $f(0) = 0^3 - 3(0^2) - 0 + 2 = 2$; $f(1) = 1^3 - 3(1^2) - 1 + 2 = -1$
   The root is possibly closer to 1 than 0. Let $x_1 = 0.6$:

n	x_n	$f(x_n)$	$f'(x_n)$	$x_n - \dfrac{f(x_n)}{f'(x_n)}$
1	0.6	0.536	-3.52	0.7522727
2	0.7522727	-0.0242935	-3.8158936	0.7459063
3	0.7459063	-0.0000304	-3.8063092	0.7458983

   $x_4 = x_3 = 0.7458983$

13M. $x_4 = x_3 = 0.745\ 898\ 3$

17. $y = 4x^2 + 16x$
   (1) The graph is continuous for all x.
   (2) The intercepts are (0,0) and (-4,0).
   (3) As $x \to +\infty$ and $-\infty$, $y \to +\infty$.
   (4) The graph is not symmetrical about
       either axis or the origin.
   (5) $y' = 8x + 16$; $y' = 0$ at $x = -2$.
       (-2,-16) is a critical point.
   (6) $y'' = 8 > 0$ for all x; the graph
       is concave up and (-2,-16) is a
       minimum.

21. $y = x^4 - 32x$
    (1) The graph is continuous for all x.
    (2) The intercepts are $(0,0)$, $(2\sqrt[3]{4},0)$.
    (3) As $x \to -\infty$, $y \to +\infty$; as $x \to +\infty$, $y \to +\infty$.
    (4) The graph is not symmetrical about either axis or origin.
    (5) $y' = 4x^3 - 32$;
    (6) $y'' = 12x^2$; $y'' = 0$ at $x = 0$;
    $(0,0)$ is a point of inflection.
    $f''(2) > 0$; the graph is concave up and $(2,-48)$ is a minimum.

25. $y = x^2 + 2$ and $y = 4x - x^2$
    $y' = 2x$; $y' = 4 - 2x$
    $2x = 4 - 2x$; $4x = 4$; $x = 1$
    The point $(1,3)$ belongs to both graphs; the slope of the tangent line is 2.
    $y - y_1 = 2(x - x_1)$; $y - 3 = 2(x - 1)$; $y - 3 = 2x - 2$
    $2x - y + 1 = 0$ is the tangent line.

29. $x = 8t$ $y = -0.15t^2$ $v = \sqrt{8^2 + (-3.6)^2}$

    $\dfrac{dx}{dt} = 8$ $\dfrac{dy}{dx} = -0.30t$ $v = \sqrt{64 + 12.96}$

    $v_x \Big|_{t = 12} = 8$ $v_y \Big|_{t = 12} = -3.6$ $v \Big|_{t = 12} = \sqrt{76.96} = 8.8$

    $\tan \theta = \dfrac{-3.6}{8} = -0.45$; $\theta = 336°$

33. $p = -0.25x^2 + 80x - 4000$

    $p' = -0.50x + 80 = 0$

    $-0.50x = -80$

    $x = 160$ calculators

    $p = -0.25(160)^2 + 80(160) - 4000 = \$2400$

37. Length of third side $y = \sqrt{3}x$

    $\dfrac{dx}{dt} = 0.750$

    $\dfrac{dy}{dt} = \sqrt{3}\dfrac{dx}{dt} = \sqrt{3}(0.750) = 1.30$

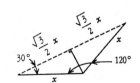

41. $y + 2x = 200;\ y = -2x + 200$

$A = xy = x(-2x + 200) = -2x^2 + 200x$

$A' = -4x + 200 = 0$

$x = \dfrac{-200}{-4} = 50;\ y = 200 - 2(50) = 100$

$A = 50(100) = 5000$

45. 8000 ft = 1.52 mi

$\dfrac{dz}{dt} = 680;\ \dfrac{dy}{dt} = 0;\ \text{find } \dfrac{dx}{dt}$

$z^2 = x^2 + y^2;\ z = \sqrt{5.00^2 + 1.52^2} = \sqrt{27.31} = 5.23$

$2z\dfrac{dz}{dt} = 2x\dfrac{dx}{dt} + 2y\dfrac{dy}{dt}$

$2(5.23)(680) = 2(5.00)\dfrac{dx}{dt} + 2(1.52)(0)$

$7107 = 10.00\dfrac{dx}{dt}$

$\dfrac{dx}{dt} = 711 \text{ mi/hr}$

45M.

2400 m = 2.40 km

$\dfrac{dz}{dt} = 1100;\ \dfrac{dy}{dt} = 0;\ \text{find } \dfrac{dx}{dt}$

$z^2 = x^2 + y^2;\ = \sqrt{(8.00)^2 + (2.40)^2} = \sqrt{69.76} = 8.35$

$2z\dfrac{dz}{dt} = 2x\dfrac{dx}{dt} + 2y\dfrac{dy}{dt}$

$2(8.35)(1100) = 2(8.00)\dfrac{dx}{dt} + 2(2.40)(0)$

$18370 = 16.00\dfrac{dx}{dt}$

$\dfrac{dx}{dt} = 1150 \text{ km/hr}$

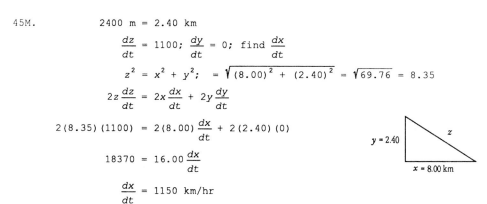

CHAPTER 24

Integration

1. $y = x^5 + x; \frac{dy}{dx} = 5x^4 + 1; dy = (5x^4 + 1) \ dx$
    Eq. (22-9), (22-11), (24-1)

5. $y = (x^2 - 1)^4; \frac{dy}{dx} = 4(x^2 - 1)^3(2x) = 8x(x^2 - 1)^3$
    $dy = 8x(x^2 - 1)^3 \ dx$

9. $y = x^2(1 - x)^3; \frac{dy}{dx} = x^2(3)(1 - x)^2(-1) + (1 - x)^3(2x)$
    $$= (1 - x)^2[-3x^2 + 2x(1 - x)]$$
    $$= x(1 - x)^2(-5x + 2) \quad \text{Eq. (22-15)}$$
    $$dy = x(1 - x)^2(-5x + 2) \ dx \quad \text{Eq. (24-1)}$$

13. $y = 7x^2 + 4x; \ x = 4; \ \Delta x = 0.2$
    $y + \Delta y = 7(x + \Delta x)^2 + 4(x + \Delta x)$
    $y + \Delta y - y = 7(x + \Delta x)^2 + 4(x + \Delta x) - (7x^2 + 4x) \quad \text{Eq. (22-4)}$
    $\Delta y = 7(x^2 + 2x\Delta x + \Delta x^2) + 4x + 4\Delta x - 7x^2 - 4x$
    $\Delta y = 7x^2 + 14x\Delta x + 7\Delta x^2 + 4x + 4\Delta x - 7x^2 - 4x$
    $\quad = 14x\Delta x + 7\Delta x^2 + 4\Delta x$

    Let $x = 4; \ \Delta x = 0.2$
    $\Delta y = 14(4)(0.2) + 7(0.2)^2 + 4(0.2) = 11.2 + 0.28 + 0.8 = 12.28$

    $\frac{dy}{dx} = 14x + 4; \ dy = (14x + 4) \ dx \quad \text{Eq. (22-9), (24-1)}$
    Let $x = 4, \ \Delta x = 0.2; \ dy = [14(4) + 4](0.2) = 12$

305

17. $y + \Delta y = [1 - 3(x + \Delta x)]^5$; $y + \Delta y - y = [1 - 3(x + \Delta x]^5 - (1 - 3x)^5$

$\Delta y = [1 - 3(x + \Delta x)]^5 - (1 - 3x)^5$ Eq. (22-5)

Let $x = 1$; $\Delta x = 0.01$

$\Delta y = [1 - 3(1 + 0.01)]^5 - [1 - 3(1)]^5$
$= (-2.03)^5 - (-2)^5 = -34.473 + 32 = -2.473$

$\dfrac{dy}{dx} = 5(1 - 3x)^4(-3) = -15(1 - 3x)^4$

$dy = -15(1 - 3x)^4 \, dx$ Eq. (22-9)

For $x = 1$, $dx = 0.01$

$dy = -15[1 - 3(1)]^4(0.01) = -15(16)(0.01) = -2.4$ Eq. (24-1)

21. $A = s^2$; $dA = 2s\,ds$

$dA = 2(0.95)(0.002) = 0.0038$

25. $V = \frac{1}{3}\pi r^2 h$ $dV = \frac{2}{3}(8.20)\pi\,rdr$

$= \frac{1}{3}(8.20)\pi r^2$ $= \frac{2}{3}(8.20)\pi(6.50)(0.25)$

$= 28$ ft

25M. $V = \frac{1}{3}\pi r^2 h$ $dV = \frac{2}{3}(2.50)\pi\,rdr$

$= \frac{1}{3}(2.50)\pi r^2$ $= \frac{2}{3}(2.50)\pi(2.00)(0.10)$

$= 1.05$ m

29. $A = s^2$; $dA = 2s\,ds = 2s(0.02) = 0.04s$
$= 4\% \; s$

Exercises 24-2, page 738

1. $3x^2$; the power of x required in the antiderivative is 3. Therefore, we must multiply by $\frac{1}{3}$. The antiderivative of $3x^2$ is $\frac{1}{3}(3x^3) = x^3$.
Checking: The derivative of x^3 is $3x^2$.

5. The power of x required in the antiderivative of $6x^3$ is 4. Therefore, we multiply by $\frac{1}{4}$. The antiderivative of $6x^3$ is $\frac{1}{4}(6x^4) = \frac{3}{2}x^4$. The power of x required in the antiderivative of $1x^0$ is 1. Therefore, we multiply by $\frac{1}{1}$. The antiderivative of $1x^0$ is $\frac{1}{1}(1x^1) = x$. The antiderivative of $6x^3 + 1$ is $\frac{3}{2}x^4 + x$.
Checking: The derivative of $\frac{3}{2}x^4 + x$ is $\frac{3}{2}(4x^3) + 1 = 6x^3 + 1$.

9. The power of x required in the antiderivative of $\frac{5}{2}x^{3/2}$ is $\frac{5}{2}$. Multiply by $\frac{2}{5}$. The antiderivative of $\frac{5}{2}x^{3/2}$ is $\frac{2}{5}(\frac{5}{2})x^{5/2} = x^{5/2}$.

13. $-\frac{1}{x^2} = -x^{-2}$; the power of x required in the antiderivative is $-2 + 1 = -1$. We multiply by -1. The antiderivative of $-x^{-2}$ is $(-1)(-x)^{-1} = \frac{1}{x}$. Checking: The derivative of $\frac{1}{x}$ is $-\frac{1}{x^2}$.

17. The power of x required for the antiderivative of $2x^4$ is $4 + 1 = 5$; multiply by $\frac{1}{5}$. The power of x required for the antiderivative of 1, or x^0, is $0 + 1 = 1$. The antiderivative of $2x^4 + 1$ is $\frac{1}{5}(2x^{4+1}) + x^1$ $= \frac{2}{5}x^5 + x$.

21. The power of x required for the antiderivative of x^2 is 3, so it will be multiplied by $\frac{1}{3}$; the power of x required for $2 = 2x^0$ is 1, and it will be multiplied by 1. The power of x required for x^{-2} is -1, and it will be multiplied by -1. The antiderivative of $x^2 + 2 + x^{-2}$ is $\frac{1}{3}x^3 + 2x - \frac{1}{x}$.

25. The antiderivative requires $(x^2 - 1)^4$. We multiply by $\frac{1}{4}$. Thus we have $\frac{1}{4}[4(x^2 - 1)^4]$. The derivative of $(x^2 - 1)^4$ is $4(x^2 - 1)^3 (2x)$. This is the correct antiderivative.

29. The antiderivative requires $(6x + 1)^{3/2}$. We multiply by $\frac{2}{3}$. Thus we have $\frac{2}{3}(\frac{3}{2})(6x + 1)^{3/2} = (6x + 1)^{3/2}$. The derivative of $(6x + 1)^{3/2}$ is $\frac{3}{2}(6x + 1)^{1/2} (6)$. This is the correct antiderivative.

Exercises 24-3, page 744

1. $\int 2x \, dx = 2\int x \, dx$ Eq. (24-3) $u = x, \; du = dx; \; n = 1$
   $2\int x \, dx = 2(\frac{x^{1+1}}{1 + 1}) + C = x^2 + C$ Eq. (24-5)

5. $\int x^{3/2} \, dx; \; u = x, \; du = dx; \; n = \frac{3}{2}$
   $\int x^{3/2} \, dx = \frac{x^{3/2+1}}{\frac{3}{2} + 1} + C = \frac{x^{5/2}}{\frac{5}{2}} + C = \frac{2}{5}x^{5/2} + C$ Eq. (24-5)

9. $\int(x^2 - x^5)\ dx = \int x^2\ dx - \int x^5\ dx$

$= \dfrac{x^3}{3} - \dfrac{x^6}{6} + C = \dfrac{1}{3}x^3 - \dfrac{1}{6}x^6 + C$ Eq. (24-4), (24-5)

13. $\int(\dfrac{1}{x^3} + \dfrac{1}{2})\ dx = \int x^{-3}\ dx + \dfrac{1}{2}\int\ dx$

$= \dfrac{1}{-2}x^{-2} + \dfrac{1}{2}x + C$

$= -\dfrac{1}{2x^2} + \dfrac{1}{2}x + C$

17. $\int(2x^{-2/3} + 3^{-2})\ dx = \int 2x^{-2/3} + \int 3^{-2}x^0\ dx$

$= 2\int x^{-2/3}\ dx + 3^{-2}\int x^0\ dx$

$= 2 + 1/3\,(3x^{1/3}) + 3^{-2}(x^1)$

$= 6x^{1/3} + \dfrac{1}{9}x + C$

21. $\int(x^2 - 1)^5(2x\ dx);\ u = x^2 - 1;\ du = 2x\ dx;\ n = 5$

$\int(x^2 - 1)^5(2x\ dx) = \dfrac{(x^2 - 1)^6}{6} + C$

$= \dfrac{1}{6}(x^2 - 1)^6 + C$ Eq. (24-5)

25. $\int(x^5 + 4)^7 x^4\ dx;\ u = x^5 + 4;\ du = 5x^4;\ n = 7$

$\int(x^5 + 4)^7 x^4\ dx = \dfrac{1}{5}\int(x^5 + 4)^7(5x^4\ dx)$

$= \dfrac{1}{5} \times \dfrac{(x^5 + 4)^8}{8} + C = \dfrac{1}{40}(x^5 + 4)^8 + C$

29. $\int\dfrac{x\ dx}{\sqrt{6x^2 + 1}} = \int(6x^2 + 1)^{-1/2}x\ dx$

$u = 6x^2 + 1;\ du = 12x;\ n = -\dfrac{1}{2}$

$\int(6x^2 + 1)^{-1/2}x\ dx = \dfrac{1}{12}\int(6x^2 + 1)^{-1/2}(12x\ dx)$

$= \dfrac{1}{12}\,\dfrac{(6x^2 + 1)^{1/2}}{\dfrac{1}{2}} + C = \dfrac{1}{6}\sqrt{6x^2 + 1} + C$

33. $\dfrac{dy}{dx} = 6x^2;\ dy = 6x^2\ dx$

$y = \int 6x^2\ dx = 6\int x^2\ dx = \dfrac{6x^3}{3} + C = 2x^3 + C$

The curve passes through $(0,2)$. $2 = 2(0^3) + C,\ C = 2;\ y = 2x^3 + 2$

37. slope: $\dfrac{dy}{dx} = -x\sqrt{1 - 4x^2}$ Eq. (22-4); $dy = -x\sqrt{1 - 4x^2}\, dx$

$y = \int -x\sqrt{1 - 4x^2}\, dx = \int (1 - 4x^2)^{1/2}(-x\, dx)$

$u = 1 - 4x^2;\ du = -8x;\ n = \dfrac{1}{2}$

$y = \dfrac{1}{8}\int (1 - 4x^2)^{1/2}(-8x\, dx) = \dfrac{1}{8}\,\dfrac{(1 - 4x^2)^{3/2}}{\dfrac{3}{2}} + C$

$\quad = \dfrac{1}{12}(1 - 4x^2)^{3/2} + C$

The curve passes through (0,7).

$7 = \dfrac{1}{12}[1 - 4(0^2)]^{3/2} + C$

$7 = \dfrac{1}{12}(1)^{3/2} + C;\ 7 = \dfrac{1}{12} + C;\ C = \dfrac{83}{12}$

$y = \dfrac{1}{12}(1 - 4x^2)^{3/2} + \dfrac{83}{12};\ 12y = 83 + (1 - 4x^2)^{3/2}$

41. $\dfrac{df}{dA} = \dfrac{0.005}{\sqrt{0.01A + 1}} = 0.005(0.01A + 1)^{-1/2}$

$f(A) = \int 0.005(0.01A + 1)^{-1/2}\, dA + C$

$\quad = \dfrac{1}{2}\int (0.01A + 1)^{-1/2}0.01\, dA = (0.01A + 1)^{1/2} + C$

$f(0) = 0 = (0.01A + 1)^{1/2} + C$

$C = -[0.01(0) + 1]^{1/2} = -1$

$f(A) = (0.01A + 1)^{1/2} - 1 = \sqrt{0.01A + 1} - 1$

Exercises 24-4, page 750

1.

x	y
1	3
2	6
3	9

$n = 3$
$\Delta x = 1$

$y = 3x$ between $x = 0$ and $x = 3$

x	y
0	0
0.3	0.9
0.6	1.8
0.9	2.7
1.2	3.6
1.5	4.5
1.8	5.4
2.1	6.3
2.4	7.2
2.7	8.1
3.0	9.0

$n = 10$
$\Delta x = 0.3$

$A = 1(0 + 3 + 6) = 9$ (first rectangle has 0 height)
$A = 0.3(0 + 0.9 + 1.8 + 2.7 + 3.6 + 4.5 + 5.4 + 6.3 + 7.2 + 8.1)$
$\quad = 0.3(40.5) = 12.15$
(See page 745.)

5. $y = 4x - x^2$ between $x = 1$ and $x = 4$

x	y
1.0	3.00
1.5	3.75
2.0	4.00
2.5	3.75
3.0	3.00
3.5	1.75
4.0	0.00

$n = 6$
$\Delta x = 0.5$

$A = 0.5(3.00 + 3.75 + 4.00 + 3.75$
$\qquad\qquad + 3.00 + 1.75 + 0.00)$

$\quad = 7.625$

x	y
1.0	3.00
1.3	3.51
1.6	3.84
1.9	3.99
2.2	3.96
2.5	3.75
2.8	3.36
3.1	2.79
3.4	2.04
3.7	1.11
4.0	0.00

$n = 10$
$\Delta x = 0.3$

$A = 0.3(3.00 + 3.51 + 3.84$
$\qquad\qquad + 3.99 + 3.96 + 3.75$
$\qquad\qquad + 3.36 + 2.79 + 2.04$
$\qquad\qquad + 1.11)$

$\quad = 8.208$

9. $F(x) = \int 3x\ dx = \dfrac{3x^2}{2} + C$

$A_{0,3} = F(3) - F(0) = [\dfrac{3(3^2)}{2} + C] - [\dfrac{3(0^2)}{2} + C] = \dfrac{27}{2} = 13.5$

13. $F(x) = \int (4x - x^2)\ dx = \int 4x\ dx - \int x^2\ dx$

$\qquad = \dfrac{4x^2}{2} - \dfrac{x^3}{3} + C = 2x^2 - \dfrac{x^3}{3} + C$

$A_{1,4} = [\dfrac{4(4^2)}{2} - \dfrac{4^3}{3} + C] - [\dfrac{4(1^2)}{2} - \dfrac{1^3}{3} + C]$

$\qquad = [32 - \dfrac{64}{3} + C] - [2 - \dfrac{1}{3} + C] = \dfrac{32}{3} - \dfrac{5}{3} = \dfrac{27}{3} = 9$

Exercises 24-5, page 753

1. $\displaystyle\int_0^1 2x\ dx = \dfrac{2x^2}{2}\bigg|_0^1 = x^2\bigg|_0^1 = 1^2 - 0^2 = 1$ Eq. (24-11)

5. $\displaystyle\int_3^6 (\frac{1}{\sqrt{x}} + 2)\ dx = \int_3^6 (\frac{1}{\sqrt{x}})\ dx + \int_3^6 2\ dx$

$\displaystyle = \int_3^6 x^{-1/2}\ dx + \int_3^6 2x^0\ dx$

$\displaystyle = 2x^{1/2}\Big|_3^6 + 2x\Big|_3^6$

$= [2(6)^{1/2} - 2(3)^{1/2}] + [2(6) - 2(3)]$

$= 6 + 2\sqrt{6} - 2\sqrt{3}$

9. $\displaystyle\int_0^3 (x^4 - x^3 + x^2)\ dx = (\frac{1}{5}x^5 - \frac{1}{4}x^4 - \frac{1}{3}x^3)\Big|_0^3$

$= [\frac{1}{5}(3)^5 - \frac{1}{4}(3)^4 + \frac{1}{3}(3)^3] - [\frac{1}{5}(0)^5 - \frac{1}{4}(0)^4 + \frac{1}{3}(0)^3]$

$= \dfrac{243}{5} - \dfrac{81}{4} + 9 - 0$

$= \dfrac{747}{20}$

13. $\displaystyle\int_0^4 (1 - \sqrt{x})^2\ dx = \int_0^4 (1 - 2\sqrt{x} + x)\ dx$

$\displaystyle = \int_0^4 1\ dx - \int_0^4 2x^{1/2}\ dx + \int_0^4 x\ dx$

$= x - \dfrac{2x^{3/2}}{\frac{3}{2}} + \dfrac{x^2}{2}\Big|_0^4 = x - \dfrac{4}{3}x^{3/2} + \dfrac{x^2}{2}\Big|_0^4$

$= [4 - \frac{4}{3}(4^{3/2}) + \frac{4^2}{2}] - 0 = 4 - \frac{4}{3}(8) + 8 - 0$

$= \dfrac{12}{3} - \dfrac{32}{3} + \dfrac{24}{3} = \dfrac{4}{3}$

17. $\displaystyle\int_0^4 \dfrac{x\ dx}{\sqrt{x^2 + 9}} = \int_0^4 (x^2 + 9)^{-1/2}\ x\ dx = \frac{1}{2}\int_0^4 (x^2 + 9)^{-1/2}\ 2x\ dx$

$= \dfrac{1}{2} \times \dfrac{(x^2 + 9)^{1/2}}{\frac{1}{2}}\Big|_0^4 = (x^2 + 9)^{1/2}\Big|_0^4$

$= (4^2 + 9)^{1/2} - (0^2 + 9)^{1/2}$

$= 25^{1/2} - 9^{1/2} = 5 - 3 = 2$

21. If $u = (2x^2 + 1)$, $n = 3$, and $du = 4x \; dx$, then

$$\int_1^3 \frac{2x \; dx}{(2x^2 + 1)^3} = \frac{1}{2}\int_1^3 \frac{4x \; dx}{(2x^2 + 1)^3} = \frac{1}{2}\int_1^3 \frac{du \; dx}{u^n} = \frac{1}{2}\int_1^3 u^{-n} \; du$$

$$= \frac{1}{2}[\frac{1}{1-n}(u)^{-n+1}]\Big|_1^3 = \frac{1}{2}[-\frac{1}{2}(2x^2 + 1)^{-2}]\Big|_1^3$$

$$= \frac{1}{2}[-\frac{1}{2}(2(3)^2 + 1)^{-2}] - \frac{1}{2}[-\frac{1}{2}(2(1)^2 + 1)^{-2}]$$

$$= -\frac{1}{4}(\frac{1}{19^2}) + \frac{1}{4}(\frac{1}{3^2}) = -\frac{1}{1444} + \frac{1}{36} = 0.0271$$

25. $\int_0^2 2x(9 - 2x^2)^2 \; dx$; $u = (9 - 2x^2)$; $du = -4x$; $n = 2$

$$-\frac{1}{2}\int_0^2 (9 - 2x^2)^2(-2)(2x \; dx) = -\frac{1}{2}\int_0^2 (9 - 2x^2)^2(-4x \; dx)$$

$$= -\frac{1}{2}\frac{(9 - 2x^2)^3}{3}\Big|_0^2 = -\frac{1}{6}(9 - 2x^2)^3\Big|_0^2$$

$$= -\frac{1}{6}[9 - 2(2)^2]^3 - (-\frac{1}{6})[9 - 2(0)^2]^3$$

$$= -\frac{1}{6}(1)^3 - (-\frac{1}{6})(9)^3 = -\frac{1}{6}(1) - (-\frac{1}{6})(729)$$

$$= -\frac{1}{6} + \frac{729}{6} = \frac{728}{6} = \frac{364}{3}$$

29. $\int_{-1}^2 \frac{8x - 2}{(2x^2 - x + 1)}dx = \int_{-1}^2 (8x - 2)[2x^2 - x + 1]^{-1} = 2\int_{-1}^2 (4x - 1)[2x^2 - x + 1]^{-3}$

$$= \frac{2(2x^2 - x + 1)^{-2}}{-2}\Big|_{-1}^2 = -(2x^2 - x + 1)^{-2}\Big|_{-1}^2 = -\frac{1}{7^2} - \left(-\frac{1}{4^2}\right) = 0.0421$$

33. $W = \int_0^{80} (1000 - 5x)\,dx = (1000x - \frac{5}{2}x^2)\Big|_0^{80}$

$$= 1000(80) - \frac{5}{2}(80)^2 - [0 - 0]$$

$$= 80000 - 16000 = 64000$$

<u>Exercises 24-6,</u> page 757

1. $\int_0^2 2x^2\,dx$; $n = 4$; $\Delta x = \dfrac{2 - 0}{4} = \dfrac{1}{2}$; $y_0 = 2(0^2) = 0$

   $y_1 = 2(\tfrac{1}{2})^2 = \dfrac{1}{2}$; $y_2 = 2(1)^2 = 2$; $y_3 = 2(\tfrac{3}{2})^2 = \dfrac{9}{2}$; $y_4 = 2(2)^2 = 8$

   $A_T = [\tfrac{1}{2}(0) + \tfrac{1}{2} + 2 + \tfrac{9}{2} + \tfrac{1}{2}(8)](\tfrac{1}{2}) = 11(\tfrac{1}{2}) = \dfrac{11}{2} = 5.50$

   Eq. (24-13)

   Check: $\displaystyle\int_0^2 2x^2\,dx = \dfrac{2x^3}{3}\Big|_0^2 = \dfrac{16}{3} - 0 = \dfrac{16}{3} = 5.33$

5. $\int_2^3 \dfrac{1}{2x}\,dx$; $n = 2$; $\Delta x = \dfrac{3 - 2}{2} = \dfrac{1}{2}$; $y_0 = \dfrac{1}{2(2)} = \dfrac{1}{4}$

   $y_1 = \dfrac{1}{2(\tfrac{5}{2})} = \dfrac{1}{5}$; $y_2 = \dfrac{1}{2(3)} = \dfrac{1}{6}$

   $A_T = [\tfrac{1}{2}(\tfrac{1}{4}) + \tfrac{1}{5} + \tfrac{1}{2}(\tfrac{1}{6})](\tfrac{1}{2}) = (\dfrac{49}{120})(\tfrac{1}{2}) = \dfrac{49}{240} = 0.2042$

9. $\int_1^5 \dfrac{1}{x^2 + x}\,dx$; $n = 10$; $\Delta x = \dfrac{5 - 1}{10} = 0.4$; $y_0 = \dfrac{1}{1 + 1} = \dfrac{1}{2} = 0.5000$

   $y_1 = \dfrac{1}{1.4^2 + 1.4} = \dfrac{1}{3.36} = 0.2976$; $y_2 = \dfrac{1}{1.8^2 + 1.8} = \dfrac{1}{5.04} = 0.1984$

   $y_3 = \dfrac{1}{2.2^2 + 2.2} = \dfrac{1}{7.04} = 0.1420$; $y_4 = \dfrac{1}{2.6^2 + 2.6} = \dfrac{1}{9.36} = 0.1068$

   $y_5 = \dfrac{1}{3.0^2 + 3.0} = \dfrac{1}{12.00} = 0.0833$; $y_6 = \dfrac{1}{3.4^2 + 3.4} = \dfrac{1}{14.96} = 0.0668$

   $y_7 = \dfrac{1}{3.8^2 + 3.8} = \dfrac{1}{18.24} = 0.0548$; $y_8 = \dfrac{1}{4.2^2 + 4.2} = \dfrac{1}{21.84} = 0.0458$

   $y_9 = \dfrac{1}{4.6^2 + 4.6} = \dfrac{1}{25.76} = 0.0388$; $y_{10} = \dfrac{1}{5.0^2 + 5.0} = \dfrac{1}{30.00} = 0.0333$

   $A_T = [\tfrac{1}{2}(0.5000) + 0.2976 + 0.1984 + 0.1420 + 0.1068 + 0.0833 + 0.0668$
   $+\ 0.0548 + 0.0458 + 0.0388 + \tfrac{1}{2}(0.0333)](0.4) = 0.5205$

13.

x	2	4	6	8	10	12	14
y	0.67	2.34	4.56	3.67	3.56	4.78	6.87

$\Delta x = 2$

$A_T \approx \left[\tfrac{1}{2}(0.67) + 2.34 + 4.56 + 3.67 + 3.56 + 4.78 + \tfrac{1}{2}(6.87) \right]2$

$A_T \approx (22.68)2 = 45.36$

Exercises 24-7, page 760

1. (a) $\displaystyle\int_0^2 (1 + x^3)\, dx$, $n = 2$; $x_0 = 0$, $y_0 = 1$; $\Delta x = h = \dfrac{2 - 0}{n} = 1$

$x_1 = 1$, $y_1 = 2$

$x_3 = 2$, $y_3 = 9$

$\displaystyle\int_0^2 (1 + x^3)\, dx = \frac{1}{3}[1 + 4(2) + 9] = \frac{1}{3}(18) = 6$

(b) $\displaystyle\int_0^2 (1 + x^3)\, dx = x + \frac{1}{4}x^4 \ \Big|_0^2 = 2 + \frac{1}{4}(16) = 6$

5. $\displaystyle\int_2^3 \frac{1}{2x}\, dx$, $n = 2$; $x_0 = 2$, $y_0 = \frac{1}{4}$ $\qquad \Delta x = h = \dfrac{3 - 2}{n} = \dfrac{1}{2}$

$x_1 = \dfrac{5}{2}$, $y_1 = \dfrac{1}{5}$

$x_2 = 3$, $y_2 = \dfrac{1}{6}$

$\displaystyle\int_2^3 \frac{1}{2x}\, dx = \frac{1/2}{3}[\frac{1}{4} + 4(\frac{1}{5}) + \frac{1}{6}]$

$\qquad\qquad = \frac{1}{6}(\frac{1}{4} + \frac{4}{5} + \frac{1}{6}) = \frac{1}{6}(\frac{73}{60}) = \frac{73}{360} = 0.2027$

9. $\displaystyle\int_1^5 \frac{dx}{x^2 + x}$, $n = 10$, $\Delta x = h = \dfrac{5 - 1}{10} = 0.4$

Using values from Exercise 9 of Section 24-6:

$(x_0, y_0) = (1, 0.5000)$ $\qquad (x_5, y_5) = (3.0, 0.0833)$
$(x_1, y_1) = (1.4, 0.2976)$ $\qquad (x_6, y_6) = (3.4, 0.0668)$
$(x_2, y_2) = (1.8, 0.1984)$ $\qquad (x_7, y_7) = (3.8, 0.0548)$
$(x_3, y_3) = (2.2, 0.1420)$ $\qquad (x_8, y_8) = (4.2, 0.0458)$
$(x_4, y_4) = (2.6, 0.1068)$ $\qquad (x_9, y_9) = (4.6, 0.0388)$
$\qquad\qquad\qquad\qquad\qquad\qquad (x_{10}, y_{10}) = (5.0, 0.0333)$

$\displaystyle\int_1^5 \frac{dx}{x^2 + x} = \frac{0.4}{3}[y_0 + 4y_1 + 2y_2 + 4y_3 + 2y_4 + 4y_5 + 2y_6 + 4y_7 + 2y_8$
$\qquad\qquad\qquad\qquad + 4y_9 + y_{10}]$

$\qquad = \frac{0.4}{3}[0.5000 + 1.1904 + 0.3968 + 0.5680 + 0.2136 + 0.3332$
$\qquad\qquad\quad + 0.1336 + 0.2192 + 0.0916 + 0.1552 + 0.0333]$

$\qquad = \frac{0.4}{3}(3.8349)$

$\qquad = 0.5114$

13. $\Delta x = 2$

$\displaystyle\int_2^{14} y\, dx \approx \frac{2}{3}[0.67 + 4(2.34) + 2(4.56) + 4(3.67) + 2(3.56) + 4(4.78)$
$\qquad + 6.87]$

$\qquad\qquad \approx 44.63$

Review Exercises for Chapter 24, page 761

1. $\int (4x^3 - x)\ dx = \int 4x^3\ dx - \int x\ dx = \dfrac{4x^4}{4} - \dfrac{x^2}{2} + C$ Eq. (24-4),(24-5)

$$= x^4 - \tfrac{1}{2}x^2 + C$$

5. $\displaystyle\int_1^4 (\sqrt{x} + \dfrac{1}{\sqrt{x}})\ dx = \int_1^4 x^{1/2}\ dx + \int_1^4 x^{-1/2}\ dx$

$$= \dfrac{x^{3/2}}{\tfrac{3}{2}} + \dfrac{x^{1/2}}{\tfrac{1}{2}}\Big|_1^4 = \tfrac{2}{3}x^{3/2} + 2x^{1/2}\Big|_1^4$$

$$= [\tfrac{2}{3}(4)^{3/2} + 2(4)^{1/2}] - [\tfrac{2}{3}(1)^{3/2} + 2(1)^{1/2}]$$

$$= [\tfrac{2}{3}(8) + 2(2)] - [\tfrac{2}{3}(1) + 2(1)] = \dfrac{16}{3} + \dfrac{12}{3} - \dfrac{2}{3} - \dfrac{6}{3} = \dfrac{20}{3}$$

9. $\int (3 + \dfrac{2}{x^3})\ dx = \int 3\ dx + \int \dfrac{2}{x^3}\ dx$

$$= \int 3\ dx + \int 2x^{-3}\ dx$$

$$= (3x) + (-\tfrac{2}{2}x^{-2})$$

$$= 3x - x^{-2} = 3x - \dfrac{1}{x^2} + C$$

13. $\displaystyle\int \dfrac{dx}{(2 - 5x)^2} = \int (2 - 5x)^{-2}\ dx$

$u = 2 - 5x;\ du = -5\ dx;\ n = -2$

$-\dfrac{1}{5}\int (2 - 5x)^{-2}(-5\ dx) = -\dfrac{1}{5} \times \dfrac{(2 - 5x)^{-1}}{-1} + C = \dfrac{1}{5} \times \dfrac{1}{(2 - 5x)} + C$

$$= \dfrac{1}{5(2 - 5x)} + C \qquad \text{Eq. (24-5)}$$

17. $\displaystyle\int_0^2 \dfrac{3x\ dx}{\sqrt[3]{1 + 2x^2}} = \int_0^2 (1 + 2x^2)^{-1/3}(3x)\ dx$

$u = 1 + 2x^2;\ du = 4x\ dx;\ n = -\dfrac{1}{3}$

$\dfrac{3}{4}\displaystyle\int_0^2 (1 + 2x^2)^{-1/3}(4x)\ dx = \dfrac{3}{4} \times \dfrac{(1 + 2x^2)^{2/3}}{\tfrac{2}{3}}\Big|_0^2$

$$= \tfrac{9}{8}(1 + 2x^2)^{2/3}\Big|_0^2$$

$$= \tfrac{9}{8}[1 + 2(2^2)]^{2/3} - \tfrac{9}{8}[1 + 2(0)^2]^{2/3}$$

$$= \tfrac{9}{8}(9)^{2/3} - \tfrac{9}{8}(1)^{2/3}$$

$$= \tfrac{9}{8}(\sqrt[3]{81} - 1) = \tfrac{9}{8}(3\sqrt[3]{3} - 1) \qquad \text{Eq. (24-5), (24-10)}$$

21. $\int \dfrac{(2 - 3x^2)\ dx}{(2x - x^3)^2} = \int (2x - x^3)^{-2}(2 - 3x^2)\ dx;\ u = 2x - x^3;\ du = 2 - 3x^2;$

$$n = -2$$

$$\int (2x - x^3)^{-2}(2 - 3x^2)\ dx = \dfrac{(2x - x^3)^{-1}}{-1} + C$$

$$= -\dfrac{1}{(2x - x^3)} + C \quad \text{Eq. (24-5)}$$

25. $y = \dfrac{1}{(x^2 - 1)^3} = (x^2 - 1)^{-3}$

$\dfrac{dy}{dx} = -3(x^2 - 1)^{-4}(2x) = -6x(x^2 - 1)^{-4} = \dfrac{-6x\ dx}{(x^2 - 1)^4}$

$dy = \dfrac{-6x\ dx}{(x^2 - 1)^4} \quad \text{Eq. (22-9), (24-1)}$

29. $y = x^3;\ x = 2;\ \Delta x = 0.1;\ y + \Delta y = (x + \Delta x)^3$

$\Delta y = (x + \Delta x)^3 - y = x^3 + 3x^2\Delta x + 3x(\Delta x)^2 + (\Delta x)^3 - x^3$

$\quad = 3x^2\Delta x + 3x(\Delta x)^2 + (\Delta x)^3$

for $x = 2,\ \Delta x = 0.1,\ \Delta y = 3(2)^2(0.1) + 3(2)(0.1)^2 + (0.1)^3$

$\quad = 1.2 + 0.06 + 0.001 = 1.261$

$\dfrac{dy}{dx} = \dfrac{dx^3}{dx} = 3x^2;\ dy = 3x^2\ dx$

for $x = 2,\ \Delta x = 0.1\ dy = 3(2)^2(0.1) = 1.2$

$\Delta y - dy = 1.261 - 1.2 = 0.061$

33a. $\int (1 - 2x)\ dx = x - x^2 + C_1$

33b. $\int (1 - 2x)\ dx = -\dfrac{1}{2}\int (1 - 2x)(-2)\ dx = -\dfrac{1}{2}[\dfrac{(1 - 2x)^2}{2}] + C_2$

$\quad = -\dfrac{1}{2}[\dfrac{1 - 4x + 4x^2}{2}] + C_2 = -\dfrac{1}{2}[\dfrac{1}{2} - 2x + 2x^2] + C_2$

$\quad = -\dfrac{1}{4} + x - x^2 + C_2 = x - x^2 - \dfrac{1}{4} + C_2$

$\quad C_1 = -\dfrac{1}{4} + C_2$

37. $\int_1^3 \dfrac{dx}{2x - 1},\ n = 4;\ \Delta x = \dfrac{3 - 1}{n} = \dfrac{2}{4} = 0.5$

$x_0 = 1.0,\ y_0 = \dfrac{1}{2(1.0) - 1} = 1.0$

$x_1 = 1.5,\ y_1 = \dfrac{1}{2(1.5) - 1} = 0.50$

$x_2 = 2.0,\ y_2 = \dfrac{1}{2(2.0) - 1} = 0.33$

$x_3 = 2.5,\ y_3 = \dfrac{1}{2(2.5) - 1} = 0.25$

$x_4 = 3.0,\ y_4 = \dfrac{1}{2(3.0) - 1} = 0.20$

$\Delta_T = \int_1^3 \dfrac{dx}{2x - 1} = [\dfrac{1}{2}(1.0) + 0.50 + 0.33 + 0.25 + \dfrac{1}{2}(0.20)]0.5$

$\quad = [1.68]0.5 = 0.842$

41.

x	y
0	0 = 0.000
1	$\frac{1}{3}$ = 0.333
2	$\frac{1}{3}$ = 0.333
3	$\frac{3}{11}$ = 0.273
4	$\frac{2}{9}$ = 0.222
5	$\frac{5}{27}$ = 0.185

$y = \dfrac{x}{x^2 + 2}$

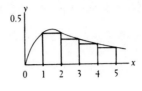

$A = 1(0.000 + 0.333 + 0.273 + 0.222 + 0.185) = 1.01$
(See page 745.)

45. $v = \dfrac{4}{3}\pi r^3; \quad \dfrac{dv}{dr} = 4\pi r^2; \quad dv = 4\pi r^2 dr$

For $r = 3.500$ m, $dr = 0.012$ m

$dV = 4\pi (3.500)^2 (0.012) = 1.85$ m^3

49. $\dfrac{dy}{dx} = k(2L^3 - 12Lx + 2x^4)$

$dy = k(2L^3 x^0 - 12Lx + 2x^4)\,dx$

$y = \displaystyle\int k(2L^3 x^0 - 12Lx + 2x^4)\,dx$

$= k \displaystyle\int (2L^3 x^0 - 12Lx + 2x^4)\,dx$

$= k\left(2L^3 x^0 - \dfrac{12Lx}{2} + \dfrac{2x^5}{5}\right) + C$

$y = 0$ for $x = 0; \quad 0 = k(0 - 0 + 0) + C$

$C = 0; \quad y = k\left(2L^3 x - 6Lx^2 + \dfrac{2}{5}x^5\right)$

CHAPTER 25

Applications of Integration

1. $v = \int a\ dt = \int_0^{2.5} (-32)\ dt = -32t\ \Big|_0^{2.5} = 80$

1M. $v = \int a\ dt = \int_0^{2.5} (-9.8)\ dt = -9.8t\ \Big|_0^{2.5} = 24.5$

5. $a = 90(1 - 4t)^{1/2};\ v = \int 90(1 - 4t)^{1/2} dt = 90 \int (1 - 4t)^{1/2} dt$

$= -\dfrac{1}{4}(90) \int (1 - 4t)^{1/2}(-4dt) = \dfrac{-45}{2}\ \dfrac{(1 - 4t)^{3/2}}{\dfrac{3}{2}} + C$

$= -15(1 - 4t)^{3/2} + C;\ v = 0\ \text{for}\ t = 0;$

$0 = -15(1 - 0)^{3/2} + C;\ 0 = -15 + C;\ C = 15$

$v = -15(1 - 4t)^{3/2} + 15;\ \text{for}\ t = 0.25\ \text{s},$

$v = -15(1 - 1)^{3/2} + 15;\ v = 15\ \text{ft/s}$

5M. $a = 30(1 - 4t)^{1/2};\ v = \int 30(1 - 4t)^{1/2} dt = 30 \int (1 - 4t)^{1/2} dt$

$= -\dfrac{1}{4}(30) \int (1 - 4t)^{1/2}(-4dt) = \dfrac{-15}{2}\ \dfrac{(1 - 4t)^{3/2}}{\dfrac{3}{2}} + C$

$= -5(1 - 4t)^{3/2} + C;\ v = 0\ \text{for}\ t = 0;$

$0 = -5(1 - 0)^{3/2} + C;\ 0 = -5 + C;\ C = 5$

$v = -5(1 - 4t)^{3/2} + 5;\ \text{for}\ t = 0.25\ \text{s},$

$v = -5(1 - 1)^{3/2} + 5;\ v = 5\ \text{m/s}$

9. $s = \int 32t \quad = 90 \qquad v = 32t = 32(2.37) = 76 \ \text{ft/s}$

$\quad = 16t^2 \quad = 90$

$\quad t = 2.37s$

9M. $s = \int 9.8t = 30 \qquad v = 9.8t = 9.8(6.12) = 60 \ \text{m/s}$

$\quad = \dfrac{9.8t^2}{2} = 30$

$\quad t = 6.12$

13. $q = \int i \ dt = i \int dt = it = 2.3 \times 10^{-7} \left(1.5 \times 10^{-3}\right) = 3.45 \times 10^{-10}$

$\quad = 0.345 \ nC$

17. $V_c = \dfrac{1}{c} \int i \ dt = \dfrac{1}{3.0 \times 10^{-6}} \int 0.20 \ dt = \dfrac{1}{3.0 \times 10^{-6}} (0.20)t$

$\quad = \dfrac{1}{3.0 \times 10^{-6}} (0.20)(1.0 \times 10^{-2})$

$\quad = \dfrac{2 \times 10^3}{3} = 667$

21. $\omega = \dfrac{d\theta}{dt} = 0.40t; \ d\theta = 0.40t \ dt; \ \theta = \int 0.40t \ dt = 0.2 + C$

$\quad 0.10 = 0.20(0)^2 + C; \ C = 0.10; \ \theta = 0.20t^2 + 0.10$

$\quad \theta = 0.20(1.50)^2 + 0.10 = 0.55$

25. $\dfrac{dV}{dx} = \dfrac{-k}{x^2}; \ V = - \int \dfrac{k}{x^2} \ dx = kx^{-1} + C = \dfrac{k}{x} + C$

$\quad \lim\limits_{V \to 0} V = \lim\limits_{x \to \infty} \dfrac{k}{x} + C; \ 0 = 0 + C; \ C = 0; \ \text{therefore } V \Big|_{x = x_1} = \dfrac{k}{x_1}.$

Exercises 25-2, page 775

1. $y = 4x; \ y = 0, \ x = 1$
Using vertical elements,

$\quad A = \int_0^1 y \ dx = \int_0^1 4x \ dx = 2x^2 \Big|_0^1$

$\quad = 2(1)^2 - 2(0) = 2.$

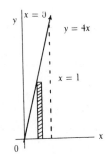

5. $y = 6 - 4x$, $x = 0$, $y = 0$, $y = 3$

$y - 6 = -4x$, $x = -\frac{1}{4}y + \frac{3}{2}$

$$A = \int_0^3 x \, dy = \int_0^3 \left(-\frac{1}{4}y + \frac{3}{2}\right) dy$$

$$= -\frac{1}{8}y^2 + \frac{3}{2}y \ \Big|_0^3$$

$$= -\frac{1}{8}(3)^2 + \frac{3}{2}(3) + \frac{1}{8}(0)^2 - \frac{3}{2}(0)$$

$$= -\frac{9}{8} + \frac{9}{2} = -\frac{9}{8} + \frac{36}{8} = \frac{27}{8}$$

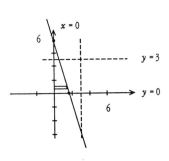

9. $y = 2 - x$; $y = 0$, $x = 1$
    Using vertical elements,

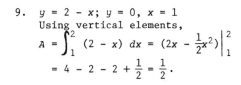

$$= 4 - 2 - 2 + \frac{1}{2} = \frac{1}{2}.$$

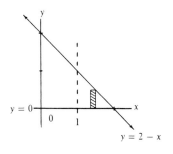

13. $y = \sqrt{x}$, $x = 0$, $y = 1$, $y = 3$

$y^2 = x$

$$A = \int_1^3 x \, dy = \int_1^3 y^2 dy = \frac{1}{3}y^3 \ \Big|_1^3$$

$$= \frac{1}{3}(3)^3 - \frac{1}{3}(1)^3$$

$$= \frac{27}{3} - \frac{1}{3} = \frac{26}{3}$$

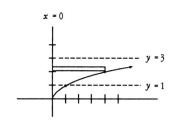

17. $y = 4 - 2x$; $x = 0$, $y = 0$, $y = 3$
$A = \int x \, dy$; $x = -\frac{1}{2}y + 2$
$$A = \int_0^3 \left(-\frac{1}{2}y + 2\right) dy = -\frac{1}{4}y^2 + 2y \ \Big|_0^3$$
$$= -\frac{9}{4} + 6 = \frac{15}{4}$$

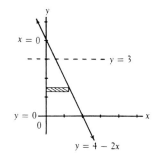

21. $y = x^4$; $y = 16$

$$\frac{1}{2}A = \int_0^{16} x\, dy = \int_0^{16} y^{1/4}\, dy = \frac{4}{5}y^{5/4}\Big|_0^{16}$$

$$= \frac{4}{5}(16)^{5/4} - 0 = \frac{4}{5}(2)^5 = \frac{128}{5}$$

$$A = 2\left(\frac{128}{5}\right) = \frac{256}{5}$$

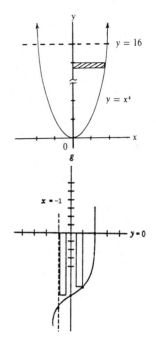

25. $y = x^3 - 8$, $x = -1$, $y = 0$

$$A = -\int_{-1}^{2} y\, dx = -\int_{-1}^{2} (x^3 - 8)\, dx$$

$$= \frac{1}{4}x^4 - 8x \Big|_2^{-1}$$

$$= \frac{1}{4}(-1)^4 - 8(-1) - \left[\frac{1}{4}(2)^4 - 8(2)\right]$$

$$= \frac{1}{4} + 8 - [4 - 16]$$

$$= \frac{1}{4} + 8 + 12 = 81\frac{1}{4}$$

29. $dw = p\, dt$; $p = 12t - 4t^2$

$$w = \int_0^3 (12t - 4t^2)\, dt$$

$$= 6t^2 - \frac{4}{3}t^3 \Big|_0^3$$

$$= 6(3)^2 - \frac{4}{3}(3)^3 - 0$$

$$= 54 - 36 = 18$$

$$w = 18.0 \text{ J}$$

33. $y = x^3 - 2x^2 - x + 2$ and $y = x^2 - 1$

Find the points of intersection:

$$x^3 - 2x^2 - x + 2 = x^2 - 1$$
$$x^3 - 3x^2 - x + 3 = 0$$
$$(x^2 - 1)(x - 3) = 0$$
$$x^2 = 1; \quad x = 3$$
$$x = \pm 1$$

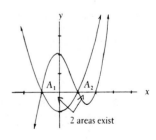

2 areas exist

$$A_1 = \int_{-1}^{1} [(x^3 - 2x^2 - x + 2) - (x^2 - 1)]\, dx$$

$$= \int_{-1}^{1} (x^3 - 3x^2 - x + 3)\, dx$$

$$= \frac{1}{4}x^4 - x^3 - \frac{1}{2}x^2 + 3x \Big|_{-1}^{1}$$

$$= \left(\frac{1}{4} - 1 - \frac{1}{2} + 3\right) - \left(\frac{1}{4} + 1 - \frac{1}{2} - 3\right)$$

$$= 4 \text{ cm}^2$$

Exercises 25-3, page 782

1.

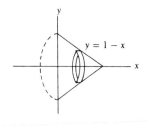

$y = 1 - x$, $x = 0$, $y = 0$
The function has intercepts $(0,1)$ and $(1,0)$.
By Eq. (25-8),

$$V = \pi \int_0^1 y^2 \, dx = \pi \int_0^1 (1 - x)^2 \, dx$$

$$= \pi [-\frac{1}{3}(1 - 1)^3 + \frac{1}{3}(1 - 0)^3]$$

$$= \frac{1}{3}\pi$$

5.

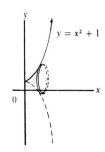

$y = 3\sqrt{x}$, $y = 0$, $x = 4$; $y^2 = 9x$

$$V = \pi \int_0^4 y^2 \, dx = \pi \int_0^4 9x \, dx = \pi (\frac{9}{2}x^2) \Big|_0^4$$

$$= \pi [\frac{9}{2}(4)^2 - 0] = 72\pi$$

7.

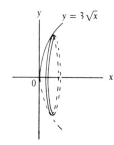

$y = x^2 + 1$, $x = 0$, $x = 3$, $y = 0$
By Eq. (25-8),

$$V = \pi \int_0^3 (x^2 + 1)^2 \, dx$$

$$= \pi \int_0^3 (x^4 + 2x^2 + 1) \, dx = \pi (\frac{1}{5}x^5 + \frac{2}{3}x^3 + x) \Big|_0^3$$

$$= \pi [\frac{1}{5}(3)^5 + \frac{2}{3}(3)^3 + 3 - 0] = \frac{348}{5}\pi$$

13.

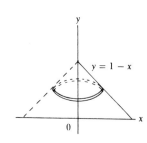

$y = 1 - x$, $x = 0$, $y = 0$
By Eq. (25-9),

$$V = \pi \int_0^1 x^2 \, dy; \quad x = 1 - y$$

$$V = \pi \int_0^1 (1 - y)^2 \, dy = \pi \int_0^1 (1 - 2y + y^2) \, dy$$

$$= \pi (y - y^2 + \frac{1}{3}y^3) \Big|_0^1 = \pi (1 - 1 + \frac{1}{3} - 0) = \frac{1}{3}\pi$$

17.

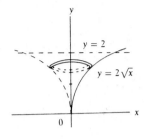

$y = 2\sqrt{x}$, $x = 0$, $y = 2$, $V = \pi \int_0^2 x^2\ dy$; $x = \frac{1}{4}y^2$

$V = \pi \int_0^2 (\frac{1}{4}y^2)^2\ dy = \pi \int_0^2 \frac{1}{16}y^4\ dy$

$= \pi (\frac{1}{80}y^5) \Big|_0^2 = \pi (\frac{32}{80} - 0) = \frac{2}{5}\pi$

21.

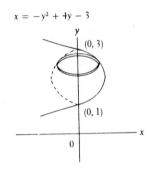

$x = -y^2 + 4y - 3$; $0 = y^2 - 4y + 3$
$\qquad\qquad\qquad = (y - 3)(y - 1)$
$\qquad\qquad y = 3,\ y = 1$

By Eq. (25-9),

$V = \pi \int_1^3 x^2\ dy$

$= \pi \int_1^3 (-y^2 + 4y - 3)^2\ dy$

$= \pi \int_1^3 (y^4 - 8y^3 + 22y^2 - 24y + 9)\ dy$

$= \pi (\frac{1}{5}y^5 - 2y^4 + \frac{22}{3}y^3 - 12y^2 + 9y) \Big|_1^3$

$= \pi [\frac{1}{5}(3)^5 - 2(3)^4 + \frac{22}{3}(3)^3 - 12(9) + 9(3)]$

$\quad -\pi (\frac{1}{5} - 2 + \frac{22}{3} - 12 + 9) = \frac{16}{15}\pi$

25.

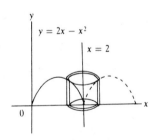

$y = 2x - x^2$, $y = 0$, rotated around $x = 2$,
using shells $r = 2 - x$, $h = y$, $t = dx$
By Eq. (25-10),

$dV = 2\pi(2 - x)y\ dx$; $V = 2\pi \int_0^2 (2 - x)(2 - x^2)\ dx$

$= 2\pi \int_0^2 (4x - 4x^2 + x^3)\ dx$

$= 2\pi [2x^2 - \frac{4}{3}x^3 + \frac{1}{4}x^4] \Big|_0^2$

$= 2\pi [8 - \frac{32}{3} + 4] = \frac{8}{3}\pi$

29. $dV = 2\pi(\text{radius})(\text{height})(\text{thickness})$; $y = x^4 + 1.5$
limits of integration are 0 and 1.1
$r = x$, $h = y$, and $t = dx$

$V = 2\pi \int_0^{1.1} xy\ dx$

$= 2\pi \int_0^{1.1} x(x^4 + 1.5)\ dx = 2\pi \int_0^{1.1} (x^5 + 1.5x)\ dx$

$= 2\pi (\frac{x^6}{6} + \frac{3x^2}{4} \Big|_0^{1.1}) = 2\pi (\frac{1.1^6}{6} + \frac{3(1.1)^2}{4}) = 7.56\ \text{mm}^3$

Exercises 25-4, page 789

1. Using the origin as reference, $m_1 = 5$, $d_1 = 1$; $m_2 = 10$, $d_2 = 4$; $m_3 = 3$, $d_3 = 5$

    By Eq. (25-15), $m_1 d_1 + m_2 d_2 + m_3 d_3 = (m_1 + m_2 + m_3)\bar{d}$

    $5(1) + 10(4) + 3(5) = (5 + 10 + 3)\bar{d}$

    $5 + 40 + 15 = 18\bar{d}$: $60 = 18\bar{d}$

    $\bar{d} = \dfrac{10}{3}$; center of mass is $(\dfrac{10}{3}, 0)$.

5. The area of the left rectangle is $4(2) = 8$. The center is $(-2, 0)$. The area of the right rectangle is $2(4) = 8$. The center is $(1, 1)$; $8(-2) + 8(1) = (8 + 8)\bar{x}$; $-8 = 16\bar{x}$
    $\bar{x} = -\dfrac{1}{2}$; $8(0) + 8(1) = (8 + 8)\bar{y}$; $8 = 16\bar{y}$; $\bar{y} = \dfrac{1}{2}$
    The centroid is $(\bar{x}, \bar{y}) = (-\dfrac{1}{2}, \dfrac{1}{2})$.

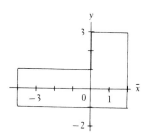

9.

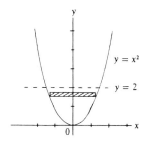

$y = x^2$, $y = 2$
The curve is symmetrical to the y-axis.
Therefore $\bar{x} = 0$.

$$ y = \frac{\int_0^2 y(2x)\,dy}{\int_0^2 2x\,dy} = \frac{\int_0^2 y(2\sqrt{y})\,dy}{\int_0^2 2\sqrt{y}\,dy} = \frac{\int_0^2 2y^{3/2}\,dy}{\int_0^2 2y^{1/2}\,dy} = \frac{\frac{4}{5}y^{5/2}\Big|_0^2}{\frac{4}{3}y^{3/2}\Big|_0^2} $$

$$ = \frac{\frac{4}{5}\sqrt{32}}{\frac{4}{3}\sqrt{8}} = \frac{6}{5}; \text{ the centroid is } (0, \frac{6}{5}). $$

13. $y = x^2$, $y = x^3$

$$\bar{x} = \frac{\int_0^1 x(x^2 - x^3)\, dx}{\int_0^1 (x^2 - x^3)\, dx} = \frac{\int_0^1 (x^3 - x^4)\, dx}{\int_0^1 (x^2 - x^3)\, dx}$$

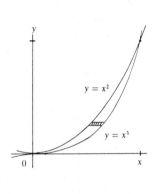

$$= \frac{\frac{1}{4}x^4 - \frac{1}{5}x^5 \Big|_0^1}{\frac{1}{3}x^3 - \frac{1}{4}x^4 \Big|_0^1} = \frac{\frac{1}{4} - \frac{1}{5}}{\frac{1}{3} - \frac{1}{4}} = \frac{\frac{1}{20}}{\frac{1}{12}} = \frac{3}{5}$$

$$\bar{y} = \frac{\int_0^1 y(y^{1/3} - y^{1/2})\, dy}{\int_0^1 (y^{1/3} - y^{1/2})\, dy} = \frac{\int_0^1 (y^{4/3} - y^{3/2})\, dy}{\int_0^1 (y^{1/3} - y^{1/2})\, dy}$$

$$= \frac{\frac{3}{7}y^{7/3} - \frac{2}{5}y^{5/2} \Big|_0^1}{\frac{3}{4}y^{4/3} - \frac{2}{3}y^{3/2} \Big|_0^1} = \frac{\frac{3}{7} - \frac{2}{5}}{\frac{3}{4} - \frac{2}{3}} = \frac{\frac{1}{35}}{\frac{1}{12}} = \frac{12}{35}$$

The centroid is $(\frac{3}{5}, \frac{12}{35})$.

17. $y^2 = 4x$, $x = 1$; $\bar{y} = 0$
By Eq. (25-18),

$$\bar{x} = \frac{\int_0^1 xy^2\, dx}{\int_0^1 y^2\, dx} = \frac{\int_0^1 x(4x)\, dx}{\int_0^1 (4x)\, dx}$$

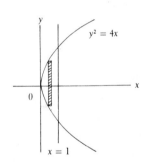

$$= \frac{\frac{4}{3}x^3 \Big|_0^1}{2x^2 \Big|_0^1} = \frac{\frac{4}{3}}{2} = \frac{2}{3}$$

The centroid is $(\frac{2}{3}, 0)$.

21. $\dfrac{x^2}{5^2} + \dfrac{y^2}{1} = 1$; $y^2 = 1 - \dfrac{x^2}{25}$;

$$x^2 = 25 - 25y^2$$

$$\int_0^1 (yx^2)\, dy = \int_0^1 y(25 - 25y^2)\, dy = \int_0^1 (25y - 25y^3)\, dy$$

$$= \left(\frac{25}{2}y^2 - \frac{25}{4}y^4\right)\Big|_0^1 = \frac{25}{2} - \frac{25}{4} = \frac{25}{4}$$

$$\int_0^1 x^2\, dy = \int_0^1 (25 - 25y^2)\, dy = (25y - \frac{25}{3}y^3)\Big|_0^1$$

$$= (25 - \frac{25}{3}) = \frac{50}{3}$$

$$\bar{y} = \frac{25}{4} \div \frac{50}{3} = \frac{25}{4} \cdot \frac{3}{50} = \frac{3}{8} = 0.375$$

1. 5 units at $(2,0)$ has $d_1 = 2$ with respect to the origin, $m_1 = 5$
    3 units at $(6,0)$ has $d_2 = 6$ with respect to the origin, $m_2 = 3$
    By Eq. (25-20), $I = (5)(2)^2 + (3)(6)^2 = 128$; $(5 + 3)R^2 = 128$; $R = 4$

5.

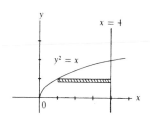

$y^2 = x$, $x = 4$, x-axis, with respect to the x-axis
By Eq. (25-19),

$$I_x = k \int_0^2 y^2 (4 - y^2) \, dy$$

$$= k \int_0^2 (4y^2 - y^4) \, dy = k\left(\frac{4}{3}y^3 - \frac{1}{5}y^5\right)\Big|_0^2$$

$$= k\left(\frac{32}{3} - \frac{32}{5}\right) = k\left(\frac{64}{15}\right)$$

9.

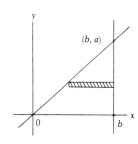

(b, a)

$$y = \frac{a}{b}x, \quad I_x = \frac{ba^3}{12} \text{ for } k = 1; \quad m = \frac{1}{2}ab$$

$$ab = 2m; \quad I_x = ab\frac{a^2}{12} = 2m\frac{a^2}{12} = \frac{1}{6}ma^2$$

13.

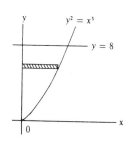

$y^2 = x^3$

$y = 8$

$y^2 = x^3$, $y = 8$, y-axis, with respect to the x-axis

$$I_x = k \int_0^8 y^2 x \, dy = k \int_0^8 y^2 (y^{2/3}) \, dy = k \int_0^8 y^{8/3} \, dy$$

$$= k\frac{3}{11}y^{11/3}\Big|_0^8 = \frac{3}{11}(8)^{11/3}k = \frac{3}{11}(2)^{11}k = \frac{6144}{11}k$$

$$m = k \int_0^8 x \, dy = k \int_0^8 y^{2/3} \, dy = k\left(\frac{3}{5}y^{5/3}\right)\Big|_0^8$$

$$= \frac{3}{5}(8)^{5/3}k = \frac{3}{5}(2)^5 k = \frac{96k}{5}$$

$$R^2 = \frac{I_x}{m} = \frac{6144k}{11} \div \frac{96k}{5} = \frac{64(5)}{11}$$

$$R = \sqrt{\frac{64(5)}{11}} = \frac{8}{11}\sqrt{55}$$

17.

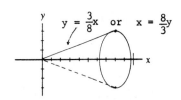

$y = 2x - x^2$, $y = 0$, rotated about y-axis

By Eq. (25-22), $I_y = 2\pi k \int_0^2 (2x - x^2)(x^3)\ dx$

$= 2\pi k \int_0^2 (2x^4 - x^5)\ dx = 2\pi k [\frac{2}{5}x^5 - \frac{1}{6}x^6 \Big|_0^2]$

$= 2\pi k [\frac{2}{5}(2)^5 - \frac{1}{6}(2)^6] = 2\pi k [\frac{64}{5} - \frac{64}{6}] = \frac{64\pi k}{15}$

$m = 2\pi k \int_0^2 (2x - x^2)(x)\ dx = 2\pi k \int_0^2 (2x^2 - x^3)\ dx$

$= 2\pi k [\frac{2}{3}x^3 - \frac{1}{4}x^4 \Big|_0^2] = 2\pi k [\frac{2}{3}(2)^3 - \frac{1}{4}(2)^4]$

$= 2\pi k [\frac{16}{3} - \frac{16}{4}] = \frac{8\pi k}{3}$

$R_y^2 = \frac{I_y}{m} = \frac{64\pi k}{15} \div \frac{8\pi k}{3} = \frac{8}{5}$

$R_y = \sqrt{\frac{8}{5}(\frac{5}{5})} = \frac{2}{5}\sqrt{10}$

21. $I_x = 2\pi k \int_0^{0.3} xy^3\ dy$

$y = \frac{3}{8}x$ or $x = \frac{8}{3}y$

where k = density = $\dfrac{\text{mass}}{\text{volume}}$

$= \dfrac{3}{\frac{1}{3}\pi(0.3)^2(0.8)}$

$I_x = \dfrac{2\pi(3)}{\frac{1}{3}\pi(0.09)(0.8)} \int_0^{0.3} (\frac{8}{3}y)y^3\ dy = \dfrac{18}{0.072} \int_0^{0.3} \frac{8}{3}y^4\ dy$

$= 250(\frac{8}{15}y^5 \Big|_0^{0.3}) = \frac{2000}{15}(0.3^5)$

$= 0.324\ \text{g} \cdot \text{cm}^2$

Exercises 25-6, page 799

1. $f(x) = kx$; $6.0 = k(1.5)$; $k = 4.0$

$w = \int_0^{2.0} 4.0x\ dx = 2.0x^2 \Big|_0^{2.0} = 2.0(2^2) - 0 = 8.0\ \text{lb} \cdot \text{in.}$

5. $f(x) = \dfrac{9.0 \times 10^9\ q_1 q_2}{x^2} = \dfrac{9.0 \times 10^9 \times 1.6 \times 10^{-19} \times 1.3 \times 10^{-18}}{x^2} = \dfrac{1.87 \times 10^{-27}}{x^2}$

$w = \int_{2.0 \times 10^{-6}}^{1.0} (1.87 \times 10^{-27}\ x^{-2})\ dx = -1.9 \times 10^{-27}\ x^{-1} \Big|_{2.0 \times 10^{-6}}^{1.0}$

$= -1.9 \times 10^{-27} - \dfrac{-1.9 \times 10^{-27}}{2.0 \times 10^{-6}}$

$= -1.9 \times 10^{-27} + 0.94 \times 10^{-21}$

$= -1.9 \times 10^{-27} + 9.4 \times 10^{-22} = 0.000019 \times 10^{-22} + 9.4 \times 10^{-22}$

$= 9.4 \times 10^{-22}$ N $\cdot$ m

9. $f(x) = 32.5 - (1.25 \times 10^{-3})x$

$w = \int_{0}^{12000} \left[32.5 - (1.25 \times 10^{-3})x\right] dx = 32.5x - 0.63 \times 10^{-3} \ x^2 \ \Big|_{0}^{12000}$

$= 300\ 000 - 0 = 3.00 \times 10^5$

13. $F = 62.4 \int_{0}^{2.5} (12x)\, dx = 62.4 \ (6x^2 \Big|_{0}^{2.5})$

$= 62.4 \ (37.5 - 0) = 2340$ lb

13M. $F = 9800 \int_{0}^{0.80} (4.0x)\, dx = 9800(2.0x^2 \Big|_{0}^{0.80})$

$= 9800(1.28 - 0) = 12500$

17. $F = 9800 \int_{0}^{1} (xy)(dy);\ y = \frac{1}{2}x;\ x = 2y$

$F = 9800 \int_{0}^{1} (2y)(y)\, dy = 9800 \left[\frac{2}{3} \ y^3 \ \Big|_{0}^{1}\right] = 9800 \left(\frac{2}{3}\right) = 6530$

21. $i = 4t - t^2$.   Find i_{av} with respect to time for $t = 0$ to $t =$

$i_{av} = \dfrac{\int_{0}^{4} i\, dt}{4 - 0} = \dfrac{\int_{0}^{4} (4t - t^2)\, dt}{4} = \dfrac{2t^2 - \frac{1}{3}t^3 \Big|_{0}^{4}}{4}$

$= \dfrac{2(16) - \frac{1}{3}(64)}{4} = \dfrac{10.7}{4} = 2.67$ A

25. $s = \int_{a}^{b} \sqrt{1 + \left(\dfrac{dy}{dx}\right)^2}\ dx;\ y = 0.04x^{3/2};\ \dfrac{dy}{dx} = 0.06x^{1/2}$

$= \int_{0}^{100} \sqrt{1 + (0.06x^{1/2})^2}\ dx = \int_{0}^{100} \sqrt{(1 + 0.0036x)}\ dx = \int_{0}^{100} (1 + 0.0036x)^{1/2}\ dx$

$= \dfrac{1}{0.0036} \int_{0}^{100} (1 + 0.0036x)^{1/2} (0.0036)\ (dx)$

$= \dfrac{1}{0.0036} \left[\dfrac{2}{3} (1 + 0.0036x)^{3/2}\right] \ \Big|_{0}^{100}$

$= \dfrac{1}{0.0036} \left[\dfrac{2}{3} (1.586) - \dfrac{2}{3} (1)^{3/2}\right]$

$= \dfrac{1}{0.0036} \left[\dfrac{2}{3} (0.586)\right] = \dfrac{1}{0.0036} \left[0.391\right] = 109$

Review Exercises for Chapter 25, page 801

1. $v = \int a\ dt;\ a = 32$

$140 = \int_0^t 32dt = 32t\ \Big|_0^t = 32t$

$140 = 32t;\ t = 4.4\ \text{sec}$

1M. $v = \int a\ dt;\ a = 9.8$

$42 = \int_0^t 9.8\ dt = 9.8t\ \Big|_0^t = 9.8t$

$42 = 9.8t;\ t = 4.3\ \text{sec}$

5. $i = 0.25(2\sqrt{t} - t);\ q = \int i\ dt$

$q = \int_0^2 0.25(2\sqrt{t} - t)\ dt = \int_0^2 0.25(2t^{1/2} - t)\ dt$

$= \int_0^2 (0.50t^{1/2} - 0.25t)\ dt = \left(\frac{1}{3}t^{3/2} - \frac{1}{8}t^2\right)\ \Big|_0^2 = \frac{\sqrt{8}}{3} - \frac{1}{2}$

$= 0.44$

9. $dy = (20 + \frac{1}{40}x^2)\ dx;\ y = \int (20 + \frac{1}{40}x^2)\ dx = 20x + \frac{1}{120}x^3 + C$

$y = 0$ when $x = 0$

$20(0) + \frac{1}{120}(0)^3 + C = 0;\ C = 0$

$y = 20x + \frac{1}{120}x^3$

13. $y^2 = 2x,\ y = x - 4$

(8, 4)

(2, −2)

$x_1 = \frac{1}{2}y^2;\ x_2 = y + 4;\ A = \int_{-2}^4 (x_2 - x_1)\ dy$

$= \int_{-2}^4 (y + 4 - \frac{1}{2}y^2)\ dy = -\frac{1}{6}y^3 + \frac{1}{2}y^2 + 4y\ \Big|_{-2}^4$

$= -\frac{1}{6}(64) + \frac{1}{2}(16) + 4(4) + \frac{1}{6}(-8) - \frac{1}{2}(4) - 4(-2)$

$= 18$

17.

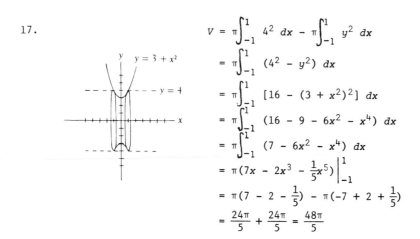

$$V = \pi \int_{-1}^{1} 4^2 \, dx - \pi \int_{-1}^{1} y^2 \, dx$$

$$= \pi \int_{-1}^{1} (4^2 - y^2) \, dx$$

$$= \pi \int_{-1}^{1} [16 - (3 + x^2)^2] \, dx$$

$$= \pi \int_{-1}^{1} (16 - 9 - 6x^2 - x^4) \, dx$$

$$= \pi \int_{-1}^{1} (7 - 6x^2 - x^4) \, dx$$

$$= \pi (7x - 2x^3 - \tfrac{1}{5}x^5) \Big|_{-1}^{1}$$

$$= \pi (7 - 2 - \tfrac{1}{5}) - \pi (-7 + 2 + \tfrac{1}{5})$$

$$= \frac{24\pi}{5} + \frac{24\pi}{5} = \frac{48\pi}{5}$$

21.

$$\frac{x^2}{a^2} - \frac{y^2}{b^2} = 1$$

$$b^2x^2 + a^2y^2 = a^2b^2$$

$$y = \sqrt{\frac{a^2b^2 - b^2x^2}{a^2}}$$

$$V = \pi \int_{-a}^{a} y^2 \, dx = \pi \int_{-a}^{a} \left(\frac{a^2b^2 - b^2x^2}{a^2}\right) dx = \pi \int_{-a}^{a} \left(b^2 - \frac{b^2}{a^2}x^2\right) dx$$

$$= \pi \left(b^2x - \frac{b^2}{3a^2}x^3\right) \Big|_{-a}^{a} = \pi \left[\left(ab^2 - \frac{ab^2}{3}\right) - \left(-ab^2 + \frac{ab^2}{3}\right)\right]$$

$$= \pi \left(2ab^2 - \frac{2ab^2}{3}\right) = \pi \left(\frac{4ab^2}{3}\right) = \frac{4\pi ab^2}{3} = \frac{4}{3}\pi ab^2$$

25. $y = \sqrt{x}$, $x = 1$, $x = 4$, $y = 0$, x-axis; $y = 0$

By Eq. (25-13), $\bar{x} = \dfrac{\int_{1}^{4} xy^2 \, dx}{\int_{1}^{4} y^2 \, dx} = \dfrac{\int_{1}^{4} x(x) \, dx}{\int_{1}^{4} x \, dx}$

$$= \frac{\int_{1}^{4} x^2 \, dx}{\int_{1}^{4} x \, dx} = \frac{\frac{1}{3}x^3 \Big|_{1}^{4}}{\frac{1}{2}x^2 \Big|_{1}^{4}} = \frac{\frac{63}{3}}{\frac{15}{2}} = \frac{14}{5}$$

Thus, $(\bar{x}, \bar{y})$ is $(\frac{14}{5}, 0)$.

29.

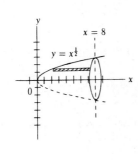

$y = x^{1/2}$, $x = 8$, $y = 0$

$r = y$; $h = 8 - x$; $t = dy$

$$I_x = 2\pi k \int_0^{\sqrt{8}} y^2 [y(8 - x)]\ dy$$

$$= 2\pi k \int_0^{\sqrt{8}} y^2 [y(8 - y^2)]\ dy$$

$$= 2\pi k \int_0^{\sqrt{8}} (8y^3 - y^5)\ dy$$

$$= 2\pi k [2y^4 - \frac{y^6}{6}]\Big|_0^{\sqrt{8}} = 2\pi k [128 - \frac{512}{6}]$$

$$= \frac{256}{3} \pi k$$

33. $y_2 = 3x^2 - x^3$; $y_1 = 0$;

$$\bar{x} = \frac{\int_0^3 x\,(y_2 - y_1)\,dx}{\int_0^3 (y_2 - y_1)\,dx} = \frac{\int_0^3 x\,(3x^2 - x^3)\,dx}{\int_0^3 (3x^2 - x^3)\,dx} = \frac{\int_0^3 (3x^3 - x^4)\,dx}{\int_0^3 (3x^2 - x^3)\,dx}$$

$$= \frac{\left(\frac{3}{4}x^4 - \frac{1}{5}x^5\right)\Big|_0^3}{\left(x^3 - \frac{1}{4}x^4\right)\Big|_0^3} = \frac{\frac{243}{4} - \frac{243}{5}}{27 - \frac{81}{4}} = \frac{12.15}{6.75} = 1.8\ \text{m}$$

37. The circumference of the bottom, $c = 2\pi r = 9\pi$, equates to l, the length of the vertical surface area.

$$F = 68.0 \int_0^{3.25} (9\pi h)\,dh = 68.0 \left[4.5\pi h^2 \Big|_0^{3.25}\right]$$

$$= 68.0 \left[4.5\pi (3.25)^2 - 0\right] = 10200\ \text{lb.}$$

37M. $F = 10.6 \int_0^{3.25} (9\pi h)\,dh = 10.6 \left[4.5\pi h^2 \Big|_0^{3.25}\right]$

$$= 10.6 \left[4.5\pi (3.25)^2 - 0\right] = 1580$$

CHAPTER 26

Differentiation of Transcendental Functions

Exercises 26-1, page 809

1. $y = \sin(x + 2)$; $\dfrac{dy}{dx} = \cos(x + 2)\dfrac{d(x + 2)}{dx} = \cos(x + 2)$

5. $y = 6 \cos\left(\dfrac{1}{2}x\right)$

$\dfrac{dy}{dx} = -6\left[\sin\left(\dfrac{1}{2}x\right)\right]\left[\dfrac{1}{2}\right] = -3 \sin\left(\dfrac{1}{2}x\right)$

9. $y = \sin^2 4x$; $\dfrac{dy}{dx} = 2 \sin 4x\dfrac{d(\sin 4x)}{dx}$

$\qquad = (2 \sin 4x)(\cos 4x)(4) = 8 \sin 4x \cos 4x$

$\qquad = 4 \sin 8x \quad$ Eq. (19-21)

13. $y = x \sin 3x$; $\dfrac{dy}{dx} = x(3 \cos 3x) + (\sin 3x)(1)$

$\qquad = \sin 3x + 3x \cos 3x$

17. $y = \sin x^2 \cos 2x$; $\dfrac{dy}{dx} = \sin x^2(-2 \sin 2x) + \cos 2x(2x \cos x^2)$

$\qquad = 2x \cos x^2 \cos 2x - 2 \sin x^2 \sin 2x$

21. $y = \dfrac{\sin 3x}{x}$; $\dfrac{dy}{dx} = \dfrac{x(\cos 3x)(3) - (\sin 3x)(1)}{x^2} = \dfrac{3x \cos 3x - \sin 3x}{x^2}$

25. $y = 2 \sin^2 3x \cos 2x$

$\dfrac{dy}{dx} = (2 \sin^2 3x)(-\sin 2x)(2) + (\cos 2x)(2)(2 \sin 3x)(\cos 3x)(3)$

$\qquad = -4 \sin^2 3x \sin 2x + 12 \cos 2x \sin 3x \cos 3x$

$\qquad = 4 \sin 3x(3 \cos 3x \cos 2x - \sin 3x \sin 2x)$

29. $y = \sin^3 x - \cos 2x$; $\dfrac{dy}{dx} = 3 \sin^2 x(\cos x) - (-\sin 2x)(2)$

$\qquad = 3 \sin^2 x \cos x + 2 \sin 2x$

33. $\dfrac{\sin 0.5}{0.5} = \dfrac{0.4794255}{0.5} = 0.958851$; $\dfrac{\sin 0.1}{0.1} = \dfrac{0.09983342}{0.1} = 0.9983342$

$\dfrac{\sin 0.05}{0.05} = \dfrac{0.04997917}{0.05} = 0.9995834$; $\dfrac{\sin 0.01}{0.01} = \dfrac{0.009999833}{0.01} = 0.9999833$

$\dfrac{\sin 0.001}{0.001} = \dfrac{0.001}{0.001} = 1$

37. $y = \sin x$

$\dfrac{dy}{dx} = \cos x$

1) $\dfrac{dy}{dx}\bigg|_{x=0} = \cos 0 = 1$

2) $\dfrac{dy}{dx}\bigg|_{x=\frac{\pi}{4}} = \cos \dfrac{\pi}{4} = \dfrac{\sqrt{2}}{2} = 0.7$

3) $\dfrac{dy}{dx}\bigg|_{x=\frac{\pi}{2}} = \cos \dfrac{\pi}{2} = 0$

4) $\dfrac{dy}{dx}\bigg|_{x=\frac{3\pi}{4}} = \cos \dfrac{3\pi}{4} = -\dfrac{\sqrt{2}}{2} = -0.7$

5) $\dfrac{dy}{dx}\bigg|_{x=\pi} = \cos \pi = -1$

6) $\dfrac{dy}{dx}\bigg|_{x=\frac{5\pi}{4}} = \cos \dfrac{5\pi}{4} = -\dfrac{\sqrt{2}}{2} = -0.7$

7) $\dfrac{dy}{dx}\bigg|_{x=\frac{3\pi}{2}} = \cos \dfrac{3\pi}{2} = 0$

8) $\dfrac{dy}{dx}\bigg|_{x=\frac{7\pi}{4}} = \cos \dfrac{7\pi}{4} = \dfrac{\sqrt{2}}{2} = 0.7$

9) $\dfrac{dy}{dx}\bigg|_{x=2\pi} = \cos 2\pi = 1$

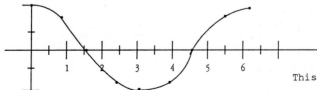

This is a cosine curve.

41. $\dfrac{d}{dx}\sin x = \cos x$

$\dfrac{d^2}{dx^2}\cos x = -\sin x$

$\dfrac{d^3}{dx^3}(-\sin x) = -\cos x$

$\dfrac{d^4}{dx^4}(-\cos x) = \sin x$

45. $y = 3\sin 2x$

$\dfrac{dy}{dx} = 3(\cos 2x)(2) = 6\cos 2x;$

$dy = (6\cos 2x)\,dx = (6\cos\dfrac{\pi}{4})(0.02)$

$= \dfrac{6\sqrt{2}}{2}(0.02)$

$= 0.084$

49. $y = 1.85\sin 36\pi t$

$v = \dfrac{dy}{dx} = (1.85\cos 36\pi t)(36\pi)$

$v\Big|_{t=0.0250} = [1.85\cos(36\pi \cdot 0.025)][36\pi] = 199 \text{ cm/s}$

Exercises 26-2, page 813

1. $y = \tan 5x;\ \dfrac{dy}{dx} = \sec^2 5x\ \dfrac{d5x}{dx} = 5\sec^2 5x$

5. $y = 3\sec 2x;\ \dfrac{dy}{dx} = 3\sec 2x\tan 2x\ \dfrac{d2x}{dx}$
             $= 6\sec 2x\tan 2x$

9. $y = 5\tan^2 3x;\ \dfrac{dy}{dx} = 5(2\tan 3x)(\sec^2 3x)\ \dfrac{d3x}{dx}$
             $= 30\tan 3x\sec^2 3x$

13. $y = \sqrt{\sec 4x};\ \dfrac{dy}{dx} = \dfrac{1}{2}(\sec 4\ x^{-1/2})\sec 4x\tan 4x\ \dfrac{d4x}{dx}$
             $= \dfrac{2\sec 4x\tan 4x}{\sqrt{\sec 4x}} = 2\tan 4x\sqrt{\sec 4x}$

17. $y = x^2\tan x;\ \dfrac{dy}{dx} = x^2\ \dfrac{d\tan x}{dx} + \tan x\ \dfrac{dx^2}{dx}$
             $= x^2\sec^2 x + (\tan x)(2x) = x^2\sec^2 x + 2x\tan x$

21. $y = \dfrac{\csc\ x}{x}$; $\dfrac{dy}{dx} = \dfrac{x(\dfrac{d\ \csc\ x}{dx}) - \csc\ x\ (\dfrac{dx}{dx})}{x^2}$

$= \dfrac{x(-\csc\ x\ \cot\ x) - \csc\ x}{x^2} = -\dfrac{\csc\ x\ (x\ \cot\ x + 1)}{x^2}$

25. $y = \dfrac{1}{3}\tan^3\ x - \tan\ x$; $\dfrac{dy}{dx} = 3(\dfrac{1}{3})\ \tan^2\ x\ \dfrac{d(\tan\ x)}{dx} - \sec^2\ x$

$= \tan^2\ x\ \sec^2\ x - \sec^2\ x = \sec^2\ x(\tan^2\ x - 1)$

29. $y = \sqrt{2x + \tan\ 4x} = (2x + \tan\ 4x)^{1/2}$

$\dfrac{dy}{dx} = \dfrac{1}{2}(2x + \tan\ 4x)^{-1/2}\ (2 + 4\ \sec^2\ 4x)$

$= \dfrac{1 + 2\ \sec^2\ 4x}{\sqrt{2x + \tan\ 4x}}$

33. $y = 4\ \tan^2\ 3x$; $\dfrac{dy}{dx} = 4(2\ \tan\ 3x)(\sec^2\ 3x)(3) = 24\ \tan\ 3x\ \sec^2\ 3x$

$dy = 24\ \tan\ 3x\ \sec^2\ 3x\ dx$

37. (a) $\sec^2\ 1.0000 = \dfrac{1}{\cos^2\ 1.0000} = 3.4255188$. This is the slope of a tangent line to the curve $f(x) = \tan\ x$ at $x = 1.0000$. It is the value of $f'(x) = \sec^2\ x$ at $x = 1.0000$ since $\dfrac{d(\tan\ x)}{dx} = \sec^2\ x$.

(b) $\dfrac{\tan\ 1.0001 - \tan\ 1.0000}{0.0001} = 3.4260524$. This is the slope of a secant line through the curve $f(x) = \tan\ x$ at $x = 1.0000$, where $\Delta x = 0.0001$

$\lim\limits_{\Delta x \to 0}\ \dfrac{\tan\ (x + \Delta x) - \tan\ x}{\Delta x} = \dfrac{d\ \tan\ x}{dx} = \sec^2\ x$

For $\Delta x = 0.0001$, the slope of the tangent line is approximately equal to the slope of the secant line. (Final digits may vary.)

41. $y = 2\ \cot\ 3x$; $\dfrac{dy}{dx} = 2(-\csc^2\ 3x)(3) = -6\ \csc^2\ 3x$

$\dfrac{dy}{dx}\Big|_{x = \frac{\pi}{12}} = -6\ \csc^2\ \dfrac{\pi}{4} = -6(\sqrt{2})^2 = -6(2) = -12$

43. $y = 2\ \tan\ x - \sec\ x$

$\dfrac{dy}{dx} = 2\ \sec^2\ x - (\sec\ x\ \tan\ x)$

$= \dfrac{2}{\cos^2\ x} - \dfrac{1}{\cos\ x} \cdot \dfrac{\sin\ x}{\cos\ x}$

$= \dfrac{2}{\cos^2} - \dfrac{\sin\ x}{\cos^2\ x} = \dfrac{2 - \sin\ x}{\cos^2\ x}$

47. $h = 1000 \tan \left(\dfrac{3t}{2t + 10}\right)$

$\dfrac{dh}{dt} = 1000\left[\sec^2 \left(\dfrac{3t}{2t + 10}\right)\right] \dfrac{(2t + 10)(3) - (3t)(2)}{(2t + 10)^2}$

$= 1000 \left[\sec^2 \left(\dfrac{3t}{2t + 10}\right)\left(\dfrac{30}{(2t + 10)^2}\right)\right]$

$\dfrac{dh}{dt}\bigg|_{t = 5.0} \; 1000\left[\sec^2 \left(\dfrac{15}{20}\right)\right]\left(\dfrac{30}{400}\right) = 1000 \,(1.87)(0.075)$

$= 140$

Exercises 26-3, page 817

1. $y = \text{Arcsin } (x^2);\; \dfrac{dy}{dx} = \dfrac{1}{\sqrt{1 - (x^2)^2}} \dfrac{dx^2}{dx} = \dfrac{2x}{\sqrt{1 - x^4}}$

5. $y = \text{Arccos } \tfrac{1}{2}x;\; \dfrac{dy}{dx} = -\dfrac{1}{\sqrt{1 - \tfrac{1}{4}x^2}} \dfrac{d(\tfrac{1}{2}x)}{dx}$

$= \dfrac{-\tfrac{1}{2}}{\sqrt{1 - \tfrac{1}{4}x^2}} = \dfrac{-1}{2\sqrt{1 - \tfrac{1}{4}x^2}} = -\dfrac{1}{\sqrt{4 - x^2}}$

9. $y = \text{Arctan } \sqrt{x} = \text{Arctan } x^{1/2};\; \dfrac{dy}{dx} = \dfrac{1}{1 + (\sqrt{x})^2} \dfrac{dx^{1/2}}{dx}$

$= \dfrac{1}{1 + x}(\tfrac{1}{2}x^{-1/2}) = \dfrac{1}{2\sqrt{x}(1 + x)}$

13. $y = x \text{ Arcsin } x;\; \dfrac{dy}{dx} = x\,\dfrac{1}{\sqrt{1 - x^2}} + (\text{Arcsin } x)(1) = \dfrac{x}{\sqrt{1 - x^2}} + \text{Arcsin } x$

17. $y = \dfrac{3x}{\text{Arcsin } 2x};\; \dfrac{dy}{dx} = \dfrac{(\text{Arcsin } 2x)(3) - 3x\,\dfrac{1}{\sqrt{1 - 4x^2}}(2)}{(\text{Arcsin } 2x)^2}$

$= (3 \text{ Arcsin } 2x - \dfrac{6x}{\sqrt{1 - 4x^2}}) \times \dfrac{1}{(\text{Arcsin}^2 2x)}$

$= \dfrac{3\sqrt{1 - 4x^2} \text{ Arcsin } 2x - 6x}{\sqrt{1 - 4x^2} \text{ Arcsin}^2 2x}$

21. $y = 2 \text{ Arccos}^3 4x = 2(\text{Arccos } 4x)^3$

$\dfrac{dy}{dx} = 6(\text{Arccos}^2 4x) \dfrac{d(\text{Arccos } 4x)}{dx} = 6(\text{Arccos}^2 4x)(-\dfrac{1}{\sqrt{1 - (4x)^2}})(4)$

$= -\dfrac{24 \text{ Arccos}^2 4x}{\sqrt{1 - 16x^2}}$ Eq. (22-15), (26-11)

25. $y = 3 \text{ Arctan}^3 \ x; \ \dfrac{dy}{dx} = 3(3) \text{ Arctan}^2 \ x \ \dfrac{d \text{ Arctan } x}{dx}$

$$= 9(\text{Arctan}^2 \ x) \ \frac{1}{1 + x^2} = \frac{9 \text{ Arctan}^2 \ x}{1 + x^2}$$

29. $y = 3(4 - \text{Arccos } 2x)^3$

$$\frac{dy}{dx} = 9(4 - \text{Arccos } 2x)^2 \left(+ \frac{1}{\sqrt{1 - (2x)^2}} \right)(2)$$

$$= \frac{18(4 - \text{Arccos } 2x)^2}{\sqrt{1 - 4x^2}}$$

33. (a) $\dfrac{1}{\sqrt{1 - 0.5^2}} = 1.1547005$

(b) $\dfrac{\text{Arcsin } 0.5001 - \text{Arcsin } 0.5000}{0.0001} = 1.1546621$ (Final digits may vary.)

$\dfrac{d \text{ Arcsin } 0.5}{dx} = \dfrac{1}{\sqrt{1 - 0.5^2}} = 1.15470$, which is approximately equal to

1.15466.

Let $x = 0.4999$; $\Delta x = 0.0001$; $\dfrac{\Delta y}{\Delta x} = 1.15466 \approx 1.15470$

33M. (a) 1.154 700 5, (b) 1.154 662 1

37. $y = \text{Arctan } 2x$

$$\frac{dy}{dx} = \frac{1}{1 + (2x)^2}(2) = \frac{2}{1 + 4x^2}$$

$$\frac{d^2 y}{dx^2} = \frac{(1 + 4x^2)(0) - 2(8x)}{(1 + 4x^2)^2} = \frac{-16x}{(1 + 4x^2)^2} \quad \text{Eq. (22-13), (26-12)}$$

41. $t = \dfrac{1}{\omega} \text{Arcsin} \ \dfrac{A - E}{mE} = \dfrac{1}{\omega} \text{Arcsin} \left(\dfrac{A - E}{E} \right) \left(\dfrac{1}{m} \right) = \dfrac{1}{\omega} \text{Arcsin} \left(\dfrac{A - E}{E} \right) m^{-1}$

$u = \left(\dfrac{A - E}{E} \right) m^{-1} \qquad \dfrac{du}{dm} = - \left(\dfrac{A - E}{E} \right) m^{-2}$

$$\frac{dt}{dm} = \frac{1}{\omega \sqrt{1 - \left(\dfrac{-A + E}{E} \right)^2 m^{-2}}} \left(\frac{-A + E}{Em^2} \right) = \frac{E - A}{\omega E m^2 \sqrt{1 - \dfrac{(A - E)^2}{E^2 m^2}}}$$

$$= \frac{E - A}{\omega E m^2 \sqrt{\dfrac{E^2 m^2 - (A - E)^2}{E^2 m^2}}}$$

$$= \frac{E - A}{\omega m \sqrt{E^2 m^2 - (A - E)^2}}$$

Exercises 26-4, page 821

1. Points of intersection occur when $\sin x = \cos x$. $y_1 = \sin x$; $y_2 = \cos x$;
$\dfrac{dy_1}{dx} = \cos x$; $\dfrac{dy_2}{dx} = -\sin x$. At points of intersection, $\dfrac{dy_1}{dx} = -\dfrac{dy_2}{dx}$.

5. $y = x - \tan x (-\frac{\pi}{2} < x < \frac{\pi}{2}); \quad \frac{dy}{dx} = 1 - \sec^2 x; \quad \frac{d^2y}{dx^2} = 0 - 2 \sec x (\sec x \tan x)$
    $= -2 \sec^2 x \tan x.$

   (a) $\frac{dy}{dx} < 0$ for $\frac{-\pi}{2} < x < 0; \quad \frac{dy}{dx} = 0$ for $x = 0; \quad \frac{dy}{dx} < 0$ for $0 < x < \frac{\pi}{2}$. See
       Figure 9-18, p. 273. The function decreases from $-\frac{\pi}{2}$ to $\frac{\pi}{2} (x \neq 0)$ and
       has one critical point at $x = 0$.

   (b) Since $0 - \tan 0 = 0$, there is an intercept at $(0,0)$.

   (c) $\frac{d^2y}{dx^2} = 0$ when $\sec^2 x = 0$ or $\tan x = 0$. Since $|\sec x| \geq 1$ for all x,
       $\sec^2 x \geq 1$ for all x, and $\sec^2 x = 0$ has no solution. Between
       $-\frac{\pi}{2}$ and $\frac{\pi}{2}$, $\tan x = 0$ when $x = 0$.

   (d) $-2 \sec^2 x \tan x > 0$ when $\tan x < 0$. $\tan x < 0$ for $-\frac{\pi}{2} < x < 0$. The
       graph is concave up in this region. $-2 \sec^2 x \tan x < 0$ when
       $\tan x > 0$ or x is between 0 and $\frac{\pi}{2}$. The graph is concave down in this
       region. Since there is a change of sign in the second derivative, the
       only critical point is an inflection point, and there are no maximum
       or minimum points.

   (e) As $x \to -\frac{\pi}{2}$ from the right, $x - \tan x \to -\frac{\pi}{2} - (-\infty)$ or $+\infty$; $x = -\frac{\pi}{2}$ is an
       asymptote. Summarizing, the function
       decreases, intersects the x-axis at 0,
       is concave up for $-\frac{\pi}{2} < x < 0$, concave
       down for $0 < x < \frac{\pi}{2}$, has point of inflec-
       tion at $x = 0$, and asymptotes at
       $x = -\frac{\pi}{2}$ and $\frac{\pi}{2}$.

   Dec., $x > 0, x < 0$
   Infl. $(0,0)$
   Asym., $x = \pi/2, x = -\pi/2$

9. By a rough sketch of the graph using $f(0) = 0$, $f(1) = -2.37$, and
    $f(2) = 0.36$, the root is between 1 and 2 and is possibly closer to 2.
    Let $x_1 = 1.7$:

n	x_n	$f(x_n)$	$f'(x_n)$	$x_n - \dfrac{f(x_n)}{f'(x_n)}$
1	1.7	-1.0766592	3.915378	1.9749822
2	1.9749822	0.2228632	5.5230459	1.9346307
3	1.9346307	0.0046391	5.2927023	1.9337542
4	1.9337542	0.0000023	5.2876722	1.9337538

   $x_3 = x_4 = 1.9337538$

9M. 1.933 753 8

13. $y = 0.50 \sin 2t + 0.30 \cos t$

               $v = \frac{dy}{dt} = 1.00 \cos 2t - 0.30 \sin t$

$$v\bigg|_{t=0.40s} = 1.00 \cos 0.80 - 0.30 \sin 0.40$$

$$= 0.58$$

$$a = \frac{d^2y}{dyt^2} = -2.00 \sin 2t - 0.30 \cos t$$

$$a\bigg|_{t=0.40s} = -2.00 \sin 0.80 - 0.30 \cos 0.40$$

$$= -1.7$$

17. $x = 2.625 \cos 12\pi t$ $y = 2.625 \sin 12\pi t$

$$\frac{dx}{dt} = -2.625 \,(12\pi) \sin 12\pi t \qquad \frac{dy}{dt} = 2.625 \,(12\,\pi) \cos 12\pi t$$

$$= -31.50\pi \sin 12\pi t \qquad\qquad = 31.50\pi \cos 12\pi t$$

$$\frac{dx}{dt}\bigg|_{t=1.250} v = -31.50\pi \sin 47.12 \qquad \frac{dy}{dt}\bigg|_{t=1.250} = 31.50\pi \cos 47.12$$

$$= -31.50\pi\,(0.0039) \qquad\qquad = 31.50\pi\,(-1.000)$$

$$= -0.3859 \qquad\qquad\qquad = -98.96$$

$$v = \sqrt{(-0.3859)^2 + (-98.96)^2} = 98.96$$

$$\tan\theta = \frac{-98.96}{0.3859} = -90° = 270°$$

21.

$$s = 16t^2, \quad \tan\theta = \frac{200-s}{100} = \frac{200-16t^2}{100} = 2 - 0.16t^2$$

$$\theta = \arctan(2 - 0.16t^2); \quad \frac{d\theta}{dt} = \frac{1}{1 + (2 - 0.16t^2)^2}\,(-0.32t)$$

$$= \frac{-0.32t}{5 - 0.64t^2 + 0.0256t^4}$$

$$\frac{d\theta}{dt}\bigg|_{t=1} = \frac{-0.32(1)}{5 - 0.64(1)^2 + 0.0256(1)^4} = -0.0730 \text{ rad/s}$$

21M.

$$s = 4.9t^2, \quad \tan\theta = \frac{60-s}{40} = \frac{60-4.9t^2}{40} = 1.5 - 0.1225t^2$$

$$\theta = \arctan(1.5 - 0.1225t^2)$$

$$\frac{d\theta}{dt} = \frac{1}{1 + (1.5 - 0.1225t^2)^2}\,(-0.245t)$$

$$= \frac{-0.245t}{3.25 - 0.3675t^2 + 0.015t^4}$$

$$\frac{d\theta}{dt}\bigg|_{t=1} = \frac{-0.245(1)}{3.25 - 0.3675(1)^2 + 0.015(1)^4} = -0.0846 \text{ rad/s}$$

25. $u = \tan \theta$, $\theta = 20° = \dfrac{\pi}{9} = 0.349$ rad; $d\theta = 1° = \dfrac{\pi}{180} = 0.0175$ rad

$\dfrac{du}{d\theta} = \sec^2 \theta$; $du = \sec^2 \theta \, d\theta$

$du\big|_{\theta = 0.349} = (\sec^2 0.349)(0.0175)$

$\qquad\qquad = (1.064)^2(0.0175) = 0.02$

29. Let x = length of rectangle; y = width of rectangle.

$2x + 2y = 40$; $y = 20 - x$; $\tan \theta = \dfrac{20 - x}{x}$

$x = \dfrac{20}{\tan \theta + 1}$; $y = 20 - \dfrac{20}{\tan \theta + 1} = \dfrac{20 \tan \theta}{\tan \theta + 1}$

$A = xy = \left(\dfrac{20}{\tan \theta + 1}\right)\left(\dfrac{20 \tan \theta}{\tan \theta + 1}\right) = \dfrac{400 \tan \theta}{(\tan \theta + 1)^2}$

The maximum area will occur when $\dfrac{dA}{d\theta} = 0$

$\dfrac{dA}{d\theta} = \dfrac{(\tan \theta + 1)^2 (400 \sec^2\theta) - (400 \tan \theta)(2)(\tan \theta + 1)\,\sec^2\theta}{(\tan \theta + 1)^4}$

$\qquad = \dfrac{400(1 + \tan \theta)\sec^2\theta[1 - \tan \theta]}{(\tan \theta + 1)^4}$

Maximum area occurs when $\dfrac{dA}{d\theta} = 0$

$\dfrac{400(1 + \tan \theta)\sec^2\theta[1 - \tan \theta]}{(\tan \theta + 1)^4} = 0$

$400(1 + \tan \theta)\sec^2\theta[1 - \tan \theta] = 0$

$1 + \tan \theta = 0$; $\sec^2\theta = 0$; $1 - \tan \theta = 0$

$\quad \tan \theta = -1 \qquad\qquad\qquad \tan \theta = 1$

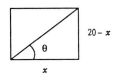

First two roots are rejected. $\tan \theta = 1$

$x = \dfrac{20}{1 + 1} = 10$; $y = 20 - \dfrac{20}{1 + 1} = 10$

Rectangle is a square. Area $= 10 \times 10 = 100$ mm^2

Exercises 26-5, page 826

1. $y = \log x^2$; $u = x^2$, $\dfrac{du}{dx} = 2x$, $\dfrac{dy}{dx} = \dfrac{1}{x^2}(\log e)(2x) = \dfrac{2 \log e}{x}$ Eq. (26-13)

5. $y = \ln(1 - 3x)$; $u = 1 - 3x$, $\dfrac{du}{dx} = -3$, $\dfrac{dy}{dx} = \dfrac{1}{1 - 3x}(-3) = \dfrac{-3}{1 - 3x}$

Eq. (26-14)

9. $y = \ln \sqrt{x} = \ln(x^{1/2}) = \dfrac{1}{2} \ln x$; $u = x$; $\dfrac{du}{dx} = 1$; $\dfrac{dy}{dx} = \dfrac{1}{2}\left(\dfrac{1}{x}\right) = \dfrac{1}{2x}$ Eq. (26-14)

13.　$y = x \ln x^2$

$\dfrac{dy}{dx} = 1(\ln x^2) + x\left(\dfrac{1}{x^2}\right)(2x) = \ln x^2 + 2$

$\qquad = 2 \ln x + 2 = 2(\ln x + 1)$

17.　$y = \ln(\ln x)$; let $u = \ln x$

$\dfrac{dy}{dx} = \dfrac{1}{\ln x}\left(\dfrac{d(\ln x)}{dx}\right) = \dfrac{1}{\ln x}\left(\dfrac{1}{x}\right) = \dfrac{1}{x \ln x}$　　Eq. (26-14)

21.　$y = \sin \ln x$; $\dfrac{dy}{dx} = \cos \ln x \dfrac{d(\ln x)}{dx} = \cos \ln x\left(\dfrac{1}{x}\right) = \dfrac{\cos \ln x}{x}$

Eq. (26-4), (26-14)

25.　$y = \ln(x \tan x)$; $u = x \tan x$; $\dfrac{du}{dx} = x(\sec^2 x) + \tan x$

$\dfrac{dy}{dx} = \dfrac{1}{x \tan x}[x(\sec^2 x) + \tan x] = \dfrac{x \sec^2 x + \tan x}{x \tan x}$

Eq. (22-12), (26-6), (26-14)

29.　$y = \sqrt{x^2 + 1} - \ln \dfrac{1 + \sqrt{x^2 + 1}}{x} = (x^2 + 1)^{1/2} - \ln \dfrac{1 + (x^2 + 1)^{1/2}}{x}$

$\dfrac{dy}{dx} = \dfrac{1}{2}(x^2 + 1)^{-1/2}(2x) - \dfrac{x}{1 + (x^2 + 1)^{1/2}} \cdot \dfrac{x\left[\frac{1}{2}(x^2 + 1)^{-1/2}(2x)\right] - \left[1 + (x^2 + 1)^{1/2}\right]}{x^2}$

$= \dfrac{x}{(x^2 + 1)^{1/2}} - \dfrac{1}{1 + (x^2 + 1)^{1/2}} \cdot \dfrac{x\left[\frac{x}{(x^2 + 1)^{1/2}}\right] - \left[1 + (x^2 + 1)^{1/2}\right]}{x}$

$= \dfrac{x}{(x^2 + 1)^{1/2}} - \dfrac{1}{1 + (x^2 + 1)^{1/2}} \cdot \dfrac{\frac{x^2}{(x^2 + 1)^{1/2}} - \frac{(x^2 + 1)^{1/2}}{(x^2 + 1)^{1/2}} - \frac{x^2 + 1}{(x^2 + 1)^{1/2}}}{x}$

$= \dfrac{x}{(x^2 + 1)^{1/2}} - \dfrac{1}{1 + (x^2 + 1)^{1/2}} \cdot \dfrac{\frac{-1\left[(x^2 + 1)^{1/2} + 1\right]}{(x^2 + 1)^{1/2}}}{x}$

$= \dfrac{x}{(x^2 + 1)^{1/2}} - \dfrac{1}{\left[1 + (x^2 + 1)^{1/2}\right]} \cdot \dfrac{-1\left[(x^2 + 1)^{1/2} + 1\right]}{(x^2 + 1)^{1/2} x}$

$= \dfrac{x}{(x^2 + 1)^{1/2}} + \dfrac{1}{(x^2 + 1)^{1/2} x} = \dfrac{x^2 + 1}{(x^2 + 1)^{1/2}(x)} = \dfrac{(x^2 + 1)^{1/2}}{x} = \dfrac{\sqrt{x^2 + 1}}{x}$

33M. 0.499 987 5, 0.5

37. $y = \text{Arctan } 2x + \ln(4x^2 + 1)$

$$\frac{dy}{dx} = \frac{1}{1 + (2x)^2} (2) + \frac{1}{4x^2 + 1} (8x) = \frac{2}{4x^2 + 1} + \frac{8x}{4x^2 + 1} = \frac{2 + 8x}{4x^2 + 1}$$

$$\frac{dy}{dx}\bigg|_{x = 0.625} = \frac{2 + 8(0.625)}{4(0.625)^2 + 1} = 2.73 \quad \text{Eq. (26-12), (26-14)}$$

41. $y = \ln \cos x$

$$\frac{dy}{dx} = \frac{1}{\cos x} (-\sin x) = -\frac{\sin x}{\cos x} = -\tan x$$

$$\frac{dy}{dx}\bigg|_{x = \frac{\pi}{4}} - \tan x = -\tan \frac{\pi}{4} = -1$$

43. $y = x^x$; $\ln y = \ln(x^x)$; $\ln y = x \ln x$

$$\frac{d(\ln y)}{dx} = x(\frac{1}{x}) + \ln x(1)$$

$$\frac{1}{y} \frac{dy}{dx} = 1 + \ln x$$

$$\frac{dy}{dx} = y(\ln x + 1) = x^x(\ln x + 1)$$

45. $b = 10 \log (I/I_0) = 10 \log (I_0^{-1} I)$ $\qquad u = (I_0^{-1} I)$

$$\frac{db}{dt} = 10 \left(\frac{I_0}{I}\right) (\log e) \left(\frac{I}{I_0}\right) \frac{dI}{dt} = \frac{10}{I} \log e \frac{dI}{dt}$$

Exercises 26-6, page 829

1. $y = 3^{2x}$; $\frac{dy}{dx} = 3^{2x} \ln 3 \frac{d2x}{dx} = (2 \ln 3) 3^{2x}$

5. $y = e^{6x}$; $\frac{dy}{dx} = e^{6x}(6) = 6e^{6x}$ $\quad$ Eq. (26-16)

9. $y = xe^{-x}$

$$\frac{dy}{dx} = x(e^{-x})(-1) + (1)(e^{-x}) = e^{-x} - xe^{-x} = e^{-x}(1 - x)$$

13. $y = \frac{3e^{2x}}{x + 1}$; $\frac{dy}{dx} = \frac{(x + 1)(3e^{2x})(2) - 3e^{2x}(1)}{(x + 1)^2}$

$$= \frac{(2x + 2)(3e^{2x}) - 3e^{2x}}{(x + 1)^2} = \frac{3e^{2x}[(2x + 2) - 1]}{(x + 1)^2}$$

$$= \frac{3e^{2x}(2x + 1)}{(x + 1)^2} \quad \text{Eq. (22-13), (26-16)}$$

17. $y = \dfrac{2e^{3x}}{4x + 3}$; $\dfrac{dy}{dx} = \dfrac{(4x + 3)(2e^{3x})(3) - (2e^{3x})(4)}{(4x + 3)^2}$

$\qquad = \dfrac{(12x + 9)(2e^{3x}) - 8e^{3x}}{(4x + 3)^2} = \dfrac{2e^{3x}(12x + 5)}{(4x + 3)^2}$

Eq. (22-13), (26-16)

21. $y = (2e^{2x})^3 \sin x^2$

$\qquad \dfrac{dy}{dx} = (2e^{2x})^3 \dfrac{d \sin x^2}{dx} + \sin x^2 \dfrac{d(2e^{2x})^3}{dx}$ Eq. (22-12)

$\qquad\quad = (2e^{2x})^3 (\cos x^2)(2x) + (\sin x^2)3(2e^{2x})^2 \dfrac{d(2e^{2x})}{dx}$ Eq. (22-16)

$\qquad\quad = 2x(2e^{2x})^3 (\cos x^2) + 3(2e^{2x})^2(\sin x^2)(2e^{2x})(2)$

$\qquad\quad = 2x(2e^{2x})^3 (\cos x^2) + 6(2e^{2x})^3(\sin x^2)$

$\qquad\quad = 2(2e^{2x})^3 (x \cos x^2 + 3 \sin x^2) = 16e^{6x}(x \cos x^2 + 3 \sin x^2)$

25. $y = xe^{xy} + \sin y$

$\qquad \dfrac{dy}{dx} = x\left(e^{xy}\right)\left(x \dfrac{dy}{dx} + y\right) + (1)\ e^{xy} + \cos y \dfrac{dy}{dx}$

$\qquad\quad = x\left(e^{xy}\right)\left(x \dfrac{dy}{dx}\right) + x\left(e^{xy}\right)(y) + e^{xy} + \cos y \dfrac{dy}{dx}$

$\qquad \dfrac{dy}{dx} - x\left(e^{xy}\right)\left(x\dfrac{dy}{dx}\right) - \cos y \dfrac{dy}{dx} = x\left(e^{xy}\right)y + e^{xy}$

$\qquad \dfrac{dy}{dx}\left(1 - x\left(e^{xy}\right)(x) - \cos y\right) = x\left(e^{xy}\right)y + e^{xy}$

$\qquad \dfrac{dy}{dx} = \dfrac{xy\left(e^{xy}\right) + e^{xy}}{1 - x^2 e^{xy} - \cos y} = \dfrac{e^{xy}(xy + 1)}{1 - x^2 e^{xy} - \cos y}$

29. $y = \ln \sin 2e^{6x}$; $\dfrac{dy}{dx} = \dfrac{1}{\sin 2e^{6x}}(\cos 2e^{6x})(2e^{6x})(6)$

$\qquad\qquad = \dfrac{12e^{6x} \cos 2e^{6x}}{\sin 2e^{6x}} = 12e^{6x} \cot 2e^{6x}$

33. (a) $e = e^x = 2.7182818$ when $x = 1.0000$. This is the slope of a tangent
     line to the curve $f(x) = e^x$ when $x = 1.0000$. It is the value of
     $f'(x) = e^x$, since $\dfrac{d\ e^x}{dx} = e^x$.

   (b) $\dfrac{e^{1.0001} - e^{1.0000}}{0.0001} = 2.7184178$ (Final digits may vary.)

     This is the slope of a secant line through the curve $f(x) = e^x$
     at $x = 1.0000$, where $\Delta x = 0.0001$.

     $\lim\limits_{\Delta x \to 0} \dfrac{e^{(x+\Delta x)} - e^x}{\Delta x} = \dfrac{d\ e^x}{dx} = e^x$

     For $\Delta x = 0.0001$, the slope of the tangent line is approximately equal
     to the slope of the secant line.

33M.   (a) 2.718 281 8, (b) 2.718 417 8

37. $y = xe^{-x}; \dfrac{dy}{dx} = x(e^{-x})(-1) + (e^{-x})(1) = -xe^{-x} + e^{-x}$

Substituting, $\dfrac{dy}{dx} + y = (-xe^{-x} + e^{-x}) + (xe^{-x}) = e^{-x}$

41. $i = 2te^{-0.5t}$

$\dfrac{di}{dt} = 2t\left(e^{-0.5t}\right)(-0.5) + 2\left(e^{-0.5t}\right)$

$= -t\left(e^{-0.5t}\right) + 2e^{-0.5t}$

$= e^{-0.5t}(-t + 2) = e^{-0.5t}(2 - t) = (2 - t)e^{-0.5t}$

45. $\cosh^2 u - \sinh^2 u = \dfrac{1}{4}\left(e^u + e^{-u}\right)^2 - \dfrac{1}{4}\left(e^u - e^{-u}\right)^2$

$= \dfrac{1}{4}\left(e^{2u} + 2e^0 + e^{-2u}\right) - \dfrac{1}{4}\left(e^{2u} - 2e^0 + e^{-2u}\right)$

$= \dfrac{1}{4}\left[0 + 2e^0 + 2e^0 + 0\right] = \dfrac{1}{4}\left(4e^0\right) = \dfrac{1}{4}(4) = 1$

Exercises 26-7, page 833

1. $y = \ln \cos x; \dfrac{dy}{dx} = \dfrac{1}{\cos x}(-\sin x)$

$= \dfrac{-\sin x}{\cos x} = -\tan x; \dfrac{d^2y}{dx^2} = -\sec^2 x$

(a)   Since $\ln(1) = 0$ and $\cos 0 = 1$, there is an intercept at $x = 0$, $y = 0$.

Int. (0,0), max. (0,0), not defined for $\cos x < 0$, asym. $x = -\dfrac{1}{2}\pi, \dfrac{1}{2}\pi, \ldots$

(b)   Since ln functions are not defined for negatives, y is undefined for $\cos x < 0$. The function is defined for x between $-\dfrac{\pi}{2}$ and $\dfrac{\pi}{2}$, etc.

(c)   As $\cos x$ approaches 0, $\ln \cos x$ approaches negative infinity.

Cos $x = 0$ when $x = -\dfrac{\pi}{2}, \dfrac{\pi}{2}$, and their odd multiples, so $x = -\dfrac{\pi}{2}$, $x = \dfrac{\pi}{2}$, $x = \dfrac{3\pi}{2}$, etc., are asymptotes.

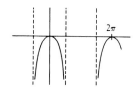

(d)   Critical points exist where $-\tan x = 0$; i.e., where $x = 0$, 2π, 4π, etc.

(e) $-\sec^2 x$ is negative at all critical points so the graph is concave down, and all critical points are maximum points.

(f)   Maximum points are (0,0), $(2\pi,0)$, $(4\pi,0)$, etc.

5. $y = \ln \dfrac{1}{x^2 + 1} = -\ln (x^2 + 1); \dfrac{dy}{dx} = -\dfrac{1}{x^2 + 1}(2x) = \dfrac{-2x}{x^2 + 1}$

$\dfrac{d^2y}{dx^2} = \dfrac{(x^2 + 1)(-2) - (-2x)(2x)}{(x^2 + 1)^2} = \dfrac{2x^2 - 2}{(x^2 + 1)^2}$ Int. (0,0), max. (0,0),

infl. (-1,-ln 2),(1,-ln 2)

(a) Since $\dfrac{1}{x^2 + 1} > 0$ for all numbers,

$\ln \dfrac{1}{x^2 + 1}$ is defined for all numbers;

there are no asymptotes.

(b) When $x = 0$, $y = \ln \dfrac{1}{0 + 1} = \ln 1 = 0$;

(0,0) is an intercept.

(c) Critical points; $\dfrac{dy}{dx} = \dfrac{-2x}{x^2 + 1} = 0$ when

$-2x = 0$; (0,0) is a critical point.

(d) $\dfrac{d^2y}{dx^2} = \dfrac{2x^2 - 2}{(x^2 + 1)^2} = 0$ when $2x^2 - 2 = 0$; $x^2 = 1$, $x = -1$, $x = 1$.

Inflections occur at $x = -1$, $y = \dfrac{\ln}{(-1)^2 + 1} = \ln \dfrac{1}{2} = \ln(2^{-1}) = -\ln 2$

and $x = 1$, $y = \ln \dfrac{1}{1^2 + 1} = -\ln 2$

(e) Since $\dfrac{d^2y}{dx^2}$ is negative at the critical point (0,0), it is a maximum.
The graph is concave down between the points (-1,-ln 2) and
(1,-ln 2). It is concave up for $x < -1$ and $x > 1$. Since the
second derivative goes through a change of sign at $x = -1$ and
again at $x = 1$, (-1,-ln 2) and (1,-ln 2) are inflection points.

9. $y = \ln x - x; \dfrac{dy}{dx} = \dfrac{1}{x} - 1; \dfrac{d^2y}{dx^2} = -x^{-2} = -\dfrac{1}{x^2}$

(a) $x \not= 0$ since ln is undefined for those   Max. (1,-1), asym. $x = 0$
values. $y \not= 0$ since $\ln x \not= x$ for any
number.

(b) There is an asymptote at $x = 0$.

(c) Critical points occur at $\dfrac{1}{x} - 1 = 0$
or $x = 1$, $y = -1$.

(d) Since $\dfrac{1}{x^2} \not= 0$ for any x, there is no
point of inflection.

(e) Since $-\dfrac{1}{x^2} < 0$ for all x, the graph
is concave down, and the critical
point (1,-1) is a maximum.

13. $y = x^2 \ln x; \dfrac{dy}{dx} = x^2 (\dfrac{1}{x}) + (\ln x)(2x) = x + 2x \ln x$ Eq. (22-12), (26-14)

$\dfrac{dy}{dx}\bigg|_{x = 1} = 1 + 2 \ln 1 = 1 + 2(0) = 1$

Slope is 1, $x = 1$, $y = 0$; using slope intercept form of the equation and
substituting gives $0 = 1(1) + b$ or $b = 1$. The equation is $y = (1)x - 1$
or $y = x - 1$.

17. $f(x) = x^2 - 2 + \ln x$; $f'(x) = 2x + \dfrac{1}{x}$

$f(1) = 1^2 - 2 + \ln 1 = -1$; $f(2) = 2^2 - 4 + \ln 2 = 0.69$

Therefore we choose $x_1 = 1.5$.

n	x_n	$f(x_n)$	$f'(x_n)$	$x_n - \dfrac{f(x_n)}{f'(x_n)}$
1	1.5	0.6554651	3.6666667	1.3212368
2	1.3212368	0.0242349	3.3993402	1.3141075
3	1.3141075	0.0000362	3.3891878	1.3140968

Therefore, the root is 1.3140968, which is correct to the number of decimal places shown.

17M. 1.314 096 8

21. $\ln p = \dfrac{a}{T} + b \ln T + C$

$P = e^{\left(\frac{a}{T} + b \ln T + c\right)}$

$\dfrac{dP}{dT} = e^{\left(\frac{a}{T} + b \ln T + c\right)}\left(-aT^{-2} + \dfrac{b}{T}\right) = e^{\left(\frac{a}{T} + b \ln T + c\right)}\left(\dfrac{-a}{T^2} = \dfrac{b}{T}\right)$

$= P\left(\dfrac{-a}{T^2} + \dfrac{bT}{T^2}\right) = P\left(\dfrac{-a + bT}{T^2}\right)$

25. $y = \ln \sec x$, $-1.5 \le x \le 1.5$; $u = \sec x$

$\dfrac{dy}{dx} = \dfrac{1}{\sec x} \cdot \sec x \tan x = \tan x = 0$ at $x = 0$

$x = 0$ is a critical value; also, multiples of 2π.

$\dfrac{d^2y}{dx^2} = \sec^2 x$; $\sec^2 (0) = 1$ so the curve is concave up and there is a

minimum point at $x = 0$, $y = \ln \sec 0$

$= \ln 1 = 0$

recurring at multiples of $x = 2\pi$; $(2\pi, 0)$ $(4\pi, 0)$ $\cdots$

$(0, 0)$ is an intercept.

Asymptotes occur where $\dfrac{dy}{dx} = \tan x$ is undefined.

These values are odd multiples of $\dfrac{\pi}{2}$; $-\dfrac{\pi}{2}$, $\dfrac{\pi}{2}$, $\dfrac{3\pi}{2}$ $\cdots$

29. $y = e^{-0.5t}(0.4 \cos 6t - 0.2 \sin 6t)$

$v = \dfrac{dy}{dt} = e^{-0.5t}(-2.4 \sin 6t - 1.2 \cos 6t) + (0.4 \cos 6t - 0.2 \sin 6t)$

$\times (e^{-0.5t})(-0.5)$

$= -2.4e^{-0.5t} \sin 6t - 1.2e^{-0.5t} \cos 6t + 0.1e^{-0.5t} \sin 6t$

$- 0.2e^{-0.5t} \cos 6t$

$= -2.3e^{-0.5t} \sin 6t - 1.4e^{-0.5t} \cos 6t = -e^{-0.5t}(1.4 \cos 6t$

$+ 2.3 \sin 6t)$

$$\left.\frac{dy}{dt}\right|_{t = \frac{\pi}{12}s} = -e^{-0.5(\pi/12)}[1.4 \cos 6(\tfrac{\pi}{12}) + 2.3 \sin 6(\tfrac{\pi}{12})]$$
$$= -e^{-\pi/24}[1.4 \cos \tfrac{\pi}{2} + 2.3 \sin \tfrac{\pi}{2}]$$
$$= -e^{-\pi/24}[1.4(0) + 2.3(1)]$$
$$= 2.3e^{-\pi/24} = -2.02 \text{ cm/s}$$

Review Exercises for Chapter 26, page 835

1. $y = 3 \cos(4x - 1)$; $\frac{dy}{dx} = [-3 \sin(4x - 1)][4] = -12 \sin(4x - 1)$

5. $y = \csc^2(3x + 2)$; $\frac{dy}{dx} = 2 \csc(3x + 2)[-\csc(3x + 2) \cot(3x + 2)](3)$
$$= -6 \csc^2(3x + 2) \cot(3x + 2)$$

9. $y = (e^{x-3})^2$; $\frac{dy}{dx} = 2(e^{x-3})(e^{x-3})(1) = 2e^{2(x-3)}$

13. $y = 3 \arctan(\frac{x}{3})$; $\frac{dy}{dx} = 3\left[\dfrac{1}{1 + (\frac{x}{3})^2}\right]\dfrac{1}{3}$
$$= \frac{1}{1 + (\frac{x}{3})^2} = \frac{1}{1 + \frac{x^2}{9}} = \frac{9}{9 + x^2}$$

17. $y = \sqrt{\csc 4x + \cot 4x} = (\csc 4x + \cot 4x)^{1/2}$
$$\frac{dy}{dx} = \frac{1}{2}(\csc 4x + \cot 4x)^{-1/2}(-4 \csc 4x \cot 4x - 4 \csc^2 4x)$$
$$= \frac{1}{2}(\csc 4x + \cot 4x)^{-1/2}(-4 \csc 4x)(\csc 4x + \cot 4x)$$
$$= -2 \csc 4x(\csc 4x + \cot 4x)^{1/2} = (-2 \csc 4x)\sqrt{\csc 4x + \cot 4x}$$

21. $y = \dfrac{\cos^2 x}{e^{3x} + 1}$
$$\frac{dy}{dx} = \frac{(e^{3x} + 1)[2 \cos x(-\sin x)] - (\cos^2 x)(e^{3x})(3)}{(e^{3x} + 1)^2}$$
$$= \frac{(e^{3x} + 1)[-2 \sin x \cos x] - 3e^{3x} \cos^2 x}{(e^{3x} + 1)^2}$$
$$= \frac{-\cos x[(e^{3x} + 1)(2 \sin x) + 3e^{3x} \cos x]}{(e^{3x} + 1)^2}$$
$$= \frac{-\cos x[2e^{3x} \sin x + 2 \sin x + 3e^{3x} \cos x]}{(e^{3x} + 1)^2}$$
$$= \frac{-\cos x(2e^{3x} \sin x + 3e^{3x} \cos x + 2 \sin x)}{(e^{3x} + 1)^2}$$

25. $y = \ln(\csc x^2)$; $\dfrac{dy}{dx} = \dfrac{1}{\csc x^2}(-\csc x^2 \cot x^2)(2x) = -2x \cot x^2$

29. $y = e^{-2x} \sec x$

$\dfrac{dy}{dx} = e^{-2x}(\sec x \tan x) + \sec x\ e^{-2x}(-2)$

$\qquad = e^{-2x} \sec x \tan x - 2e^{-2x} \sec x$

$\qquad = e^{-2x} \sec x\ (\tan x - 2)$

33.

$\text{Arctan } \dfrac{y}{x} = x^2 e^y$; $u = \dfrac{y}{x} = yx^{-1}$; $\dfrac{du}{dx} = -yx^{-2} + x^{-1}\dfrac{dy}{dx}$

$\dfrac{1}{1 + \left(yx^{-1}\right)^2}\left(-yx^{-2} + x^{-1}\dfrac{dy}{dx}\right) = x^2 e^y \dfrac{dy}{dx} + 2xe^y$

$\dfrac{\dfrac{-y}{x^2} + \dfrac{1}{x}\dfrac{dy}{dx}}{1 + y^2 x^{-2}} = x^2 e^y \dfrac{dy}{dx} + 2xe^y$

$\dfrac{-y}{x^2} + \dfrac{1}{x}\dfrac{dy}{dx} = \left(x^2 e^y \dfrac{dy}{dx} + 2xe^y\right)\left(1 + y^2 x^{-2}\right)$

$\dfrac{-y}{x^2} + \dfrac{1}{x}\dfrac{dy}{dx} = x^2 e^y \dfrac{dy}{dx} + 2xe^y + y^2 e^y \dfrac{dy}{dx} + 2x^{-1} y^2 e^y$

$\dfrac{1}{x}\dfrac{dy}{dx} - x^2 e^y \dfrac{dy}{dx} - y^2 e^y \dfrac{dy}{dx} = 2xe^y + 2x^{-1}y^2 e^y + \dfrac{y}{x^2}$

$\dfrac{dy}{dx}\left(\dfrac{1}{x} - x^2 e^y - y^2 e^y\right) = 2xe^y + \dfrac{2y^2 e^y}{x} + \dfrac{y}{x^2}$

$\dfrac{dy}{dx}\left(\dfrac{1 - x^3 e^y - xy^2 e^y}{x}\right) = \dfrac{2x^3 e^y + 2xy^2 e^y + y}{x^2}$

$\dfrac{dy}{dx} = \dfrac{2x^3 e^y + 2xy^2 e^y + y}{x^2} \cdot \dfrac{x}{1 - x^3 e^y - xy^2 e^y}$

$\dfrac{dy}{dx} = \dfrac{2x^3 e^y + 2xy^2 e^y + y}{x - x^4 e^y - x^2 y^2 e^y}$

37. $\ln xy + ye^{-x} = 1$

Using implicit differentiation, $\dfrac{d \ln xy}{dx} + \dfrac{d\ ye^{-x}}{dx} = \dfrac{d(1)}{dx}$

$\dfrac{1}{xy}\left(x\dfrac{dy}{dx} + y\dfrac{dx}{dx}\right) + y\dfrac{de^{-x}}{dx} + e^{-x}\dfrac{dy}{dx} = 0$

$\dfrac{1}{xy}\left(x\dfrac{dy}{dx} + y\right) + ye^{-x}(-1) + e^{-x}\dfrac{dy}{dx} = 0$

$\dfrac{1}{y}\dfrac{dy}{dx} + \dfrac{1}{x} - ye^{-x} + e^{-x}\dfrac{dy}{dx} = 0$

$\dfrac{1}{y}\dfrac{dy}{dx} + e^{-x}\dfrac{dy}{dx} = ye^{-x} - \dfrac{1}{x}$

$\dfrac{dy}{dx}\left(\dfrac{1}{y} + e^{-x}\right) = ye^{-x} - \dfrac{1}{x}$

$$\frac{dy}{dx}(\frac{1 + ye^{-x}}{y}) = \frac{x\ ye^{-x} - 1}{x}$$

$$\frac{dy}{dx} = (\frac{x\ ye^{-x} - 1}{x})(\frac{y}{1 + ye^{-x}}) = \frac{y(x\ ye^{-x} - 1)}{x(1 + ye^{-x})}$$

41. $y = x - \cos x$; $\frac{dy}{dx} = 1 + \sin x$; $\frac{d^2y}{dx^2} = \cos x$ Infl.: $(\frac{1}{2}\pi, \frac{1}{2}\pi)$, $(\frac{3}{2}\pi, \frac{3}{2}\pi)$

(a) $x = 0$, $y = 0 - \cos 0 = 0 - 1$
     $= -1$; $(0,-1)$ is an intercept.
     $x - \cos x = 0$ when $x = \cos x$;
     $x = 0.74$ (see Table 3); $(0.74,0)$
     is an intercept.

(b) y is defined for all x; no asymp-
     totes.

(c) Critical points occur at $1 + \sin x = 0$, $\sin x = -1$, $x = -\frac{\pi}{2}$, $\frac{3\pi}{2}$, $\frac{7\pi}{2}$,
     etc.

(d) Inflections occur at $\cos x = 0$, $x = -\frac{\pi}{2}$, $\frac{\pi}{2}$, $\frac{3\pi}{2}$, $\frac{5\pi}{2}$, etc., since the
     second derivative undergoes a change of sign at each of these points.

(e) All critical points are inflections; no maximum or minimum points.

(f) Checking concavity at $x = 0$, $-\cos 0 = -1$; the graph is concave up at
     $(0,-1)$ and on each side of this point up to the inflection points
     at $(-\frac{\pi}{2},-\frac{\pi}{2})$ and $(\frac{\pi}{2},\frac{\pi}{2})$. It will switch concavity again at each subse-
     quent inflection point.

x	$-\frac{3\pi}{2}$	$-\pi$	$-\frac{\pi}{2}$	0	0.7	$\frac{\pi}{2}$	π	$\frac{3\pi}{2}$
y	-4.7	-4.1	-1.6	-1	0	1.6	4.1	4.7

45. $y = 4\cos^2(x^2)$; slope $= \frac{dy}{dx} = 2[4\cos(x^2)][-\sin(x^2)](2x) = -16x\cos x^2 \sin x^2$

$\frac{dy}{dx}\Big|_{x = 1} = -16\cos(1^2)\sin(1^2) = -16(0.5403)(0.8415) = -7.27$

$f(1) = 4\cos^2(1^2) = 4(0.5403)^2 = 1.168$

By Eq. (20-9), $y = -7.27x + b$; $1.168 = -7.27(1) + b$, $b = 8.44$

$y = -7.27x + 8.44$; $7.27x + y - 8.44 = 0$

49. $\sin^2 x + \cos^2 x = 1$

$$\frac{d(\sin^2 x + \cos^2 x)}{dx} = \frac{d(1)}{dx}$$

$$\frac{d \sin^2 x}{dx} + \frac{d \cos^2 x}{dx} = 0$$

$2 \sin x \cos x + 2 \cos x(-\sin x) = 0$

$2 \sin x \cos x - 2 \cos x \sin x = 0; \; 0 = 0$

53. $y = 3.5 \sin (0.75 \pi t + 0.50); \; u = 0.75\pi t + 0.50$

$$\frac{dy}{dt} = 3.5 \cos u \frac{du}{dt} \qquad\qquad \frac{du}{dt} = 0.75\pi$$

$$= 3.5[\cos (0.75\pi t + 0.50)] [0.75\pi]$$

$$= 8.25 \cos (0.75\pi t + 0.50)$$

$$\frac{dy}{dt} \Big|_{t = 1.50} = 8.25 \cos \left[0.75\pi (1.50) + 0.50\right] = -5.17$$

57. $t = a \ln \dfrac{x}{b - x} - c; \; u = \dfrac{x}{b - x}; \; \dfrac{du}{dx} = \dfrac{(b - x)(1) - x(-1)}{(b - x)^2}$

$$\frac{dt}{dx} = a \frac{1}{u} \frac{du}{dx} \qquad\qquad\qquad = \frac{b - x + x}{(b - x)^2} = \frac{b}{(b - x)^2}$$

$$= a\left(\frac{b - x}{x}\right)\left(\frac{b}{(b - x)^2}\right) = \frac{ab}{x(b - x)}$$

61. $\theta = \text{Arcsin} \dfrac{Ff}{R} \qquad\qquad u = \dfrac{Ff}{R} = \dfrac{f}{R} F$

$$\frac{d\theta}{dF} = \frac{1}{\sqrt{1 - u^2}} \frac{du}{dx} \qquad\qquad \frac{du}{dF} = \frac{f}{R}$$

$$= \frac{1}{\sqrt{1 - \left(\frac{Ff}{R}\right)^2}} \cdot \frac{f}{R} = \frac{1 \cdot f}{\sqrt{1 - \left(\frac{Ff}{R}\right)^2} \sqrt{R^2}} = \frac{f}{\sqrt{R^2 - F^2 f^2}}$$

65. $$T = 80 + 120 \ (0.5)^{0.2t} \qquad u = 0.2t, \ \frac{du}{dt} = 0.2$$

$$T = 80 + 120 \ (0.5)^{u}$$

$$\frac{dT}{dt} = 120 \ (0.5)^{0.2t} \big(\ln \ (0.5)\big) \ (0.2) \qquad \text{Eq. (26-15)}$$

$$\frac{dT}{dt}\bigg|_{t = 5.00} = 120 \ (0.5)^{0.2(5.00)} \ (-0.693)(0.2) = 120 \ (0.5)(-0.693)(0.2)$$

$$= -8.32°\text{F/min}$$

69. $A = xy = 3xe^{-0.5x^2}$ (Working with $\frac{1}{2}$ the actual area)

$$\frac{dA}{dx} = 3x \ d\big(e^{-0.5x^2}\big) + e^{-0.5x^2} \left(\frac{d \ 3x}{dx}\right)$$

$$= 3x \ \big(e^{-0.5x^2}\big)(-1.0x) + e^{-0.5x^2} \ (3)$$

$$= -3x^2 e^{-0.5x^2} + 3e^{-0.5x^2}$$

The maximum value will occur when $\frac{dA}{dx} = 0$

$$-3x^2 e^{-0.5x^2} + 3e^{-0.5x^2} = 0$$

$$\big(e^{-0.5x^2}\big) \big(-3x^2 + 3\big) = 0$$

$$-3x^2 + 3 = 0; \ x^2 = 1, \ x = 1.00$$

$$e^{-0.5x^2} = 0 \text{ has no real solution}$$

$$y = 3e^{-0.5(1)^2} = 3e^{-0.5} = 1.82$$

$$W = 2x = 2.00; \ H = 1.82$$

73. $x = r(\theta - \sin \theta); \ y = r(1 - \cos \theta); \ r = 5.500 \text{ cm}; \ \frac{d\theta}{dt} = 0.12 \text{ rad/s}; \ \theta = 35°$

horizontal component of velocity, $v_x = \dfrac{dx}{dt} = \dfrac{d[5.5 (\theta - \sin \theta)]}{dt}$

$$= 5.5 \left(\frac{d\theta}{dt} - \cos \theta \frac{d\theta}{dt}\right) = 5.5(0.12 - .12 \cos 35°) = 0.119 \text{ cm/s}$$

vertical component of velocity, $v_y = \dfrac{dy}{dt} = d[5.5(1 - \cos \theta)]$

$$= 5.5 \sin \theta \frac{d\theta}{dt} = (5.5)(0.12) \sin 35° = 0.379$$

$$v = \sqrt{.119^2 + 0.379^2} = 0.4 \text{ cm/s}$$

77.

$$\cos \theta = \frac{y}{4}; \quad y = 4 \cos \theta$$

$$\sin \theta = \frac{x}{4}; \quad x = 4 \sin \theta$$

$$A = (4 + x)y = 4y + xy = 16 \cos \theta + 16 \sin \theta \cos \theta$$

$$= 16 \cos \theta (1 + \sin \theta)$$

$$\frac{dA}{d\theta} = -16 \sin \theta + 16 \sin \theta (-\sin \theta) + 16 \cos \theta (\cos \theta)$$

$$= -16 \sin \theta - 16 \sin^2\theta + 16 \cos^2\theta$$

$$= 16(\cos^2\theta - \sin^2\theta - \sin \theta) = 16(1 - 2 \sin^2\theta - \sin \theta)$$

1) Not valid for negative θ or A. Domain and range are positive real numbers.

2) A-intercept at $\theta = 0$, $A = 16$; To find θ-intercept, $A = 0$

$$16 \cos \theta (1 + \sin \theta) = 0$$

$$16 \cos \theta = 0 \qquad 1 + \sin \theta = 0$$

$$\theta = \frac{\pi}{2} \qquad \sin \theta = -1, \quad \theta = -\frac{\pi}{2} \text{ (not in domain)}$$

3) Critical value is $16(1 - 2 \sin^2\theta - \sin \theta) = 0$

$$1 - 2 \sin^2\theta - \sin \theta = 0$$

$$-1 + 2 \sin^2\theta + \sin \theta = 0$$

$$(2 \sin \theta - 1)(\sin \theta + 1) = 0$$

$$2 \sin \theta = 1 \qquad \sin \theta = -1$$

$$\sin \theta = \frac{1}{2} \qquad \theta = -\frac{\pi}{2} \text{ (not in domain)}$$

$$\theta = \frac{\pi}{6}$$

4) $\dfrac{d^2A}{d\theta^2} = 16(-4 \sin \theta \cos \theta - \cos \theta) = -16 \cos \theta (4 \sin \theta + 1)$

$$= -16(4 \sin \theta + 1) \Big|_{\theta = \frac{\pi}{6}} = -16[4(0.5) + 1] = -16(3) = -48$$

Curve is concave down at $\frac{\pi}{6}$ and $\theta = \frac{\pi}{6}$ is a maximum.

77M.

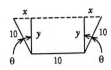

$\cos \theta = \dfrac{y}{10}$; $y = 10 \cos \theta$

$\sin \theta = \dfrac{x}{10}$; $x = 10 \sin \theta$

$A = (10 + x)(y) = 10y + xy = 100 \cos \theta + 100 \sin \theta \cos \theta$
$$= 100 \cos \theta \, (1 + \sin \theta)$$

Analysis of curve follows that of Exercise 77.

CHAPTER 27

Methods of Integration

1. $u = \sin x$, $du = \cos x\ dx$. The integral is of the form $\int u^4\ du$.

    $\int \sin^4 x \cos x\ dx = \dfrac{1}{5} \sin^5 x + C$ Eq. (26-2), (27-1)

5. $\displaystyle\int 4 \tan^2 x\ \sec^2 x\ dx = 4\int \tan^2 x\ \sec^2 x\ dx$

    $u = \tan x$, $du = \sec^2 x\ dx$.

    The integral is of the form $\displaystyle\int u^2\ du$.

    $4\displaystyle\int \tan^2 x\ \sec^2 x\ dx = 4\left(\dfrac{1}{3}\ \tan^3 x + c\right) = \dfrac{4}{3}\ \tan^3 x + c$

9. $u = \arcsin x$, $du = \dfrac{1}{\sqrt{1 - x^2}}\ dx = \dfrac{dx}{\sqrt{1 - x^2}}$. The integral is of the form $\int u^3\ du$.

    $\int (\arcsin x)^3 \left(\dfrac{dx}{\sqrt{1 - x^2}}\right) = \dfrac{1}{4}(\arcsin x)^4 + C$ Eq. (26-10), (27-1)

13. $u = \ln(x + 1)$, $du = \dfrac{1}{x + 1}(1)\ dx = \dfrac{dx}{x + 1}$. The integral is of the form $\int u^2\ du$.

    $\int [\ln(x + 1)]^2\ \dfrac{dx}{x + 1} = \dfrac{1}{3}[\ln(x + 1)]^3 + c$ Eq. (26-14), (27-1)

17. $u = 4 + e^x$, $du = e^x\ dx$. The integral is of the form $\int u^3\ du$.

    $\int (4 + e^x)^3 e^x\ dx = \dfrac{1}{4}(4 + e^x)^4 + c$ Eq. (26-16), (27-1)

21. $u = 1 + \sec^2 x$, $\dfrac{du}{dx} = 2 \sec x \dfrac{d(\sec x)}{dx} = 2 \sec x \sec x \tan x$;

   $du = 2 \sec^2 x \tan x \, dx$.

   du needs a factor of 2. $\int (1 + \sec^2 x)^4 (\sec^2 x \tan x \, dx)$

   $= \dfrac{1}{2} \int (1 + \sec^2 x)^4 \, 2 \sec^2 x \tan x \, dx$

   $= \dfrac{1}{2} \times \dfrac{1}{5} (1 + \sec^2 x)^5 + C = \dfrac{1}{10} (1 + \sec^2 x)^5 + C$

25. $A = \displaystyle\int_0^2 \dfrac{1 + \text{Arctan } 2x}{1 + 4x^2} \, dx$; $u = \text{Arctan } 2x$, $du = \dfrac{1}{1 + (2x)^2} \times 2$

   $A = \dfrac{1}{2} \displaystyle\int_0^2 1 + u \, du = \dfrac{1}{2} (u + \dfrac{1}{2} u^2) \Big|_0^2$

   $= \dfrac{1}{2} \Big[\text{Arctan } 2x + \dfrac{1}{2} (\text{Arctan } 2x)^2 \Big] \Big|_0^2$

   $= \dfrac{1}{2} \Big[\text{Arctan } 4 + \dfrac{1}{2} (\text{Arctan } 2x)^2 - \text{Arctan } 0 - \dfrac{1}{2} (\text{Arctan } 0)^2 \Big]$

   $= \dfrac{1}{2} \Big[1.326 + 0.879 - 0 - 0 \Big] = 1.102$

29. $P = mnv^2 \displaystyle\int_0^{\pi/2} \sin \theta \cos^2 \theta \, d\theta$; $n = 2$, $\mu = \cos \theta$, $du = -\sin \theta \, d\theta$

   $P = -mnv^2 \displaystyle\int_0^{\pi/2} \cos^2 \theta (-\sin \theta \, d\theta) = -mnv^2 [\dfrac{\cos^3 \theta}{3} \Big|_0^{\pi/2}]$

   $= -mnv^2 [\dfrac{1}{3} (\cos^3 \dfrac{\pi}{2} - \cos^3 0)] = -mnv^2 [\dfrac{1}{3} (0 - 1)]$

   $= -mnv^2 (-\dfrac{1}{3}) = \dfrac{1}{3} mnv^2$

Exercises 27-2, page 846

1. $u = 1 + 4x$, $du = 4 \, dx$. Introduce a factor of 4.

   $\displaystyle\int \dfrac{dx}{1 + 4x} = \dfrac{1}{4} \int \dfrac{4dx}{1 + 4x} = \dfrac{1}{4} \ln |1 + 4x| + C$ Eq. (27-2)

5. $u = 8 - 3x$; $du = -3 \, dx$

   $\displaystyle\int_0^2 \dfrac{dx}{8 - 3x} = -\dfrac{1}{3} \int_0^2 \dfrac{-3dx}{8 - 3x} = -\dfrac{1}{3} \ln |8 - 3x| \Big|_0^2$

   $= -\dfrac{1}{3} \ln 2 + \dfrac{1}{3} \ln 8 = -0.231 + 0.693 = 0.462$

9. $u = 1 + \sin x;\ du = \cos x\ dx$

$$\int_{0}^{\pi/2} \frac{\cos x\ dx}{1 + \sin x} = \ln |1 + \sin x|\ \Big|_{0}^{\pi/2}$$

$$= \ln \left|1 + \sin \frac{\pi}{2}\right| - \ln |1 + \sin 0|$$

$$= \ln |2| - \ln |1| = 0.693 - 0 = 0.693$$

13. $u = x + e^x;\ du = (1 + e^x)\ dx$

$$\int \frac{1 + e^x}{x + e^x}\ dx = \ln |x + e^x| + C$$

17. $u = 4x + 2x^2;\ du = (4 + 4x)\ dx$

$$\int_{1}^{3} \frac{1 + x}{4x + 2x^2}\ dx = \frac{1}{4} \int_{1}^{3} \frac{4 + 4x}{4x + 2x^2}\ dx = \frac{1}{4} \ln |4x + 2x^2|\ \Big|_{1}^{3}$$

$$= \tfrac{1}{4} \ln 30 - \tfrac{1}{4} \ln 6 = 0.850 - 0.448 = 0.402$$

21. $u = 2x + \tan x;\ du = (2 + \sec^2 x)\ dx$

$$\int \frac{2 + \sec^2 x}{2x + \tan x}\ dx = \ln |2x + \tan x| + c$$

25. $\displaystyle \int \frac{x + 2}{x^2}\ dx = \int \frac{1}{x}\ dx + \int \frac{2}{x^2}\ dx = \int \frac{1}{x}\ dx + \int (2x^{-2})\ dx$

$$= \ln |x| - 2x^{-1} + c = \ln |x| - \frac{2}{x} + c$$

29.

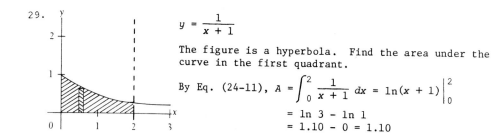

$$y = \frac{1}{x + 1}$$

The figure is a hyperbola. Find the area under the curve in the first quadrant.

By Eq. (24-11), $A = \displaystyle\int_{0}^{2} \frac{1}{x + 1}\ dx = \ln(x + 1)\ \Big|_{0}^{2}$

$$= \ln 3 - \ln 1$$
$$= 1.10 - 0 = 1.10$$

33. $m = \dfrac{dy}{dx} = \dfrac{\sin x}{3 + \cos x}$

$y = \displaystyle\int \dfrac{1}{3 + \cos x} \times \sin x \; dx$

$y = -\displaystyle\int \dfrac{1}{3 + \cos x}(-\sin x)dx;$ let $u = 3 + \cos x$

$\hspace{6cm} du = -\sin x \; dx$

$y = -\displaystyle\int \dfrac{1}{u} \; du = -\ln |u| + C$

$y = -\ln(3 + \cos x) + C$

$2 = -\ln(3 + \cos \frac{\pi}{3}) + C;$ substitute values of x and y

$2 = -\ln(3 + 0.5) + C$

$C = 2 + \ln 3.5$

$y = -\ln(3 + \cos x) + \ln 3.5 + 2;$ substituting for C

$y = \ln \dfrac{3.5}{3 + \cos x} + 2$

37. $t = L\displaystyle\int \dfrac{di}{E - iR};\; u = E - iR,\; du = -R\; di$

$t = -\dfrac{L}{R}\displaystyle\int \dfrac{-R\; di}{E - iR} = \dfrac{-L}{R} \ln |E - iR| + C;\; t = 0$ for $i = 0$

$0 = -\dfrac{L}{R} \ln E + C;\; C = \dfrac{L}{R} \ln |F|;\; t = \dfrac{L}{R}(-\ln |E - iR| + \ln |E|)$

$\hspace{1.5cm} = \dfrac{L}{R} \ln \dfrac{E}{E - iR};\; \dfrac{R}{L}t = \ln\dfrac{E}{E - iR};\; e^{Rt/L} = \dfrac{E}{E - iR};\; i = \dfrac{E}{R} - \dfrac{E}{R}e^{-Rt/L};$

$i = \dfrac{E}{R}\left(1 - e^{-Rt/L}\right)$

Exercises 27-3, page 853

1. $u = 7x,\; du = 7\; dx.$ By Eq. (27-3), $\displaystyle\int e^{7x}(7\; dx) = e^{7x} + C$

5. $u = \dfrac{x}{2},\; du = \dfrac{1}{2}\; dx.$ Introduce a factor of $\dfrac{1}{2}$.

$\displaystyle\int_{0}^{2} e^{x/2}\; dx = 2\displaystyle\int_{0}^{2} e^{x/2}(\tfrac{1}{2}\; dx)$

$\hspace{2cm} = 2e^{x/2}\Big|_{0}^{2} = 2e - 2e^{0} = 2(e - 1)$

$\hspace{2cm} = 2(2.718 - 1) = 3.44$

9. $u = \sqrt{x} = x^{1/2},\; du = \dfrac{1}{2}x^{-1/2}\; dx = \dfrac{dx}{2\sqrt{x}}.$ Introduce a factor of $\dfrac{1}{2}$.

$\displaystyle\int \dfrac{e^{\sqrt{x}}}{\sqrt{x}}\; dx = 2\displaystyle\int e^{\sqrt{x}}\dfrac{dx}{2\sqrt{x}} = 2e^{\sqrt{x}} + C$

13. $\displaystyle\int \frac{(3 - e^x)\ dx}{e^{2x}} = \int e^{-2x}(3 - e^x)\ dx$

$\displaystyle = \int (3e^{-2x} - e^{-x})\ dx = \int 3e^{-2x}\ dx - \int e^{-x}\ dx$

$\displaystyle \int 3e^{-2x}\ dx; \quad u = -2x, \quad du = -2\ dx. \quad \text{Introduce a factor of } -2.$

$\displaystyle -\frac{3}{2}\int e^{-2x}(-2\ dx) = -\frac{3}{2}e^{-2x}$

$\displaystyle \int e^{-x}\ dx; \quad u = -x, \quad du = -1\ dx. \quad \text{Introduce a factor of } -1.$

$\displaystyle -\int e^{-x}(-dx) = -e^{-x}$

$\displaystyle \int \frac{(3 - e^x)\ dx}{e^{2x}} = -\frac{3}{2}e^{-2x} - (-e^{-x}) + C$

$\displaystyle = -\frac{3}{2}e^{-2x} + e^{-x} + C = \frac{1}{e^x} - \frac{3}{2e^{2x}} + C$

$\displaystyle = \frac{2e^x - 3}{2e^{2x}} + C$

17. $\displaystyle\int \frac{2\ dx}{\sqrt{x}e^{\sqrt{x}}} = 2\int \frac{dx}{x^{1/2}e^{x^{1/2}}} = 2\int e^{-x^{1/2}}x^{-1/2}\ dx$

$u = -x^{1/2}; \quad du = -\frac{1}{2}x^{-1/2}\ dx.$

We place $-\frac{1}{2}$ with dx, and -2 before the integral. Therefore,

$\displaystyle 2(-2)\int e^{-x^{1/2}}(-\frac{1}{2}x^{-1/2}) = -4e^{-x^{1/2}} + C$

$\displaystyle = -\frac{4}{e^{x^{1/2}}} + C = -\frac{4}{e^{\sqrt{x}}} + C$

21. $u = \cos 3x\ dx, \quad du = -\sin 3x(3\ dx) = \dfrac{-3}{\csc 3x}\ dx.$ Eq. (19-1)

Introduce a factor of -3.

$\displaystyle \int \frac{e^{\cos 3x}\ dx}{\csc 3x} = -\frac{1}{3}\int e^{\cos 3x}(\frac{-3}{\csc 3x}\ dx) = -\frac{1}{3}e^{\cos 3x} + C$

25. $A = \displaystyle\int_0^2 e^x\ dx = e^x\Big|_0^2 = e^2 - e^0$

$= 2.718^2 - 1 = 6.389$

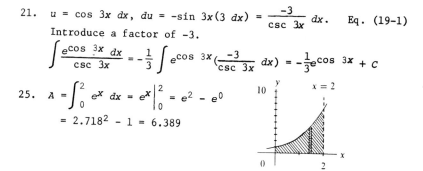

29. From Eq. (25-28), $\displaystyle y_{av} = \frac{\int_0^4 e^{2x}\ dx}{4 - 0} = \frac{\frac{1}{2}\int_0^4 e^{2x}(2\ dx)}{4}$

$\displaystyle = \frac{\frac{1}{2}e^{2x}\Big|_0^4}{4} = \frac{1}{8}e^{2x}\Big|_0^4$

$\displaystyle = \frac{1}{8}(e^8 - 1) = \frac{1}{8}(2977) = 372$

33. $qe^{t/RC} = \dfrac{E}{R} \displaystyle\int e^{t/RC} \, dt; \quad u = t/RC; \quad du = \dfrac{1}{RC} \, dt$

$qe^{t/RC} = RC \cdot \dfrac{E}{R} \displaystyle\int e^{t/RC} \left(\dfrac{1}{RC}\right) dt$

$qe^{t/RC} = EC \, (e^{t/RC}) + C_1$ where C_1 is the constant of integration.

$\quad q = 0$ for $t = 0$; $\; 0 = EC + C_1$; $C_1 = -EC$; $\; qe^{t/RC} = EC(e^{t/RC}) - EC$

$\quad q = EC - \dfrac{EC}{e} t/RC; \; q = EC\left(1 - e^{-t/RC}\right)$

<u>Exercises 27-4</u>, page 846

1. $u = 2x$, $du = 2 \, dx$. Introduce a factor of 2.

$\displaystyle\int \cos 2x \, dx = \dfrac{1}{2} \displaystyle\int \cos 2x \, (2 \, dx)$

$\qquad\qquad = \dfrac{1}{2} \sin 2x + C$ Eq. (27-5)

5. $u = \dfrac{1}{2}x$, $du = \dfrac{1}{2} \, dx$. Introduce a factor of $\dfrac{1}{2}$.

$\displaystyle\int \sec \dfrac{1}{2}x \, \tan \dfrac{1}{2}x \, dx = 2 \displaystyle\int \sec \dfrac{1}{2}x \, \tan \dfrac{1}{2}x \left(\dfrac{1}{2} \, dx\right) = 2 \sec \dfrac{1}{2}x + C$ Eq. (27-8)

9. $\qquad\qquad\qquad u = x^2; \; du = 2x \, dx$

$\displaystyle\int 3x \sec x^2 \, dx = \dfrac{3}{2} \displaystyle\int 2x \sec x^2 \, dx$

$\qquad\qquad = \dfrac{3}{2} \ln \left| \sec x^2 + \tan x^2 \right| + c$ Eq. (27-12)

13. $u = 2x$, $du = 2 \, dx$. Introduce a factor of 2.

$\displaystyle\int_0^{\pi/6} \dfrac{dx}{\cos^2 2x} = \dfrac{1}{2} \displaystyle\int_0^{\pi/6} \sec^2 2x \, (2 \, dx)$

$\qquad\qquad = \dfrac{1}{2} \tan 2x \Big|_0^{\pi/6} = \dfrac{1}{2}\left(\tan \dfrac{\pi}{3} - \tan 0\right)$

$\qquad\qquad = \dfrac{1}{2}(\sqrt{3} - 0) = \dfrac{1}{2}\sqrt{3}$ Eq. (27-6)

17. By Eq. (19-7), $\displaystyle\int \sqrt{\tan^2 2x + 1} \, dx = \displaystyle\int \sqrt{\sec^2 2x} \, dx = \displaystyle\int \sec 2x \, dx$

$u = 2x$, $du = 2 \, dx$. Introduce a factor of 2.

$\displaystyle\int \sec 2x \, dx = \dfrac{1}{2} \displaystyle\int \sec 2x \, (2 \, dx) = \dfrac{1}{2} \ln \left| \sec 2x + \tan 2x \right| + C$ Eq. (27-12)

21. $\displaystyle\int \frac{1 - \sin x}{1 + \cos x}\, dx = \int \frac{1 - \sin x}{1 + \cos x} \times \frac{1 - \cos x}{1 - \cos x}\, dx$

$\displaystyle = \int \frac{1 - \sin x - \cos x + \sin x \cos x}{1 - \cos^2 x}\, dx$

$\displaystyle = \int \frac{1 - \sin x - \cos x + \sin x \cos x}{\sin^2 x}\, dx$

$\displaystyle = \int (\frac{1}{\sin^2 x} - \frac{\sin x}{\sin^2 x} - \frac{\cos x}{\sin^2 x} + \frac{\sin x \cos x}{\sin^2 x})\, dx$

$\displaystyle = \int (\csc^2 x - \csc x - \cot x \csc x + \cot x)\, dx$

$= -\cot x - \ln \left| \csc x - \cot x \right| - (-\csc x) + \ln \left| \sin x \right| + C$

$= \csc x - \cot x - \ln \left| \csc x - \cot x \right| + \ln \left| \sin x \right| + C$

25. $\displaystyle A = \int_0^{\pi/4} y\, dx = \int_0^{\pi/4} \tan x = -\ln \left| \cos x \right| \Big|_0^{\pi/4}$

$= -[\ln \left| \cos \frac{\pi}{4} \right| - \ln \left| \cos 0 \right|] = -(\ln 0.7071 - \ln 1)$

$= -(-0.347) = 0.347$

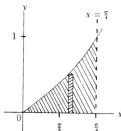

29. $\omega = -0.25 \sin 2.5t$

$\theta = \displaystyle\int -0.25 \sin 2.5t\, dt;\ u = 2.5t,\ du = 2.5\, dt$

$\theta = -0.10 \displaystyle\int \sin 2.5t\ (2.5dt) = -0.10(-\cos 2.5t) + c$

$\theta = 0.10 \cos 2.5t + c$

$0.10 = 0.10 \cos 0 + c;\ c = 0$

$\theta = 0.10 \cos 2.5t$

Exercises 27-5, page 858

1. $u = \sin x,\ du = \cos x\, dx$

$\displaystyle\int \sin^2 x \cos x\, dx = \frac{1}{3} \sin^3 x + C$

5. $\int 2 \sin^2 x \cos^3 x \, dx = 2\int \sin^2 x \cos^3 x \, dx$

$2\int \cos^3 x \sin^2 x \, dx = 2\int \cos^2 x \cos x \sin^2 x \, dx$

$$= 2\int (1 - \sin^2 x) \cos x \sin^2 x \, dx$$

$$= 2\int (\sin^2 x - \sin^4 x) \cos x \, dx$$

$$= 2\int \sin^2 x \cos x \, dx - \int \sin^4 x \cos x \, dx$$

$$= 2\left(\frac{1}{3} \sin^3 x - \frac{1}{5} \sin^5 x\right) + c = \frac{2}{3} \sin^3 x - \frac{2}{5} \sin^5 x + c$$

9. $\int \sin^2 x \, dx = \int [\frac{1}{2}(1 - \cos 2x)] \, dx = \frac{1}{2} \int dx - \frac{1}{2} \int \cos 2x \, dx$

$$= \frac{1}{2} \int dx - \frac{1}{4} \int \cos 2x (2 \, dx) = \frac{1}{2}x - \frac{1}{4} \sin 2x + C$$

13. $\int \tan x^3 \, dx = \int \tan^2 x \tan x \, dx = \int (\sec^2 x - 1)\tan x \, dx$

$$= \int \tan x \sec^2 x \, dx - \int \tan x \, dx$$

$$= \frac{1}{2} \tan^2 x - (-\ln |\cos x|) + C$$

$$= \frac{1}{2} \tan^2 x + \ln |\cos x| + C \quad \text{Eq. (27-15)}$$

17. $\int \tan^4 2x \, dx = \int (\tan^2 2x)(\tan^2 2x) \, dx = \int (\tan^2 2x)(\sec^2 2x - 1) \, dx$

$$= \int (\tan^2 2x \sec^2 2x - \tan^2 2x) \, dx$$

$$= \int \tan^2 2x \sec^2 2x \, dx - \int \tan^2 2x \, dx$$

$$= \int \tan^2 2x \sec^2 2x \, dx - \int (\sec^2 2x - 1) \, dx$$

$$= \frac{1}{2} \int (\tan^2 2x \sec^2 2x)(2 \, dx) - \int \sec^2 2x \, dx + \int 1 \, dx$$

$$= \frac{1}{2} \int (\tan^2 2x \sec^2 2x)(2 \, dx) - \frac{1}{2} \int \sec^2 2x (2 \, dx) + \int 1 \, dx$$

$$= \frac{1}{2} \times \frac{1}{3} \tan^3 2x - \frac{1}{2} \tan 2x + x + C$$

$$= \frac{1}{6} \tan^3 2x - \frac{1}{2} \tan 2x + x + C$$

21. $\int (\sin x + \cos x)^2 \, dx = \int (\sin^2 x + \cos^2 x) \, dx + \int 2 \sin x \cos x \, dx$

$$= \int 1 \, dx + \int \sin 2x \, dx = \int 1 \, dx + \frac{1}{2} \int \sin 2x (2 \, dx)$$

$$= x - \frac{1}{2} \cos 2x + C$$

25. $\displaystyle\int_{\pi/6}^{\pi/4} \cot^5 x\, dx = \int_{\pi/6}^{\pi/4} \cot^3 x(\csc^2 x - 1)\, dx$

$\displaystyle = \int_{\pi/6}^{\pi/4} \cot^3 x\, \csc^2 x\, dx - \int_{\pi/6}^{\pi/4} \cot^3 x\, dx$

$\displaystyle = -\int_{\pi/6}^{\pi/4} \cot^3 x(-\csc^2 x\, dx) - \int_{\pi/6}^{\pi/4} \cot x(\csc^2 x - 1)\, dx$

$\displaystyle = -\frac{1}{4}\cot^4 x\Big|_{\pi/6}^{\pi/4} - \int_{\pi/6}^{\pi/4} \cot x\, \csc^2 x\, dx + \int_{\pi/6}^{\pi/4} \cot x\, dx$

$\displaystyle = -\frac{1}{4}\cot^4 x + \frac{1}{2}\cot^2 x + \ln\,|\sin x|\,\Big|_{\pi/6}^{\pi/4}$

$\displaystyle = -\frac{1}{4}\cot^4\frac{\pi}{4} + \frac{1}{2}\cot^2\frac{\pi}{4} + \ln\,\left|\sin\frac{\pi}{4}\right| - \left(-\frac{1}{4}\cot^4\frac{\pi}{6} + \frac{1}{2}\cot^2\frac{\pi}{6}\right.$

$\displaystyle \left. + \ln\,\left|\sin\frac{\pi}{6}\right|\right)$

$\displaystyle = -\frac{1}{4} + \frac{1}{2} + \ln 0.7071 + 2.25 - 1.5 - \ln 0.5000$

$\displaystyle = 1 - 0.347 + 0.693 = 1.347$

29. By Eq. (25-11), the disk method, $V = \pi\displaystyle\int_0^\pi y^2\, dx = \pi\int_0^\pi \sin^2 x\, dx$

$\displaystyle = \pi\int_0^\pi \frac{1}{2}(1 - \cos 2x)\, dx = \frac{\pi}{2}\int_0^\pi dx - \frac{\pi}{2}\int_0^\pi \cos 2x\, dx$

$\displaystyle = \frac{\pi}{2}x - \frac{\pi}{2} \times \frac{1}{2}\sin 2x\Big|_0^\pi = \frac{\pi^2}{2} - \frac{\pi}{4}\sin 2\pi - 0 + \frac{\pi}{4}\sin 0$

$\displaystyle = \frac{1}{2}\pi^2 = 4.935$ Eq. (27-18)

33. $\displaystyle\int \sin x\, \cos x\, dx;\ u = \sin x,\ du = \cos x\, dx$

$\displaystyle\int u\, du = \tfrac{1}{2}\, u^2 + C = \tfrac{1}{2}\sin^2 x + C_1$

Let $u = \cos x;\ du = -\sin x\, dx$

$\displaystyle -\int \cos x\, (-\sin x)\, dx = -\frac{1}{2}\cos^2 x + C_2$

$\displaystyle \frac{1}{2}\sin^2 x + C_1 = \frac{1}{2}(1 - \cos^2 x + C_1) = \frac{1}{2} - \frac{1}{2}\cos^2 x + C_1$

$\displaystyle -\frac{1}{2}\cos^2 x + C_2 = \frac{1}{2} - \frac{1}{2}\cos^2 x + C_1$

$\displaystyle C_2 = C_1 + \frac{1}{2}$

37.
$$V_{rms} = \sqrt{\frac{1}{2\pi} \int_0^{2\pi} (170 \sin 120\pi t)^2 \, dt}$$

$$\int_0^{2\pi} (170 \sin 120\pi t)^2 dt = 170^2 \int_0^{2\pi} \sin^2 120\pi t \, dt = 170^2 \int_0^{2\pi} \sin^2 2\pi t \, dt$$

$$= 170^2 \int_0^{2\pi} \sin^2 t \, dt = \frac{170^2}{2} \int_0^{2\pi} 2 \sin^2 t \, dt$$

$$= \frac{170^2}{2} \int_0^{2\pi} (1 - \cos 2t) \, dt$$

$$= \frac{170^2}{2} \int_0^{2\pi} 1 \, dt - \frac{170^2}{2} \int_0^{2\pi} \cos 2t \, dt$$

$$= \frac{170^2}{2} \int_0^{2\pi} 1 \, dt - \frac{170^2}{4} \int_0^{2\pi} 2 \cos 2t \, dt$$

$$= \frac{170^2}{2} \left[t \, \Big|_0^{2\pi} \right] - \frac{170^2}{4} \left[- \sin 2t \, \Big|_0^{2\pi} \right]$$

$$= \frac{170^2}{2} \left[2\pi - 0 \right] - \frac{170^2}{4} \left[-\sin 4\pi + \sin 0 \right]$$

$$= 90800 - \frac{170^2}{4} (0) = 90800$$

$$\sqrt{\frac{1}{2\pi} (90800)} = 120V$$

Exercises 27-6, page 862

1. $a = 2; \ u = x; \ \displaystyle\int \frac{dx}{\sqrt{4 - x^2}} = \int \frac{dx}{\sqrt{2^2 - x^2}}$

   $= \arcsin \frac{x}{2} + C = \arcsin \frac{1}{2}x + C$ Eq. (27-20)

5. $a = 1; \ u = 4x, \ du = 4 \, dx. \ \displaystyle\int \frac{dx}{\sqrt{1 - (4x)^2}} = \frac{1}{4} \int \frac{4 \, dx}{\sqrt{1 - (4x)^2}}$

   $= \frac{1}{4} \arcsin \frac{4x}{1} + C = \frac{1}{4} \arcsin 4x + C$ Eq. (27-20)

9. $\int_0^{0.4} \dfrac{2\ dx}{\sqrt{4-5x^2}} = 2\int_0^{0.4} \dfrac{dx}{\sqrt{4-5x^2}}$; $a = 2,\ u = \sqrt{5}x,\ du = \sqrt{5}$

$\dfrac{2}{\sqrt{5}} \int_0^{0.4} \dfrac{\sqrt{5}\ dx}{\sqrt{4-5x^2}} = \dfrac{2}{\sqrt{5}} \left(\text{Arcsin}\ \dfrac{\sqrt{5}x}{2}\ \Big|_0^{0.4} \right)$

$= \dfrac{2\sqrt{5}}{5} \Big[\text{Arcsin}\ (0.2\sqrt{5}) - \text{Arcsin}\ 0 \Big]$

$= \dfrac{2\sqrt{5}}{5} \Big[0.463 - 0 \Big] = 0.415$

13. $a = \sqrt{7};\ u = \sqrt{5}x,\ du = \sqrt{5}\ dx.$ $\int_1^2 \dfrac{dx}{5x^2+7} = \int_1^2 \dfrac{dx}{7+5x^2}$

$= \dfrac{1}{\sqrt{5}} \int_1^2 \dfrac{\sqrt{5}\ dx}{(\sqrt{7})^2 + (\sqrt{5}x)^2} = \dfrac{1}{\sqrt{5}} \times \dfrac{1}{\sqrt{7}}\ \text{arctan}\ \dfrac{\sqrt{5}x}{\sqrt{7}}\Big|_1^2$

$= \dfrac{\sqrt{35}}{35}\ \text{arctan}\ \dfrac{1}{7}\sqrt{35}x\ \Big|_1^2 = \dfrac{1}{35}\sqrt{35}\ (\text{arctan}\ \dfrac{2}{7}\sqrt{35} - \text{arctan}\ \dfrac{1}{7}\sqrt{35})$

$= 0.169(\text{arctan}\ 1.690 - \text{arctan}\ 0.845)$

$= 0.169(1.036 - 0.702) = 0.057$ Eq. (27-21)

17. $a = 1;\ u = x + 1,\ du = dx.$ $\int \dfrac{dx}{x^2 + 2x + 2} = \int \dfrac{dx}{(x^2 + 2x + 1) + 1}$

$= \int \dfrac{dx}{(x+1)^2 + 1^2} = \dfrac{1}{1}\ \text{arctan}\ \dfrac{(x+1)}{1} + C$

$= \text{arctan}\ (x + 1) + C$ Eq. (27-21)

21. $a = 1;\ u = \sin 2x,\ du = \cos 2x(2\ dx).$

$\int_{\pi/6}^{\pi/2} \dfrac{\cos 2x}{1 + \sin^2 2x}\ dx = \dfrac{1}{2} \int_{\pi/6}^{\pi/2} \dfrac{\cos 2x(2\ dx)}{1 + \sin^2 2x}$

$= \dfrac{1}{2} \times \dfrac{1}{1}\ \text{arctan}\ \dfrac{\sin 2x}{1}\Big|_{\pi/6}^{\pi/2} = \dfrac{1}{2}(\text{arctan}\ \sin \pi - \text{arctan}\ \sin \dfrac{\pi}{3})$

$= \dfrac{1}{2}\ \text{arctan}\ 0 - \text{arctan}\ 0.8660) = \dfrac{1}{2}(0 - 0.714) = -0.357$

25. (a) $\int \dfrac{2\ dx}{4 + 9x^2} = 2\int \dfrac{dx}{2^2 + (3x)^2}$; $u = 3x$; $du = 3\ dx$; $a = 2$.

Therefore, the integral is an inverse tangent, Eq. (27-21).

(b) $\int \dfrac{2\ dx}{4 + 9x} = 2\int \dfrac{dx}{4 + 9x}$; $u = 4 + 9x$; $du = 9\ dx$

Therefore, the integral is logarithmic, Eq. (27-2).

(c) $\int \dfrac{2x\ dx}{\sqrt{4 + 9x^2}} = 2\int (4 + 9x^2)^{-1/2}(x\ dx)$; $u = 4 + 9x^2$; $du = 18x\ dx$

Therefore, the form of the integral is general power.

29. $y = \dfrac{1}{1 + x^2}$; $A = \displaystyle\int_0^2 \dfrac{1}{1 + x^2}\ dx$ Eq. (25-7)

$a = 1$; $u = x$, $du = dx$

$A = \dfrac{1}{1} \arctan \dfrac{x}{1}\Big|_0^2 = \arctan 2 - \arctan 0$

$= 1.11 - 0 = 1.11$ Eq. (27-21)

33. $\int \dfrac{dx}{\sqrt{A^2 - x^2}} = \int \sqrt{\dfrac{k}{m}}\ dt$. $\arcsin \dfrac{x}{A} + C_1 = \sqrt{\dfrac{k}{m}}\, t + C_2$

$\arcsin \dfrac{x}{A} = \sqrt{\dfrac{k}{m}}\, t + C$.   Solve for C by letting $x = x_0$ and $t = 0$.

$\arcsin \dfrac{x_0}{A} = \sqrt{\dfrac{k}{m}}(0) + C$; $C = \arcsin \dfrac{x_0}{A}$

Therefore, $\arcsin \dfrac{x}{A} = \sqrt{\dfrac{k}{m}}\, t + \arcsin \dfrac{x_0}{A}$.

Exercises 27-7, page 864

1. $u = x$, $du = dx$, $dv = \cos x\ dx$; $v = \int \cos x\ dx = \sin x$

$\int(x)(\cos x\ dx) = x(\sin x) - \int \sin x\ dx$

$= x(\sin x) - (-\cos x) + C = \cos x + x \sin x + C$

Eq. (27-22)

5. $u = x$, $du = dx$, $dv = \sec^2 x\ dx$; $v = \int \sec^2 x\ dx = \tan x$

$\int(x)(\sec^2 x\ dx) = x \tan x - \int \tan x\ dx$

$= x \tan x - (-\ln|\cos x| + C)$

$= x \tan x + \ln|\cos x| + C$

Eq. (27-22)

9. $$\int \frac{4x \; dx}{\sqrt{1 - x}} = 4 \int \frac{x \; dx}{\sqrt{1 - x}} \; ; \; u = x; \; du = dx$$

$$dv = \frac{1}{\sqrt{1 - x}} \; dx; \; v = \int (1 - x)^{-1/2} \; dx$$

$$v = - \int (1 - x)^{-1/2} \; (-dx) = -(1 - x)^{1/2} (2) = -2(1 - x)^{1/2}$$

Using Eq. (27-22),

$$4 \int (x) \; (1 - x)^{-1/2} \; dx = 4x \left[-2(1-x)^{1/2} \right] - 4 \int -2(1 - x)^{1/2} \; dx$$

$$= -8x(1 - x)^{1/2} + 8 \int (1 - x)^{1/2} \; dx$$

$$= -8x(1 - x)^{1/2} = 8(1 - x)^{1/2} \; (-dx)$$

$$= -8x(1 - x)^{1/2} - 8(1 - x)^{3/2} \left(2/3 \right)$$

$$= -8x\sqrt{1 - x} - 16/3 (1 - x)^{3/2} + C$$

13. $u = x^2$, $du = 2x \; dx$, $dv = \sin 2x \; dx$

$$v = \int \sin 2x \; dx = \frac{1}{2} \int \sin 2x \; (2 \; dx) = -\frac{1}{2} \cos 2x$$

By Eq. (27-22), $\int (x^2)(\sin 2x \; dx) = x^2 (-\frac{1}{2} \cos 2x) - \int -\frac{1}{2} \cos 2x(2x \; dx)$

$$= -\frac{x^2}{2} \cos 2x + \int x \cos 2x \; dx$$

Integrating $\int x \cos 2x \; dx$, $u = x$, $du = dx$, $dv = \cos 2x \; dx$

$$v = \int \cos 2x \; dx = \frac{1}{2} \int \cos 2x(2 \; dx) = \frac{1}{2} \sin 2x.$$

By Eq. (27-22), $\int x \cos 2x \; dx = x(\frac{1}{2} \sin 2x) - \int \frac{1}{2} \sin 2x \; dx$

$$= \frac{x}{2} \sin 2x - \frac{1}{4} \int \sin 2x(2 \; dx) = \frac{x}{2} \sin 2x - \frac{1}{4}(-\cos 2x)$$

$$= \frac{x}{2} \sin 2x + \frac{1}{4} \cos 2x$$

$$\int x^2 \sin 2x \; dx = -\frac{x^2}{2} \cos 2x + \frac{x}{2} \sin 2x + \frac{1}{4} \cos 2x$$

$$= \frac{1}{2}x \sin 2x - \frac{2x^2}{4} \cos 2x + \frac{1}{4} \cos 2x$$

$$= \frac{1}{2}x \sin 2x - \frac{1}{4}(2x^2 - 1) \cos 2x + C$$

17. $A = \displaystyle\int_0^2 xe^{-x} \; dx$; $u = x$, $du = dx$, $dv = e^{-x} \; dx$, $v = \int e^{-x} \; dx = -e^{-x}$

By Eq. (27-22), $A = -xe^{-x} \Big|_0^2 - \displaystyle\int_0^2 -e^{-x} \; dx = -xe^{-x} - e^{-x} \Big|_0^2$

$$= -2e^{-2} - e^{-2} - (0 - 1) = -3e^{-2} + 1 = 1 - \frac{3}{e^2}$$

$$= 1 - \frac{3}{2.718^2} = 0.594$$

21. $y_{rms} = \sqrt{\dfrac{1}{T}\displaystyle\int_0^T y^2\ dx},\ y = \sqrt{\text{Arcsin } x}$

$(y_{rms})^2 = \dfrac{1}{1}\displaystyle\int_0^1 (\sqrt{\text{Arcsin } x})^2\ dx = \int_0^1 (\text{Arcsin } x)\ dx$

Let $u = \text{Arcsin } x$, $dv = dx$, $du = \dfrac{dx}{\sqrt{1 - x^2}}$, $v = x$

By Eq. (27-22), $(y_{rms})^2 = x\text{ Arcsin } x + \sqrt{1 - x^2}\ \Big|_0^1$

$\qquad = [1(\text{Arcsin } 1) + \sqrt{1 - 1}] - [0(\text{Arcsin } 0) + \sqrt{1 - 0}]$

$\qquad = [1(\tfrac{\pi}{2}) + 0] - [0(0) + 1] = \dfrac{\pi}{2} - 1 = 0.756$

Exercises 27-8, page 870

1. Let $x = \sin \theta$, $dx = \cos \theta\ d\theta$; $\displaystyle\int \dfrac{\sqrt{1 - x^2}}{x^2}\ dx$

$\displaystyle\int \dfrac{\sqrt{1 - \sin^2 \theta}}{\sin^2 \theta}\cos \theta\ d\theta = \int \dfrac{\cos^2 \theta}{\sin^2 \theta}\ d\theta = \int \cot^2 \theta\ d\theta$

$= \displaystyle\int (\csc^2 \theta - 1)\ d\theta = \int \csc^2 \theta\ d\theta - \int d\theta$

$= -\cot \theta - \theta + C = \dfrac{-\sqrt{1 - x^2}}{x} - \text{Arcsin } x + C$

Eq. (27-6)

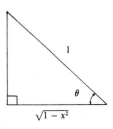

5. Let $x = 3 \tan \theta$, $dx = 3 \sec^2 \theta\ d\theta$

$\displaystyle\int \dfrac{dx}{x^2\sqrt{x^2 + 9}} = \int \dfrac{3 \sec^2 \theta\ d\theta}{9 \tan^2 \theta\ \sqrt{9 \tan^2 \theta + 9}}$

$= \displaystyle\int \dfrac{3 \sec^2 \theta\ d\theta}{27 \tan^2 \theta\ \sqrt{\tan^2 \theta + 1}} = \dfrac{1}{9}\int \dfrac{\sec \theta\ d\theta}{\tan^2 \theta} = \dfrac{1}{9}\int \dfrac{\cos \theta\ d\theta}{\sin^2 \theta}$

$= \dfrac{1}{9}\displaystyle\int \csc \theta \cot \theta\ d\theta = -\dfrac{1}{9}\csc \theta + C = -\dfrac{1}{9 \sin \theta} + C$

$\tan \theta = \dfrac{x}{3}$, $\sin \theta = \dfrac{x}{\sqrt{9 + x^2}}$

$\dfrac{-1}{9 \sin \theta} + C = \dfrac{-1}{\dfrac{9x}{\sqrt{9 + x^2}}} + C = -\dfrac{\sqrt{x^2 + 9}}{9x} + C$

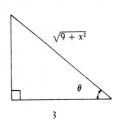

9. Let $x = \sin \theta$, $dx = \cos \theta \; d\theta$

$$\int_0^{0.5} \frac{x^3 \; dx}{\sqrt{1 - x^2}} = \int \frac{\sin^3 \theta \cos \theta \; d\theta}{\sqrt{1 - \sin^2 \theta}} = \int \sin^3 \theta \; d\theta$$

$$= \int \sin \theta \sin^2 \theta \; d\theta = \int \sin \theta (1 - \cos^2 \theta) \; d\theta$$

$$= \int \sin \theta \; d\theta - \int \cos^2 \theta \sin \theta \; d\theta = -\cos \theta + \frac{\cos^3 \theta}{3}$$

$$\cos \theta = \sqrt{1 - x^2}; \quad -\sqrt{1 - x^2} + \frac{1}{3}(\sqrt{1 - x^2})^3 \Big|_0^{0.5}$$

$$= -\sqrt{1 - 0.5^2} + \frac{1}{3}(\sqrt{1 - 0.5^2})^3 + \sqrt{1} - \frac{1}{3}\sqrt{1} = 0.017$$

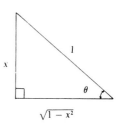

13. We have the form $\sqrt{x^2 - a^2}$; let $2x = 3 \sec \theta$; $x = \frac{3}{2} \sec \theta$,

$dx = \frac{3}{2} \sec \theta \tan \theta \; d\theta$.

$$\int \frac{\frac{3}{2} \sec \theta \tan \theta \; d\theta}{\frac{3}{2} \sec \theta \sqrt{4(\frac{3}{2} \sec \theta)^2 - 9}} = \int \frac{\tan \theta \; d\theta}{\sqrt{9 \sec^2 \theta - 9}}$$

$$= \int \frac{\tan \theta \; d\theta}{3\sqrt{\sec^2 \theta - 1}} = \int \frac{\tan \theta \; d\theta}{3 \tan \theta}$$

$$= \frac{1}{3} \int d\theta = \frac{1}{3} \theta + C = \frac{1}{3} \operatorname{arcsec} \frac{2}{3}x + C$$

17. The equation of the circle with the center at the origin is $x^2 + y^2 = 1^2$;

$y = \sqrt{1 - x^2}$. By Eq. (25-7), $\frac{1}{4}A = \int_0^1 \sqrt{1 - x^2} \; dx$.

Let $x = \sin \theta$, $dx = \cos \theta \; d\theta$. $\int \sqrt{1 - x^2} \; dx$

$$= \int \sqrt{1 - \sin^2 \theta} \cos \theta \; d\theta = \int \cos^2 \theta \; d\theta$$

$$= \int \frac{1}{2}(1 + \cos 2\theta) \; d\theta = \frac{1}{2} \int d\theta + \frac{1}{2} \int \cos 2\theta (2 \; d\theta)$$

$$= \frac{1}{2} \theta + \frac{1}{2} \sin 2\theta = \frac{1}{2} \arcsin x + \frac{1}{2} \sin(2 \arcsin x)$$

$$\frac{1}{4}A = \frac{1}{2} \arcsin x + \frac{1}{2} \sin(2 \arcsin x) \Big|_0^1 = \frac{1}{2} \arcsin 1 + \frac{1}{2} \sin(2 \arcsin 1)$$

$$= \frac{1}{2}(\frac{\pi}{2}) + \frac{1}{2} \sin(2)(\frac{\pi}{2}) = A = \frac{4\pi}{4} = \pi \quad \text{Eq. (27-17)}$$

21. By Eq. (25-10), $V = 2\pi \int_4^5 \dfrac{x(\sqrt{x^2 - 16})}{x^2}\, dx.$

(The limits of integration are $x = 4$
since $x = 4$ when $y = 0$, and $x = 5$.)
Let $x = 4 \sec\theta$, $dx = 4 \sec\theta \tan\theta\, d\theta$;

$V = 2\pi \int_4^5 \dfrac{\sqrt{x^2 - 16}}{x}\, dx.$

$2\pi \int \dfrac{\sqrt{x^2 - 16}}{x}\, dx$

$\quad = 2\pi \int \dfrac{\sqrt{16 \sec^2\theta - 16}}{4 \sec\theta}\, 4 \sec\theta \tan\theta\, d\theta$

$\quad = 2\pi \int \sqrt{16 \sec^2\theta - 16}\, \tan\theta\, d\theta$

$\quad = 2\pi \int 4\sqrt{\sec^2\theta - 1}\, \tan\theta\, d\theta$

$\quad = 8\pi \int (\tan\theta) \tan\theta\, d\theta = 8\pi \int \tan^2\theta\, d\theta$

$\quad = 8\pi \int (\sec^2\theta - 1)\, d\theta = 8\pi(\tan\theta - \theta)$

Since $x = 4\sec\theta$, $\sec\theta = \dfrac{x}{4}$, and $\tan\theta = \dfrac{\sqrt{x^2 - 16}}{4}$, and

$2\pi \int \dfrac{\sqrt{x^2 - 16}}{x}\, dx = 8\pi(\tan\theta - \theta) = 8\pi\left(\dfrac{\sqrt{x^2 - 16}}{4} - \text{arcsec } \dfrac{x}{4}\right)$

$V = 2\pi \int_4^5 \dfrac{x\sqrt{x^2 - 16}}{x^2}\, dx = 8\pi\left(\dfrac{\sqrt{x^2 - 16}}{4} - \text{arcsec } \dfrac{x}{4}\right)\Big|_4^5$

$\quad = 8\pi(0.75 - 0.643) = 2.68$

Exercises 27-9, page 872

1. Use formula 1, where $u = x$, $a = 2$, $b = 5$; $du = dx$.

$\displaystyle\int \dfrac{3x\, dx}{2 + 5x} = 3\int \dfrac{x\, dx}{2 + 5x} = 3\left[\dfrac{1}{25}[(2 + 5x) - 2 \ln|2 + 5x|]\right] + C$

$\quad = \dfrac{3}{25}[2 + 5x - 2 \ln|2 + 5x|] + C$

5. Use formula 15, where $a = 2$, $u = x$, $du = dx$.

$\displaystyle\int \sqrt{4 - x^2}\, dx = \int \sqrt{2^2 - x^2}\, dx$

$\quad = \dfrac{1}{2}x\sqrt{4 - x^2} + 2 \arcsin\dfrac{1}{2}x + C$

9. Use formula 17, where $u = 2x$, $du = 2\, dx$, $a = 3$.

$\displaystyle\int \dfrac{\sqrt{4x^2 - 9}}{x}\, dx = \int \dfrac{\sqrt{(2x)^2 - 3^2}}{2x}\, 2\, dx$

$\quad = \sqrt{4x^2 - 9} - 3 \text{ arcsec }\left(\dfrac{2x}{3}\right) + C$

13. Use formula 52, where $u = x^2$, $du = 2x\ dx$.

$$\int \text{Arctan } x^2\,(x\ dx) = \frac{1}{2}\int \text{Arctan } x^2\,(2x\ dx)$$

$$= \frac{1}{2}[x^2 \text{ Arctan } x^2 - \frac{1}{2}\ln(1 + x^4)] + C$$

$$= \frac{1}{2}x^2 \text{ Arctan } x^2 - \frac{1}{4}\ln(1 + x^4) + C$$

17. Use formula 11, where $u = 2x$, $du = 2\ dx$, $a = 1$.

$$\int \frac{dx}{x\sqrt{4x^2 + 1}} = \int \frac{2\ dx}{2x\sqrt{(2x)^2 + 1^2}} = -\ln(\frac{1 + \sqrt{4x^2 + 1}}{2x}) + C$$

21. Use formula 40, where $a = 1$, $u = x$, $du = dx$, $b = 5$.

$$\int \sin x \cos 5x\ dx = -\frac{\cos(-4x)}{2(-4)} - \frac{\cos 6x}{12} + C$$

$$= \frac{1}{8}\cos 4x - \frac{1}{12}\cos 6x + C$$

25. Use formula 25 where $u = x^2$, $du = 2x\ dx$, $a = 1$

$$\int \frac{2x\ dx}{(1 - x^2)^{3/2}} = \frac{x^2}{\sqrt{1 - x^4}} + C$$

29. Use formula 46, where $u = x^2$, $du = 2x\ dx$, $n = 1$.

$$\int x^3 \ln x^2\ dx = \frac{1}{2}\int x^2 \ln x^2\,(2x\ dx)$$

$$= \frac{1}{2}[(x^2)^2(\frac{\ln x^2}{2} - \frac{1}{4})] = \frac{1}{2}[\frac{x^4}{2}(\ln x^2 - \frac{1}{2})]$$

$$= \frac{1}{4}x^4(\ln x^2 - \frac{1}{2}) + C$$

33. From Exercise 17 of Exercises 25-6, $s = \int_a^b \sqrt{1 + (\frac{dy}{dx})^2}\ dx$; $y = x^2$, $\frac{dy}{dx} = 2x$

$$s = \int_0^1 \sqrt{1 + (2x)^2}\ dx = \frac{1}{2}\int_0^1 \sqrt{(2x)^2 + 1}\,(2\ dx)$$

Use formula 14, where $u = 2x$, $du = 2\ dx$.

$$s = \frac{1}{2}[\frac{2x}{2}\sqrt{4x^2 + 1} + \frac{1}{2}\ln(2x + \sqrt{4x^2 + 1})]\Big|_0^1$$

$$= \frac{1}{2}([1\sqrt{5} + \frac{1}{2}\ln(2 + \sqrt{5})] - \frac{1}{2}\ln 1)$$

$$= \frac{1}{4}[2\sqrt{5} + \ln(2 + \sqrt{5})] = \frac{1}{4}(4.472 + 1.444)$$

$$= 1.479$$

37.
$$F = w\int_0^3 \ell h\ dh = w\int_0^3 x(3-y)\ dy = w\int_0^3 \frac{3-y}{\sqrt{1+y}}\ dy \quad [\text{Formula 6}]$$

$$\int \frac{3-y}{\sqrt{1+y}}\ dy = 3\int \frac{dy}{\sqrt{1+y}} - \int \frac{y\ dy}{\sqrt{1+y}} = 3\frac{(1+y)^{\frac{1}{2}}}{\frac{1}{2}} - \left[\frac{-2(2-y)\sqrt{1+y}}{3(1)^2}\right] + C$$

$$F = w\int_0^3 \frac{3-y}{\sqrt{1+y}}\ dy = w\left[6(1+y)^{\frac{1}{2}} + \frac{2}{3}(2-y)(1+y)^{\frac{1}{2}}\right]\ \Big|_0^3$$

$$= w\left[6(2) + \frac{2}{3}(-1)(2) - 6(1) - \frac{2}{3}(2)(1)\right]$$

$$= w\left(12 - \frac{4}{3} - 6 - \frac{4}{3}\right) = \frac{10w}{3} = \frac{10(62.4)}{3} = 208\ \text{lb}$$

37M. $F = \frac{10w}{3} = 10(9800) = 98000$

Review Exercises for Chapter 27, page 874

1. $u = -2x,\ du = -2\ dx$
$$\int e^{-2x}\ dx = -\frac{1}{2}\int e^{-2x}(-2\ dx) = -\frac{1}{2}e^{-2x} + C \quad \text{Eq. (27-3)}$$

5. $\int \frac{4\cos x\ dx}{1+\sin x} = 4\int \frac{\cos x\ dx}{1+\sin x}$; $u = 1 + \sin x,\ du = \cos x$
$$= 4\ln(1+\sin x) + c$$

9. $\int_0^{\pi/2} \cos^3 2x\ dx = \int_0^{\pi/2} \cos^2 2x\ \cos 2x\ dx$

$$= \int_0^{\pi/2} (1 - \sin^2 2x)\ \cos 2x\ dx$$

$$= \int_0^{\pi/2} \cos 2x\ dx - \int_0^{\pi/2} \sin^2 2x\ \cos 2x\ dx$$

$$= \frac{1}{2}\int_0^{\pi/2} \cos 2x(2\ dx) - \frac{1}{2}\int_0^{\pi/2} \sin^2 2x\ \cos 2x(2\ dx)$$

$$= \frac{1}{2}[\sin 2x - \frac{1}{3}\sin^3 2x]\ \Big|_0^{\pi/2}$$

$$= \frac{1}{2}[(\sin \pi - \frac{1}{3}\sin^3 \pi) - (\sin 0 - \frac{1}{3}\sin^3 0)]$$

$$= \frac{1}{2}(0) = 0$$

13. Let $u = \sec 3x,\ du = \sec 3x\ \tan 3x(3\ dx)$
$$\int \sec^4 3x\ \tan 3x\ dx = \frac{1}{3}\int \sec^3 3x\ \sec 3x\ \tan 3x(3\ dx)$$

$$= \frac{1}{3}(\frac{\sec^4 3x}{4}) + C = \frac{1}{12}\sec^4 3x + C \quad \text{Eq. (26-8), (27-1)}$$

17. $\int \sec^4 3x\ dx = \int \sec^2 3x \sec^2 3x\ dx = \int (1 + \tan^2 3x) \sec^2 3x\ dx$

$= \frac{1}{3} \int \sec^2 3x\,(3\ dx) + \frac{1}{3} \int \tan^2 3x \sec^2 3x\,(3\ dx)$

$= \frac{1}{3} \tan 3x + \frac{1}{3}\frac{\tan^3 3x}{3} + C = \frac{1}{9} \tan^3 3x + \frac{1}{3} \tan 3x + C$

21. $\int \frac{3x\ dx}{4 + x^4} = 3\int \frac{x\ dx}{4 + x^4} = 3 \int \frac{1}{2^2 + (x^2)^2}\ x\,dx = \frac{3}{2} \int \frac{1}{2^2 + (x^2)^2}\ 2\ x\,dx$

$= \frac{3}{2}(\frac{1}{2}\arctan\frac{x^2}{2} + C) = \frac{3}{4}\arctan\frac{x^2}{2} + C$

　　Use Eq. (27-21); $a = 2$ and $u = x^2$

25. $u = e^{2x},\ du = e^{2x}(2\ dx)$

$\int \frac{e^{2x}\ dx}{\sqrt{e^{2x} + 1}} = \frac{1}{2} \int (e^{2x} + 1)^{-1/2} e^{2x}(2\ dx)$

$= \frac{1}{2}(e^{2x} + 1)^{1/2}(2) + C = \sqrt{e^{2x} + 1} + C \qquad$ Eq. (27-1)

29. By Eq. (27-22), $u = x,\ du = dx;\ dv = \csc^2 2x\ dx$

$v = \int \csc^2 2x\ dx = \frac{1}{2} \int \csc^2 2x\,(2\ dx) = -\frac{1}{2}\cot 2x$

$\int x \csc^2 2x\ dx = x(-\frac{1}{2}\cot 2x) - \int -\frac{1}{2}\cot 2x\ dx$

$= -\frac{1}{2}x \cot 2x + \frac{1}{4}\int \cot 2x\,(2\ dx)$

$= -\frac{1}{2}x \cot 2x + \frac{1}{4}\ln|\sin 2x| + C$

33. Let $u = \ln x,\ du = \frac{1}{x}\ dx$.

$\int_1^e \frac{(\ln x)^2\ dx}{x} = \int_1^e (\ln x)^2 \times \frac{1}{x}\ dx = \frac{(\ln x)^3}{3}\Big|_1^e$

$= \frac{(\ln e)^3}{3} - \frac{(\ln 1)^3}{1} = \frac{1}{3} \qquad$ Eq. (26-14), (27-1)

37. Use Eq. (27-1), the general power formula. Let $u = e^x + 1,\ du = e^x\ dx$, $n = 2$.

$\int e^x (e^x + 1)^2\ dx = \int (e^x + 1)^2 e^x\ dx = \frac{(e^x + 1)^3}{3} + C_1$

The second method used is to multiply the factors before integrating:

$\int e^x (e^x + 1)^2\ dx = \int e^x (e^{2x} + 2e^x + 1)\ dx$

$= \int (e^{3x} + 2e^{2x} + e^x)\ dx = \int e^{3x}\ dx + \int 2e^{2x}\ dx + \int e^x\ dx$

$= \frac{1}{3} \int e^{3x}(3\ dx) + 2(\frac{1}{2}) \int e^{2x}(2\ dx) + \int e^x\ dx$

$= \frac{1}{3}e^{3x} + e^{2x} + e^x + C_2; \qquad C_2 = C_1 + \frac{1}{3}$

The value of the constant of integration is different in the different answers.

41. $\int \sec^4 x \, dx = \dfrac{\sec^2 \tan x}{3} + \dfrac{2}{3} \int \sec^2 x \, dx$ (form 37)

$\qquad y = \dfrac{\sec^2 x \tan x}{3} + \dfrac{2}{3} \tan x + C$

$\qquad\qquad = \dfrac{1}{3}(1 + \tan^2 x)(\tan x) + \dfrac{2}{3}\tan x + C$

$\qquad\qquad = \dfrac{1}{3}\tan x + \dfrac{1}{3}\tan^3 x + \dfrac{2}{3}\tan x + C$

$\qquad\qquad = \dfrac{1}{3}\tan^3 x + \dfrac{2}{3}\tan x + C$

$\qquad 0 = \dfrac{1}{3}\tan^3(0) + \dfrac{2}{3}\tan 0 + C$

$\qquad 0 = 0 + 0 + C; \quad C = 0; \quad y = \dfrac{1}{3}\tan^3 x + \tan x$

45. $x^2 + y^2 = 5^2; \quad y = \sqrt{25 - x^2} \qquad A = 2\displaystyle\int_3^5 \sqrt{25 - x^2}\, dx$

Using integral form 15,

$$A = 2\left[\dfrac{x}{2}\sqrt{25 - x^2} + \dfrac{25}{2}\,\text{Arcsin}\,\dfrac{x}{5}\right]\Big|_3^5$$

$$A = 2\left[\dfrac{5}{2}\sqrt{0} + \dfrac{25}{2}\,\text{Arcsin}\,1\right] - 2\left[\dfrac{3}{2}\sqrt{16} + \dfrac{25}{2}\,\text{Arcsin}\,\dfrac{3}{5}\right]$$

$$= 2\left[\dfrac{25}{2}\left(\dfrac{\pi}{2}\right)\right] - 2\left[6 + \dfrac{25}{2}(0.6435)\right]$$

$$= 2\left[19.63\right] - 2\left[14.04\right] = 11.18$$

49. $y = e^x \sin x; \quad x = 0; \quad x = \pi$

$V = \pi\displaystyle\int \sin^2 x \, e^{2x} dx = \dfrac{\pi}{2}\int 2\sin^2 x \, e^{2x} dx = \dfrac{\pi}{2}\int (1 - \cos 2x)\, e^{2x} dx$

$\quad = \dfrac{\pi}{2}\displaystyle\int (e^{2x} - e^{2x}\cos 2x)\, dx = \dfrac{\pi}{2}\left[\int e^{2x} dx - \int e^{2x}\cos 2x \, dx\right]$

$\quad = \dfrac{\pi}{2}\left[\dfrac{1}{2}\displaystyle\int e^{2x} dx - \int e^{2x}\cos 2x \, dx\right]$ using formula 50 in table of integrals

$\quad = \dfrac{\pi}{2}\left[\dfrac{1}{2}e^{2x} - \dfrac{e^{2x}(2\cos 2x + 2\sin 2x)}{2^2 + 2^2}\right]\Big|_0^\pi$

$\quad = \dfrac{\pi}{2}\left[\dfrac{1}{2}e^{2x} - e^{2x}\dfrac{(\cos 2x + \sin 2x)}{4}\right]\Big|_0^\pi$

$\quad = \dfrac{\pi}{2}\left[\dfrac{2e^{2x} - e^{2x}(\cos 2x + \sin 2x)}{4}\right]\Big|_0^\pi$

$\quad = \dfrac{\pi}{8}\left[2e^{2x} - e^{2x}(\cos 2x + \sin 2x)\right]\Big|_0^\pi$

$\quad = \dfrac{\pi}{8}\left[2e^{2\pi} - e^{2\pi}(\cos 2\pi + \sin 2\pi) - 2e^0 + e^0(\cos 0 + \sin 0)\right]\Big|_0^\pi$

$$= \frac{\pi}{8}\left[2e^{2\pi} - e^{2\pi}(1) - 2 + 1\right] = \frac{\pi}{8}\left[e^{2\pi} - 1\right] = 209.9$$

53. $\Delta s = \int (C_v/T)\, dT = \int \frac{a + bT + cT^2}{T}\, dt$

$$= \int \frac{a}{T}dt + \int b\, dt + \int cT\, dt$$

$$= a \ln T + bT + \tfrac{1}{2}cT^2 + C$$

57. $y_{rms} = \sqrt{\frac{1}{T} \int i^2 dt} = \sqrt{\frac{1}{T} \int (2 \sin t)^2\, dt}$

$$\int_0^{2\pi} (2 \sin t)^2\, dt = \int_0^{2\pi} 4 \sin^2 t\, dt = 4 \int_0^{2\pi} \sin^2 t\, dt$$

$$= 4\, (\frac{t}{2} - \frac{1}{2} \sin t \cos t)\,\Big|_0^{2\pi} \qquad \text{(Form 29)}$$

$$= 2t - 2 \sin t \cos t\,\Big|_0^{2\pi} = 4\pi;\ \ T = 2\pi$$

$$y_{rms} = \sqrt{\frac{1}{2\pi}\,(4\pi)} = \sqrt{2}$$

61. $V = \pi \int_2^4 y^2\, dx = \pi \int_2^4 e^{-0.2x}dx;\ \ u = -0.2x\ \ du = -0.2dx$

$$= -\frac{\pi}{0.2} \int_2^4 e^{-0.2x}\, (-0.2)\, dx = \frac{-\pi}{0.2}\, e^{-0.2x}\,\Big|_2^4$$

$$= -\frac{\pi}{2}\left[e^{-0.8} - e^{-0.4}\right] = \frac{-\pi}{0.2}\left[0.4493 - 0.6703\right] = 3.47$$

CHAPTER 28

Expansion of Functions in Series

1. $a_n = n^2$, $n = 1, 2, 3...$

$a_1 = (1)^1 = 1$ $a_3 = (3)^2 = 9$

$a_2 = (2)^2 = 4$ $a_4 = (4)^2 = 16$

5. $a_n = \left(-\dfrac{2}{5}\right)^n$

   (a) $a_1 = \left(-\dfrac{2}{5}\right)^1 = -\dfrac{2}{5}$ $a_3 = \left(-\dfrac{2}{5}\right)^3 = -\dfrac{8}{25}$

   $a_2 = \left(-\dfrac{2}{5}\right)^2 = \dfrac{4}{25}$ $a_4 = \left(-\dfrac{2}{5}\right)^4 = \dfrac{16}{625}$

   (b) $S = -\dfrac{2}{5} + \dfrac{4}{25} - \dfrac{8}{125} + \dfrac{16}{625}$

9. $\dfrac{1}{2} + \dfrac{1}{3} + \dfrac{1}{4} + \dfrac{1}{5} + ...$

   $n = 1, \quad a_1 = \dfrac{1}{2} = \dfrac{1}{2 + (n - 1)}$

   $n = 2, \quad a_2 = \dfrac{1}{3} = \dfrac{1}{2 + (n - 1)}$

   $n^{th} \text{ term} = \dfrac{1}{2 + (n - 1)} = \dfrac{1}{m + 1}$

17. $\displaystyle\sum_{n=0}^{\infty} (-n)$ The first five terms are: $0, -1, -2, -3, -4$

   The first five partial sums are: $0, -1, -3, -6, -10$

   The series is divergent.

21. $1 + 2 + 4 + \ldots + 2^n + \ldots$

This is a geometric series with $a = 1$ and $r = 2$. Since $|r| > 1$, $\lim_{n \to \infty} r^n$ is unbounded. Therefore the series is divergent.

25. $10 + 9 + 8.1 + 7.29 + 6.561 + \ldots$

This is a geometric series where $a = 10$, $x = \dfrac{9}{10}$

The sum by (28-3) is: $s = \dfrac{a}{1 - r} = \dfrac{10}{1 - \dfrac{9}{10}} = 100$

29. Successive square roots of 2 are approximately: 1.414214, 1.189207, 1.090508, 1.044274, 1.021897, 1.010889, 1.005430, 1.002711, 1.001355, 1.000677, 1.000339, 1.000169, 1.000085, 1.000042, 1.000021, 1.000011, 1.000005, 1.000003, 1.000001, 1.000000

(a) It appears that the value of $\lim_{n \to \infty} 2^{1/2n}$ is 1.

(b) The infinite series will diverge since each successive term will increase the sum by approximately 1.

Successive square roots of 0.01 are approximately: 0.100000, 0.316228, 0.562341, 0.749894, 0.865964, 0.930572, 0.964662, 0.982172, 0.991046, 0.995513, 0.997754, 0.998876, 0.999438, 0.999719, 0.999860, 0.999930, 0.999964, 0.999982, 0.999991, 0.999996

(a) It appears that the value of $\lim 0.01^{1/n}$ is 1.

(b) The series will diverge $n \to \infty$

Successive square roots of 100 are approximately: 10.000000, 3.162278, 1.778279, 1.333521, 1.154782, 1.074608, 1.036633, 1.018152, 1.009035, 1.004507, 1.002251, 1.001125, 1.000562, 1.000281, 1.000141, 1.000070, 1.000035, 1.000018, 1.000009, 1.000004

(a) It appears that the value of $\lim 100^{1/n}$ is 1.

(b) The series will diverge $n \to \infty$

Exercises 28-2, page 885

1. $f(x) = e^x$ $f(0) = e^0 = 1$
    $f'(x) = e^x$ $f'(0) = e^0 = 1$
    $f''(x) = e^x$ $f''(0) = e^0 = 1$

    By Eq. (28-2), $f(x) = 1 + (1)x + \dfrac{(1)x^2}{2!} + \ldots = 1 + x + \dfrac{1}{2}x^2 + \ldots$

5. $f(x) = (1 + x)^{1/2}$ $f(0) = 1$

    $f'(x) = \dfrac{1}{2}(1 + x)^{-1/2}$ $f'(0) = \dfrac{1}{2}$

    $f''(x) = -\dfrac{1}{4}(1 + x)^{-3/2}$ $f''(0) = -\dfrac{1}{4}$

    By Eq. (28-2), $f(x) = 1 + \dfrac{1}{2}x - \dfrac{1}{4}\dfrac{x^2}{2!} + \ldots = 1 + \dfrac{1}{2}x - \dfrac{1}{8}x^2 + \ldots$

9. $f(x) = \cos 4x$ $f(0) = 1$
    $f'(x) = -\sin 4x(4) = -4 \sin 4x$ $f'(0) = 0$
    $f''(x) = -4 \cos 4x(4) = -16 \cos 4x$ $f''(0) = -16$
    $f'''(x) = 16 \sin 4x(4) = 64 \sin 4x$ $f'''(0) = 0$
    $f^{iv}(x) = 64 \cos 4x(4) = 256 \cos 4x$ $f^{iv}(0) = 256$

    By Eq. (28-2), $f(x) = 1 + (0)x - \dfrac{16x^2}{2!} + \dfrac{0x^3}{3!} + \dfrac{256x^4}{4!} + \ldots$

                    $= 1 - 8x^2 + \dfrac{32}{3}x^4 - \ldots$

13. $f(x) = \ln(1 - 2x)$ $f(0) = \ln 1 = 0$

    $f'(x) = \dfrac{1}{1 - 2x}(-2) = \dfrac{-2}{1 - 2x}$ $f'(0) = -2$

    $f''(x) = \dfrac{0 - (-2)(-2)}{(1 - 2x)^2} = \dfrac{-4}{(1 - 2x)^2}$ $f''(0) = -4$

    $f'''(x) = \dfrac{0 - (-4)2(1 - 2x)(-2)}{(1 - 2x)^4} = \dfrac{-16(1 - 2x)}{(1 - 2x)^4}$ $f'''(0) = -16$

    $\ln(1 - 2x) = 0 + (-2)x + (-4)\dfrac{x^2}{2!} + (-16)\dfrac{x^3}{3!} + \ldots$

    $\ln(1 - 2x) = -2x - 2x^2 - \dfrac{8}{3}x^3 - \ldots$

17. $f(x) = \text{Arctan } x$ $f(0) = 0$

    $f'(x) = \dfrac{1}{1 + x^2} = (1 + x^2)^{-1}$ $f'(0) = 1$

    $f''(x) = -(1 + x^2)^{-2}2x = -2x(1 + x^2)^{-2}$ $f''(0) = 0$

    $f'''(x) = -2x[-2(1 + x^2)^{-3}(2x)] + (1 + x^2)^{-2}(-2)$ $f'''(0) = -2$

    By Eq. (28-2), $f(x) = 0 + 1x + \dfrac{0x^2}{2!} - \dfrac{2x^3}{3!} + \ldots = x - \dfrac{1}{3}x^3 + \ldots$

21. $f(x) = \ln \cos x$ $f(0) = \ln 1 = 0$

$f'(x) = -\dfrac{1}{\cos x} \sin x = -\tan x$ $f'(0) = 0$

$f''(x) = -\sec^2 x$ $f''(0) = -1$

$f'''(x) = -2 \sec x \sec x \tan x = -2 \sec^2 x \tan x$ $f'''(0) = 0$

$f^{iv}(x) = -2 \sec^2 x \sec^2 x - 2 \tan x (2 \sec x \sec x \tan x)$ $f^{iv}(0) = -2 - 0 = -2$

By Eq. (28-2), $f(x) = 0 + 0x - \dfrac{1x^2}{2!} + \dfrac{0x^3}{3!} - \dfrac{2x^4}{4!} + \ldots$

$= -\dfrac{1}{2}x^2 - \dfrac{1}{12}x^4 - \ldots$

25. (a) It is not possible to find a MacLaurin expansion for $f(x) = \csc x$ since the function is not defined when $x = 0$.

(b) $f(x) = \ln x$ is not defined when $x = 0$.

29. $f(x) = x^3;$ $f(0) = 0$

$f'(x) = 3x^2;$ $f'(0) = 0$

$f''(x) = 6x$ $f''(0) = 0$

$f'''(x) = 6$ $f'''(0) = 6$

$f(x) = 0 + 0x + \dfrac{0x^2}{2!} + \dfrac{6x^3}{3!} = \dfrac{6x^3}{6} = x^3$

Exercises 28-3, page 891

1. From Eq. (28-4), $f(x) = 1 + x + \dfrac{x^2}{2!} + \dfrac{x^3}{3!} + \ldots$

$e^{3x} = f(3x) = 1 + 3x + \dfrac{(3x)^2}{2!} + \dfrac{(3x)^3}{3!} + \ldots$

$= 1 + 3x + \dfrac{9}{2}x^2 + \dfrac{9}{2}x^3 + \ldots$

5. From Eq. (28-6), $f(x) = 1 - \dfrac{x^2}{2!} + \dfrac{x^4}{4!} - \dfrac{x^6}{6!} + \ldots$

$\cos 4x = f(4x) = 1 - \dfrac{(4x)^2}{2!} + \dfrac{(4x)^4}{4!} - \dfrac{(4x)^6}{6!} + \ldots$

$= 1 - 8x^2 + \dfrac{32}{3}x^4 - \dfrac{256}{45}x^6 + \ldots$

9. By Eq. (28-5), $\displaystyle\int_0^1 \sin x^2\, dx = \int_0^1 \left(x^2 - \dfrac{(x^2)^3}{3!} + \dfrac{(x^2)^5}{5!} + \ldots\right) dx$

$= \displaystyle\int_0^1 \left(x^2 - \dfrac{x^6}{6} + \dfrac{x^{10}}{120}\right) dx$

$= \dfrac{x^3}{3} - \dfrac{x^7}{42} + \dfrac{x^{11}}{1320} + \ldots \Big|_0^1 = \dfrac{1}{3} - \dfrac{1}{42} + \dfrac{1}{1320} + \ldots$

$= 0.33333 - 0.02381 + 0.00076$

$= 0.31028 = 0.3103$ (rounded off)

13. $f(x) = \frac{1}{2}(e^x + e^{-x}); \quad e^x = 1 + x + \frac{x^2}{2!} + \frac{x^3}{3!} + \frac{x^4}{4!} + \frac{x^5}{5!} + \frac{x^6}{6!} + \ldots$ Eq. (28-4)

$$e^{-x} = 1 - x + \frac{x^2}{2!} - \frac{x^3}{3!} + \frac{x^4}{4!} - \frac{x^5}{5!} + \frac{x^6}{6!} + \ldots$$

$$e^x + e^{-x} = 2 + \frac{2x^2}{2!} + \frac{2x^4}{4!} + \frac{2x^6}{6!}$$

$$= 2 + x^2 + \frac{1}{12}x^4 + \frac{1}{360}x^6 + \ldots$$

$$f(x) = \frac{1}{2}(2 + x^2 + \frac{1}{12}x^4 + \frac{1}{360}x^6 + \ldots)$$

$$= 1 + \frac{1}{2}x^2 + \frac{1}{24}x^4 + \frac{1}{720}x^6 + \ldots$$

17. $\dfrac{d(\sin x)}{dx} = \dfrac{d}{dx}\left(x - \dfrac{x^3}{3!} + \dfrac{x^5}{5!} + \ldots\right)$

$$= 1 - \frac{3x^2}{3!} + \frac{5x^4}{5!} + \ldots \quad = 1 - \frac{x^2}{2!} + \frac{x^4}{4!} + \ldots \quad = \cos x$$

21. $\displaystyle\int_0^1 e^x\,dx = e^x\Big|_0^1 = e - e^0 = e - 1 \approx 2.718 - 1 \approx 1.718$

$f(x) = e^x; \quad f(0) = e^0 = 1$

$f'(x) = e^x; \quad f'(0) = 1$

$f''(x) = e^x; \quad f''(0) = 1$

$f'''(x) = e^x; \quad f'''(0) = 1$

$$e^x = 1 + x + \frac{x^2}{2!} + \frac{x^3}{3!} + \ldots$$

$$\int_0^1 \left(1 + x + \frac{x^2}{2} + \frac{x^3}{6}\right)dx = x + \frac{x^2}{2} + \frac{x^3}{6} + \frac{x^4}{24}\ \Big|_0^1$$

$$= 1 + \frac{1}{2} + \frac{1}{6} + \frac{1}{24} = 1.7083333$$

23. $A = \displaystyle\int_0^{0.2} x^2 e^x\,dx = \int_0^{0.2} x^2\left(1 + x + \frac{x^2}{2!} + \ldots\right) dx$

$$= \int_0^{0.2} \left(x^2 + x^3 + \frac{x^4}{2} + \ldots\right) dx = \frac{x^3}{3} + \frac{x^4}{4} + \frac{x^5}{10}\ \Big|_0^{0.2}$$

$$= 0.002667 + 0.000400 + 0.000032 = 0.003099$$

23M.   0.003 099

<u>Exercises 28-4,</u> page 894

1. By Eq. (28-6), $e^x = 1 + x + \dfrac{x^2}{2!} + \ldots$

    Let $x = 0.2$, $e^{0.2} = 1 + 0.2 + \dfrac{(0.2)^2}{2!} = 1.22$; 1.2214028 (calculator)

1M. 1.22; 1.221 402 8

5. By Eq. (28-6), $e^x = 1 + x + \dfrac{x}{2!} + \dfrac{x}{3!} + \dfrac{x}{4!} + \dfrac{x}{5!} + \dfrac{x}{6!} + \ldots$

    Let $x = 1$; $e^1 = 1 + 1 + \dfrac{1}{2} + \dfrac{1}{6} + \dfrac{1}{24} + \dfrac{1}{120} + \dfrac{1}{720} + \ldots = 2.7180556$; 2.7182818 (calculator)

5M. 2.718; 2.718 281 8

9. By Eq. (28-9), $\ln(1 + x) = x - \dfrac{x^2}{2} + \dfrac{x^3}{3} - \dfrac{x^4}{4} + \ldots$

    Let $x = 0.4$; $\ln(1 + 0.4) = 0.4 - \dfrac{(0.4)^2}{2} + \dfrac{(0.4)^3}{3} - \dfrac{(0.4)^4}{4} + \ldots$

    $= 0.3349333$; 03364722 (calculator)

9M. 0.3349333; 0.336 472 2

13. ln 0.9861

    $\ln(1 + x) = x - \dfrac{x^2}{2} + \dfrac{x^3}{3} - \dfrac{x^4}{4} + \ldots$ Eq. (28-9)

    Let $x = -0.0139$

    $\ln[1 + (-0.0139)] = -0.0139 - \dfrac{(-0.0139)^2}{2} + \dfrac{(-0.0139)^3}{3} + \ldots$

    $= -0.0139 - 0.0000966 - 0.0000009 + \ldots$

    $= -0.0139957$; -0.0139975 (calculator)

13M. $-0.013\ 995\ 7$; $-0.013\ 997\ 5$

17. $\sqrt{1.1076} = 1.1076^{1/2} = (1 + 0.1076)^{1/2}$

    Using Eq. (28-10), $(1 + x)^n = 1 + nx + \dfrac{n(n - 1)x^2}{2!} + \ldots$

    $x = 0.1076$ and $n = \dfrac{1}{2}$

    $\sqrt{1.1076} = 1 + \dfrac{1}{2}(0.1076) + \dfrac{\frac{1}{2}(-\frac{1}{2})(0.1076)^2}{2} + \ldots$

    $= 1 + 0.0538000 - 0.0014472 + \ldots = 1.0523528$

17M. 1.052 352 8

21. From Exercise 3, $\sin(0.1) = 0.1 + \dfrac{0.1^3}{6} = 0.1001667$

    The maximum possible error is the value of the first term omitted, $\dfrac{x^5}{5!} = \dfrac{0.1^5}{120} = 8.333 \times 10^{-8}$. See Eq. (28-7).

25. $\arctan \frac{1}{2} = \frac{1}{2} - \frac{1}{3}(\frac{1}{2})^3 + \frac{1}{5}(\frac{1}{2})^5 = 0.4646$

   $\arctan \frac{1}{3} = \frac{1}{3} - \frac{1}{3}(\frac{1}{3})^3 + \frac{1}{5}(\frac{1}{3})^5 = 0.3218$

   $\pi = 4(0.4646 + 0.3218) = 3.146$

29. $f(t) = \frac{E}{R}(1 - e^{-Rt/L}); \quad e^x = 1 + x + \frac{x^2}{2} + \ldots$

   $e^{-Rt/L} = 1 - \frac{Rt}{L} + \frac{R^2 t^2}{2L^2} + \ldots$

   $i = \frac{E}{R}[1 - (1 - \frac{Rt}{L} + \frac{R^2 t^2}{2L^2})] = \frac{E}{L}(t - \frac{Rt^2}{2L})$

   The approximation will be valid for small values of t.

Exercises 28-5, page 895

1. From Example A, $e^x = e[1 + (x - 1) + \frac{(x-1)^2}{2} + \frac{(x-1)^3}{6} + \ldots]$

   Let $x = 1.2$

   $e^{1.2} = 2.718[1 + (1.2 - 1) + \frac{(1.2 - 1)^2}{2} + \frac{(1.2 - 1)^3}{6} + \ldots]$

   $= 2.718(1.2227) = 3.32$

5. Let $a = 30° = \frac{\pi}{6}$; from Example D, $\sin x = \frac{1}{2} + \frac{\sqrt{3}}{2}(x - \frac{\pi}{6}) - \frac{1}{4}(x - \frac{\pi}{6})^2 - \ldots$

   $31° = 0.5410$ radians

   $\sin 31° = \frac{1}{2} + \frac{\sqrt{3}}{2}(0.5410 - 0.5236) - \frac{1}{4}(0.5410 - 0.5236)^2 = 0.5150$

9. $f(x) = e^{-x}$ $f(2) = e^{-2}$

   $f'(x) = -e^{-x}$ $f'(2) = -e^{-2}$

   $f''(x) = e^{-x}$ $f''(2) = e^{-2}$

   By Eq. (28-16), $f(x) = e^{-2} - e^{-2}(x - 2) + \frac{e^{-2}(x - 2)^2}{2!} + \ldots$

   $= e^{-2}[1 - (x - 2) + \frac{(x - 2)^2}{2!} + \ldots]$

13. $f(x) = x^{1/3}$ $f(8) = 8^{1/3} = 2$

   $f'(x) = \frac{1}{3}x^{-2/3}$ $f'(8) = \frac{1}{3(8)^{2/3}} = \frac{1}{12}$

   $f''(x) = -\frac{2}{9}x^{-5/3}$ $f''(8) = \frac{-2}{9(8^{5/3})} = -\frac{1}{144}$

   By Eq. (28-16), $f(x) = 2 + \frac{1}{12}(x - 8) - \frac{1}{288}(x - 8)^2 + \ldots$

17. From Exercise 9, $f(x) = e^{-2}[1 - (x - 2) + \frac{(x - 2)^2}{2!} + \ldots]$; let $x = 2.2$

   $f(2.2) = 0.1353[1 - (2.2 - 2) + \frac{(2.2 - 2)^2}{2} + \ldots] = 0.111$

21. $\sqrt[3]{8.3}$; $f(x) = x^{1/3}$ $f(8) = 2$

    $f'(x) = \frac{1}{3}x^{-2/3}$ $f'(8) = \dfrac{1}{3\sqrt[3]{8^2}} = \dfrac{1}{12} = 0.083333$

    $f''(x) = -\frac{2}{9}x^{-5/3}$ $f''(8) = -\dfrac{2}{9\sqrt[3]{8^5}} = -\dfrac{2}{9(32)} = -0.0069444$

    $a = 8$; $\sqrt[3]{8.3} = 2 + 0.083333(8.3 - 8) - 0.0069444(8.3 - 8)^2 = 2.025$

25. $f(x) = c_0 + c_1(x - a) + c_2(x - a)^2 + c_3(x - a)^3 + c_4(x - a)^4 + c_5(x - a)^5$
     $+ \ldots c_n(x - a)^n$

     $f'(x) = c_1 + 2c_2(x - a) + 3c_3(x - a)^2 + 4c_4(x - a)^3 + 5c_5(x - a)^4$
     $+ \ldots nc_n(x - a)^{n-1}$

     $f''(x) = 2c_2 + 2 \times 3c_3(x - a) + 3 \times 4c_4(x - a)^2 + 4 \times 5c_5(x - a)^3$
     $+ \ldots (n - 1)nc_n(x - a)^{n-2}$

     $f'''(x) = 2 \times 3c_3 + 2 \times 3 \times 4c_4(x - a) + 3 \times 4 \times 5c_5(x - a)^2$
     $+ \ldots (n - 2)(n - 1)nc_n(x - a)^{n-3}$

     $f^{iv}(x) = 2 \times 3 \times 4c_4 + 2 \times 3 \times 4 \times 5c_5(x - a)$
     $+ \ldots (n - 3)(n - 2)(n - 1)nc_n(x - a)^{n-4}$

     Let $x = a$; $f(a) = c_0$

     $f'(a) = c_1$

     $f''(a) = 2c_2$; $c_2 = \dfrac{f''(a)}{2!}$

     $f'''(a) = 2 \times 3c_3$; $c_3 = \dfrac{f'''(a)}{3!}$

     $f^{iv}(a) = 2 \times 3 \times 4c_4$; $c_4 = \dfrac{f^{iv}(a)}{4!}$

     $f(x) = f(a) + f'(a)(x - a) + \dfrac{f''(a)(x - a)^2}{2!} + \dfrac{f'''(a)(x - a)^3}{3!}$
     $+ \dfrac{f^{iv}(a)(x - a)^4}{4!} + \ldots$

Exercises 28-6, page 905

1. $a_0 = \dfrac{1}{2\pi}\displaystyle\int_{-\pi}^{0} 1\, dx + \dfrac{1}{2\pi}\displaystyle\int_{0}^{\pi} 0\, dx = \dfrac{1}{2\pi}x \Big|_{-\pi}^{0}$

    $= 0 - \dfrac{1}{2\pi}(-\pi) = \dfrac{1}{2}$ Eq. (28-25)

    $a_1 = \dfrac{1}{\pi}\displaystyle\int_{-\pi}^{0} (1)\cos x\, dx + \dfrac{1}{\pi}\displaystyle\int_{0}^{\pi} 0\, dx = \dfrac{1}{\pi}\sin x \Big|_{-\pi}^{0} = 0$ Eq. (28-24)

    $a_n = 0$ for all values of n since $\sin n\pi = 0$

$$b_1 = \frac{1}{\pi}\int_{-\pi}^{0} 1 \sin x\, dx + \int_{0}^{\pi} 0\, dx = \frac{1}{\pi}(-\cos x)\Big|_{-\pi}^{0}$$

$$= \frac{1}{\pi}(-1 - 1) = -\frac{2}{\pi} \quad \text{Eq. (28-26)}$$

$$b_2 = \frac{1}{\pi}\int_{-\pi}^{0} 1 \sin 2x\, dx + \int_{0}^{\pi} 0\, dx = \frac{1}{2\pi}\int_{-\pi}^{0} \sin 2x\,(2\, dx)$$

$$= \frac{1}{2\pi}\int_{-\pi}^{0} \sin 2x\,(2\, dx) = \frac{1}{2\pi}(-\cos 2x)\Big|_{-\pi}^{0} = \frac{1}{2\pi}(-1 + 1) = 0$$

$$b_3 = \frac{1}{\pi}\int_{-\pi}^{0} 1 \sin 3x\, dx + \int_{0}^{\pi} 0\, dx = \frac{1}{3\pi}\int_{-\pi}^{0} \sin 3x\,(3\, dx)$$

$$= \frac{1}{3\pi}(-\cos 3x)\Big|_{-\pi}^{0} = \frac{1}{3\pi}(-1 - 1) = \frac{-2}{3\pi}$$

By Eq. (28-17), $f(x) = \frac{1}{2} - \frac{2}{\pi} \sin x - \frac{2}{3\pi} \sin 3x - \cdots$

5. $\quad a_0 = \frac{1}{2\pi}\int_{-\pi}^{0} 0\, dx + \frac{1}{2\pi}\int_{0}^{\pi} x\, dx = \frac{1}{2\pi}\left(\frac{x^2}{2}\right)\Big|_{0}^{\pi} = \frac{\pi}{4} \quad$ Eq. (28-25)

$$a_1 = \frac{1}{\pi}\int_{-\pi}^{0} 0 \cos x\, dx + \frac{1}{\pi}\int_{0}^{\pi} x \cos x\, dx$$

$$= \frac{1}{\pi}(\cos x + x \sin x)\Big|_{0}^{\pi} = \frac{1}{\pi}(-1 - 1) = -\frac{2}{\pi}$$

(Using Eq. (28-24) and formula 48)

$$a_2 = \frac{1}{\pi}\int_{-\pi}^{0} 0\, dx + \frac{1}{\pi}\int_{0}^{\pi} x \cos 2x\, dx = \frac{1}{4\pi}\int_{0}^{\pi} 2x \cos 2x\,(2\, dx)$$

$$= \frac{1}{4\pi}(\cos 2x + 2x \sin 2x)\Big|_{0}^{\pi} = \frac{1}{4\pi}(\cos 2\pi - \cos 0) = 0$$

$$a_3 = \frac{1}{\pi}\int_{-\pi}^{0} 0\, dx + \frac{1}{\pi}\int_{0}^{\pi} x \cos 3x\, dx = \frac{1}{9\pi}\int_{0}^{\pi} 3x \cos 3x\,(3\, dx)$$

$$= \frac{1}{9\pi}(\cos 3x + 3x \sin 3x)\Big|_{0}^{\pi} = \frac{1}{9\pi}[\cos 3\pi + 3\pi \sin 3\pi) - (\cos 0 + 0)]$$

$$= \frac{1}{9\pi}(-1 - 1) = -\frac{2}{9\pi}$$

$$b_1 = \frac{1}{\pi}\int_{-\pi}^{0} 0\, dx + \frac{1}{\pi}\int_{0}^{\pi} x \sin x\, dx = \frac{1}{\pi}(\sin x - x \cos x)\Big|_{0}^{\pi}$$

$$= \frac{1}{\pi}[(\sin \pi - \pi \cos \pi) - (\sin 0 - 0)]$$

$$= \frac{1}{\pi}(-\pi)(-1) = 1 \quad \text{Eq. (28-26) and formula 47}$$

$$b_2 = \frac{1}{\pi}\int_{-\pi}^{0} 0\, dx + \frac{1}{\pi}\int_{0}^{\pi} x \sin 2x\, dx$$

$$= \frac{1}{4\pi}\int_{0}^{\pi} 2x \sin 2x\,(2\, dx) = \frac{1}{4\pi}(\sin 2x - 2x \cos 2x)\Big|_{0}^{\pi}$$

$$= \frac{1}{4\pi}[(\sin 2\pi - 2\pi \cos 2\pi) - (\sin 0 - 0)]$$

$$= \frac{1}{4\pi}(0 - 2\pi - 0) = -\frac{1}{2}$$

By Eq. (28-17), $f(x) = \frac{\pi}{4} - \frac{2}{\pi} \cos x + 0 \cos 2x - \frac{2}{9\pi} \cos 3x + \ldots$

$$+ \sin x - \frac{1}{2} \sin 2x + \ldots$$

$$= \frac{\pi}{4} - \frac{2}{\pi}(\cos x + \frac{1}{9} \cos 3x + \ldots)$$

$$+ (\sin x - \frac{1}{2} \sin 2x + \ldots)$$

9. $a_0 = \frac{1}{2\pi}\int_{-\pi}^{0} -x\,dx + \frac{1}{2\pi}\int_{0}^{\pi} x\,dx = -\frac{1}{2\pi}(\frac{x^2}{2})\Big|_{-\pi}^{0} + \frac{1}{2\pi}(\frac{x^2}{2})\Big|_{0}^{\pi}$

$$= 0 - (-\frac{\pi}{4}) + \frac{\pi^2}{4\pi} - 0 = \frac{\pi}{4} + \frac{\pi}{4} = \frac{\pi}{2} \quad \text{Eq. (28-22)}$$

$a_1 = \frac{1}{\pi}\int_{-\pi}^{0} -x \cos x\,dx + \frac{1}{\pi}\int_{0}^{\pi} x \cos x\,dx$

$$= -\frac{1}{\pi}(\cos x + x \sin x)\Big|_{-\pi}^{0} + \frac{1}{\pi}(\cos x + x \sin x)\Big|_{0}^{\pi}$$

$$= -\frac{1}{\pi}[\cos 0 + 0 \sin 0 - \cos(-\pi) + \pi \sin(-\pi)]$$

$$+ \frac{1}{\pi}(\cos \pi + \pi \sin \pi - \cos 0 - 0 \sin 0)$$

$$= -\frac{4}{\pi}$$

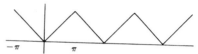

Using Eq. (28-24) and formula 48,

$a_2 = \frac{1}{\pi}\int_{-\pi}^{0} -x \cos 2x\,dx + \frac{1}{\pi}\int_{0}^{\pi} x \cos 2x\,dx$

$$= -\frac{1}{4\pi}\int_{-\pi}^{0} 2x \cos 2x(2\,dx) + \frac{1}{4\pi}\int_{0}^{\pi} 2x \cos 2x\,2\,dx$$

$$= -\frac{1}{4\pi}(\cos 2x + 2x \sin 2x)\Big|_{-\pi}^{0} + \frac{1}{4\pi}(\cos 2x + 2x \sin 2x)\Big|_{0}^{\pi} = 0$$

$a_3 = \frac{1}{\pi}\int_{-\pi}^{0} -x \cos 3x\,dx + \frac{1}{\pi}\int_{0}^{\pi} x \cos 3x\,dx$

$$= -\frac{1}{9\pi}\int_{-\pi}^{0} 3x \cos 3x(3\,dx) + \frac{1}{9\pi}\int_{0}^{\pi} 3x \cos 3x(3\,dx)$$

$$= -\frac{1}{9\pi}(\cos 3x + 3x \sin 3x)\Big|_{-\pi}^{0} + \frac{1}{9\pi}(\cos 3x + 3x \sin 3x)\Big|_{0}^{\pi}$$

$$= -\frac{4}{9\pi}$$

$b_1 = \frac{1}{\pi}\int_{-\pi}^{0} -x \sin x\,dx + \frac{1}{\pi}\int_{0}^{\pi} x \sin x\,dx$

$$= -\frac{1}{\pi}(\sin x - x \cos x)\Big|_{-\pi}^{0} + \frac{1}{\pi}(\sin x - x \cos x)\Big|_{0}^{\pi}$$

$$= -1 + 1 = 0$$

$$b_2 = \frac{1}{\pi}\int_{-\pi}^{0} -x \sin 2x \, dx + \frac{1}{\pi}\int_{0}^{\pi} x \sin 2x \, dx$$

$$= -\frac{1}{4\pi}\int_{-\pi}^{0} 2x \sin 2x (2 \, dx) + \frac{1}{4\pi}\int_{0}^{\pi} 2x \sin 2x (2 \, dx)$$

$$= -\frac{1}{4\pi}(\sin 2x - 2x \cos 2x)\Big|_{-\pi}^{0} + \frac{1}{4\pi}(\sin 2x - 2x \cos 2x)\Big|_{0}^{\pi}$$

$$= \frac{1}{2} - \frac{1}{2} = 0 \qquad b_n = 0 \text{ for all values of } n$$

By Eq. (28-17), $f(x) = \frac{\pi}{2} - \frac{4}{\pi} \cos x + 0 \cos 2x - \frac{4}{9\pi} \cos 3x + \dots$

$$= \frac{\pi}{2} - \frac{4}{\pi} \cos x - \frac{4}{9\pi} \cos 3x - \dots$$

13. $a_n = \frac{1}{\pi}\int_{-\pi}^{0} -\sin t \cos nt \, dt + \frac{1}{\pi}\int_{0}^{\pi} \sin t \cos nt \, dt$

$$= \frac{1}{2\pi}[\frac{\cos(1-n)t}{1-n} + \frac{\cos(1+n)t}{1+n}]_{-\pi}^{0} + -\frac{1}{2\pi}[\frac{\cos(1-n)t}{1-n} + \frac{\cos(1+n)t}{1+n}]_{0}^{\pi}$$

$$= \frac{1}{2\pi}(\frac{1}{1-n} + \frac{1}{1+n} - \frac{\cos(1-n)(-\pi)}{1-n} - \frac{\cos(1+n)(-\pi)}{1+n})$$

$$-\frac{1}{2\pi}(\frac{\cos(1-n)\pi}{1-n} + \frac{\cos(1+n)\pi}{1+n} - \frac{1}{1-n} - \frac{1}{1+n})$$

$a_1 = \frac{1}{\pi}\int_{-\pi}^{0} -\sin t \cos t \, dt + \int_{0}^{\pi} \sin t \cos t \, dt$

$$= -\frac{1}{2\pi} \sin^2 t \Big|_{-\pi}^{0} + \frac{1}{2\pi} \sin^2 t \Big|_{0}^{\pi} = 0$$

$a_2 = \frac{1}{2\pi}[\frac{1}{-1} + \frac{1}{3} - \frac{(-1)}{-1} - \frac{(-1)}{3}] - \frac{1}{2\pi}[\frac{-1}{-1} + \frac{(-1)}{3} - \frac{1}{-1} - \frac{1}{3}]$

$$= \frac{1}{2\pi}(\frac{-4}{3}) - \frac{1}{2\pi}(\frac{4}{3}) = -\frac{4}{3\pi}$$

$a_3 = \frac{1}{2\pi}(\frac{1}{-2} + \frac{1}{4} - \frac{1}{-2} - \frac{1}{4}) - \frac{1}{2\pi}(\frac{1}{-2} + \frac{1}{4} - \frac{1}{-2} - \frac{1}{4})$

$$= \frac{1}{2\pi}(0) - \frac{1}{2\pi}(0) = 0$$

$a_4 = \frac{1}{2\pi}(\frac{1}{-3} + \frac{1}{5} - \frac{(-1)}{-3} - \frac{(-1)}{5}) - \frac{1}{2\pi}(\frac{-1}{-3} + \frac{-1}{5} - \frac{1}{-3} - \frac{1}{5})$

$$= \frac{1}{2\pi}(-\frac{2}{5}) - \frac{1}{2\pi}(\frac{2}{5}) = -\frac{4}{15\pi}$$

$b_n = \frac{1}{\pi}\int_{-\pi}^{0} -\sin t \sin nt \, dt + \frac{1}{\pi}\int_{0}^{\pi} \sin t \sin nt$

$$= -\frac{1}{2\pi}\left[\frac{\sin(1-n)t}{1-n} - \frac{\sin(1+n)t}{1+n}\right]_{-\pi}^{0} + \frac{1}{2\pi}\left[\frac{\sin(1-n)t}{1-n} - \frac{\sin(1+n)t}{1+n}\right]_{0}^{\pi}$$

$$= -\frac{1}{2\pi}\left[\frac{-\sin(1-n)(-\pi)}{1-n} - \frac{\sin(1+n)(-\pi)}{1+n}\right]$$

$$+ \frac{1}{2\pi}\left[\frac{\sin(1-n)\pi}{1-n} - \frac{\sin(1+n)\pi}{1+n}\right]$$

$$b_1 = \frac{1}{\pi}\int_{-\pi}^{0} -\sin^2 t\, dt + \frac{1}{\pi}\int_{0}^{\pi} \sin^2 t\, dt$$

$$= -\frac{1}{2\pi}(t - \sin t \cos t)\Big|_{-\pi}^{0} + \frac{1}{2\pi}(t - \sin t \cos t)\Big|_{0}^{\pi}$$

$$= -\frac{1}{2\pi}(\pi) + \frac{1}{2\pi}(\pi) = 0$$

$b_n = 0$ if $n > 1$, since each is evaluated in terms of the sine of a multiple of π.

By Eq. (28-17), $f(t) = \frac{2}{\pi} - \frac{4}{3\pi}\cos 2t - \frac{4}{15\pi}\cos 4t - \dots$

Review Exercises for Chapter 28, page 907

1. $f(x) = \frac{1}{1+e^x} = (1 + e^x)^{-1}$ $f(0) = \frac{1}{1+1} = \frac{1}{2}$

$f'(x) = -(1 + e^x)^{-2}(e^x) = -e^x(1 + e^x)^{-2}$ $f'(0) = -1(2^{-2}) = -\frac{1}{4}$

$f''(x) = -(1 + e^x)^{-2}e^x + e^x(2)(1 + e^x)^{-3}(e^x)$ $f''(0) = -\frac{1}{4} + \frac{1}{4} = 0$

$f'''(x) = -(1 + e^x)^{-2}e^x + e^x(2)(1 + e^x)^{-3}(e^x)$ $f'''(0) = -\frac{1}{4} + \frac{1}{4} - \frac{3}{8} + \frac{1}{2}$

$\quad\quad\quad + 2e^{2x}(-3)(1 + e^x)^{-4}e^x$ $\quad\quad = \frac{1}{8}$

$\quad\quad\quad + (1 + e^x)^{-3}(2e^{2x})(2)$

By Eq. (28-5), $f(x) = \frac{1}{2} - \frac{1}{4}x + \frac{0x^2}{2!} + (\frac{1}{8})\frac{x^3}{3!} + \dots$

$\quad\quad\quad\quad\quad\quad = \frac{1}{2} - \frac{1}{4}x + \frac{1}{48}x^3 - \dots$

5. $f(x) = (x + 1)^{1/3}$ $f(0) = 1$

$f'(x) = \frac{1}{3}(x + 1)^{-2/3}$ $f'(0) = \frac{1}{3}$

$f''(x) = -\frac{2}{9}(x + 1)^{-5/3}$ $f''(0) = -\frac{2}{9}$

By Eq. (28-5), $f(x) = 1 + \frac{1}{3}x - \frac{2x^2}{9(2)} + \dots$

$\quad\quad\quad\quad\quad\quad = 1 + \frac{1}{3}x - \frac{1}{9}x^2 + \dots$

9. By Eq. (28-6), $e^x = 1 + x + \dfrac{x^2}{2} + \ldots$ Let $x = -0.2$

$e^{-0.2} = 1 - 0.2 + \dfrac{(0.2)^2}{2} + \ldots = 0.82$

13. $\sqrt{1.07} = 1.07^{1/2} = (1 + 0.07)^{1/2}$

$(1 + x)^n = 1 + nx + \dfrac{n(n - 1)x^2}{2!} + \ldots$

$x = 0.07$ and $n = \dfrac{1}{2}$

$\sqrt{1.07} = 1 + \dfrac{1}{2}(0.07) + \dfrac{\frac{1}{2}(-\frac{1}{2})(0.07)^2}{2} + \ldots$

$= 1 + 0.035 - 0.0006125 + \ldots = 1.0344$

17. $f(x) = \tan x; \ a = \dfrac{\pi}{4}$ $\qquad\qquad f(\dfrac{\pi}{4}) = 1$

$f'(x) = \sec^2 x = 1 + \tan^2 x$ $\qquad f'(\dfrac{\pi}{4}) = 2$

$f''(x) = 2 \tan x \sec^2 x$ $\qquad\qquad f''(\dfrac{\pi}{4}) = 4$

$f(x) = 1 + 2(x - \dfrac{\pi}{4}) + \dfrac{4(x - \frac{\pi}{4})^2}{2!} + \ldots$

$\tan 43.62° = \tan(45° - 1.38°) = \tan(\dfrac{\pi}{4} - \dfrac{1.38\pi}{180})$

$= 1 + 2(\dfrac{\pi}{4} - \dfrac{1.38\pi}{180} - \dfrac{\pi}{4}) + 2(\dfrac{\pi}{4} - \dfrac{1.38\pi}{180} - \dfrac{\pi}{4})^2 + \ldots$

$= 1 + 2(-0.0240855) + 2(0.0005801) = 0.953$

21. By Eq. (28-8), $\displaystyle\int_{0.1}^{0.2} \dfrac{1 - \frac{x^2}{2} + \frac{x^4}{24} + \ldots}{\sqrt{x}} \, dx$

$= \displaystyle\int_{0.1}^{0.2} (x^{-1/2} - \dfrac{x^{3/2}}{2} + \dfrac{x^{7/2}}{24} + \ldots) \, dx$

$= \displaystyle\int_{0.1}^{0.2} x^{-1/2} \, dx - \dfrac{1}{2}\int_{0.1}^{0.2} x^{3/2} \, dx + \dfrac{1}{24}\int_{0.1}^{0.2} x^{7/2} \, dx$

$= 2x^{1/2} - \dfrac{1}{5}x^{5/2} + \dfrac{1}{108}x^{9/2} + \ldots \Big|_{0.1}^{0.2} = 0.259$

25. $a_0 = \dfrac{1}{2\pi} \displaystyle\int_{-\pi}^{-\pi/2} 0 \, dx + \int_{-\pi/2}^{\pi/2} 1 \, dx + \int_{\pi/2}^{\pi} 0 \, dx + \dfrac{1}{2\pi}x \Big|_{-\pi/2}^{\pi/2} = \dfrac{1}{2}$ Eq. (28-25)

$a_1 = \dfrac{1}{\pi} \displaystyle\int_{-\pi}^{-\pi/2} 0 \cos x \, dx + \dfrac{1}{\pi}\int_{-\pi/2}^{\pi/2} 1 \cos x \, dx + \dfrac{1}{\pi}\int_{\pi/2}^{\pi} 0 \cos x \, dx$

$= \dfrac{1}{\pi} \sin x \Big|_{-\pi/2}^{\pi/2} = \dfrac{2}{\pi}$

$$a_2 = \frac{1}{\pi} \int_{-\pi}^{\pi/2} 0 \cos 2x \, dx + \frac{1}{\pi} \int_{-\pi/2}^{\pi/2} 1 \cos 2x \, dx + \frac{1}{\pi} \int_{\pi/2}^{\pi} 0 \cos 2x \, dx$$

$$= \frac{1}{2\pi} \int_{-\pi/2}^{\pi/2} \cos 2x \,(2 \, dx) = \frac{1}{2\pi} \sin 2x \bigg|_{-\pi/2}^{\pi/2} = 0$$

$$a_3 = \frac{1}{\pi} \int_{-\pi}^{\pi/2} 0 \cos 3x \, dx + \frac{1}{\pi} \int_{-\pi/2}^{\pi/2} 1 \cos 3x \, dx + \frac{1}{\pi} \int_{\pi/2}^{\pi} 0 \cos 3x \, dx$$

$$= \frac{1}{3\pi} \int_{-\pi/2}^{\pi/2} \cos 3x \,(3 \, dx) = \frac{1}{3\pi} \sin 3x \bigg|_{-\pi/2}^{\pi/2} = -\frac{2}{3\pi}$$

$$b_1 = \frac{1}{\pi} \int_{-\pi}^{\pi/2} 0 \sin x \, dx + \frac{1}{\pi} \int_{-\pi/2}^{\pi/2} \sin x \, dx + \frac{1}{\pi} \int_{\pi/2}^{\pi} 0 \sin x \, dx$$

$$= \frac{1}{\pi}(-\cos x) \bigg|_{-\pi/2}^{\pi/2} = 0$$

$$b_2 = \frac{1}{\pi} \int_{-\pi}^{\pi/2} 0 \sin 2x \, dx + \frac{1}{\pi} \int_{-\pi/2}^{\pi/2} \sin 2x \, dx + \frac{1}{\pi} \int_{\pi/2}^{\pi} 0 \sin 2x \, dx$$

$$= \frac{1}{2\pi} \int_{-\pi/2}^{\pi/2} \sin 2x \,(2 \, dx) = -\frac{1}{2\pi} \frac{\cos 2x}{2\pi} \bigg|_{-\pi/2}^{\pi/2} = -\frac{1}{2\pi}(-1 + 1) = 0$$

$b_n = 0$ for all n

By Eq. (28-17), $f(x) = \frac{1}{2} + \frac{2}{\pi} \cos x + 0 \cos 2x - \frac{2}{3\pi} \cos 3x + \ldots$

$$= \frac{1}{2} + \frac{2}{\pi}(\cos x - \frac{1}{3} \cos 3x + \ldots)$$

29. $\sin(x + h) - \sin(x - h) = (x + h) - \frac{(x + h)^3}{3!} + \ldots$

$$-(x - h) + \frac{(x - h)^3}{3!} + \ldots = (x + h) - (x - h) - \frac{(x^3 + 3hx^2 + 3h^2x + h^3)}{3!}$$

$$+ \frac{(x^3 - 3hx^2 + 3h^2x - h^3)}{3!} + \ldots$$

$$= 2h + \frac{(-6x^2h - 2h^3)}{6!} + \ldots = 2h - \frac{2x^2h}{2!} + \ldots$$

$$= 2h(1 - \frac{x^2}{2!} + \ldots) = 2h \cos x$$

For small values of h, the term $-2h^3$ is approximately zero.

33.

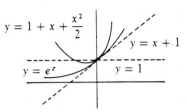

$$y = 1 + x + \frac{x^2}{2}$$

$$y = x + 1$$

$$y = e^z \qquad y = 1$$

37. $f(x) = \cos x$ $\qquad\qquad\qquad$ $f(0) = 1$
$f'(x) = -\sin x$ $\qquad\qquad\qquad$ $f'(0) = 0$
$f''(x) = -\cos x$ $\qquad\qquad\qquad$ $f''(0) = -1$
$f'''(x) = \sin x$ $\qquad\qquad\qquad$ $f'''(0) = 0$
$f^{iv}(x) = \cos x$ $\qquad\qquad\qquad$ $f^{iv}(0) = 1$

$$f(x) = 1 + 0x - \frac{1}{2}x^2 + \frac{0x^3}{3!} + \frac{1x^4}{4!} = 1 - \frac{1}{2}x^2 + \frac{1}{24}x^4 + \cdots$$

$$\sec x = \frac{1}{\cos x} = \frac{1}{1 - \frac{1}{2}x^2 - \frac{1}{24}x^4} = 1 + \frac{1}{2}x^2 + \frac{5}{24}x^4 + \cdots$$

The long division process is shown below:

$$
\begin{array}{r}
1 + \frac{1}{2}x^2 + \frac{5}{24}x^4 \\
1 - \frac{1}{2}x^2 + \frac{1}{24}x^4 \overline{\smash{\big)}\, 1 + 0 \quad + 0 \quad\ + 0\ + 0} \\
1 - \frac{1}{2}x^2 + \frac{1}{24}x^4 \\
\hline
\frac{1}{2}x^2 - \frac{1}{24}x^4 \\
\frac{1}{2}x^2 - \frac{1}{4}x^4 \\
\hline
\frac{5}{24}x^4 + 0 \quad + 0 \\
\frac{5}{24}x^4 - \frac{5}{48}x^2 + \frac{5}{576}x^4
\end{array}
$$

41. $\displaystyle\int_{0.1}^{0.2} = A \frac{x - \sin x}{x^2}\, dx = \int_{0.1}^{0.2} \frac{x - \left(x - \frac{x^3}{3!} + \frac{x^5}{5!}\right)}{x^2}\, dx$

$\qquad = \displaystyle\int_{0.1}^{0.2} \frac{\frac{x^3}{3!} + \frac{x^5}{5!}}{x^2}\, dx = \int_{0.1}^{0.2} \left(\frac{x}{6} - \frac{x^3}{120}\right) dx$

$\qquad = \left(\dfrac{x^2}{12} - \dfrac{x^4}{480}\right) \Big|_{0.1}^{0.2} =$

$\qquad = \left(\dfrac{0.04}{12} - \dfrac{0.0016}{480}\right) - \left(\dfrac{0.01}{12} - \dfrac{0.0001}{480}\right)$

$\qquad = 0.0025$

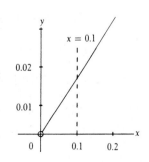

$x = 0.1$

45. $\dfrac{N_0}{1 - e^{-kt}} = N_0\left(\dfrac{1}{1 - e^{-kt}}\right)$; Let $x = e^{-kt}$

$N_0\left(\dfrac{1}{1 - e^{-kt}}\right) = N_0\left(\dfrac{1}{1 - x}\right)$

The MacLaurin expansion for $f(x) = \dfrac{1}{1 - x}$ is:

$f(x) = \dfrac{1}{1 - x}$; $f(0) = 1$

$f'(x) = \dfrac{1}{(1 - x)^2}$; $f'(0) = 1$

$f''(x) = \dfrac{2}{(1 - x)^3}$; $f''(0) = 2$

$f(x) = 1 + x + \dfrac{2x^2}{2!} + \dots = 1 + x + x^2 + \dots$

Substituting $e^{-\frac{k}{t}}$ for x; $f(x) = 1 + e^{-kt} + e^{-2kt} + \dots$;

$\therefore \dfrac{N_0}{1 - e^{-kt}} = N_0\left(1 + e^{-kt} + e^{-2kt} + \dots\right)$

CHAPTER 29

Differential Equations

1. $y = e^{-x^2}$; $y' = -2 \times e^{-x^2}$

    Substituting into the differential equation; $\frac{dy}{dx} + 2xy = 0$,

    $-2xe^{-x^2} + 2x\left(e^{-x^2}\right) = 0$; $0 = 0$ particular solution

5. $\frac{dy}{dx} = 1$, $y = x + 3$ Substituting,

    $\qquad\qquad \frac{dy}{dx} = 1$ $\frac{dy}{dx} = 1$, $1 = 1$

9. $xy' = 2y$, $y = cx^2$ Substituting,

    $\qquad\qquad y' = 2cx$ $xy' = x(2cx) = 2(cx^2) = 2y$

13. $\frac{d^2y}{dx^2} + 4y = 0$, $y = 3 \cos 2x$ Substituting,

    $\qquad\qquad \frac{dy}{dx} = 3(2)(-\sin 2x)$ $\frac{d^2y}{dx^2} + 4y = (-12 \cos 2x) + 4(3 \cos 2x)$

    $\qquad\qquad\qquad = -6 \sin 2x$ $= -12 \cos 2x + 12 \cos 2x$

    $\qquad\qquad \frac{d^2y}{dx^2} = -6(2)(\cos 2x)$ $= 0$

    $\qquad\qquad\qquad = -12 \cos 2x$

17. $x^2y' + y^2 = 0$, $xy = cx + cy$ Substituting,

    $\qquad\qquad y = \frac{cx}{x - c}$ $x^2y' + y^2 = x^2\left(\frac{-c^2}{(x - c)^2}\right) + \left(\frac{cx}{x - c}\right)^2$

    $\qquad\qquad y' = \frac{(x - c)(c) - cx(1)}{(x - c)^2}$ $= \frac{-c^2x^2}{(x - c)^2} + \frac{c^2x^2}{(x - c)^2}$

    $\qquad\qquad y' = \frac{-c^2}{(x - c)^2}$ $= 0$

21. $y' + y = 2 \cos x$, $y = \sin x + \cos x - e^{-x}$ Substituting,

    $\qquad\qquad y' = \cos x - \sin x - e^{-x}(-1)$ $y' + y = \cos x - \sin x$

    $\qquad\qquad y' = \cos x - \sin x + e^{-x}$ $+ e^{-x} + \sin x$

    $\qquad\qquad\qquad\qquad\qquad\qquad\qquad\qquad\qquad\qquad\qquad + \cos x - e^{-x}$

    $\qquad\qquad\qquad\qquad\qquad\qquad\qquad\qquad\qquad\qquad\qquad = 2 \cos x$

25. $\cos x \dfrac{dy}{dx} + \sin x = 1 - y$, $y = \dfrac{x + c}{\sec x + \tan x}$

$\dfrac{dy}{dx} = \dfrac{(\sec x + \tan x)(1) - (x + c)(\sec x \tan x + \sec^2 x)}{(\sec x + \tan x)^2}$

$= \dfrac{(\sec x + \tan x) - (x + c)(\sec x)(\tan x + \sec x)}{(\sec x + \tan x)^2}$

$= \dfrac{(\sec x + \tan x)[1 - (x + c)(\sec x)]}{(\sec x + \tan x)(\sec x + \tan x)}$

$= \dfrac{1 - (x + c)(\sec x)}{(\sec x + \tan x)}$

Substituting,

$\cos x \dfrac{dy}{dx} + \sin x = \cos x[\dfrac{1 - (x + c)(\sec x)}{\sec x + \tan x}] + \sin x$

$= \dfrac{\cos x - (x + c)}{\sec x + \tan x} + \sin x$

$= \dfrac{\cos x - (x + c)}{\sec x + \tan x} + \dfrac{\sin x \sec x + \sin x \tan x}{\sec x + \tan x}$

$= \dfrac{\cos x - x - c + \dfrac{\sin x}{\cos x} + \dfrac{\sin^2 x}{\cos x}}{\sec x + \tan x}$

$= \dfrac{\dfrac{\cos^2 x}{\cos x} + \dfrac{\sin^2 x}{\cos x} + \tan x - x - c}{\sec x + \tan x}$

$= \dfrac{\dfrac{1}{\cos x} + \tan x - x - c}{\sec x + \tan x}$

$= \dfrac{\sec x + \tan x - x - c}{\sec x + \tan x} = 1 - \dfrac{(x + c)}{\sec x + \tan x} = 1 - y$

Exercises 29-2, page 917

1. $2x\,dx + dy = 0$; integrating, $x^2 + y = c$; $y = c - x^2$

5. $x^2 + (x^3 + 5)\dfrac{dy}{dx} = 0$

Multiply by dx and divide by $x^3 + 5$: $\dfrac{x^2}{x^3 + 5}\,dx + dy = 0$

Integrating, $\dfrac{1}{3}\int \dfrac{3x^2}{x^3 + 5} + \int dy = 0$

$\dfrac{1}{3}\ln(x^3 + 5) + y = c_1$; $\ln(x^3 + 5) + 3y = c$; $(c = 3c_1)$

9. $e^{x+y} dx + dy = 0$; $e^x e^y dx + dy = 0$

Divide by e^y: $e^x dx + \dfrac{dy}{e^y} = 0$; $e^x dx + e^{-y} dy = 0$

Integrating, $e^x - e^{-y} = c$

13. $x \dfrac{dy}{dx} = y^2 + y^2 \ln x$; divide by xy^2 and multiply by dx

$\dfrac{dy}{y^2} = \dfrac{dx}{x} + \dfrac{1}{x} \ln x \, dx$

$\dfrac{dy}{y^2} = (1 + \ln x) \dfrac{dx}{x}$.

Integrating, $-\dfrac{1}{y} = \dfrac{1}{2}(1 + \ln x)^2 + \dfrac{c}{2}$

$-2 = y(1 + \ln x)^2 + cy$

$y(1 + \ln x)^2 + cy + 2 = 0$

17. $yx^2 \, dx = y \, dx - x^2 \, dy$; divide by y and by x^2: $dx = \dfrac{dx}{x^2} - \dfrac{dy}{y}$

Integrating, $x + c = \dfrac{x^{-1}}{-1} - \ln y$

$x + \dfrac{1}{x} + \ln y = c$; $x^2 + 1 + x \ln y + cx = 0$

21. $2 \ln x \, dx + x \, dy = 0$; $2 \ln x \dfrac{dx}{x} + dy = 0$

Integrating, $\dfrac{2(\ln x)^2}{2} + y = c$

$(\ln x)^2 + y = c$; $y = c - (\ln x)^2$

25. $\dfrac{dy}{dx} + yx^2 = 0$; multiply by dx and divide by y: $\dfrac{dy}{y} + x^2 \, dx = 0$

Integrating, $\ln y + \dfrac{x^3}{3} + c$

Substitute $x = 0$ when $y = 1$; $\ln 1 = c$; $c = 0$

Therefore, $3 \ln y + x^3 = 0$.

29. $\dfrac{dy}{dx} = (1 - y) \cos x$, $x = \dfrac{\pi}{6}$ when $y = 0$

$dy = (1 - y) \cos x \, dx$

Divide by $(1 - y)$: $\dfrac{1}{1 - y} \, dy = \cos x \, dx$

Integrating, $-\ln(1 - y) = \sin x + C$

$\sin x + \ln(1 - y) = C$

Substituting $x = \dfrac{\pi}{6}$ when $y = 0$, $\sin \dfrac{\pi}{6} + \ln 1 = c$

$c = \dfrac{1}{2}$

$\sin x + \ln(1 - y) = \dfrac{1}{2}$; $2 \ln(1 - y) = 1 - 2 \sin x$

Exercises 29-3, page 919

1. $x\,dy + y\,dx + x\,dx = 0$; by Eq. (29-3), $d(xy) + x\,dx = 0$

    Integrating, $xy + \dfrac{x^2}{2} = c$; $2xy + x^2 = c$

5. $x^3\,dy + x^2y\,dx + y\,dx - x\,dy = 0$

    Divide by x^2: $x\,dy + y\,dx - \left(\dfrac{x\,dy - y\,dx}{x^2}\right) = 0$

    By Eq. (29-3) and (29-5), $d(xy) - d\dfrac{y}{x} = 0$

    Integrating, $xy - \dfrac{y}{x} = c$; $x^2y - y = cx$

9. $\sqrt{x^2 + y^2}\,dx - 2y\,dy = 2x\,dx$

    By Eq. (29-4), $\sqrt{x^2 + y^2}\,dx = d(x^2 + y^2)$

    $dx = (x^2 + y^2)^{-1/2}d(x^2 + y^2)$

    Integrating, $x = 2(x^2 + y^2)^{1/2} + c$; $2\sqrt{x^2 + y^2} = x + c$

13. $y\,dy - x\,dx + (y^2 - x^2)\,dx = 0$; $\dfrac{y\,dy - x\,dx}{y^2 - x^2} + dx = 0$

    $\dfrac{2y\,dy - 2x\,dx}{y^2 - x^2} + 2\,dx = 0$

    Integrating, $\ln(y^2 - x^2) + 2x = c$

17. $2(x\,dy + y\,dx) + 3x^2\,dx = 0$; $x\,dy + y\,dx + \dfrac{3}{2}x^2\,dx = 0$

    By Eq. (29-3), $d(xy) + \dfrac{3}{2}x^2\,dx = 0$

    Integrating, $xy + \dfrac{1}{2}x^3 + c = 0$; $2xy + x^3 + c = 0$

    Substituting $x = 1$, $y = 2$, $4 + 1 + c = 0$; $c = -5$

    $2xy + x^3 = 5$

Exercises 29-4, page 922

1. $dy + y\,dx = e^{-x}\,dx$; from Eq. (29-7), $P = 1$, $Q = e^{-x}$; $e^{\int dx} = e^x$

    Substituting into Eq. (29-8), $ye^x = \int e^{-x}e^x\,dx + c$

    $ye^x = \int dx + c$; $ye^x = x + c$; $y = e^{-x}(x + c)$

5. $\dfrac{dy}{dx} - 2y = 4$; $dy - 2y\,dx = 4\,dx$

    From Eq. (29-7), $P = -2$, $Q = 4$; $e^{\int -2\,dx} = e^{-2x}$

    Substituting into Eq. (29-8), $ye^{-2x} = \int 4e^{-2x}\,dx + c$

    $ye^{-2x} = 4\left[-\dfrac{1}{2}\int e^{-2x}(-2\,dx)\right] + c$

    $ye^{-2x} = -2e^{-2x} + c$ or $y = -2 + ce^{2x}$

9. $2x \, dy + y \, dx = 8x^3 \, dx$; $dy + \frac{1}{2x}y \, dx = 4x^2 \, dx$

From Eq. (29-7), $P = \frac{1}{2x}$, $Q = 4x^2$

$e^{\int 1/2x \, dx} = e^{1/2 \, \ln x} = e^{\ln x^{1/2}} = x^{1/2}$

Substituting into Eq. (29-8), $yx^{1/2} = \int 4x^2(x^{1/2}) \, dx + c$

$= \int 4x^{5/2} \, dx + c = 4x^{7/2}\left(\frac{2}{7}\right) + c = \frac{8}{7}x^{7/2} + c$

$y = \frac{8}{7}x^3 + \frac{c}{\sqrt{x}}$

13. $\sin x \, \frac{dy}{dx} = 1 - y \cos x$

$\sin x \, dy + y \cos x \, dx = dx$

$dy + y\frac{\cos x}{\sin x} \, dx = \frac{dx}{\sin x}$

$dy + \cot xy \, dx = \csc x \, dx$

From Eq. (29-7), $P = \cot x$, $Q = \csc x$

$e^{\int \cot x \, dx} = e^{\ln \sin x} = \sin x$

Substituting into Eq. (29-8), $y \sin x = \int \csc x \sin x \, dx$

$y \sin x = \int dx + c = x + c$; $y = \frac{x + c}{\sin x} = (x + c) \csc x$

17. $\frac{dy}{dx} = xe^{4x} + 4y$; $dy - 4y \, dx = xe^{4x} \, dx$

From Eq. (29-7), $P = -4$, $Q = xe^{4x}$; $e^{\int -4 \, dx} = e^{-4x}$

Substituting into Eq. (29-8), $ye^{-4x} = \int xe^{4x} + c$

$ye^{-4x} = \int x + c$; $ye^{-4x} = \frac{x^2}{2} + c$

$2y = e^{4x}(x^2 + c)$

21. $x\frac{dy}{dx} = y + (x^2 - 1)^2$; $x\frac{dy}{dx} = y + x^4 - 2x^2 + 1$

$dy = \frac{1}{x}y \, dx + \frac{1}{x}(x^4 - 2x^2 + 1) \, dx$

$dy - \frac{1}{x}y \, dx = (x^3 - 2x + \frac{1}{x}) \, dx$

From Eq. (29-7), $P = -\frac{1}{x}$; $Q = (x^3 - 2x + \frac{1}{x})$; $e^{\int (-1/x) \, dx} = e^{-\ln x}$

$= e^{\ln x^{-1}} = \frac{1}{x}$

Substituting into Eq. (29-8), $\frac{y}{x} = \int (x^3 - 2x + \frac{1}{x})(\frac{1}{x}) \, dx + c$

$\frac{y}{x} = \int (x^2 - 2 + x^{-2}) \, dx + c$; $\frac{y}{x} = \frac{x^3}{3} - 2x - x^{-1} + c$

$y = \frac{x^4}{3} - 2x^2 - 1 + cx$; $3y = x^4 - 6x^2 - 3 + cx$

25. $\frac{dy}{dx} + 2y = e^{-x}$; $dy + 2y\ dx = e^{-x}\ dx$

From Eq. (29-7), $P = 2$, $Q = e^{-x}$; $e^{\int 2\ dx} = e^{2x}$

Substituting into Eq. (29-8), $ye^{2x} = \int e^{x}\ dx + c$

$ye^{2x} = e^{x} + c$; $y = e^{-x} + ce^{-2x}$

Substituting $x = 0$ when $y = 1$, $1 = 1 + c$; $c = 0$; $y = e^{-x}$

29. $\sqrt{x}\ y' + \frac{1}{2}y = e^{\sqrt{x}}$; $dy + \frac{1}{2\sqrt{x}}y\,dx = \frac{1}{\sqrt{x}}e^{\sqrt{x}}$; $P = \frac{1}{2}x^{-1/2}$, $Q = \frac{1}{\sqrt{x}}e^{\sqrt{x}}$

$e^{\int P dx} = e^{\int (1/2)x^{1/2}dx} = e^{x^{1/2}} = e^{\sqrt{x}}$

$ye^{\sqrt{x}} = \int \frac{1}{\sqrt{x}}e^{\sqrt{x}}e^{\sqrt{x}}dx = \int \frac{1}{\sqrt{x}}e^{2\sqrt{x}}dx = e^{2\sqrt{x}} + c$; $y = e^{\sqrt{x}} + ce^{-\sqrt{x}}$

$x = 1$ when $y = 3$; $3 = e + ce^{-1}$, $c = 3e - e^{2}$; $y = e^{\sqrt{x}} + (3e - e^{2})e^{-\sqrt{x}}$

Exercises 29-5, page 927

1. Slope $= \frac{dy}{dx} = \frac{2x}{y}$; $y\ dy = 2x\ dx$. Integrating, $\frac{1}{2}y^{2} = x^{2} + c$

   Substitute $x = 2$ when $y = 3$. $\frac{1}{2}(9) = 4 + c$; $c = 0.5$

   $\frac{1}{2}y^{2} = x^{2} + 0.5$; $y^{2} = 2x^{2} + 1$

5. See Example B. $\frac{dy}{dx} = ce^{x}$. $y = ce^{x}$; $c = \frac{y}{e^{x}}$

   Substitute for c in the equation for the derivative.

   $\frac{dy}{dx} = \frac{y}{e^{x}}e^{x}$, $\frac{dy}{dx} = y$; $\frac{dy}{dx}\Big|_{OT} = -\frac{1}{y}$; $y\ dy = -dx$

   Integrating, $\frac{y^{2}}{2} = -x + \frac{c}{2}$; $y^{2} = c - 2x$

9. See example C. $N = N_0\ e^{kt}$

   Use the condition that half this isotope decays in 40s.

   $N = \frac{N_0}{2}$ when $t = 40s$. $\frac{N_0}{2} = N_0\ e^{40k}$; $0.5 = e^{40k}$

   $0.5^{1/40} = e^{k}$; $0.5^{0.025} = e^{k}$; $N = N_0\ (0.5)^{0.025t}$

   Evaluating when $t = 60s$, $N = N_0\ (0.5)^{0.025(60)} = N_0\ (0.5)^{1.5} = 0.35N_0$. Therefore 35.4% remains after 60s.

13. $r \dfrac{ds}{dr} = 2(a - s)$; $r\,ds = 2a\,dr - 2s\,dr$; $ds + \dfrac{2}{r}s\,dx = 2a\dfrac{dr}{r}$

From Eq. (29-7) and Eq. (29-8); $y = s$, $x = r$, $p = \dfrac{2}{r}$, $Q = \dfrac{2a}{r}$

$$e^{\int P\,dx} = e^{\int \frac{2\,dr}{r}} = e^{2 \ln x} = e^{\ln r^2} = r^2$$

$$sr^2 = \int \left(2a\dfrac{dr}{r}\right) - 2a \ln r + C = \int 2ar\,dr$$

$$sr^2 = ar^2 + C; \quad s = a\ \dfrac{C}{r^2}$$

17. $\dfrac{dT}{dt} = k(T - 80)$; $\dfrac{dT}{T - 80} = k\,dt$; $\ln(T - 80) = kt + \ln c$

$T = 80 + ce^{kt}$; $200 = 80 + c$; $T = 80 + 120e^{kt}$

$140 = 80 + 120e^{5k}$; $e^{5k} = 0.5$; $e^k = (0.5)^{1/5}$

$T = 80 + 120(0.5)^{t/5}$

$100 = 80 + 120(0.5)^{t/5}$

$0.5^{t/5} = \dfrac{1}{6}$; $t = 5\ \dfrac{\ln \dfrac{1}{6}}{\ln \dfrac{1}{2}} = 12.9$ min

17M. $\dfrac{dT}{dt} = k(T - 25)$, $\dfrac{dT}{T - 25} = k\,dt$; $\ln(T - 25) = kt + \ln c$

$T = 25 + ce^{kt}$; $90 = 25 + c$; $T = 25 + 65e^{kt}$

$60 = 25 + 65e^{5k}$; $e^{5k} = \dfrac{7}{13}$; $e^k = (\dfrac{7}{13})^{1/5}$

$T = 25 + 65(\dfrac{7}{13})^{t/5}$; $40 = 25 + 65(\dfrac{7}{13})^{t/5}$

$(\dfrac{7}{13})^{t/5} = \dfrac{3}{13}$; $t = 5\ \dfrac{\ln \dfrac{3}{13}}{\ln \dfrac{7}{13}} = 11.8$

21. See Example D, $i = \dfrac{E}{R}(1 - e^{-(R/L)t})$

$\lim\limits_{i \to \infty} = \lim\limits_{i \to \infty}(\dfrac{E}{R} - \dfrac{E}{R}e^{-(R/L)t}) = \dfrac{E}{R} - 0 = \dfrac{E}{R}$

25. $Ri + \dfrac{q}{C} = 0$; $i = \dfrac{dq}{dt}$; $R\dfrac{dq}{dt} + \dfrac{q}{C} = 0$; $\dfrac{dq}{q} + \dfrac{1}{RC}\,dt = 0$

Integrating, $\ln q = -\dfrac{1}{RC}t + c$

Let $q = q_0$ when $t = 0$, $\ln q_0 = c$; $\ln q = -\dfrac{1}{RC}t + \ln q_0$

$\ln q - \ln q_0 = -\dfrac{1}{RC}t$; $\ln \dfrac{q}{q_0} = -\dfrac{1}{RC}t$

$\dfrac{q}{q_0} = e^{(-1/RC)t}$; $q = q_0 e^{-t/RC}$

29. $\dfrac{dv}{dt} = 32 - v$; $\dfrac{dv}{32 - v} = dt$; $-\dfrac{-dv}{32 - v} = dt$

Integrating, $(-1)\ln(32 - v) = t - \ln c$; $\ln \dfrac{32 - v}{c} = -t$

$32 - v = ce^{-t}$; $-v = -32 + ce^{-t}$; $v = 32 - ce^{-t}$

Starting from rest means $v = 0$ when $t = 0$; $0 = 32 - c$; $c = 32$

$v = 32 - 32e^{-t} = 32(1 - e^{-t})$

$\lim\limits_{t \to \infty} 32(1 - e^{-t}) = 32$ ft/s

29M. $\dfrac{dv}{dt} = 9.8 - v$; $\dfrac{dv}{9.8 - v} = dt$; $-\dfrac{-dv}{9.8 - v} = dt$

Integrating, $(-1)\ln(9.8 - v) = t - \ln c$; $\ln \dfrac{9.8 - v}{c} = -t$

$9.8 - v = ce^{-t}$; $-v = -9.8 + ce^{-t}$; $v = 9.8 - ce^{-t}$

Starting from rest, $v = 0$ when $t = 0$; $0 = 9.8 - c$; $c = 9.8$

$v = 9.8 - 9.8e^{-t} = 9.8(1 - e^{-t})$

$\lim\limits_{t \to \infty} 9.8(1 - e^{-t}) = 9.8$ m/s

33. $\dfrac{dx}{dt} = 6t - 3t^2$; $dx = \left(6t - 3t^2\right)$; $x = \dfrac{6t^2}{2} - \dfrac{3t^3}{3} + C$;

$x = 3t^2 - t^3 + C$; $x = 0$ for $t = 0$; $0 = C$;

$x = 3t^2 - t^3$; $y = 2\left(3t^2 - t^3\right) - \left(3t^2 - t^3\right)^2 = -t^6 + 6t^5 - 9t^4 - 2t^3 + 6t^2$

37. $\dfrac{dv}{dt} = kv$; $\dfrac{dv}{v} = kdt$; integrating, $\ln v = kt + \ln C$

Let $v = 8200$ when $t = 0$; $8200 = C$

$\ln v = kt + \ln 8200$; $\ln v - \ln 8200 = kt$

$\ln \dfrac{v}{8200} = kt$; $\dfrac{v}{8200} = e^{kt}$; $v = 8200kt$

Let $v = 4920$ when $t = 3$; $4920 = 8200e^{3k}$

$e^k = (0.6)^{1/3}$; $v = 8200(0.6)^{t/3}$

After 11 years, $v = 8200(0.6)^{11/3} = \$1260$

Exercises 29-6, 29-7, page 933

1. $D^2y - Dy - 6y = 0$; by Eq. (29-14), $m^2 - m - 6 = 0$

$(m - 3)(m + 2) = 0$; $m_1 = 3$ and $m_2 = -2$.

By Eq. (29-15), $y = c_1 e^{3x} + c_2 e^{-2x}$

5. $D^2y - 3Dy = 0$; by Eq. (29-14), $m^2 - m = 0$; $m_1 = 0$ and $m_2 = 3$

By Eq. (29-15), $y = c_1e^0 + c_2e^{3x}$; $y = c_1 + c_2e^{3x}$

9. $3D^2_y + 8Dy - 3y = 0$; by Eq. (29-14), $3m^2 + 8m - 3 = 0$

$(3m - 1)(m + 3) = 0$; $m_1 = \dfrac{1}{3}$ and $m_2 = -3$

By Eq. (29-15), $y = c_1e^{x/3} + c_2e^{-3x}$

13. $2\dfrac{d^2y}{dx^2} - 4\dfrac{dy}{dx} + y = 0$; $2D^2y - 4Dy + y = 0$

By Eq. (29-14), $2m^2 - 4m + 1 = 0$

By the quadratic formula, $m = \dfrac{4 \pm\sqrt{16 - 8}}{4}$; $m_1 = 1 + \dfrac{\sqrt{2}}{2}$ and $m_2 = 1 - \dfrac{\sqrt{2}}{2}$

By Eq. (29-15), $y = c_1e^{(1 + \frac{\sqrt{2}}{2})x} + c_2e^{(1 - \frac{\sqrt{2}}{2})x}$

$= c_1e^xe^{\frac{\sqrt{2}}{2}x} + c_2e^xe^{-\frac{\sqrt{2}}{2}x} = e^x(c_1e^{x\frac{\sqrt{2}}{2}} + c_2e^{-x\frac{\sqrt{2}}{2}})$

17. $y'' = 3y' + y$; $D^2y - 3Dy - y = 0$. By Eq. (29-14), $m^2 - 3m - 1 = 0$

By the quadratic formula, $m = \dfrac{3 \pm\sqrt{9 + 4}}{2}$

$m_1 = \dfrac{3}{2} + \dfrac{\sqrt{13}}{2}$, $m_2 = \dfrac{3}{2} - \dfrac{\sqrt{13}}{2}$

By Eq. (29-15), $y = c_1e^{(\frac{3}{2} + \frac{\sqrt{13}}{2})x} + c_2e^{(\frac{3}{2} - \frac{\sqrt{13}}{2})x}$

$= e^{\frac{3x}{2}}\left(c_1e^{x\frac{\sqrt{13}}{2}} + c_2e^{-x\frac{\sqrt{13}}{2}}\right)$

21. $D^2y - 4Dy - 21y = 0$; by Eq. (29-14), $m^2 - 4m - 21 = 0$

$(m - 7)(m + 3) = 0$; $m_1 = 7$ and $m_2 = -3$

By Eq. (29-15), $y = c_1e^{7x} + c_2e^{-3x}$

The derivative of the general equation is: $Dy = 7c_1e^{7x} - 3c_2e^{-3x}$.

Substituting the conditions $Dy = 0$ and $y = 2$ when $x = 0$, we have the

equations: $c_1 + c_2 = 2$ and $7c_1 - 3c_2 = 0$.

Solving this system of equations gives $c_1 = \dfrac{3}{5}$ and $c_2 = \dfrac{7}{5}$.

Therefore, $y = \dfrac{3}{5}e^{7x} + \dfrac{7}{5}e^{-3x}$; $y = \dfrac{1}{5}(3e^{7x} + 7e^{-3x})$

25. $D^3y - 2D^2y - 3Dy = 0$; by Eq. (29-14), $m^3 - 2m^2 - 3m = 0$

$m(m + 1)(m - 3) = 0$; $m_1 = 0$, $m_2 = -1$ and $m_3 = 3$

By Eq. (29-15), $y = c_1e^0 + c_2e^{-x} + c_3e^{3x}$

$y = c_1 + c_2e^{-x} + c_3e^{3x}$

Exercises 29-8, page 937

1. $D^2y - 2Dy + y = 0$; by Eq. (29-14), $m^2 - 2m + 1 = 0$; $(m - 1)^2 = 0$

The roots are 1,1.

By Eq. (29-16), $y = e^x(c_1 + c_2x)$; $y = (c_1 + c_2x)e^x$

5. $D^2y + 9y = 0$; by Eq. (29-14), $m^2 + 9 = 0$

$m_1 = 3j$ and $m_2 = -3j$, $\alpha = 0$, $\beta = 3$

By Eq. (29-17), $y = e^{0x}(c_1 \sin 3x + c_2 \cos 3x)$

$y = c_1 \sin 3x + c_2 \cos 3x$

9. $D^2y = 0$; by Eq. (29-14), $m^2 = 0$

The roots are 0,0.

By Eq. (29-16), $y = e^{0x}(c_1 + c_2x)$; $y = c_1 + c_2x$

13. $16D^2y - 24Dy + 9y = 0$; by Eq. (29-14),

$16m^2 - 24m + 9 = 0$; $(4m - 3)^2 = 0$

The roots are $\frac{3}{4}, \frac{3}{4}$.

By Eq. (29-16), $y = e^{3x/4}(c_1 + c_2x)$; $y = (c_1 + c_2x)e^{3x/4}$

17. $2D^2y + 5y = 4Dy$; $2D^2y + 5y - 4Dy = 0$

By Eq. (29-14), $2m^2 - 4m + 5 = 0$

By the quadratic formula, $m = \dfrac{4 \pm\sqrt{16 - 40}}{4} = \dfrac{4 \pm 2\sqrt{-6}}{4}$

$m_1 = 1 + \frac{\sqrt{6}}{2}j$; $m_2 = 1 - \frac{\sqrt{6}}{2}j$; $\alpha = 1$, $\beta = \frac{1}{2}\sqrt{6}$

By Eq. (29-17), $y = e^x(c_1 \cos \frac{1}{2}\sqrt{6}x + c_2 \sin \frac{1}{2}\sqrt{6}x)$

21. $2D^2y - 3Dy - y = 0$; by Eq. (29-14), $2m^2 - 3m - 1 = 0$.

By the quadratic formula, $m = \dfrac{3 \pm\sqrt{9 + 8}}{4}$

$m_1 = \frac{3}{4} + \frac{\sqrt{17}}{4}$, $m_2 = \frac{3}{4} - \frac{\sqrt{17}}{4}$

By Eq. (29-15), $y = c_1e^{\left(\frac{3}{4} + \frac{\sqrt{17}}{4}\right)x} + c_2e^{\left(\frac{3}{4} + \frac{\sqrt{17}}{4}\right)x}$

$$y = e^{\frac{3}{4}x}\left(c_1e^{x\frac{\sqrt{17}}{4}} + c_2e^{-x\frac{\sqrt{17}}{4}}\right)$$

25. $D^2y + 2Dy + 10y = 0$; by Eq. (29-14), $m^2 + 2m + 10 = 0$

By the quadratic formula, $m = \dfrac{-2 \pm\sqrt{4 - 40}}{2}$, $m_1 = -1 + 3j$, $m_2 = -1 - 3j$

$\alpha = -1$, $\beta = 3$; by Eq. (29-17), $y = e^{-x}(c_1 \sin 3x + c_2 \cos 3x)$

Substituting $y = 0$ when $x = 0$, $0 = e^0(c_1 \sin 0 + c_2 \cos 0)$, $c_2 = 0$

Substituting $y = e^{-1}$ when $x = \dfrac{\pi}{6}$, $e^{-1} = e^{-\pi/6}(c_1 \sin \dfrac{\pi}{2})$

$e^{-1} = e^{-\pi/6}c_1$; $c_1 = e^{\pi/6 - 1}$

$y = e^{-x}[e^{(\pi/6 - 1)} \sin 3x] = e^{(\pi/6 - 1 - x)} \sin 3x$

29. $y = c_1 e^{3x} + c_2 e^{-3x}$. The auxiliary equation is $(m - 3)(m + 3) = 0$.
$m^2 - 9 = 0$. The simplest form of the differential equation is
$(D^2 - 9)y = 0$.

Exercises 29-9, page 941

1. $D^2y - Dy - 2y = 4$. By Eq. (29-14), $m^2 - m - 2 = 0$

$(m - 2)(m + 1) = 0$; $m_1 = 2$, $m_2 = -1$

By Eq. (29-15), $y_c = c_1 e^{2x} + c_2 e^{-x}$

$y_p = A$; $Dy_p = 0$; $D^2 y_p = 0$

Substituting into the differential equation, $0 - 0 - 2A = 4$; $A = -2$.

Thus, $y_p = -2$. By Eq. (29-19), $y = c_1 e^{2x} + c_2 e^{-x} - 2$

5. $D^2y + 4Dy + 3y = 2 + e^x$. By Eq. (29-14), $m^2 + 4m + 3 = 0$

$(m + 1)(m + 3) = 0$; $m_1 = -1$, $m_2 = -3$

By Eq. (29-15), $y_c = c_1 e^{-x} + c_2 e^{-3x}$

$y_p = A + Be^x$; $Dy_p = Be^x$; $D^2 y_p = Be^x$

Substituting into the differential equation, $Be^x + 4Be^x + 3(A + Be^x)$
$= 2 + e^x$.

$3A + 8Be^x = 2 + e^x$; $3A = 2$

$A = \dfrac{2}{3}$; $8B = 1$; $B = \dfrac{1}{8}$; $y_p = \dfrac{2}{3} + \dfrac{1}{8}e^x$

By Eq. (29-19), $y = c_1 e^{-x} + c_2 e^{-3x} + \dfrac{1}{8}e^x + \dfrac{2}{3}$

9.
$$9D^2y - y = \sin x; \text{ let } y_p = A \sin x + B \cos x$$
$$9m^2 - 1 = 0; \ m = -\frac{1}{3}, \ \frac{1}{3}$$
$$y_c = c_1 e^{-1/3 x} + c_2 e^{1/3 x}$$
$$y_p = A \sin x + B \cos x$$
$$Dy_p = A \cos x - B \cos x$$
$$D^2 y_p = -A \sin x - B \cos x$$
$$9(-A \sin x - B \cos x) - (A \sin x + B \cos x) = \sin x$$
$$-10 \sin x - 10 \cos x = \sin x$$
$$-10A = 1; \ A = -\frac{1}{10}; \ -10B = 0; \ B = 0$$
$$y_p = -\frac{1}{10} \sin x + 0 \cos x = -\frac{1}{10} \sin x$$
$$\text{Therefore, } y = c_1 e^{1/3 x} + c_2 e^{-1/3 x} - \frac{1}{10} \sin x$$

13. $D^2 y - Dy - 30y = 10$. By Eq. (29-14), $m^2 - m - 30 = 0$

$(m - 6)(m + 5) = 0; \ m_1 = -5; \ m_2 = 6$

By Eq. (29-15), $y_c = c_1 e^{-5x} + c_2 e^{6x}; \ y_p = A; \ Dy_p = 0; \ D^2 y_p = 0$

Substituting into the differential equation, $0 = 0 - 30A = 10;$

$A = -\frac{1}{3}; \ y_p = -\frac{1}{3}.$

By Eq. (29-19), $y = c_1 e^{-5x} + c_2 e^{6x} - \frac{1}{3}$

17. $D^2 y - 4y = \sin x + 2 \cos x$. By Eq. (29-14), $m^2 - 4 = 0; \ m_1 = 2; \ m_2 = -2$

By Eq. (29-15), $y_c = c_1 e^{2x} + c_2 e^{-2x}$

$y_p = A \sin x + B \cos x; \ Dy_p = A \cos x - B \sin x$

$D^2 y_p = -A \sin x - B \cos x$

Substituting into the differential equation,

$-A \sin x - B \cos x - 4A \sin x - 4B \cos x = \sin x + 2 \cos x$

$-5A \sin x - 5B \cos x = \sin x + 2 \cos x$

$-5A = 1; \ A = -\frac{1}{5}; \ -5B = 2; \ B = -\frac{2}{5}; \ y_p = -\frac{1}{5} \sin x - \frac{2}{5} \cos x$

By Eq. (29-19), $y = c_1 e^{2x} + c_2 e^{-2x} - \frac{1}{5} \sin x - \frac{2}{5} \cos x$

21. $D^2y + 5Dy + 4y = xe^x + 4$; by Eq. (29-14), $m^2 + 5m + 4 = 0$

$(m + 1)(m + 4) = 0$; $m_1 = -1$, $m_2 = -4$

By Eq. (29-15), $y_c = c_1e^{-x} + c_2e^{-4x}$; $y_p = Ae^x + Bxe^x + C$

$Dy_p = Ae^x + B(xe^x + e^x) = Ae^x + Bxe^x + Be^x$

$D^2y_p = Ae^x + B(xe^x + e^x) + Be^x = Ae^x + 2Be^x + Bxe^x$

Substituting into the differential equation,

$Ae^x + 2Be^x + Bxe^x + 5(Ae^x + Bxe^x + Be^x) + 4(Ae^x + Bxe^x + C) = xe^x + 4$

$(10A + 7B)e^x + 10Bxe^x + 4C = xe^x + 4$

$10A + 7B = 0$; $10B = 1$, $B = \dfrac{1}{10}$

$4C = 4$; $C = 1$; $10A + 7(\dfrac{1}{10}) = 0$

$A = -\dfrac{7}{100}$; $y_p = -\dfrac{7}{100}e^x + \dfrac{1}{10}e^x + 1$

By Eq. (29-19), $y = c_1e^{-x} + c_2e^{-4x} - \dfrac{7}{100}e^x + \dfrac{1}{10}xe + 1$

25. $D^2y - Dy - 6y = 5 - e^x$. By Eq. (29-14), $m^2 - m - 6 = 0$

$(m - 3)(m + 2) = 0$; $m_1 = 3$, $m_2 = -2$

By Eq. (29-15), $y_c = c_1e^{3x} + c_2e^{-2x}$

$y_p = A + Be^x$; $Dy_p = Be^x$; $D^2y_p = Be^x$

Substituting into the differential equation, $Be^x - Be^x - 6(A + Be^x)$

$= 5 - e^x$

$-6A - 6Be^x = 5 - e^x$; $-6A = 5$, $A = -\dfrac{5}{6}$; $-6B = -1$, $B = \dfrac{1}{6}$

$y_p = -\dfrac{5}{6} + \dfrac{1}{6}e^x$; $y = c_1e^{3x} + c_2e^{-2x} + \dfrac{1}{6}e^x - \dfrac{5}{6}$

Substituting $x = 0$ when $y = 2$, $3c_1 + 3c_2 = 8$.

$Dy = 3c_1e^{3x} - 2c_2e^{-2x} + \dfrac{1}{6}e^x$

Substituting, $Dy = 4$ when $x = 0$; $18c_1 - 12c_2 = 23$.

Solving the two linear equations simultaneously, $c_1 = \dfrac{11}{6}$, $c_2 = \dfrac{5}{6}$.

$y = \dfrac{11}{6}e^{3x} + \dfrac{5}{6}e^{-2x} + \dfrac{1}{6}e^x - \dfrac{5}{6}$

$y = \dfrac{1}{6}(11e^{3x} + 5e^{-2x} + e^x - 5)$

Exercises 29-10, page 947

1. $D^2\theta + \dfrac{g}{e}\theta = 0$; $g = 9.8$, $e = 0.1$

$D^2\theta + 9.8\theta = 0$; by Eq. (29 - 14), $m^2 + 9.8 = 0$

$$m_1 = \sqrt{9.8}\,j, \quad m_2 = -\sqrt{9.8}\,j; \quad \alpha = 0, \quad \beta = \sqrt{9.8};$$

By Eq. (29-17), $x = C_1 \sin \sqrt{9.8}\,t + C_2 \cos \sqrt{9.8}\,t$

Substituting $\theta = 0.1$ when $t = 0$, $0.1 = C_1 \sin 0 + C_2 \cos 0$;

$0.1 = C_2$; $D_x = C_1 \cos \sqrt{9.8}\,t - C_2 \sin \sqrt{9.8}\,t$

Substituting $D\theta = 0$ when $t = 0$, $0 = C_1 \cos 0 - C_2 \sin 0$

$C_1 = 0$; $\theta = 0.1 \cos \sqrt{9.8}\,t$

$\sqrt{9.8}\,t$	t	$\cos\sqrt{9.8}\,t$	$0.1 \cos \sqrt{9.8}\,t$
0	0	1	0.1
$\pi/2$	0.50	0	0
π	1.00	-1	-0.1
$3\pi/2$	1.50	0	0
2π	2.00	1	0.1

5. To get spring constant: $F = kx$; $4 = k(.125)$; $k = 32$ lb/ft

To get mass of object: $F = ma$; $4 = m(32)$; $m = 0.125 \dfrac{\text{lb} \cdot s^2}{\text{ft}}$ (a slug)

Using Newton's Second Law: mass × accel. = restoring force

$0.125 \dfrac{d^2x}{dt^2} = -32x$; $D^2x + 256x = 0$

$m^2 + 256 = 0$; $m = \pm 16j$

$x = c_1 \sin 16t + c_2 \cos 16t$

$Dx = 16c_1 \cos 16t - 16c_2 \sin 16t$

Let $x = \dfrac{1}{4}$ ft when $t = 0$

$\dfrac{1}{4} = c_1 \sin 0 + c_2 \cos 0$; $c_2 = \dfrac{1}{4}$

Let $Dx = 0$ when $t = 0$, $0 = 16c_1 \cos 0 - 16c_2 \sin 0$

$c_1 = 0$; $x = \dfrac{1}{4} \cos 16t$

5M. $F = kx$; $4 = k(\dfrac{1}{20})$; $k = 80$ N/m; $F = ma$

$4 = m(9.8)$; $m = \dfrac{4}{9.8} = \dfrac{20}{49}$ kg; $\dfrac{20}{49} \dfrac{d^2x}{dt^2} = -80x$

$D^2x + 196x = 0$, $m^2 + 196 = 0$; $m = \pm 14j$

$x = c_1 \sin 14t + c_2 \cos 14t$

$Dx = 14c_1 \cos 14t - 14c_2 \cos 14t$

Let $x = 0.1$ m when $t = 0$; $0.1 = c_1 \sin 0 + c_2 \cos 0$; $c_2 = 0.1$

Let $Dx = 0$ when $t = 0$; $0 = 14c_1 \cos 0 - 14c_2 \cos 0$

$c_1 = 0$; $x = 0.1 \cos 14t$

9. To get spring constant: $F = kx$; $0.820 = k(0.250)$, $k = 3.28^{kg}/_m$

To get mass of object: $F = ma$; $0.820 = m(9.8)$; $m = 0.0837$

Using Newton's Second Law: mass × accel. = restoring force.

$$0.0837\frac{d^2x}{dt^2} = -3.28x; \quad D^2x + 39.19 = 0$$

$$m^2 + 13.12 = 0; \quad m = \pm\sqrt{39.19}\,j$$

$$x = c_1 \sin\sqrt{39.19}\,t + c_2 \cos\sqrt{39.19}\,t$$

$$Dx = \sqrt{39.19}\,c_1 \cos\sqrt{39.19}$$

Let $x = 0.150m$ when $t = 0$

$0.150 = c_1 \sin 0 + c_2 \cos 0$; $c_2 = 0.150$

Let $Dx = 0$ when $t = 0$,

$$0 = \sqrt{39.19}\,c_1 \cos 0 - \sqrt{39.19}\,\sin 0$$

$$c_1 = 0; \quad x = 0.150 \cos\sqrt{13.12}\,t$$

$$x = 0.150 \cos 6.26t$$

13. $L = 0.1H$, $R = 0$, $C = 100\mu F = 10^{-4}F$, $E = 100V$

From Eq. (29-20), $0.1\dfrac{d^2q}{dt^2} + 0\dfrac{dq}{dt} + \dfrac{q}{10^{-4}} = 100$

$$\frac{d^2q}{dt^2} + 10^5q = 1000; \quad m^2 + 10^5 = 0; \quad m = \pm 315j$$

$q_c = c_1 \sin 316t + c_2 \cos 316t$; $q_p = A$; $q_p' = 0$; $q_p'' = 0$

Substituting into the differential equation,

$$0 + 0 + 10^5A = 1000; \quad A = \frac{1}{100}$$

$q = c_1 \sin 316t + c_2 \cos 316t + \dfrac{1}{100}$; $q = 0$ when $t = 0$;

$$0 = 0 + c_2(1) + \frac{1}{100}$$

$c_2 = -\dfrac{1}{100}$; $\dfrac{dq}{dt} = 316c_1 \cos 316t + \dfrac{315}{100} \sin 316t$; $i = \dfrac{dq}{dt} = 0$ when $t = 0$

$0 = 316c_1 + 0$; $c_1 = 0$; $y = 0 \sin 316t - \dfrac{1}{100} \cos 316t + \dfrac{1}{100}$;

$$y = \frac{1}{100}(1 - \cos 316t)$$

17. $L = 8mH = 8 \times 10^{-3}H; \; R = 0; \; C = 0.5\mu F$

$= 5 \times 10^{-1} \times 10^{-6} = 5 \times 10^{-7}F;$

$E = 20e^{-200t} \; mv = 2 \times 10^{-2}e^{-200t} V.$ From Eq. (29-20),

$8 \times 10^{-3} \dfrac{d^2q}{dt^2} + 0\dfrac{dq}{dt} + \dfrac{q}{5 \times 10^{-7}} = 2 \times 10^{-2} \; e^{-200t}$

$\dfrac{d^2q}{dt^2} + 2.5 \times 10^8q = 2.5e$

$m^2 + 2.5 \times 10^8 = 0; \; m_1 = 1.58 \times 10^4 j$ and $m_2 = -1.58 \times 10^4 j$

$q_c = c_1 \sin 1.58 \times 10^4t + c_2 \cos 1.58 \times 10^4t$

$q_p = Ae^{-200t}; \; q_p' = -200Ae^{-200t}; \; q''p = 4 \times 10^4Ae^{-200t}$

Substituting into the differential equation,

$4 \times 10^4Ae^{-200t} + 2.5 \times 10^8 Ae^{-200t} = 2.5e^{-200t}$

$4 \times 10^4A + 2.5 \times 10^8A = 2.5; \; A = 10^{-8}$

$q = c_1 \sin 1.58 \times 10^4t + c_2 \cos 1.58 \times 10^4 + 10^{-8}e^{-200t}$

$q = 0$ when $t = 0$

$0 = c_1(0) + c_2(1) + 10^{-8}; \; c_2 = -10^{-8}$

$i = \dfrac{dq}{dt} = 1.58 \times 10^4c_1 \cos 1.58 \times 10^4t$

$\qquad -1.58 \times 10^4c_2 \sin 1.58 \times 10^4t - 200 \times 10^{-8}e^{-200t}$

$i = 0$ when $t = 0$

$0 = 1.58 \times 10^4c_1(1) - 2 \times 10^{-6}(1); \; c_1 = \dfrac{2 \times 10^{-6}}{1.58 \times 10^4}$

$i = 1.58 \times 10^4 \times \dfrac{2 \times 10^{-6}}{1.58 \times 10^4} \cos 1.58 \times 10^4t$

$\qquad -1.58 \times 10^4 \times (-10^{-8})\sin 1.58 \times 10^4t - 2 \times 10^{-6}e^{-200t}$

$\qquad = 2 \times 10^{-6} \cos 1.58 \times 10^4t + 1.58 \times 10^{-4} \sin 1.58 \times 10^4t - 2 \times 10^{-6}e^{-200t}$

$\qquad = 10^{-6}(2.00 \cos 1.58 \times 10^4t + 158 \sin 1.58 \times 10^4t - 2.00e - 200t$

Exercises 29-11, page 953

1. $f(t) = 1; \; L(f) = L(t)$

$F(s) = \displaystyle\int_0^\infty e^{-st} \, dt = -\dfrac{1}{s}\int_0^\infty e^{-st}(-s \, dt)$

$= \lim_{c\to\infty} \left[-\dfrac{1}{s}\int_0^c e^{-st}(-s \, dt) \right] = \lim_{c\to\infty} \left[-\dfrac{1}{s}e^{-st} \Big|_0^c \right]$

$= \lim_{c\to\infty} \left[-\dfrac{1}{s}e^{-sc} + \dfrac{1}{s} \right] = 0 + \dfrac{1}{s} = \dfrac{1}{s}$

5. $f(t) = e^{3t}$; from transform (3) of the table, $a = -3$, $L(3t) = \dfrac{1}{s - 3}$.

9. $f(t) = \cos 2t - \sin 2t$; by Eq. (29-23), $L(f) = L(\cos 2t) - L(\sin 2t)$
    By transforms (5) and (6),
    $$L(f) = \frac{s}{s^2 + 4} - \frac{2}{s^2 + 4}; \; L(f) = \frac{s - 2}{s^2 + 4}$$

13. $y'' + y'$; $f(0) = 0$; $f'(0) = 0$; by Eq. (29-23), $L[f''(y) + f'(y)]$
    $= L(f'') + L(f') = s^2 L(f) - sf(0) - f'(0) + sL(f) - f(0)$
    $s^2 L(f) - s(0) - 0 + sL(f) - 0 = s^2 L(f) + sL(f)$

17. $L^{-1}(F) = L^{-1}(\dfrac{2}{s^3}) = 2L^{-1}(\dfrac{1}{s^3})$.
    From transform (2) of the table, $L^{-1}(F) = \dfrac{2t^2}{2} = t^2$

21. $L^{-1}(F) = L^{-1}\dfrac{1}{(s + 1)^3} = L^{-1}\dfrac{1}{2}[\dfrac{2}{(s + 1)^3}]$
    $= \dfrac{1}{2}t^2 e^{-t}$, from transform (12)

Exercises 29-12, page 972

1. $y' + y = 0$; $y(0) = 1$; $L(y') + L(y) = L(0)$; $L(y') + L(y) = 0$
    By Eq. (29-24), $sL(y) - y(0) + L(y) = 0$
    $sL(y) - 1 + L(y) = 0$; $(s + 1)L(y) = 1$; $L(y) = \dfrac{1}{s + 1}$
    The inverse of $F(s) = \dfrac{1}{s + 1}$ is found from transform (3).
    $y = e^{-t}$

5. $y' + 3y = e^{-3t}$, $y(0) = 1$; $L(y') + L(3y) = L(e^{-3t})$
    $L(y') + 3L(y) = L(e^{-3t})$
    By Eq. (29-24) and transform (3), $[sL(y) - 1] + 3L(y) = \dfrac{1}{s + 3}$
    $(s + 3)L(y) = \dfrac{1}{s + 3} + 1$; $L(y) = \dfrac{1}{(s + 3)^2} + \dfrac{1}{s + 3}$
    The inverse is found from transforms (11) and (3).
    $L(y) = te^{-3t} + e^{-3t}$; $L(y) = (1 + t)e^{-3t}$

9. $y'' + 2y' = 0$, $y(0) = 0$, $y'(0) = 2$
    $L(y'') + L(2y') = 0$; $L(y'') + 2L(y') = 0$
    By Eq. (29-24), (29-25), $[s^2 L(y) - 0 - 2] + 2sL(y) - 0 = 0$
    $(s^2 + 2s)L(y) = 2$; $L(y) = \dfrac{2}{s^2 + 2s} = \dfrac{2}{s(s + 2)}$
    By transform (4), $y = 1 - e^{-2t}$

13. $y'' + y = 1$, $y(0) = 1$, $y'(0) = 1$; $L(y'') + L(y) = L(1)$

By Eq. (29-25) and transform (1), $s^2 L(y) - s - 1 + L(y) = \dfrac{1}{s}$

$(s^2 + 1)L(y) = \dfrac{1}{s} + s + 1$; $L(y) = \dfrac{1}{s(s^2 + 1)} + \dfrac{s}{s^2 + 1} + \dfrac{1}{s^2 + 1}$

By transforms (7), (5), and (6), $y = 1 - \cos t + \cos t + \sin t$

$y = 1 + \sin t$

17. $2v' = 6 - v$. Since the object starts from rest, $f(0) = 0$, $f'(0) = 0$

$2L(v') + L(v) = 6L(1)$. By Eq. (29-24) and transform (1),

$2sL(v) - 0 + L(v) = \dfrac{6}{s}$; $(2s + 1)L(v) = \dfrac{6}{s}$

$L(v) = \dfrac{6}{s(2s + 1)} = 6\left[\dfrac{\frac{1}{2}}{s(s + \frac{1}{2})}\right]$

By transform (4), $v = 6(1 - e^{-t/2})$

21. By Eq. (29-20), $10\dfrac{d^2q}{dt^2} + \dfrac{q}{4 \times 10^{-5}} = 100 \sin 50t$

It is given that $q(0) = 0$ and $q'(0) = 0$.

$10L(q'') + 2.5 \times 10^4 L(q) = L(100 \sin 50t)$

By Eq. (29-25), $10s^2 L(q) + 2.5 \times 10^4 L(q) = L(100 \sin 50t)$

$(s^2 + 2.5 \times 10^3)L(q) = 10[\dfrac{50}{s^2 + 50}]$

$L(q) = \dfrac{500}{(s^2 + 50^2)^2}$; $L(q) = \dfrac{1}{500} \dfrac{2(50)^3}{(s^2 + 50^2)^2}$

By transform (15), $q = \dfrac{1}{500}(\sin 50t - 50t \cos 50t)$

$i = \dfrac{dq}{dt} = \dfrac{1}{500}[50 \cos 50t - [50t(-50 \sin 50t) + (\cos 50t)50]]$

$= \dfrac{1}{500}(2500t \sin 50t) = 5t \sin 50t$

Review   Exercises for Chapter 29, page 956

1. $4xy^3\ dx + (x^2 + 1)\ dy = 1$; divide by y^2 and $x^2 + 1$

$\dfrac{4x}{x^2 + 1}\ dx + \dfrac{dy}{y^3} = 0$; integrating, $2 \ln (x^2 + 1) - \dfrac{1}{2y^2} = c$

5. $2D^2 y + Dy = 0$. The auxiliary equation is $2m^2 + m = 0$

$m(2m + 1) = 0$; $m_1 = 0$ and $m_2 = -\dfrac{1}{2}$

By Eq. (29-15), $y = c_1 e^0 + c_2 e^{-1/2x}$

$y = c_1 + c_2 e^{-x/2}$

9. $(x + y)\, dx + (x + y^3)\, dy = 0;\ x\, dx + y\, dx + x\, dy + y^3\, dy = 0$

By Eq. (29-3), $x\, dx + d(xy) + y^3\, dy = 0$

Integrating, $\frac{1}{2}x^2 + xy + \frac{1}{4}y^4 = c_1;\ 2x^2 + 4xy + y^4 = c$

13. $dy = (2y + y^2)\, dx;\ \dfrac{dy}{(2y + y^2)} = dx$

Integrating by formula 2, $-\frac{1}{2} \ln\left(\dfrac{2 + y}{y}\right) = x + \ln c_1$

$\ln \dfrac{2 + y}{y} = -2x - \ln c_1^2 \left(\dfrac{2 + y}{y}\right) = -2x$

$\dfrac{2 + y}{y} = \dfrac{e^{-2x}}{c_1^2};\ y = c_1^2(y + 2)e^{2x};\ y = c(y + 2)e^{2x}$

17. $y' + 4y = 2;\ \dfrac{dy}{dx} + 4y = 2;\ dy + 4y\, dx = 2\, dx$

$P = 4,\ Q = 2$

$e^{\int 4\, dx} = e^{4x};\ ye^{4x} = \int 2e^{4x}\, dx + c$

$ye^{4x} = 2(\frac{1}{4})e^{4x} + \dfrac{c}{2} = \dfrac{1}{2}e^{4x} + \dfrac{c}{2}$

$y = \dfrac{1}{2} + \dfrac{1}{2e^{4x}} + c = \dfrac{1}{2}(1 + ce^{-4x})$

21. $2D^2y + Dy - 3y = 6.$ By Eq. (29-14), $2m^2 + m - 3 = 0$

$(m - 1)(2m + 3) = 0;\ m_1 = 1,\ m_2 = -\dfrac{3}{2}$

By Eq. (29-15), $y_c = c_1e^x + c_2e^{-3x/2};\ y_p = A;\ y_p' = 0;\ y_p'' = 0$

Substituting into the differential equation, $2(0) + 0 - 3A = 6;\ A = -2;$

$y_p = -2.$

By Eq. (29-19), $y = c_1e^x + c_2e^{-3x/2} - 2$

25. $9D^2y - 18Dy + 8y = 16 + 4x$

$D^2y - 2Dy + \dfrac{8}{9}y = \dfrac{16}{9} + \dfrac{4}{9}x$

$m^2 - 2m + \dfrac{8}{9} = 0;\ \ m = \dfrac{2 \pm \sqrt{4 - 4\left(\frac{8}{9}\right)}}{2};\ m_1 = \dfrac{2}{3};\ m_2 = \dfrac{4}{3}$

$y_c = c_1e^{2x/3} + c_2e^{4x/3};\ y_p = A + Bx;\ y_p' = B;\ y''_p = 0$

Substituting into the differential equation,

$0 - 2B + \dfrac{8}{9}(A + Bx) = \dfrac{16}{9} + \dfrac{4}{9}x$

$\left(-2B + \dfrac{8}{9}A\right) + \dfrac{8}{9}Bx = \dfrac{16}{9} + \dfrac{4}{9}x;\ -2B + \dfrac{8}{9}A = \dfrac{16}{9};$

$\dfrac{8}{9}B = \dfrac{4}{9};\ B = \dfrac{1}{2};\ -2\left(\dfrac{1}{2}\right) + \dfrac{8}{9}A = \dfrac{16}{9};\ A = \dfrac{25}{8};\ y_p = \dfrac{1}{2}x + \dfrac{25}{8}$

$y = c_1e^{2x/3} + c_2e^{4x/3} + \dfrac{1}{2}x + \dfrac{25}{8}$

29.

$$3y' = 2y \cot x; \quad \frac{dy}{dx} = \frac{2}{3}y \cot x; \quad \frac{dy}{y} = \frac{2}{3}\cot x\, dx; \text{ integrating,}$$

$$\ln y = \frac{2}{3}\ln \sin x + \ln c; \quad \ln y - \ln \sin^{2/3} x = \ln c;$$

$$\ln \frac{y}{\sin^{2/3}x} = \ln c; \quad \frac{y}{\sin^{2/3}x} = c; \quad y = c \sin^{2/3} x$$

Substituting $y = 2$ when $x = \frac{\pi}{2}$, $2 = c \sin^{2/3}\frac{\pi}{2}$; $c = 2$

Therefore $y = 2 \sin^{2/3}x = 2\sqrt[3]{\sin^2 x}$; $y^3 = 8 \sin^2 x$

33. $D^2 y + Dy + 4y = 0$. By Eq. (29-14), $m^2 + m + 4 = 0$; $m = \dfrac{-1 \pm\sqrt{-15}}{2}$

$m_1 = -\dfrac{1}{2} + \dfrac{\sqrt{15}}{2}j$; $m_2 = -\dfrac{1}{2} - \dfrac{\sqrt{15}}{2}j$

$\alpha = -\dfrac{1}{2}$, $\beta = \dfrac{\sqrt{15}}{2}$; by Eq. (29-17),

$y = e^{-x/2}(c_1 \sin \dfrac{\sqrt{15}}{2}x + c_2 \cos \dfrac{\sqrt{15}}{2}x)$

$Dy = e^{-x/2}(\dfrac{\sqrt{15}}{2}c_1 \cos \dfrac{\sqrt{15}}{2}x - c_2 \sin \dfrac{\sqrt{15}}{2}x)$

$\quad + (c_1 \sin \dfrac{\sqrt{15}}{2}x + c_2 \cos \dfrac{\sqrt{15}}{2}x)(-\dfrac{1}{2}e^{-x})$

Substituting $Dy = \sqrt{15}$, $y = 0$ when $x = 0$; $c_2 = 0$; $c_1 = 2$.

$y = 2e^{-x/2}\sin(\dfrac{1}{2}\sqrt{15}x)$

37. $L(4y') - L(y) = 0$; $4L(y') - L(y) = 0$

By Eq. (29-24), $4sL(y) - 1 - L(y) = 0$

$(4s - 1)L(y) = 1$; $L(y) = \dfrac{1}{4s - 1} = \dfrac{\frac{1}{4}}{s - \frac{1}{4}}$

By transform (3), $y = e^{t/4}$.

41. $L(y'') + L(y) = 0$; by Eq. (29-25), $y(0) = 0$, $y'(0) = -4$

$s^2 L(y) + 4 + L(y) = 0$; $(s^2 + 1)L(y) = -4$; $L(y) = -4(\dfrac{1}{s^2 + 1})$

By transform (6), $y = -4 \sin t$.

45. ① $v = \pi\dfrac{4}{3}r^3$; ② $A = 4\pi r^2$; $\dfrac{dv}{dt} = kA$

From ①, $\dfrac{(d\frac{4}{3}\pi r^3)}{dt} = k(4\pi r^2)$; $4\pi r^2 \dfrac{dr}{dt} = 4\pi kr^2$

$\dfrac{dr}{dt} = k$; $r = kt + C$; $r_0 = 0 + C$

$C = r_0$; $r = r_0 + kt$

49. See Example C, p. 941. $N = N_0 e^{kt}$; $N_0 = 500$ mg, $N = 400$ mg, $t = 4.02$ years

$400 = 500\ e^{4.02k}$; $e^{4.02k} = 0.8$; $\ln e^{4.02k} = \ln 0.80$; $k = -0.0555$

$N = N_0 e^{-0.0555t}$ Half-life occurs when 250 mg remains.

$250 = 500 e^{-0.0555t}$; $0.5 = e^{-0.0555t}$; $\ln 0.5 = -0.0555t \ln e$

$-0.6931 = -0.0555t$; $t = 12.5$ years

53. $y = cx^5$; $c = \dfrac{y}{x^5}$; $\dfrac{dy}{dx} = 5cx^4$

$\dfrac{dy}{dx} = 5\left(\dfrac{y}{x^5}\right)x^4 = \dfrac{5y}{x}$; $\left.\dfrac{dy}{dx}\right|_{OT} = -\dfrac{x}{5y}$; $5y\ dy = -x\ dx$

Integrating, $\dfrac{5y^2}{2} = -\dfrac{x^2}{2}$; $5y^2 + x^2 = c$

57. $F = kx$; $40 = 0.5k$; $k = 80$ N/m; $m = 4$ kg

$4\dfrac{d^2x}{dt^2} = -16\dfrac{dx}{dt} - 80x$; $4D^2x + 16Dx + 80 = 0$

$D^2x + 4Dx + 20 = 0$; $4m^2 + 16m + 80 = 0$

$m_1 = -2 + 4j$; $m_2 = -2 - 4j$

By Eq. (29-17), $x = e^{-2t}(c_1 \sin 4t + c_2 \cos 4t)$

Let $x = 0.5$ when $t = 0$; $0.5 = c_2$

$Dx = e^{-2t}(4c_1 \cos 4t - 4c_2 \sin 4t) + (c_1 \sin 4t + c_2 \cos 4t)(-2)e^{-2t}$

Let $Dx = 0$ when $t = 0$; $0 = 4c_1 - 2c_2$; $0 = 4c_1 - 1$; $c_1 = 0.25$

$x = e^{-2t}(0.25 \sin 4t + 0.5 \cos 4t)$

$= 0.25e^{-2t}(\sin 4t + 2 \cos 4t)$ underdamped

61. $R = 20\Omega$, $L = 4H$, $C = 10^{-4}F$; $V = 100V$; $q(0) = 10^{-2}C$

$4q'' + 20q' + 10^4 q = 100$; $\dfrac{1}{25}q'' + \dfrac{1}{5}q' + 100q = 1$

$\dfrac{1}{25}\left[s^2 L(q) - s\left(10^{-2}\right)\right] + \dfrac{1}{5}\left[sL(q) - 10^{-2}\right] + 100\ L(q) = L(1)$

$\dfrac{1}{25}s^2 L(q) - \dfrac{1}{2500}s + \dfrac{1}{5}sL(q) - \dfrac{1}{500} + 100\ L(q) = \dfrac{1}{s}$

$L(q)\left(\dfrac{1}{25}s^2 + \dfrac{1}{5}s + 100\right) = \dfrac{1}{s} + \dfrac{1}{2500}s + \dfrac{1}{500}$

$L(q)\left(\dfrac{s^2 + 5s + 2500}{25}\right) = \dfrac{2500 + s^2 + 5s}{2500s}$

$L(q) = \dfrac{1}{100s}$; $q = \dfrac{1}{100}$; $i = \dfrac{dq}{dt} = 0$

65. By Eq. (29-20), $0.25 \dfrac{d^2q}{dt^2} + \dfrac{4dq}{dt} + \dfrac{q}{10^{-4}} = 0$

$q(0) = 400 \ \mu C = 4 \times 10^{-4} \ C$

$i = \dfrac{dq}{dt} = q'(0) = 0; \ 0.25L(q'') + 4L(q') + 10^4 L(q) = 0$

$L(q'') + 16L(q') + 4 \times 10^4 L(q) = 0$

By Eq. (29-24), (29-25), $s^2 L(q) - 4 \times 10^{-4}s - 0 + 16sL(q) - 4 \times 10^{-4}$
$+ 4 \times 10^4 L(q) = 0$

$(s^2 + 16s + 4 \times 10^4)L(q) = 4 \times 10^{-4}s + 64 \times 10^{-4}$

$L(q) = 4 \times 10^{-4}[\dfrac{s + 16}{s^2 + 16s + 64 + 4 \times 10^4}]$

$\quad = 4 \times 10^{-4}[\dfrac{s + 8}{(s + 8)^2 + 200^2} + \dfrac{8}{(s + 8)^2 + 200^2}]$

$\quad = 4 \times 10^{-4}[\dfrac{s + 8}{(s + 8) + 200^2} + \dfrac{8}{200} \times \dfrac{200}{(s + 8)^2 + 200^2}]$

$q = 4 \times 10^{-4}(e^{-8t} \cos 200t + 0.04e^{-8t} \sin 200t)$

$\quad = 10^{-4}e^{-8t}(4.0 \cos 200t + 0.16 \sin 200t)$

69. $\dfrac{d^2y}{dx^2} = \dfrac{1}{EI}m = \dfrac{1}{EI}(2000x - 40x^2);$ Integrating

$\dfrac{dy}{dx} = \dfrac{1}{EI}\left[\left(2000\dfrac{x^2}{2} - \dfrac{40x^3}{3}\right) + c_1\right]$

$\dfrac{dy}{dx} = \dfrac{1}{EI}\left(1000x^2 - \dfrac{40x^3}{3} + c_1\right);$ integrating

$y = \dfrac{1}{EI}\left(\dfrac{1000x^3}{3} - \dfrac{10x^4}{3} + c_1x + c_2\right) = \dfrac{10}{3EI}(100x^3 - x^4) + c_1x + c_2$

$y = 0$ when $x = 0; \ 0 = c_2; \ y = \dfrac{10}{3EI}(100x^3 - x^4) + c_1x$

$y = 0$ when $x = L; \ 0 = \dfrac{10}{3EI}(100L^3 - L^4) + c_1L$

$c_1L = -\dfrac{10}{3EI}(100L^3 - L^4); \ c_1 = -\dfrac{10}{3EI}(100L^2 - L^3)$

$y = \dfrac{10}{EI}(100x^3 - x^4) - \dfrac{10}{3EI}(100L^2 - L^3)x = \dfrac{10}{3EI}[100x^3 - x^4 - (100L^2 - L^3)x]$

$\quad = \dfrac{10}{3EI}[100x^3 - x^4 + xL^2(L - 100)]$